Philosophy of Biology

BLACKWELL PHILOSOPHY ANTHOLOGIES

Each volume in this outstanding series provides an authoritative and comprehensive collection of the essential primary readings from philosophy's main fields of study. Designed to complement the *Blackwell Companions to Philosophy* series, each volume represents an unparalleled resource in its own right, and will provide the ideal platform for course use.

PHILOSOPHY OF BIOLOGY
AN ANTHOLOGY

Edited by

Alex Rosenberg

and

Robert Arp

WILEY-BLACKWELL

A John Wiley & Sons, Ltd., Publication

Blackwell Publishing was acquired by John Wiley & Sons in February 2007. Blackwell's publishing program has been merged with Wiley's global Scientific, Technical, and Medical business to form Wiley-Blackwell.

Registered Office
John Wiley & Sons Ltd, The Atrium, Southern Gate, Chichester, West Sussex, PO19 8SQ, United Kingdom

Editorial Offices
350 Main Street, Malden, MA 02148-5020, USA
9600 Garsington Road, Oxford, OX4 2DQ, UK
The Atrium, Southern Gate, Chichester, West Sussex, PO19 8SQ, UK

For details of our global editorial offices, for customer services, and for information about how to apply for permission to reuse the copyright material in this book please see our website at www.wiley.com/wiley-blackwell.

The right of Alex Rosenberg and Robert Arp to be identified as the author of the editorial material in this work has been asserted in accordance with the Copyright, Designs and Patents Act 1988.

Library of Congress Cataloging-in-Publication Data
Philosophy of biology : an anthology / edited by Alex Rosenberg and Robert Arp.
        p. cm. — (Blackwell philosophy anthologies)
    Includes bibliographical references and index.
    ISBN 978-1-4051-8317-8 (hardcover : alk. paper) — ISBN 978-1-4051-8316-1 (pbk. : alk. paper)   1. Biology—Philosophy.   2. Evolution (Biology)—Philosophy.   I. Rosenberg, Alexander, 1946–   II. Arp, Robert.

QH331.P4686 2010
570.1—dc22

                                                                      2008054086

A catalogue record for this book is available from the British Library.

Set in 9.5/11.5pt Minion by Graphicraft Limited, Hong Kong
Printed in Singapore by Fabulous Printers Pte Ltd

01   2010

Contents

Rob Arp would like to thank his mom and dad, LaVerne and Jerry Arp, and his sister and brother-in-law, Laura and Jim McDonald, for their encouragement to study the biological sciences early in his life. He owes his accomplishments to his family's love and support.

Source Acknowledgments

The editor and publisher gratefully acknowledge the permission granted to reproduce the copyright material in this book:

1. Charles Darwin, Chapters 3 and 4 of *The Origin of Species* (1859). Reprinted by permission of Random House, Inc.
2. Eugenie Scott, Chapters 1 and 2 of *Evolution vs. Creationism: An Introduction* (University of Carolina Press, 2004). Reproduced with permission of Greenwood Publishing Group, Inc., Westport, CT.
3. Michael Denton, "Beyond the Reach of Chance," Chapter 13 of *Evolution: A Theory in Crisis* (Burnett, 1985). Reprinted by permission of Adler & Adler Publishers, Inc.
4. Richard Dawkins, *The Blind Watchmaker: Why the Evidence of Evolution Reveals a Universe Without Design* (Norton, 1986), Chapter 2, "Accumulating Small Change." Copyright © 1996, 1987, 1986, by Richard Dawkins. Used by permission of W. W. Norton & Company, Inc. Reproduced by permission of Penguin Books Ltd.
5. John Beatty, "Chance and Natural Selection," *Philosophy of Science* (1984) 51, 183–211. Reprinted by permission of the publisher, the University of Chicago Press.
6. Robert Brandon, "The Principle of Drift: Biology's First Law," *Journal of Philosophy* (2006) 102, 319–335. Reprinted by permission of the *Journal of Philosophy* and by Robert Brandon.
7. Stephen Gould, "This View of Life: Darwin's Untimely Burial," *Natural History* (1976) 85, 24–30.
8. Robert Brandon, "Adaptation and Evolutionary Theory." Reprinted from *Studies in the History and Philosophy of Science*, vol. 9, Adaptation and Evolutionary Theory, pp. 9, 181–206. Copyright 1978, with permission from Elsevier.
9. Stephen Gould and Richard Lewontin. "The Spandrels of San Marco and the Panglossian Paradigm: A Critique of the Adaptationist Programme," *Proceedings of the Royal Society of London* (1978) 205, 581–598. Reprinted by permission of the Royal Society
10. Ernst Mayr, "How to Carry Out the Adaptationist Program?" *The American Naturalist* (1983) 121, 324–334. Reprinted by permission of the publisher, the University of Chicago Press.
11. Mark Perlman, "The Modern Philosophical Resurrection of Teleology," *The Monist* (2004) 87, 3–51. Copyright © 2004, *The Monist: An International Quarterly Journal of General Philosophical Inquiry*, Peru, Illinois, 61354. Reprinted by permission.
12. Robert Cummins, "Neo-Teleology," in Cummins, Ariew, Perlman (eds.) *Functions*. Oxford: OUP (2002), pp. 157–172.
13. Peter Godfrey-Smith, "A Modern History Theory of Functions," *Nôus* (1994) 28, 244–362. Reproduced with permission of Blackwell Publishing Ltd.

14. Sean Carroll, "Endless Forms: The Evolution of Gene Regulation and Morphological Diversity." Reprinted from *Cell*, vol. 101, Sean Carroll, Endless Forms: The Evolution of Gene Regulation and Morphological Diversity, pp. 101, 577–580, Copyright 2000, with permission from Elsevier.

15. Casper Breuker, Vincent Debat, and Christian Klingenberg, "Functional Evo-Devo." Reprinted from *Trends in Ecology and Evolution*, vol. 21, Casper Breuker, Vincent Debat and Christian Klingenberg, Functional Evo Devo, pp. 21, 488–492, Copyright 2006, with permission from Elsevier.

16. Philip Kitcher, "1953 and All That: A Tale of Two Sciences," *The Philosophical Review*, vol. 93, pp. 335–373. Copyright 1984, The Sage School of Philosophy at Cornell University. All rights reserved. Used by permission of the publisher, Duke University Press.

17. Elliot Sober, "The Multiple Realizability Argument against Reductionism," *Philosophy of Science* (Philosophy of Science Association, 1999), 66, 542–564. Reprinted by permission of the publisher, the University of Chicago Press.

18. Marc Ereshefsky, "Species, Taxonomy, and Systematics," in Mohan Matthen and Christopher Stephens (eds.) *Handbook for the Philosophy of Science, Philosophy of Biology* (Elsevier, 2007), pp. 403–427.

19. Jerry Coyne and H. Allen Orr, Appendix from *Speciation* (Sinauer, 2004), pp. 27, 447–472.

20. Elliott Sober and Richard Lewontin, "Artifact, Cause, and Genic Selection," *Philosophy of Science* (Philosophy of Science Association, 1982) 49, 157–180. Reprinted by permission of the publisher, the University of Chicago Press.

21. Kim Sterelny and Philip Kitcher, "The Return of the Gene," *Journal of Philosophy* (1988) 85, 339–361. Reprinted by permission of the *Journal of Philosophy* and by Philip Kitcher.

22. Samir Okasha, "The Levels of Selection Debate: Philosophical Issues," *Philosophy Compass* (2006) 1, 74–85. Reproduced with permission of Blackwell Publishing Ltd.

23. Edward Wilson, Chapters 1 and 27 from *Sociobiology: The New Synthesis* by Edward O. Wilson, pp. 3–6, 547, 562–564 (Cambridge, Mass: The Belknap Press of Harvard University Press). Reprinted by permission of the publisher. Copyright © 1975, 2000 by the President and Fellows of Harvard College.

24. Robert Axelrod and William D. Hamilton, "The Evolution of Cooperation," *Science* (American Association for the Advancement of Science, 1981), 211, 1390–1396. Reprinted with permission from AAAS.

25. Alex Rosenberg, "Darwinism in Moral Philosophy and Social Theory," in Hodge & Radick (eds.) *The Cambridge Companion to Darwin* (Cambridge University Press, 2003), pp. 310–332. Reprinted with permission.

26. John Tooby and Leda Cosmides, "Conceptual Foundations of Evolutionary Psychology," in D. Buss (ed.) *The Handbook of Evolutionary Psychology* (Wiley, 2005), pp. 5–67.

27. Robert Arp, "The Environments of Our Hominin Ancestors, Tool Usage, and Scenario Visualization," *Biology & Philosophy* (2006), 1, 95–117. With kind permission from Springer Science and Business Media.

28. Donald Prothero, Chapter 2 from *Evolution: What the Fossils Say and Why It Matters* (Columbia, 2007).

29. Michael Behe, "Irreducible Complexity: Obstacle to Darwinian Evolution," in W. Dembski and M. Ruse (eds.) *Debating Design: From Darwin to DNA Black Box: The Biochemical Challenge to Evolution* (Cambridge, 2004), pp. 352–370.

30. Kenneth Miller, "The Flagellum Unspun: The Collapse of 'Irreducible Complexity'," in W. Dembski and M. Ruse (eds.) *Debating Design: From Darwin to DNA Black Box: The Biochemical Challenge to Evolution* (Cambridge, 2004), pp. 81–97.

General Introduction: A Short History of Philosophy of Biology

Alex Rosenberg and Robert Arp

One useful way to introduce the agenda of problems in the philosophy of biology is by tracing the history of biology since Darwin and showing how its developments raised problems that have concerned both biologists and philosophers.

On the Origin of Species

Charles Darwin's (1809–82) work in evolution and natural selection, complete with its own history, would challenge every biological principle put forward in the two millennia since Aristotle (384–322 BCE) began the systematic study of the subject (Barnes, 1995), and would have an impact on virtually every topic in the philosophy of biology since Darwin's time as well.

Darwin made the term 'natural selection' a household name, and in his famous work, *The Origin of Species by Natural Selection: Or, the Preservation of Favoured Races in the Struggle for Life* (1859/1999), described it as a "principle by which each slight variation, if useful, is preserved" in the individuals of a population who have been able to survive long enough to bear offspring of their own. In fact, we have included significant parts of the *Origin* in this anthology, as the reader will see. Darwin mounts what he calls "one long argument" for natural selection, complete with numerous examples as evidence, and his insights concerning evolution by natural selection – ones that still hold today – are summarized here:

1. there is variation in organisms such that they differ from each other in ways that are inherited;
2. there is a struggle or competition for existence, since more organisms are born than can survive;
3. there is a natural selection of the traits that are most fit in an organism, given a particular environment in which the organism inhabits;
4. organisms fortunate enough to have the variation in traits that fit a particular environment will have an increased chance of surviving to pass those traits on to their progeny (survival of the fittest); and
5. natural selection leads to the accumulation of favored variants, which may produce new species (evolution), given the right environmental conditions and a certain amount of time.

Although we have been able to determine several other factors and principles at work in evolution since Darwin (and Darwin himself recognized other factors and principles), *natural selection* and *variation* remain core principles in the explanation of how species emerge and go extinct, as well as develop adaptive traits.

Darwin (1859/1999) himself was aware that variability was "governed by many unknown laws" (p. 37). Since Darwin's time, we have been able to determine that a major source of variation in organisms has to do with genetic mutation. A *gene* is a functional segment of DNA located at a particular site on a chromosome in the nucleus of all cells. Basically, DNA is the template from which RNA copies are made that transmits genetic information concerning an organism's physical and behavioral traits (phenotypic traits) to synthesis sites in the cytoplasm of the cell. RNA takes this information to ribosomes in a cell where amino acids, and then proteins, are formed according to that information. The proteins are the so-called building blocks of life, since they ultimately determine the physical characteristics of organisms (Carroll, 2005; Audesirk, Audesirk, & Byers, 2008; Campbell & Reece, 2007; Strickberger, 1985, 2000).

DNA and RNA are composed of nucleic acids. These nucleic acids specify the amino-acid sequences of all the proteins needed to make up the physical characteristics of an organism, much like a code or cryptogram. This code consists of specific sequences of nucleotides that are composed of a sugar (deoxyribose in DNA, ribose in RNA), a phosphate group, and one of four different nitrogen-containing bases, namely, adenine, guanine, cytosine, and thymine in DNA (uracil replaces thymine in RNA). These four bases are like a four-letter alphabet, and triplets of bases form three-letter words or *codons* that identify an amino acid or signal a function.

There are 64 possible permutations of the four bases, and if one of the nucleotides in a sequence is either deleted or substituted, or if an alternate nucleotide is inserted, then a *mutation* is said to occur. A mutation is nothing other than an alteration in the nucleotide sequence of a DNA or RNA molecule. Mutations can result from a variety of environmental sources, including certain chemicals, radiation from X rays, and ultraviolet rays in sunlight. Mutations also can occur spontaneously. However, the most common source of mutations occurs regularly in base pairing during replication, as a cell prepares for cell division. In other words, mutations are occurring all of the time, since cell division is occurring in organisms all of the time.

Now, the genetic makeup of an organism directly affects its phenotypic characteristics. Whether an animal will have all of its limbs, or be stronger than another member of its species, or look more appealing to the opposite sex –

all of these phenotypic characteristics are under genetic control to greater and lesser extents (Carroll, 2005; Mayr, 2001; Lewontin, 1992). When an organism exists in a particular environment, the chance of it being naturally selected to survive depends upon whether its genetic makeup happens to have produced the phenotypic characteristics necessary for optimal survival in that particular environment. To a certain extent, the randomness of a mutation makes the business of life kind of chancy. If your wolf genes coded you to have three legs instead of four, then it is likely you will not survive in the wolf pack out in the forests of Colorado. And if your rabbit genes coded you to have poor eyesight, then it is likely you will not survive in the same forests, where good eyesight is essential for avoiding such packs of wolves. The phenotypic effects of mutations need only be slight so that, for example, one wolf may be just a little stronger, or a little faster, or a little more aggressive than the rest of the pack. This small genotypic variation leads to a slight phenotypic benefit, giving the wolf an advantage in hunting, mating, and passing its genes on to future generations.

From the previous information, taken all together, we can define *natural selection* as a mechanism of evolution by which the environment favors the reproductive success of individuals possessing desirable genetic variants with greater phenotypic fitness, increasing the chance that those genotypes for the phenotypic traits will predominate in succeeding generations.

Inspired by Dawkins (1986), the evolutionary principles of *genetic variation* and the *natural selection* of the traits most fit in a particular environment are illustrated in figure I.1. In this figure, we try to show how natural selection acts like a sieve that allows for a certain phenotypic characteristic to pass through to a subsequent generation. The various shapes represent organisms having certain phenotypic traits that are genetically controlled. The sieves themselves (the rectangular planes) represent the certain environments in which these organisms live. The preformed slot or hole represents the optimal survival of organisms possessing a desirable phenotypic trait in that particular environment.

The point of this illustration is to represent pictorially what biologists such as Audesirk *et al.* (2008) and Berra (1990) have claimed about

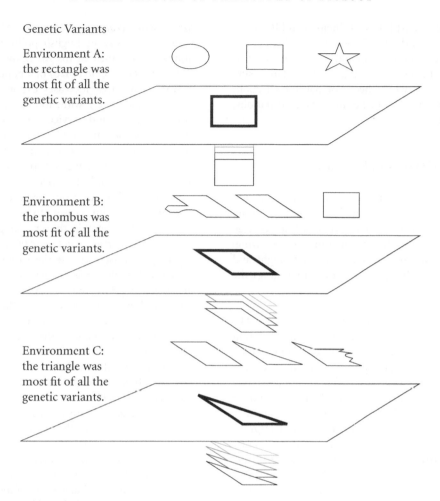

Figure I.1 The evolutionary sieve.

genetic variability and natural selection. According to Audesirk *et al.*:

Mutations are essential for evolution, because these random changes in DNA sequence are the ultimate source of all genetic variation. New base sequences undergo natural selection as organisms compete to survive and reproduce. Occasionally, a mutation proves beneficial in the organism's interactions with its environment. The mutant base sequence may spread throughout the population and become common as organisms that possess it outcompete rivals that bear the original, unmutated base sequence. (p. 175)

In Berra's words:

Some genetic variants will be better adapted to their environment than others of their sort, and will therefore tend to survive to maturity and to leave more offspring than will organisms with less favorable variations. . . . The environment is the selecting agent, and because the environment changes over time and from one region to another, different variants will be selected under different environmental conditions. (p. 8)

Stated simply, the various species around us today are those organisms that have made it through one of these environmental sieves, the result of some fortunate mutation in combination with the traits that were most fit for some environment. The reader will discover that the material in the first two sections of this anthology – Part I: Basic Principles and Proofs of Darwinism and Part II: Evolution and Chance – will underscore what we have communicated so far about the principles, arguments, and ideas surrounding Darwin's theory of natural selection, as well as provide further insights, explanations, and examples.

A Few of Darwin's Influences

Just like any idea, argument, hypothesis, theory, or stroke of genius, Darwin's ideas concerning evolution and natural selection did not emerge out of thin air. He was analyzing, synthesizing, and systematizing several lines of thought and research, which included, among other things, the following:

1. *Lamarckism.* Jean-Baptiste Lamarck (1744–1829) believed in a version of the evolution of species and put forward the idea that each generation of offspring could acquire certain characteristics from their parents based upon whether the parents had used a trait or not. Through interaction with the organism's environment, the organism could improve and adapt to the environment and pass that improvement/adaptation on to its offspring. For example, Lamarck hypothesized that giraffes on the African savannah used to have short necks at a time in the past when leaves were easier to get at because they were lower to the ground. As time passed on the African savannah, and the leaves of trees started growing at higher levels (or the giraffes ate all of the lower-level leaves), the giraffes would stretch their necks to get at the leaves growing higher up on the tree. The giraffes with stretched-out, longer necks lived and would pass that improved/adapted trait onto their progeny (somehow). Simple tests like cutting off the tails of parent mice and breeding children of those tailless mice *with* tails show Lamarck's hypothesis to be false; however, the kernel of Lamarck's insight regarding the organism and its adapted relationship to the environment was on the right track, and contributed to Darwin's thinking (see McKinney, 1971; Radick, 2003; Hodge & Radick, 2003).

2. *Malthusian economics.* Thomas Malthus (1766–1834) was an English political economist who argued in his *An Essay on the Principle of Population* (1826) that population growth in an area increases exponentially ($2 \times 2, \times 2, \times 2$, etc.), while food supply and other resources and services in that same area increases arithmetically ($2 + 2, + 2, +2$, etc.). As one can imagine, after a short time there would be many people with no food, resources, and/or services, and people would be dead of starvation, exposure, disease, and so on, thus ceasing to exist in that area. Yet (and here is the crucial point), populations in areas *do not cease to exist* (not regularly, at least) because the strongest or slyest people in that area – namely, those most fit and able to get the resources for themselves, their children, their friends – get the food, resources, and services, while the weakest and/or stupidest people in that area do not, and die off. Thus, despite the population exponential number and the resource arithmetic number, the population survives because there is a naturally selected survival of the fittest that *actually* comprises a smaller number to keep pace with the smaller resource number. Darwin (1883) probably had an "a-ha" or eureka moment when he "saw, on reading Malthus on Population that natural selection was the inevitable result of the rapid increase of all organic beings." According to Darwin (1859/1999) in the *Origin*, the struggle for existence is the "doctrine of Malthus applied with manifold force to the whole animal and vegetable kingdoms" (p. 55; also see Paul, 2003; Glick & Kohn, 1996).

3. *Geology and fossil records.* Charles Lyell (1797–1875) was an early naturalist and geologist who studied stratifications of rocks and fossils, arguing in his *Principles of Geology* (1830/1998) that species had gone extinct, as well as occupied new niches, throughout time, principally through climate change. Darwin was a bit of a geologist himself, including chapters on geology in the *Origin* where he follows the findings and hypotheses of Lyell, especially Lyell's stratigraphic investigations and his ideas concerning geologic eons of time. In the selection from the *Origin* contained in this anthology (chapter 1),

Darwin mentions Lyell's "philosophical arguments" showing that "all organic beings are exposed to severe competition." A few pages later, Darwin would combine Lamarck, Malthus, and Lyell's ideas and note that, "after any physical change, such as of climate or elevation of the land . . . new places in the natural economy of the country are left open for the old inhabitants to struggle for, and become adapted to, through modifications in their structure and constitution" (p. 59). Both Lyell and Darwin drew upon the work of Georges Cuvier (1769–1832), who established himself as one of the progenitors of modern comparative anatomy and paleontology with his influential work titled (in English) *The Animal Kingdom: Arranged in Conformity with Its Organization* (Cuvier, 1831; Rudwick, 1997).

4. *Artificial selection.* Darwin readily admitted that the way in which animal and plant breeders had been able to manipulate the morphology and behavior of organisms was one of the influences for natural selection. In the *Origin*, he notes: "I have collected so large a body of fact, showing, in accordance with the almost universal belief of breeders, that with animals and plants a cross between different varieties, or between individuals of the same variety but of another strain, gives vigor and fertility to the offspring" (p. 81). And, a few pages later he talks about how "an analogous principle" – that is, an analogous principle to artificial selection – can "apply in nature" (also Darwin, 1883; Wilner, 2006; Ruse, 1975). Artificial selection was a key piece of evidence in favor of natural selection for Darwin, most likely because the process and effects of artificial selection can be witnessed by anyone who took the time to investigate. For example, we can observe the morphology and behavior of a male German shepherd and a female border collie. Then, we can mate the two and observe the morphology and behavior of their offspring. The same can be done for any number of species.

5. *The success and unity of science.* The Scientific Revolution brought about rapid discoveries, changes, and successes in the chemical and physical sciences. Like other scientists and naturalists of his time, Darwin felt the need to explain and systematize the biosphere so as to achieve similar success in the biological sciences. In the selection from the *Origin* contained in this anthology, Darwin mentions Carl Linnaeus, and Darwin was impressed with the way in which Linnaeus was such a rigorous and detailed classifier who had made great strides in putting biology on the same footing with other sciences. In humble fashion, of course, Darwin most assuredly felt that he was offering the scientific world a unifying principle for living things with natural selection, a "power incessantly ready for action, and . . . as immeasurably superior to man's feeble efforts, as the works of Nature are to those of Art" (p. 53).

6. *A natural, empirical science.* Nowadays, most people recognize that science is concerned with empirically observable objects (either directly or indirectly) in the phenomenally experienced world. If you can't experience "it" directly or indirectly with one of the five senses, then subject it to a multitude of tests in controlled environments so that another multitude of researchers can sense it directly or indirectly with one of their five senses, and then use it, predict it, and control it somehow, then "it" is neither in the purview of science, nor is it (or should it be) countenanced by scientists (Gauch, 2003; Solomon & Hunt, 1992). Thus, the divine simply does not fit into this picture, and this was something that Darwin – feeling the effects of the scientism of his time – most probably realized. Like Pierre Laplace (1749–1827), when asked about why his account of the system of the universe lacked a god as an ultimate explanatory cause, with respect to the Tree of Life's multitude of species, Darwin likewise would have responded, "I have no need for that hypothesis" (Shapiro, 2006, p. 443). In the *Origin*, for example, it is not a god who acts; rather *nature* (understood as a set of natural principles) becomes responsible for biological parts and processes. Actually, the fact that a god or gods infiltrated every aspect of reality and acted as a kind of ultimate answer (note, ultimate) to every "why"

question – Why did it happen, ultimately? god; Why did it *not* happen, ultimately? god; Why *this* rather than that? god; Why *that* rather than this? god – most likely *prompted* Darwin to come up with a naturalistic explanation for the diversity of life. Having said this, by the end of his life Darwin seemed to remain agnostic about the existence of some god or gods (see his autobiography and letters; Darwin, 1958; Darwin, 1887); although, he may have felt that a god was necessary as a kind of first cause setting up the conditions for evolution to take place. In the final paragraph of the first edition of the *Origin*, Darwin does make the claim that life, "with its several powers, having been originally breathed into a few forms or into one . . ." which can be interpreted as a nod to the existence of a first cause. And Darwin's principles actually were used by, and continue to be used by, religious folks who think that evolution is completely compatible with some kind of divine being (Ruse, 2003). For example, the Catholic Church's position is that science and evolution account for everything in the natural world except for mind, meaning, and morality, which God takes care of and are beyond the purview of science anyway (*Humani generis*; John Paul II, 1996; *Catechism of the Catholic Church*, 1994). Evolution is just another manifestation of the Creator's *co*-creative work. In the words of Arthur Peacocke (2004), which Darwin might not have objected to:

> We now see in a new way the role in evolution of the interplay between random chance micro-events and the necessity which arises from the stuff of this world having its particular "given" properties. These potentialities a theist must regard as written into creation by the creator himself in order that they may be unveiled by chance exploring their gamut . . . the creator may be imagined to unfold the potentialities of the universe that he himself has given it, selecting and shaping by his providential action those that are to come to fruition. (pp. 46–7)

There are more influences – both positive and negative ones – that helped shape Darwin's thinking

(see, for example, Quammen, 2007), and the literature cited throughout this anthology will lead the reader to other sources of inspiration for Darwin.

1953 and All That

This is the title of Philip Kitcher's famous paper (the subtitle is: "A Tale of Two Sciences"), included in this anthology (chapter 16), which discusses the possibility of Mendelian genetics being reduced to molecular genetics, given James Watson and Francis Crick's elucidation of the molecular structure of DNA that occurred in a *Nature* paper from 1953. In fact, the middle of the twentieth century was a significant period that shaped the basic topics that would soon form the sub-discipline of philosophy of biology.

First, the publication of Watson and Crick's *Nature* (1953a) paper, and a quick follow-up paper (1953b), got scientists and philosophers of science to think about what principles, parts, and processes were really at work in heritability. Is it all just complex molecules in motion? Given that chemistry and physics were making incredible strides at explaining the real workings of other parts of the universe – and, thereby, reducing things to chemistry, then physics – it would only seem to make sense that living things could be explainable in terms of these basic sciences as well. Thus, the possibility of biology being reduced to chemistry and/or physics became solidified as a topic in philosophy of biology with papers such as J. J. C. Smart's "Can Biology Be an Exact Science?" (1959), Kenneth Schaffner's "Approaches to Reduction" (1967), Alexander Rosenberg's "The Supervenience of Biological Concepts" (1978), Ernst Mayr's "The Autonomy of Biology: The Position of Biology among the Sciences" (1996), and Elliot Sober's "The Multiple Realizability Argument Against Reductionism" (1999), which is reprinted in this anthology (chapter 17).

Not only did Smart's 1959 paper assist in generating discussion about the reducibility of biology to chemistry or physics, in that paper Smart also challenged one of the central principles of evolution: survival of the fittest. Smart wondered if, in attempting to generate laws of evolution, "we seem to reduce to tautologies." An example of a tautology is "either you are running or you are not running." In this context, tautologies have negative connotations because they are explanatorily bankrupt; consider that nothing new is learned about running in the above example (for more on this, see Sober, 2000). Smart questioned if Darwin's fitness principle was tautological because "suppose we say that even in Andromeda 'the fittest will survive' we say nothing, for 'fittest' has to be defined in terms of 'survival'" (p. 69). This facilitated another topic central to the philosophy of biology which has come to be regarded as *fitness and the tautology problem*: we need to be able to define and explain adaptive fitness (those who survive . . .) *without* appealing to adaptive fitness in our definition and explanation (. . . survive because they were able to survive). There have been a variety of responses to this problem, ranging from Alexander Rosenberg's (1978, 1983) treatment of adaptive fitness as an irreducible, primitive supervenient property, to Susan Mills and John Beatty's (1979) generally recognized solution of explaining fitness in terms of the propensity of an organism to leave a large number of offspring. In a paper that preceded the Mills & Beatty paper by one year titled "Adaptation and Evolutionary Theory," Robert Brandon argued for a similar propensity view of relative adaptedness, which we have included in this anthology.

By the 1950s, Darwin's theory of natural selection had been combined with Mendelian genetics (the "Neo-Darwinian" Evolutionary Synthesis) through the work of Theodosius Dobzhansky (1937), Julian Huxley (1942), Ernst Mayr (1942), and others, and researchers were beginning to unlock more of the secrets of the gene. As research progressed toward the end of the twentieth century, it was becoming apparent that many phenotypic traits were dependent upon genetic control. In 1971, for example, Ronald Konopka and Seymour Benzer published a ground-breaking paper on the typical fruit fly (*Drosophila melanogaster*) demonstrating that much of its behavior, and the development of a few of its body parts, could be altered through genetic manipulation. Biologists and philosophers of biology began to wonder which biological entity *exactly* was being selected in natural selection: Is it the gene, individual, population, or species? Or, which of these entities might be

most influenced by natural selection? After all, in the following paragraph of chapter 4 of the *Origin*, Darwin seems to indicate that natural selection works on individuals, populations, and species:

> Natural selection will modify the structure of the young in relation to the parent, and of the parent in relation to the young. In social animals it will adapt the structure of each individual for the benefit of the community; if each in consequence profits by the selected change. What natural selection cannot do, is to modify the structure of one species, without giving it any advantage, for the good of another species. (pp. 73–4)

The seminal work of George C. Williams (1966) and Richard Dawkins (1976, 1982), with their position that the gene is the principal unit of selection, helped facilitate the emergence of the *unit of selection debate* as another topic in the philosophy of biology. Critics of this position have included Stephen Gould (2002) as well as Elliott Sober and Richard Lewontin, whose paper titled "Artifact, Cause, and Genic Selection" we have included in this anthology (chapter 20).

In the previous paragraph, we questioned whether a species could be the unit of selection. Many people take it for granted that species are really things, existing in their own right: there are cats and dogs that really exist as our pets, but then there are the species "*felis catus*" and "*canis lupus familiaris*" that really exist, somehow, too. Further, people take it for granted that there are obvious ways to define a species and distinguish one species from another. However, one of the biggest quagmires for biologists and philosophers of biology that exists today has to do with not only demonstrating with universal acceptability the *nature* of species (Are species real? Are they concepts? or just words referring to collections of organisms?), but also coming up with necessary and sufficient conditions for the *definition* of species. As the reader will find out when reading Jerry Coyne and H. Allen Orr's material from their text *Speciation*, included in this anthology (chapter 19), there are multiple species concepts presenting us with multiple conceptual and empirical problems. Thus, the *species problem* is yet another topic in philosophy of biology.

Darwin claimed that natural selection "has been the main but not exclusive means of modification" of species over time (p. 7). As we tried to show in our diagram of the evolutionary sieve above (figure I.1), the organisms in a population are naturally selected to survive because they have the good fortune to be fit or *adapted to* the environment they inhabit. Thus, for many biologists and philosophers throughout the years, natural selection has become tantamount to adaptation. Further, given the importance, power, and simplicity of natural selection, many researchers since Darwin have looked on natural selection (and adaptation) as if it were the primary and, sometimes, *only* evolutionary principle worth considering. This has given rise to yet another significant topic in philosophy of biology known as the *adaptationist debate*, where biologists and philosophers of biology argue over the extent to which natural selection and adaptation are relevant factors in speciation (the emergence of a new species), extinction, and the solidification of some phenotypic trait in an organism or population. Adaptationists like Richard Dawkins (1996) and Daniel Dennett (1995) see natural selection as the most significant factor in evolutionary processes (Dennett calls natural selection a *universal acid*, cutting through and affecting virtually everything in nature), while non-adaptationists like Stephen Gould (2002), Richard Lewontin (1978), and Motoo Kimura (1968, 1983) have argued for other factors such as random genetic drift and exaptation (a trait that is a by-product of adaptation and that has become useful for an organism, as in flight in birds) that have enabled organisms to survive and evolve.

Function is an idea that is presupposed in adaptationism: consider that certain traits such as opposable thumbs, colored wings of birds, the gag reflex, and others, have become solidified in organisms as adaptations because they perform beneficial functions in the lives of these organisms. For example, Darwin maintains in the *Origin*:

> When we see leaf-eating insects green, and bark-feeders mottled-grey; the alpine ptarmigan white in winter, the red-grouse the colour of heather, and the black-grouse that of peaty earth, we must believe that these tints are of service to these birds and insects in preserving them from danger. (p. 72)

In fact, it is arguable that it is the notion of function that principally distinguishes (and has always distinguished) the biological sciences from the chemical and physical sciences (see FitzPatrick, 2000; Nissen, 1997; Hartwell, Hopfield, Leibner, & Murray, 1999). Try and describe a biological trait without talking about its function, and you will likely be *under-describing* that trait. Beginning in the 1960s with Ernest Nagel (1961) and Carl Hempel (1965), philosophers of science and biology have not only proposed a variety of definitions for the term *function*, but also have attempted to specify the conditions where it may be appropriate to predicate functions of natural and artificial things, and to appeal to these functions in explanations. In chapter 11 of our anthology, in his paper titled "The Modern Philosophical Resurrection of Teleology," Mark Perlman offers a survey of the history and current literature concerning biological function. There, he mentions the work of Robert Cummins and Peter Godfrey-Smith, who offer two of the most popular and viable approaches to defining biological function today, and we have included an important paper from each of them in this anthology (chapters 12 and 13).

Near the close of the twentieth century, the principle of natural selection continued to play a key factor in explaining not only animal and human anatomy and physiology, but also psychology and behavior. In this anthology (chapter 23), we have included sections from Edward Wilson's *Sociobiology: The New Synthesis* (1975), where Wilson uses natural selection to explain many animal and human traits, including altruistic behavior. This work would help facilitate the emergence of other topics in the philosophy of biology as they exist today, including *evolutionary ethics* and *evolutionary psychology*.

Throughout his book, Wilson argues for the evolution of cooperation and altruism in animals via natural selection which, although, on the face of it seems counterintuitive to the idea that an individual is naturally selected because of its own fortunate survival traits, still seems sensible to most people. For example, one would think that an animal like a worker bee would *not* have evolved a stinger that ends its life when used; however, since stinging some predator or intruder assists in saving the entire hive (and its own species), then we can see how the worker bee has made it through the evolutionary sieve. The same goes for many other species, especially those group or cooperative species that form tight, interconnected networks.

Now, here's what many considered to be the rub (and, possibly, the genius). Wilson suggests that *human*, ethical, altruistic behavior should be considered as "biologicized" since we humans basically do the exact same kinds of things as animals, at a basic evolutionary biological level. So, consider the possibility that the *only* or *real* reason why a mother takes care of her child, or Mother Theresa helped AIDS victims, or a Good Samaritan comes to your aid, or you don't just randomly take whatever you want from people whenever you want to, etc., is *not* because of a fully conscious, rational choice but, rather, because of basic, biologically based, kin selection. Nowadays, this may not strike us as shocking, but back then it downright angered a lot of people (in fact, at a conference in 1978 someone actually threw water on Wilson, accusing him of being racist; Thacker, 2001). Since Wilson's publication, several thinkers have used plenty of ink debating both (a) whether ethics is biologically based, and (b) the extent to which ethics is affected by our *biology* as well as our conscious, rational decisions to act in social settings through the erection of norms, values, moral principles, and civil laws. Michael Ruse (1986, 1995) has been a well-known proponent of ethics as being biologically based, and has had his share of responders (see Maienschein & Ruse, 1999; Nitecki & Nitecki, 1993; Ayala, 1995). In Part X of this anthology, Sociobiology and Ethics, we have included papers by Robert Axelrod and William Hamilton and Alexander Rosenberg (chapters 24 and 25) that deal with this debate.

Wilson's suggestion to biologicize human behavior in *Sociobiology*, along with other significant research in evolutionary theory, also got people to take seriously the idea that evolution is responsible not only for human anatomy and physiology, but also for certain human psychological and behavioral characteristics that evolved in our past to solve specific problems of survival. The logic here is straightforward: Traits (e.g., organs, capacities, or behaviors) develop in evolutionary history to function as a result of chance mutations and the natural selection of the trait that is most fit, given the environment in which the trait exists. For example, eyes developed in order to see food, prey, mate, or predator; webbed appendages developed to allow an organism to swim more efficiently; and physiological systems in the body developed to serve each specific end – digestion, circulation, and so forth – with the greater and overall end of survival and reproduction. Just as other traits developed functions in some specified evolutionary history, so too, the human brain has developed psychological modules that function so as to react to, interpret, or integrate information presented to it in an environment and produce behavior. Thus, the science of *evolutionary psychology* was born in the late 1980s with its own research program (for example, see the journal *Evolutionary Psychology* at: www.epjournal.net), and it has become a topic in the philosophy of biology. John Tooby and Leda Cosmides have been at the forefront of this science, and we have included sections of their work in this anthology (chapter 26).

It is now a serious suggestion that we might be able to biologicize not only ethics and human psychology, but also any and all human products (like culture and thoughts) as well as the universe itself. The explanatory power of natural selection and evolution has seeped into virtually every one of the sciences. Consider that physicists seriously consider the possibility of evolving universes (Steinhardt & Turok, 2002), economists recognize naturally selected markets (Berg & Kalleberg, 2001), and information scientists consider search engines such as Google to be just like biological traits resulting from a kind of evolutionary "good trick" or Baldwin Effect of usability (Cho & Roy, 2004).

There are a variety of ways in which Darwinism has affected the world, both in positive and negative ways (see Ruse, 2001). Thus, it is understandable that people who still want to believe in a god that is necessary to explain certain aspects of nature might be threatened by the explanatory success of evolutionary theory and science in general. Another important topic in the philosophy of biology – one that has both theoretical and practical importance and impact – is the *evolution–creationism debate*. Darwin gave us design without the proximate need for an intelligent designer and, since some kind of a designer-god has been advocated by serious thinking individuals in the history of Western philosophy (and East, too) since the pre-Socratics, he knew that his ideas would cause a stir amongst believers – this probably accounted for his delay in publishing the *Origin*. Creationists of one sort or another believe that an intelligent designer is necessary to explain the incredible complexity found in nature (see Pennock, 2000; Berra, 1990). The most recent creationists are known as Intelligent Design Creationists, and we have included a paper from one of these creationists, Michael Behe, in this anthology (chapter 29) along with material from Kenneth Miller and Donald Prothero that disputes Behe's arguments (chapters 30 and 28).

One of the more recent topics to be discussed in the philosophy of biology has been the extent to which the new science of evolutionary developmental biology ("evo-devo") has affected, or will affect, the biological sciences. Evo-devo's basic claim is broadly threefold: (a) through our work in mapping genes and manipulating the genetic toolkit that controls the phenotypic traits of animals (e.g., in fruit flies, mice, worms, and others), we have discovered that all animals are constructed out of the same genes; (b) many genes can produce a variety of different traits in different organisms, depending upon the switching on or off of genes (genes being expressed or not) in the very development of the organism; (c) the various species we see around us today actually are the result of genetic switches being turned on or off at various points in the development of an organism throughout evolutionary history. Thus – and here is where there is probably controversy – evolution is not so much the result of wholesale genetic variation in terms of *mutation*, as we talked about earlier in this general introduction, as it is a matter of changing when and where genetic switches will be turned on and off in the development of an organism (Carroll, 2005). One of the fathers of evo-devo, Sean Carroll, has popularized it in publications, and we have included one of his early journal articles in our anthology (chapter 14).

In this section of our general introduction, we have tried to touch upon the topics that comprise much of the subject-matter of philosophy of biology, subject-matter that has emerged in the "and all that" years following 1953. Of course,

the philosophy of biology has a long, varied, and complex history that would take several lifetimes to ingest completely. For virtually up-to-the-minute coverage of present debates in philosophy of biology, the reader can look at *Contemporary Debates in Philosophy of Biology*, also published by Wiley-Blackwell (Ayala & Arp, 2009). Besides the actual selections included in this anthology themselves, the reader is encouraged to look at the further reading material located at the end of this introduction and the section introductions.

References

Audesirk, T., Audesirk, G., & Beyers, B. (2008). *Biology: Life on earth*. Upper Saddle River, NJ: Prentice Hall, Inc.

Ayala, F. (1995). The difference of being human: Ethical behavior as an evolutionary byproduct. In H. Rolston (ed.). *Biology, ethics, and the origins of life* (pp. 113–135). London: Jones and Bartlett.

Ayala, F., & Arp, R. (eds.). (2009). *Contemporary debates in philosophy of biology*. Malden, MA: Wiley-Blackwell.

Barnes, J. (ed.). (1995). History of animals, Parts of animals, Movement of animals, Progression of animals, Generation of animals. In *The complete works of Aristotle*. Princeton, NJ: Princeton University Press.

Berg, I., & Kalleberg, A. (eds.). (2001). *Sourcebook of labor markets: Evolving structures and processes*. London: Springer.

Berra, T. (1990). *Evolution and the myth of creationism*. Stanford: Stanford University Press.

Campbell, N., & Reece, M. (2007). *Biology*. Menlo Park, CA: Benjamin-Cummings.

Carroll, S. (2005). *Endless forms most beautiful: The new science of evo-devo and the making of the animal kingdom*. New York: Norton.

Catechism of the Catholic Church. (1994). New York: Doubleday Books.

Cho, J., & Roy, S. (2004). Impact of search engines on page popularity. *Proceedings of the 13th International Conference on the World Wide Web*. Available at: http://portal. acm.org/citation.cfm?id=988676.

Cuvier, G. (1831). *The animal kingdom: Arranged in conformity with its organization* (trans.P. Latrielle). New York: Carvill Publishers.

Darwin, C. (1839). *Narrative of the surveying voyages of His Majesty's ships Adventure and Beagle between the years 1826 and 1836, describing their examination of the southern shores of South America, and the*

Beagle's circumnavigation of the globe. Journal and remarks. 1832–1836. London: Henry Colburn.

Darwin, C. (1859/1999). *The origin of species by natural selection: Or, the preservation of favoured races in the struggle for life.* New York: Bantam Books.

Darwin, C. (1883). *The variations of animals and plants under domestication.* New York: D. Appleton & Company.

Darwin, C. (1958). *The autobiography of Charles Darwin 1809–1882. With the original omissions restored. Edited and with appendix and notes by his grand-daughter Nora Barlow.* London: Collins.

Darwin, F. (ed.). (1887). *The life and letters of Charles Darwin, including an autobiographical chapter.* London: John Murray.

Dawkins, R. (1976). *The selfish gene.* Oxford: Oxford University Press.

Dawkins, R. (1982). *The extended phenotype.* Oxford: Oxford University Press.

Dawkins, R. (1986). *The blind watchmaker: Why the evidence of evolution reveals a universe without design.* New York: Norton.

Dawkins, R. (1996). *Climbing mount improbable.* New York: Norton.

Dennett, D. (1995). *Darwin's dangerous idea: Evolution and the meanings of life.* New York: Simon & Schuster.

Dobzhansky, T. (1937). *Genetics and the origin of species.* New York: Columbia University Press.

FitzPatrick, W. (2000). *Teleology and the norms of nature.* New York: Garland Publishing.

Gauch, H. (2003). *Scientific method in practice.* Cambridge: Cambridge University Press.

Glick, T., & Kohn, D. (eds.). (1996). *Darwin and evolution: The development of the theory of evolution.* Indianapolis: Hackett Publishing Company.

Gould, S. (2002). *The structure of evolutionary theory.* Cambridge, MA: Harvard University Press.

Harms, W. (2004). *Information and meaning in evolutionary processes.* Cambridge: Cambridge University Press.

Hartwell, H., Hopfield, J., Leibner, S., & Murray, A. (1999). From molecular to modular cell biology. *Nature, 402,* C47–C52.

Hempel, C. (1965). *Aspects of scientific explanation.* New York: The Free Press.

Hodge, J., & Radick, G. (eds.). (2003). *The Cambridge companion to Darwin.* Cambridge: Cambridge University Press.

Huxley, J. (1942). *Evolution: The modern synthesis.* London: Allen & Unwin.

John Paul II, Pope. (1996). Message to the Pontifical Academy of Sciences: On evolution. Available at: www.ewtn.com/library/PAPALDOC/JP961022.HTM.

Kimura, M. (1968). Evolutionary rate at the molecular level. *Nature, 217,* 624–626.

Kimura, M. (1983). *The neutral theory of molecular evolution.* Cambridge: Cambridge University Press.

Konopka, R., & Benzer, S. (1971). Clock mutants of *Drosophila melanogaster. Proceedings of the National Academy of the Sciences, USA, 68,* 2112–2116.

Lewontin, R. (1978). Adaptation. *Scientific American, 239,* 157–169.

Lewontin, R. (1992). Genotype and phenotype. In E. Keller & E. Lloyd (eds.). *Keywords in evolutionary biology* (pp. 137–144). Cambridge, MA: Harvard University Press.

Lyell, C. (1830/1998). *Principles of geology.* New York: Penguin.

Maienschein, J., & Ruse, M. (eds.). (1999). *Biology and the foundation of ethics.* Cambridge, UK: Cambridge University Press.

Malthus, T. (1826). *An essay on the principle of population.* London: John Murray.

Mayr, E. (1942). *Systematics and the origin of species for the viewpoint of a zoologist.* Cambridge, MA: Harvard University Press.

Mayr, E. (1996). The autonomy of biology: The position of biology among the sciences. *Quarterly Review of Biology, 71,* 97–106.

Mayr, E. (2001). *What evolution is.* New York: Basic Books.

McKinney, H. (1971). *Lamarck to Darwin: Contributions to evolutionary biology, 1809–1859.* New York: Coronado Press.

Mills, S., & Beatty, J. (1979). The propensity interpretation of fitness. *Philosophy of Science, 46,* 263–286.

Nagel, E. (1961). *The structure of science.* New York: Harcourt, Brace & World.

National Research Council. (1996). *National science education standards.* Washington, DC: National Academies Press.

National Research Council. (2000). *Inquiry and national science education standards: A guide for teaching and learning.* Washington, DC: National Academies Press.

Nissen, L. (1997). *Teleological language in the life sciences.* Lanham, MD: Rowman & Littlefield.

Nitecki, M., & Nitecki, D. (eds.). (1993). *Evolutionary ethics.* Binghamton, NY: SUNY Press.

Paul, D. (2003). Darwin, social Darwinism, and eugenics. In J. Hodge & G. Radick, (eds.). *The Cambridge companion to Darwin* (pp. 214–239). Cambridge: Cambridge University Press.

Peacocke, A. (2004). *Evolution: The disguised friend of faith?* West Conshohocken, PA: Templeton Foundation Press.

Pennock, R. (2000). *Tower of Babel: The evidence against the new creationism.* Cambridge, MA: Cambridge University Press.

Quammen, D. (2007). *The reluctant Mr. Darwin: An intimate portrait of Charles Darwin and the making of his theory of evolution.* New York: W.W. Norton.

Radick, G. (2003). Is the theory of natural selection independent of its history? In J. Hodge & G. Radick (eds.). *The Cambridge companion to Darwin* (pp. 143–167). Cambridge: Cambridge University Press.

Raven, C., & Walters, S. (1986). *John Ray, naturalist: His life and works.* Cambridge: Cambridge University Press.

Rosenberg, A. (1978). The supervenience of biological concepts. *Philosophy of Science,* 45, 368–386.

Rosenberg, A. (1983). Fitness. *Journal of Philosophy,* 80, 457–473.

Rudwick, M. (1997). *Georges Cuvier, fossil bones, and geological catastrophes: New translations and interpretations of the primary texts.* Chicago: University of Chicago Press.

Ruse, M. (1975). Charles Darwin and artificial selection. *Journal of the History of Ideas,* 36, 339–350.

Ruse, M. (1986). Evolutionary ethics: A phoenix arisen. *Zygon,* 21, 95–112.

Ruse, M. (1995). Evolutionary ethics: A defense. In H. Rolston (ed.). *Biology, ethics, and the origins of life* (pp. 93–112). London: Jones & Bartlett.

Ruse, M. (2001). *The evolution wars: A guide to the debates.* New Brunswick, NJ: Rutgers University Press.

Ruse, M. (2003). *Darwin and design: Does evolution have a purpose?* Cambridge, MA: Harvard University Press.

Ruse, M. (ed.). (2008). *The Oxford handbook of philosophy of biology.* Oxford: Oxford University Press.

Schaffner, K. (1967). Approaches to reduction. *Philosophy of Science,* 34, 137–147.

Shapiro, F. (ed.). (2006). *The Yale book of quotations.* New Haven: Yale University Press.

Smart, J. (1959). Can biology be an exact science? *Synthese,* 11, 359–368.

Sober, E. (1999). The multiple realizability argument against reductionism. *Philosophy of Science,* 66, 542–564.

Sober, E. (2000). "The multiple realizability argument against reductionism." In *Philosophy of Biology.* Boulder, CO: Westview Press.

Solomon, J., & Hunt, A. (1992). *What is science?* Hatfield, UK: Association for Science Education.

Steinhardt, P., & Turok, N. (2002). Cosmic evolution in a cyclical universe. *Physical Review D,* 65, 12.

Sterelny, K., & Griffiths, P. (1999). *Sex and death: An introduction to philosophy of biology.* Chicago: University of Chicago Press.

Strickberger, M. (1985). *Genetics.* New York: Macmillan Publishing Company.

Strickberger, M. (2000). *Evolution.* Sudbury, MA: Jones and Bartlett.

Thacker, P. (2001). Interview: Edward O. Wilson. *HMS Beagle: The BioMedNet Magazine, 119.*

Watson, J., & Crick, F. (1953a). A structure for deoxyribose nucleic acid. *Nature, 171,* 737–738.

Watson, J., & Crick, F. (1953b). Genetical implications of the structure of deoxyribonucleic acid. *Nature, 171,* 964–967.

Williams, G. (1966). *Adaptation and natural selection.* Princeton, NJ: Princeton University Press.

Wilner, E. (2006). Darwin's artificial selection as an experiment. *Studies in History and Philosophy of Biology and Biomedical Sciences, 37,* 26–40.

Further Reading in Philosophy of Science

Balashov, Y. (ed.). (2002). *Philosophy of science: Contemporary readings.* London: Routledge.

Boyd, R., Gasper, R., & Trout, J. (eds.). (1991). *The philosophy of science.* Cambridge, MA: MIT Press.

Curd, M., & Cover, J. (eds.). (1998). *Philosophy of science: The central issues.* New York: W. W. Norton.

Godfrey-Smith, P. (2003). *Theory and reality: An introduction to the philosophy of science.* Chicago: University of Chicago Press.

Klemke, E., Hollinger, R., & Rudge, D. (eds.). (1998). *Introductory readings in the philosophy of science.* New York: Prometheus.

Kuipers, T. (ed.). (2007). *General philosophy of science: Focal issues.* London: Elsevier.

Ladyman, J. (2002). *Understanding philosophy of science.* London: Routledge.

Losee, J. (2001). *A historical introduction to the philosophy of science.* Oxford: Oxford University Press.

Machamer, P., & Silberstein, M. (eds.). (2002). *The Blackwell guide to the philosophy of science.* Malden, MA: Blackwell Publishers.

Newton-Smith, W. (ed.). (2001). *A companion the philosophy of science.* London: Blackwell Publishers.

Okasha, S. (2002). *Philosophy of science: A very short introduction.* Oxford: Oxford University Press.

Papineau, D. (ed.). (1996). *The philosophy of science.* Oxford: Oxford University Press.

Rosenberg, A. (2005). *Philosophy of science: A contemporary introduction.* London: Routledge.

Rosenberg, A. (2008). *Philosophy of social science.* Boulder, CO: Westview Press.

Salmon, M., Earman, J., Glymour, C., Lennox, J., Schaffner, K., Salmon, W., et al., (1999). *An introduction to the philosophy of science.* Indianapolis: Hackett Publishing Company.

Schick, T. (ed.). (1999). *Readings in the philosophy of science: From positivism to postmodernism.* New York: McGraw-Hill.

Further Reading in Philosophy of Biology

Ayala, F., & Arp, R. (eds.). (2009). *Contemporary debates in philosophy of biology*. Malden, MA: Wiley-Blackwell.

Garvey, B. (2007). *Philosophy of biology*. Montreal: McGill-Queen's University Press.

Grene, M., & Depew, D. (2004). *The philosophy of biology: An episodic history*. Cambridge: Cambridge University Press.

Hull, D., & Ruse, M. (eds.). (2007). *The Cambridge companion to the philosophy of biology*. Cambridge: Cambridge University Press.

Rosenberg, A., & McShea, D. (2008). *Philosophy of biology: A contemporary introduction*. London: Routledge.

Ruse, M. (ed.). (2008). *The Oxford handbook of philosophy of biology*. Oxford: Oxford University Press.

Sarkar, S., & Plutynski, A. (eds.). (2008). *A companion to the philosophy of biology*. Malden, MA: Wiley-Blackwell.

Further Reading in the History of Science and Biology

Bowler, P. (2003). *Evolution: The history of an idea*. Berkeley: University of California Press.

Cormack, L., & Ede, A. (2004). *A history of science in society: From philosophy to utility*. New York: Broadview Press.

Ede, A., & Cormack, L. (2007). *A history of science in society: A reader*. New York: Broadview Press.

French, R., & Greenaway, F. (eds.). (1986). *Science in the early Roman Empire: Pliny the Elder, his sources and influence*. London: Croom Helm.

Gaukroger, S. (2007). *The emergence of a scientific culture: Science and the shaping of modernity, 1210–1685*. New York: Oxford University Press.

Grant, E. (1996). *The foundations of modern science in the Middle Ages: Their religious, institutional and intellectual contexts*. Cambridge: Cambridge University Press.

Grant, E. (2007). *A history of natural philosophy: From the ancient world to the nineteenth century*. Cambridge: Cambridge University Press.

Gregory, F. (2007). *Natural science in Western history, Volumes 1 & 2*. Boston: Houghton Mifflin.

Grene, M., & Depew, D. (2004). *The philosophy of biology: An episodic history*. Cambridge: Cambridge University Press.

Gribbin, J. (2003). *Science: A history*. New York: Penguin Books.

Gribbin, J. (2004). *The scientists: A history of science told through the lives of its greatest inventors*. New York: Random House.

Hull, D. (2008). The history of the philosophy of biology. In M. Ruse (ed.). *The Oxford handbook of philosophy of biology* (pp. 11–33). Oxford: Oxford University Press.

Larson, E. (2004). *Evolution: The remarkable history of a scientific theory*. New York: Random House.

Lindberg, D. (2008). *The beginnings of Western science: The European scientific tradition in philosophical, religious, and institutional context, prehistory to AD 1450*. Chicago: University of Chicago Press.

Lindberg, D., & Numbers, R. (eds.). (2002–2009). *The Cambridge history of science, vols. 3–7*. Cambridge: Cambridge University Press.

Magner, L. (2002). *A history of the life sciences*. New York: Marcel Dekker.

McClellan, J., & Dorn, H. (1999). *Science and technology in world history: An introduction*. Baltimore: Johns Hopkins University Press.

Moore, J. (1999). *Science as a way of knowing: The foundations of modern biology*. Cambridge: MA: Harvard University Press.

Morange, M. (2000). *A history of molecular biology* (trans. M. Cobb,). Cambridge: MA: Harvard University Press.

Rubenstein, R. (2003). *Aristotle's children: How Christians, Muslims, and Jews rediscovered ancient wisdom and illuminated the Dark Ages*. New York: Harcourt.

PART I

BASIC PRINCIPLES AND PROOFS OF DARWINISM

Introduction

Several of the topics in philosophy of biology exist because of Darwin's ideas concerning evolutionary biology, many of which can be found in the *The Origin of Species by Natural Selection: Or, the Preservation of Favoured Races in the Struggle for Life* (1859/1999). Also, every topic in philosophy of biology touches upon Darwin's principles in some way, and vice versa. Further, 2009, the year of this anthology's publication, marks the one hundred and fiftieth anniversary of the publication of the *Origin*, as well as the two hundredth anniversary of Darwin's birth. So, it is perhaps fitting that we begin the material in this anthology with a general consideration of a few of the basic principles and proofs surrounding Darwinism. *Darwinism* has meant many things throughout the past one hundred and fifty years, including referring to the ideas surrounding Darwin's grandfather, Erasmus Darwin (1731–1802), who put forward his own version of evolution. For us here in this anthology (and others), Darwinism refers to the basic principles surrounding evolutionary theory such as variation, inheritance, population increase, struggle for existence, differential survival, differential production, and natural selection. Our two selections in this section are devoted mainly to a general introduction to these principles.

Every student of the philosophy of biology needs to read at least some of Darwin's great work, to see the wealth of his evidence and the force of his arguments. We include extracts from two chapters of the *Origin* that identify the main outlines of his theory. Other chapters, not included but easy to locate (on the Web for example), dispose of misunderstandings and objections – for example, chapter 6 deals with the argument that organs of extreme perfection such as the eye could not be the result of a gradual process of natural selection operating over millions of years. Darwin's reply is relevant to the last controversy taken up in this anthology, the debate about "intelligent design."

Scott's chapter treats other objections to the theory based on common misunderstandings, especially the belief that the appearance of design – adaptation – can only be explained by the reality of a designer-god. Besides describing natural selection – along with other ideas, concepts, proofs, and principles surrounding evolution – she also speaks about the nature of science, scientific practices, and how it is that hypotheses surrounding evolutionary theory might be tested.

Further Reading

Audesirk, T., Audesirk, G., & Beyers, B. (2008). *Biology: Life on earth*. Upper Saddle River, NJ: Prentice Hall, Inc.

Berra, T. (1990). *Evolution and the myth of creationism*. Stanford: Stanford University Press.

Campbell, N., & Reece, M. (2007). *Biology*. Menlo Park, CA: Benjamin-Cummings.

Darwin, C. (1859/1999). *The origin of species by nat-*
ural selection: Or, the preservation of favoured races
in the struggle for life. New York: Bantam Books.

Dennett, D. (1995). *Darwin's dangerous idea: Evolution*
and the meanings of life. New York: Simon &
Schuster.

Eldredge, N. (1995). *Reinventing Darwin: The great*
debate at the high table of evolutionary theory. New
York: John Wiley.

Eldredge, N. (2001). *The triumph of evolution and the*
failure of creationism. New York: Henry Holt.

Fry, I. (2000). *The emergence of life on earth: A histor-*
ical and scientific overview. Brunswick, NJ: Rutgers
University Press.

Gould, S. (2002). *The structure of evolutionary theory.*
Cambridge, MA: Harvard University Press.

Griesemer, J. (2008). Origins of life studies. In
M. Ruse (ed.), *The Oxford handbook of philosophy of*
biology (pp. 263–290). Oxford: Oxford University
Press.

Hodge, J., & Radick, G. (eds.) (2003). *The Cambridge*
companion to Darwin. Cambridge: Cambridge
University Press.

Hull, D. (2001). *Science and selection: Essays on*
biological evolution and the philosophy of science.
Cambridge: Cambridge University Press.

Koshland, D. (2002). The seven pillars of life. *Science,*
295, 2215–2216.

Lange, M. (1996). Life, artificial Life, and scientific
explanation. *Philosophy of Science, 63,* 225–244.

Mayr, E. (1991). *One long argument: Charles Darwin*
and the genesis of modern evolutionary thought.
Cambridge, MA: Harvard University Press.

Mayr, E. (2001). *What evolution is.* New York: Basic
Books.

Prothero, D. (2007). *Evolution: What the fossils say and*
why it matters. New York: Columbia University Press.

Rosenberg, A., & McShea, D. (2008). *Philosophy of*
biology: A contemporary introduction (Chapter 1).
London: Routledge.

Ruse, M. (2001). *The evolution wars: A guide to the*
debates. New Brunswick, NJ: Rutgers University
Press.

Ruse, M. (2006). *Darwinism and its discontents.*
Cambridge: Cambridge University Press.

Ruse, M. (2008). Darwinian evolutionary theory:
Its structure and its mechanism. In M. Ruse (ed.),
The Oxford handbook of philosophy of biology
(pp. 34–63). Oxford: Oxford University Press.

Skelton, P. (ed.) (1993). *Evolution: A biological and*
palaeontological approach. Upper Saddle River, NJ:
Prentice-Hall.

1

Struggle for Existence and Natural Selection

Charles Darwin

Before entering on the subject of this chapter, I must make a few preliminary remarks, to show how the struggle for existence bears on Natural Selection. It has been seen in the last chapter that amongst organic beings in a state of nature there is some individual variability; indeed I am not aware that this has ever been disputed. It is immaterial for us whether a multitude of doubtful forms be called species or sub-species or varieties; what rank, for instance, the two or three hundred doubtful forms of British plants are entitled to hold, if the existence of any well-marked varieties be admitted. But the mere existence of individual variability and of some few well-marked varieties, though necessary as the foundation for the work, helps us but little in understanding how species arise in nature. How have all those exquisite adaptations of one part of the organization to another part, and to the conditions of life, and of one distinct organic being to another being, been perfected? We see these beautiful co-adaptations most plainly in the woodpecker and mistletoe; and only a little less plainly in the humblest parasite which clings to the hairs of a quadruped or feathers of a bird; in the structure of the beetle which dives through the water; in the plumed seed which is wafted by the gentlest breeze; in short, we see beautiful adaptations everywhere and in every part of the organic world.

Again, it may be asked, how is it that varieties, which I have called incipient species, become ultimately converted into good and distinct species, which in most cases obviously differ from each other far more than do the varieties of the same species? How do those groups of species, which constitute what are called distinct genera, and which differ from each other more than do the species of the same genus, arise? All these results, as we shall more fully see in the next chapter, follow inevitably from the struggle for life. Owing to this struggle for life, any variation, however slight and from whatever cause proceeding, if it be in any degree profitable to an individual of any species, in its infinitely complex relations to other organic beings and to external nature, will tend to the preservation of that individual, and will generally be inherited by its offspring. The offspring, also, will thus have a better chance of surviving, for, of the many individuals of any species which are periodically born, but a small number can survive. I have called this principle, by which each slight variation, if useful, is preserved, by the term of Natural Selection, in order to mark its relation to man's power of selection. We have seen that man by selection can certainly produce great results, and can adapt organic beings to his own uses, through the accumulation of slight but useful variations, given to him by the

Charles Darwin, of *The Origin of Species* (1859), excerpts from chapters 3 and 4. Reprinted by permission of Random House, Inc.

hand of Nature. But Natural Selection, as we shall hereafter see, is a power incessantly ready for action, and is as immeasurably superior to man's feeble efforts, as the works of Nature are to those of Art.

We will now discuss in a little more detail the struggle for existence. In my future work this subject shall be treated, as it well deserves, at much greater length. The elder De Candolle and Lyell have largely and philosophically shown that all organic beings are exposed to severe competition. In regard to plants, no one has treated this subject with more spirit and ability than W. Herbert, Dean of Manchester, evidently the result of his great horticultural knowledge. Nothing is easier than to admit in words the truth of the universal struggle for life, or more difficult at least I have found it so than constantly to bear this conclusion in mind. Yet unless it be thoroughly engrained in the mind, I am convinced that the whole economy of nature, with every fact on distribution, rarity, abundance, extinction, and variation, will be dimly seen or quite misunderstood. We behold the face of nature bright with gladness, we often see superabundance of food; we do not see, or we forget, that the birds which are idly singing round us mostly live on insects or seeds, and are thus constantly destroying life; or we forget how largely these songsters, or their eggs, or their nestlings are destroyed by birds and beasts of prey; we do not always bear in mind, that though food may be now superabundant, it is not so at all seasons of each recurring year.

I should premise that I use the term Struggle for Existence in a large and metaphorical sense, including dependence of one being on another, and including (which is more important) not only the life of the individual, but success in leaving progeny. Two canine animals in a time of dearth, may be truly said to struggle with each other which shall get food and live. But a plant on the edge of a desert is said to struggle for life against the drought, though more properly it should be said to be dependent on the moisture. A plant which annually produces a thousand seeds, of which on an average only one comes to maturity, may be more truly said to struggle with the plants of the same and other kinds which already clothe the ground. The mistletoe is dependent on the apple and a few other trees, but can only in a far-fetched sense be said to struggle with these trees, for if too many of these parasites grow on the same tree, it will languish and die. But several seedling mistletoes, growing close together on the same branch, may more truly be said to struggle with each other. As the mistletoe is disseminated by birds, its existence depends on birds; and it may metaphorically be said to struggle with other fruit-bearing plants, in order to tempt birds to devour and thus disseminate its seeds rather than those of other plants. In these several senses, which pass into each other, I use for convenience sake the general term of struggle for existence.

A struggle for existence inevitably follows from the high rate at which all organic beings tend to increase. Every being, which during its natural lifetime produces several eggs or seeds, must suffer destruction during some period of its life, and during some season or occasional year, otherwise, on the principle of geometrical increase, its numbers would quickly become so inordinately great that no country could support the product. Hence, as more individuals are produced than can possibly survive, there must in every case be a struggle for existence, either one individual with another of the same species, or with the individuals of distinct species, or with the physical conditions of life. It is the doctrine of Malthus applied with manifold force to the whole animal and vegetable kingdoms; for in this case there can be no artificial increase of food, and no prudential restraint from marriage. Although some species may be now increasing, more or less rapidly, in numbers, all cannot do so, for the world would not hold them.

There is no exception to the rule that every organic being naturally increases at so high a rate, that if not destroyed, the earth would soon be covered by the progeny of a single pair. Even slow-breeding man has doubled in twenty-five years, and at this rate, in a few thousand years, there would literally not be standing room for his progeny. Linnaeus has calculated that if an annual plant produced only two seeds and there is no plant so unproductive as this and their seedlings next year produced two, and so on, then in twenty years there would be a million plants. The elephant is reckoned to be the slowest breeder of all known animals, and I have taken some pains to estimate its probable minimum rate of natural increase: it will be under the mark to assume that it breeds when thirty years old, and goes on breeding till ninety years old, bringing forth three pairs of young in this interval; if this be so, at the end of the fifth century there would be alive fifteen million elephants, descended from the first pair.

[. . .]

In looking at Nature, it is most necessary to keep the foregoing considerations always in mind never to forget that every single organic being around us may be said to be striving to the utmost to increase in numbers; that each lives by a struggle at some period of its life; that heavy destruction inevitably falls either on the young or old, during each generation or at recurrent intervals. Lighten any check, mitigate the destruction ever so little, and the number of the species will almost instantaneously increase to any amount. The face of Nature may be compared to a yielding surface, with ten thousand sharp wedges packed close together and driven inwards by incessant blows, sometimes one wedge being struck, and then another with greater force.

[. . .]

To sum up the circumstances favorable and unfavorable to natural selection, as far as the extreme intricacy of the subject permits. I conclude, looking to the future, that for terrestrial productions a large continental area, which will probably undergo many oscillations of level, and which consequently will exist for long periods in a broken condition, will be the most favorable for the production of many new forms of life, likely to endure long and to spread widely. For the area will first have existed as a continent, and the inhabitants, at this period numerous in individuals and kinds, will have been subjected to very severe competition. When converted by subsidence into large separate islands, there will still exist many individuals of the same species on each island: intercrossing on the confines of the range of each species will thus be checked: after physical changes of any kind, immigration will be prevented, so that new places in the polity of each island will have to be filled up by modifications of the old inhabitants; and time will be allowed for the varieties in each to become well modified and perfected. When, by renewed elevation, the islands shall be re-converted into a continental area, there will again be severe competition: the most favored or improved varieties will be

enabled to spread: there will be much extinction of the less improved forms, and the relative proportional numbers of the various inhabitants of the renewed continent will again be changed; and again there will be a fair field for natural selection to improve still further the inhabitants, and thus produce new species.

That natural selection will always act with extreme slowness, I fully admit. Its action depends on there being places in the polity of nature, which can be better occupied by some of the inhabitants of the country undergoing modification of some kind. The existence of such places will often depend on physical changes, which are generally very slow, and on the immigration of better adapted forms having been checked. But the action of natural selection will probably still oftener depend on some of the inhabitants becoming slowly modified; the mutual relations of many of the other inhabitants being thus disturbed. Nothing can be effected, unless favorable variations occur, and variation itself is apparently always a very slow process. The process will often be greatly retarded by free intercrossing. Many will exclaim that these several causes are amply sufficient wholly to stop the action of natural selection. I do not believe so. On the other hand, I do believe that natural selection will always act very slowly, often only at long intervals of time, and generally on only a very few of the inhabitants of the same region at the same time. I further believe, that this very slow, intermittent action of natural selection accords perfectly well with what geology tells us of the rate and manner at which the inhabitants of this world have changed.

Slow though the process of selection may be, if feeble man can do much by his powers of artificial selection, I can see no limit to the amount of change, to the beauty and infinite complexity of the coadaptations between all organic beings, one with another and with their physical conditions of life, which may be effected in the long course of time by nature's power of selection.

Extinction

[. . .] Natural selection acts solely through the preservation of variations in some way advantageous, which consequently endure. But as from the high geometrical powers of increase of all organic beings, each area is already fully stocked with inhabitants, it follows that as each selected and favored form increases in number, so will the less favored forms decrease and become rare. Rarity, as geology tells us, is the precursor to extinction. We can, also, see that any form represented by few individuals will, during fluctuations in the seasons or in the number of its enemies, run a good chance of utter extinction. But we may go further than this; for as new forms are continually and slowly being produced, unless we believe that the number of specific forms goes on perpetually and almost indefinitely increasing, numbers inevitably must become extinct. That the number of specific forms has not indefinitely increased, geology shows us plainly; and indeed we can see reason why they should not have thus increased, for the number of places in the polity of nature is not indefinitely great, not that we have any means of knowing that any one region has as yet got its maximum of species, probably no region is as yet fully stocked, for at the Cape of Good Hope, where more species of plants are crowded together than in any other quarter of the world, some foreign plants have become naturalized, without causing, as far as we know, the extinction of any natives.

Furthermore, the species which are most numerous in individuals will have the best chance of producing within any given period favorable variations. We have evidence of this, in the facts given in the second chapter, showing that it is the common species which afford the greatest number of recorded varieties, or incipient species. Hence, rare species will be less quickly modified or improved within any given period, and they will consequently be beaten in the race for life by the modified descendants of the commoner species.

From these several considerations I think it inevitably follows, that as new species in the course of time are formed through natural selection, others will become rarer and rarer, and finally extinct. The forms which stand in closest competition with those undergoing modification and improvement, will naturally suffer most. And we have seen in the chapter on the Struggle for Existence that it is the most closely-allied forms, varieties of the same species, and species

of the same genus or of related genera, which, from having nearly the same structure, constitution, and habits, generally come into the severest competition with each other. Consequently, each new variety or species, during the progress of its formation, will generally press hardest on its nearest kindred, and tend to exterminate them. We see the same process of extermination amongst our domesticated productions, through the selection of improved forms by man. Many curious instances could be given showing how quickly new breeds of cattle, sheep, and other animals, and varieties of flowers, take the place of older and inferior kinds. In Yorkshire, it is historically known that the ancient black cattle were displaced by the long-horns, and that these 'were swept away by the short-horns' (I quote the words of an agricultural writer) 'as if by some murderous pestilence.'

[. . .]

Summary

If during the long course of ages and under varying conditions of life, organic beings vary at all in the several parts of their organization, and I think this cannot be disputed; if there be, owing to the high geometrical powers of increase of each species, at some age, season, or year, a severe struggle for life, and this certainly cannot be disputed; then, considering the infinite complexity of the relations of all organic beings to each other and to their conditions of existence, causing an infinite diversity in structure, constitution, and habits, to be advantageous to them, I think it would be a most extraordinary fact if no variation ever had occurred useful to each being's own welfare, in the same way as so many variations have occurred useful to man. But if variations useful to any organic being do occur, assuredly individuals thus characterized will have the best chance of being preserved in the struggle for life; and from the strong principle of inheritance they will tend to produce offspring similarly characterized. This principle of preservation, I have called, for the sake of brevity, Natural Selection. Natural selection, on the principle of qualities being inherited at corresponding ages, can modify the egg, seed, or young, as easily as the adult. Amongst many animals, sexual selection will give its aid to ordinary selection, by assuring to the most vigorous and best adapted males the greatest number of offspring. Sexual selection will also give characters useful to the males alone, in their struggles with other males.

Whether natural selection has really thus acted in nature, in modifying and adapting the various forms of life to their several conditions and stations, must be judged of by the general tenor and balance of evidence given in the following chapters. But we already see how it entails extinction; and how largely extinction has acted in the world's history, geology plainly declares. Natural selection, also, leads to divergence of character; for more living beings can be supported on the same area the more they diverge in structure, habits, and constitution, of which we see proof by looking at the inhabitants of any small spot or at naturalized productions. Therefore during the modification of the descendants of any one species, and during the incessant struggle of all species to increase in numbers, the more diversified these descendants become, the better will be their chance of succeeding in the battle of life. Thus the small differences distinguishing varieties of the same species, will steadily tend to increase till they come to equal the greater differences between species of the same genus, or even of distinct genera.

We have seen that if is the common, the widely-diffused, and widely-ranging species, belonging to the larger genera, which vary most; and these will tend to transmit to their modified offspring that superiority which now makes them dominant in their own countries. Natural selection, as has just been remarked, leads to divergence of character and to much extinction of the less improved and intermediate forms of life. On these principles, I believe, the nature of the affinities of all organic beings may be explained. It is a truly wonderful fact the wonder of which we are apt to overlook from familiarity that all animals and all plants throughout all time and space should be related to each other in group subordinate to group, in the manner which we everywhere behold namely, varieties of the same species most closely related together, species of the same genus less closely and unequally related together, forming sections and sub-genera, species of distinct genera much less

closely related, and genera related in different degrees, forming sub-families, families, orders, sub-classes, and classes. The several subordinate groups in any class cannot be ranked in a single file, but seem rather to be clustered round points, and these round other points, and so on in almost endless cycles. On the view that each species has been independently created, I can see no explanation of this great fact in the classification of all organic beings; but, to the best of my judgment, it is explained through inheritance and the complex action of natural selection, entailing extinction and divergence of character.

The affinities of all the beings of the same class have sometimes been represented by a great tree. I believe this simile largely speaks the truth. The green and budding twigs may represent existing species; and those produced during each former year may represent the long succession of extinct species. At each period of growth all the growing twigs have tried to branch out on all sides, and to overtop and kill the surrounding twigs and branches, in the same manner as species and groups of species have tried to overmaster other species in the great battle for life. The limbs divided into great branches, and these into lesser and lesser branches, were themselves once, when the tree was small, budding twigs; and this connexion of the former and present buds by ramifying branches may well represent the classification of all extinct and living species in groups subordinate to groups. Of the many twigs which flourished when the tree was a mere bush, only two or three, now grown into great branches, yet survive and bear all the other branches; so with the species which lived during long-past geological periods, very few now have living and modified descendants. From the first growth of the tree, many a limb and branch has decayed and dropped off; and these lost branches of various sizes may represent those whole orders, families, and genera which have now no living representatives, and which are known to us only from having been found in a fossil state. As we here and there see a thin straggling branch springing from a fork low down in a tree, and which by some chance has been favored and is still alive on its summit, so we occasionally see an animal like the Ornithorhynchus or Lepidosiren, which in some small degree connects by its affinities two large branches of life, and which has apparently been saved from fatal competition by having inhabited a protected station. As buds give rise by growth to fresh buds, and these, if vigorous, branch out and overtop on all sides many a feebler branch, so by generation I believe it has been with the great Tree of Life, which fills with its dead and broken branches the crust of the earth, and covers the surface with its ever branching and beautiful ramifications.

2

Evolution

Eugenie C. Scott

Evolution Broad and Narrow

It has been my experience as both a college professor and a longtime observer of the creationism/evolution controversy that most people have a definition of evolution rather different from that of scientists. To the question "What does evolution mean?" most people will answer, "Man evolved from monkeys" or "molecules to man." Setting aside the sex-specific language (surely no one believes that only males evolved; reproduction is challenging enough without trying to do it using only one sex), both definitions are much too narrow. Evolution involves far more than just human beings and, for that matter, far more than just living things.

The broad definition of evolution is "a cumulative change through time," and refers to the fact that the universe has had a history – that if we were able to go back into time, we would find different stars, galaxies, planets, and different forms of life on Earth. Stars, galaxies, planets, and living things have changed through time. There is astronomical evolution, geological evolution, and biological evolution. Evolution, far from being "Man evolved from monkeys," is thus integral to astronomy, geology, and biology. As we will see, it is relevant to physics and chemistry as well.

Evolution needs to be defined more narrowly within each of the scientific disciplines because both the phenomena studied and the processes and mechanisms of cosmological, geological, and biological evolution are different. Astronomical evolution deals with cosmology: the origin of elements, stars, galaxies, and planets. Geological evolution is concerned with the evolution of our own planet: its origin and its cumulative changes through time. Mechanisms of astronomical and geological evolution involve the laws and principles of physics and chemistry: thermodynamics, heat, cold, expansion, contraction, erosion, sedimentation, and the like. In biology, evolution is the inference that living things share common ancestors and have, in Darwin's words, "descended with modification" from these ancestors. The main – but not the only – mechanism of biological evolution is natural selection. Although biological evolution is the most contentious aspect of the teaching of evolution in public schools, some creationists raise objections to astronomical and geological evolution as well.

Astronomical and Chemical Evolution

Cosmologists conclude that the universe as we know it today originated from the "big bang"

Eugenie Scott, of *Evolution vs. Creationism: An Introduction* (University of Carolina Press, 2004), excerpts from chapters 1 and 2. Reproduced with permission of Greenwood Publishing Group, Inc., Westport, CT.

explosion that erupted from a compact, extremely dense mass and spread outward in all directions. Astronomers have found evidence that stars evolved from gravitational effects on swirling gases left over from the big bang. Stars combined into galaxies, the total number of which is only dimly perceived (Silk 1994). Beginning with helium, heavier elements were formed in the energy-rich cores of young stars through atomic fusion. Thus the elements were formed over approximately 10–12 billion years.

Cosmologists and geologists tell us that between 4 and 5 billion years ago the planet Earth formed from the accumulation of matter encircling our sun. In earlier times Earth looked far different from what we see today: an inhospitable place scorched by radiation, bombarded by meteorites and comets, and belching noxious chemicals from volcanoes and massive cracks in the planet's crust. Yet it is hypothesized that Earth's atmosphere evolved from that outgassing, and water might well have been brought to our planet's surface by those comets crashing into it.

Earth was bombarded by meteors and comets until about 3.8 billion years ago. In such an environment, life could not have survived. Shortly after the bombardment ceased, however, primitive replicating structures appeared. Currently, there is not yet a consensus about how these first living things originated, and there are several directions of active research. Before there were living creatures, of course, there had to be organic (carbon-containing) molecules. Presently, these molecules are produced only by living things, so the question of chemical "prebiotic" evolution involves developing plausible scenarios for the emergence of organic molecules such as sugars, purines, and pyrimidines, and the building blocks of life, amino acids.

To explore this question, in the 1950s scientists began experimenting to see whether organic compounds could be formed from methane, ammonia, water vapor, and hydrogen gases that researchers believed would have been present in Earth's early atmosphere. By electrically sparking combinations of gases, they were able to produce most of the amino acids that occur in proteins – and also the same amino acids that are found in meteorites – as well as other organic molecules (Miller 1992: 19). Because the actual composition of Earth's early atmosphere is not known, investigators have tried sparking various combinations of gases as well as the original blend. These also produce amino acids (Rode 1999: 774). Apparently, organic molecules form spontaneously on Earth and elsewhere, leading one investigator to conclude, "There appears to be a universal organic chemistry, one that is manifest in interstellar space, occurs in the atmospheres of the major planets of the solar system, and must also have occurred in the reducing atmosphere of the primitive Earth" (Miller 1992: 20).

For life to emerge, some organic molecules had to be formed and then combined into amino acids and proteins, while other organic molecules had to be combined into something that could replicate: some material that could pass information from generation to generation. Modern living things are composed of cells, and cells are set off from their environments and are recognizable entities because their constituent parts are surrounded by membranes. Much origin-of-life research focuses on explaining the origin of proteins, heredity material, and membranes.

Origin-of-life researchers joke about their models falling into two camps: "heaven" and "hell." "Hellish" theories for the origin of life point to the present-day existence of some of the simplest known forms of life in severe environments, both hot and cold. Some primitive forms of life live in hot deep-sea vents where sulfur compounds and heat provide the energy used to carry on metabolism and reproduction. Could such an environment have been the breeding ground of the first primitive forms of life? Other scientists are discovering that primitive bacteria can be found in permanently or nearly permanently frozen environments in the Arctic and Antarctic. Perhaps deep in ice or deep in the sea, protected from harmful ultraviolet radiation, organic molecules assembled into primitive replicating structures.

More "heavenly" theories note that organic molecules occur spontaneously in dust clouds of space and that amino acids have been found in meteorites. Perhaps these basic components of life were brought by these rocky visitors from outer space and combined in Earth's waters to form replicating structures.

Origin of Life

Whether the proponents of hell or heaven finally convince their rivals as to the most plausible scenario for the origin of the first replicating structures, it is clear that the origin of life is not a simple issue. One problem is the definition of life itself. From the ancient Greeks up through the early nineteenth century, people from European cultures believed that living things possessed an élan vital or vital spirit – a quality that sets them apart from dead things and nonliving things such as minerals or water. Organic molecules, in fact, were thought to differ from other molecules because of the presence of this spirit. This view was gradually abandoned in science when more detailed study on the structure and functioning of living things repeatedly failed to discover any evidence for such an élan vital, and when it was realized that organic molecules could be synthesized from inorganic chemicals. Vitalistic ways of thinking persist in some East Asian philosophies, such as in the concept of *chi*.

So how do we define life, then? According to generally agreed-upon scientific definition, if something is "living," it is able to acquire and use energy and to reproduce. The simplest living things today are primitive bacteria, enclosed by a membrane and not containing very many "moving parts." But they can take in and use energy, and they can reproduce by division. Even this definition is fuzzy, though: What about viruses? The microscopic viruses are hardly more than hereditary material in a packet – a protein shell. Are they alive? Well, they reproduce. They sort of use energy, in the sense that they take over a cell's machinery to duplicate their own hereditary material. But they can also form crystals, which no living thing can do, so biologists are divided over whether viruses are "living" or not. They tend to be treated as a separate, special category.

If life itself is difficult to define, you can see why explaining its origin is also going to be difficult. The first cell would have been more primitive than the most primitive bacterium known today, which itself is the end result of a long series of events: no scientist thinks that it popped into being with all its components present and functioning! A simple bacterium is "alive": it takes in energy that enables it to function, and it reproduces (in particular, it duplicates itself through division).

We can recognize that a bacterium can do these things because the components that process the energy and that allow it to divide are enclosed within a membrane; we can recognize a bacterium as an entity: a cell that has several components that in a sense "cooperate." What if there were a single structure that was not enclosed by a membrane, but that nonetheless could conduct a primitive metabolism? Would we consider it "alive"? It is beginning very much to look like the origin of life was not a sudden event, but a continuum of events producing structures that early in the sequence we would agree are not "alive," and at the end of the sequence we would agree are alive, with a lot of iffy stuff in the middle.

We know that virtually all life on Earth today is based on DNA, a chainlike molecule that directs the construction of proteins and enzymes, which in turn produce creatures composed of one cell or trillions. DNA instructs cellular structures to link amino acids in a particular order to form a particular protein or enzyme. It also is the material of heredity, being passed from generation to generation. The structure of DNA is rather simple, considering all it does. DNA codes for amino acids use a "language" of four letters – A (adenine), T (thymine), C (cytosine), and G (guanine) which, combined three at a time, determine the amino acid order of a particular protein. For example, CCA codes for the amino acid proline, and AGU for the amino acid serine. The exception to the generalization that all life is based on DNA is viruses, which can be composed of strands of RNA, another chainlike molecule that is quite similar to DNA. Like DNA, RNA is based on A, C, and G, but uses uracil (U) rather than thymine.

The origin of DNA and proteins is thus of considerable interest to origin-of-life researchers. How did the components of RNA and DNA assemble into these structures? One theory is that clay or calcium carbonate – both latticelike structures – could have provided a foundation upon which primitive chainlike molecules could have formed (Hazen *et al.* 2001). Because RNA is one-stranded rather than two-stranded like DNA, some scientists are building theory around the possibility of a simpler RNA-based organic "world" preceding our current DNA "world" (Joyce 1991; Lewis 1997), and very recently there has been speculation that an even simpler but related chainlike molecule, PNA (peptide nucleic acid) may have preceded the evolution of RNA (Nelson *et al.* 2000). Where did RNA or PNA come from? In a series of experiments combining chemicals available on the early Earth, scientists have been able to synthesize purines and pyrimidines, which form the backbones of DNA and RNA (Miller 1992).

After a replicating structure evolved (whether it was based on PNA or RNA or DNA), the structure had to be enclosed in a membrane and to acquire other bits of machinery to process energy and perform other tasks. The origin of life is a complex but active research area with many interesting avenues being investigated, though there is not yet consensus on the sequence of events that led to living things. But at some point in Earth's early history, perhaps as early as 3.8 billion years ago, life in the form of simple single-celled organisms appeared. Once life evolved, biological evolution became possible.

Although some people confuse the origin of life itself with evolution, the two are conceptually separate. Biological evolution is defined as the descent of living things from ancestors from which they differ. Life had to precede evolution! Regardless of how the first replicating molecule appeared, we see in the subsequent historical record the gradual appearance of more complex living things, and many variations on the many themes of life. We know much more about evolution than about the origin of life.

Biological Evolution

Biological evolution is a subset of the general idea that the universe has changed through time. In the nineteenth century, Charles Darwin spoke of "descent with modification," and that phrase still nicely communicates the essence of biological evolution. "Descent" connotes heredity, and indeed, members of species pass genes from generation to generation. "Modification" connotes change, and indeed, the composition of species may change through time. Descent with modification refers to a genealogical relationship of species through time. Just as an individual genealogy theoretically can be traced back through time, so too can the genealogy of a species. And

just as an individual genealogy has "missing links" – ancestors whose names or other details are uncertain – so too the history of a species is understandably incomplete. Evolutionary biologists are concerned both with the history of life – the tracing of life's genealogy – and the processes and mechanisms that produced the tree of life. This distinction between the patterns of evolution and the processes of evolution is relevant to the evaluation of some of the criticisms of evolution that will emerge later in this book. First, let's look briefly at the history of life.

The history of life

Deep time

The story of life unfurls against a backdrop of time – deep time: the length of time the universe has existed, the length of time that Earth has been a planet, the length of time that life has been on Earth. We are better at understanding things that we can have some experience of, but we have, and can have, no experience of deep time. Most of us can relate to a period of 100 years; a person in his fifties might reflect that 100 years ago, his grandmother was a young woman. A person in her twenties might be able to imagine what life was like for a great-grandparent 100 years ago. Thinking back to the time of Jesus, 2,000 years ago, is more difficult; although we have written descriptions of people's houses, clothes, and how they made their living, there is much we don't know of official as well as everyday life. The ancient Egyptians were building pyramids 5,000 years ago, and their lives and way of life are known in only the sketchiest outlines.

And yet the biological world of 5,000 years ago was virtually identical to ours today. The geological world 5,000 years ago would be quite recognizable: the continents would be in the same places, the Appalachians and Rocky Mountains would look pretty much as they do today, and major features of coastlines would be identifiable. Except for some minor remodeling of Earth's surface due to volcanoes and earthquakes, the filling in of some deltas due to the deposition of sediments by rivers, and some other small changes, little has changed, geologically. But our planet and life on it are far, far older than 5,000 years. We need to measure the age of Earth and the time spans important to the history of life in *billions* of years, a number that we can grasp only in the abstract.

A second is a short period of time. Sixty seconds make up a minute, and 60 minutes make up an hour. There are therefore 3,600 seconds in an hour, 86,400 in a day, 604,800 in a week, and 31,536,000 in a year. But to count to a *billion* seconds at the rate of one a second, you would have to count night and day for approximately 31 years and eight months. The age of Earth is 4.5 billion *years*, not seconds. It is an enormous amount of time. As Stephen Jay Gould remarked, "An abstract, intellectual understanding of deep time comes easily enough – I know how many zeros to place after the 10 when I mean billions. Getting it into the gut is quite another matter. Deep time is so alien that we can really only comprehend it as metaphor" (Gould 1987: 3).

Figure 2.1 presents divisions of geological time used to understand geological and biological evolution. The solar system formed approximately 4.6 billion years ago; Earth formed around 4.5 billion years ago, and life may have begun appearing relatively soon after that. During our planet's early years, however, it was bombarded regularly by comets and meteorites, as was the moon. This may have impeded the development of life: the first evidence of bacteria-like fossils dates from around 3.5 billion years ago, about half a billion years after the bombardment of Earth stopped. As discussed in the section "Astronomical and Chemical Evolution," there was a period of about half a billion years of chemical evolution before the first structures that one might consider "alive" appeared on Earth: primitive bacteria.

After these first living things appeared between 3.5 and 4 billion years ago, life continued to remain simple for more than 2 billion years. Single-celled living things bumped around in water, absorbed energy, and divided – if they weren't absorbed by some other organism first. Reproduction was asexual: when a cell divided, the result was almost always two identical cells. Very slow changes occur with asexual reproduction, and this is probably an important reason that the evolution of life moved so slowly during life's first few billion years.

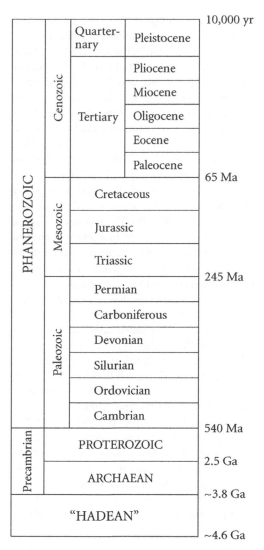

Figure 2.1   Timescale of Earth's history. Courtesy of Alan Gishlick.

Nucleated cells

The first cells lacked a nucleus. Nucleated (eukaryotic) cells didn't evolve until about 1.5 billion years ago. In eukaryotes, the nucleus contains the DNA, which is important to cell division and to directing the processes that living things need to carry out to survive. Around 2 billion years ago, great changes in Earth's surface were taking place, including movement of continents, increases in the amount of oxygen in the atmosphere, and the establishment of an ozone shield that protected living things against ultraviolet radiation (Strickberger 1996: 168). The increase and spread of photosynthesizing single-celled organisms is suggested by the appearance of large red geological deposits. Photosynthesis produces oxygen as a by-product, and the gradual increase of photosynthesizing bacteria resulted in the buildup of oxygen in the atmosphere over hundreds of millions of years. Dissolved iron would oxidize in the presence of free oxygen, thus accounting for the "red beds." In the words of researcher William Schopf, "The Earth's oceans had been swept free of dissolved iron; lowly cyanobacteria – pond scum – had rusted the world!" (Schopf 1992: 48). The increase of oxygen in the atmosphere resulted in a severe change in the environment: many organisms could not live in the new "poisonous" oxygenated environment. Others managed to survive and adapt.

Eukaryotic cells may have evolved as unnucleated cells incorporated by other cells within their membranes. Nucleated cells have structures called *organelles* within their cytoplasm which perform a variety of functions having to do with energy capture and use, cell division, predation, and other activities. Some of these structures, such as mitochondria and chloroplasts, have their own DNA. Similarities between the DNA of such organelles and that of some simple bacteria have supported the theory that early in evolution, the ancestors of eukaryotes absorbed certain bacteria and formed a cooperative or *symbiotic* relationship with them, the newcomers functioning to enhance performance of metabolism, predation, or some other task (Margulis 1993). The nucleus itself may also have been acquired in a similar fashion, from "recycled" parts obtained after the absorption of other bacteria. Evidence for these theories comes, of course, not from the fossil record but from inferences based on biochemical comparisons of living forms.

Once nucleated cells developed, sexual reproduction was not far behind. Sexual reproduction has the advantage of mixing up genetic information, allowing the organism to adapt to environmental change or challenge. Some researchers theorize that geological and atmospheric changes, together with the evolution of sexual reproduction, stimulated a burst of evolutionary activity during the late Precambrian period, about 900 million years ago, when the first multicellular animals, or *metazoans*, appear in the fossil record.

The Cambrian explosion

As might be expected, the first evidence we have of metazoa is of simple creatures like sponges, jellyfish, echinoderms, and wormlike forms. But by the Cambrian, about 500 million years ago, we have rapid divergent evolution of invertebrate groups. Structures like body segmentation, a mouth and an anus, segmented appendages, and other "inventions" characterize new forms of animal life, some of which died out but many of which continue until the present day. These new body plans appear over a geologically sudden – if not biologically sudden – period of about 10–20 million years. Crustaceans, brachiopods, mollusks, and annelid worms, as well as representatives of other groups, appear in the Cambrian.

Evolutionary biologists are studying how these groups are related to one another, and investigating whether they indeed have roots in the Precambrian period. In evolutionary biology, as in the other sciences, theory building depends on cross-checking ideas against different types of data. There are three basic types of data used to investigate the evolutionary relationships among the invertebrate groups: size and shape (morphological) comparisons among modern representatives of these groups, biochemical comparisons among modern descendants, and the fossil record. Largely because of problems of preservation of key fossils at key times – and the fact that evolution might have taken "only" tens of millions of years, an eve blink from the perspective of deep time – the fossil evidence currently does not illuminate links among most of the basic invertebrate groups. Nonetheless, much nonfossil research is being conducted to understand similarities and differences of living members of these groups, from which we may infer evolutionary relationships.

One particularly interesting area of research has to do with understanding the evolution and developmental biology (embryology) of organisms, a new field referred to as "evo-devo."

"Evo-devo"

Advances in molecular biology have permitted developmental biologists to study the genetics behind the early stages of embryological development in many groups of animals. What they are discovering is astounding. It is apparent that very small changes in genes affecting early basic structural development can cause major changes in body plans. For example, there is a group of genes operating very early in animal development that is responsible for determining the basic front-to-back, top-to-bottom, and side-to-side orientations of the body. Other early-acting genes also control such bodily components as segments and their number, and the production of structures such as legs, antennae, and wings. Major changes in body plan can come about through rather small changes in these early-acting genes. What is perhaps the most intriguing result of this research is the discovery of identical or virtually identical early genes in groups as different as insects, worms, and vertebrates. Could some of the body plan differences of invertebrate groups be the result of changes in genes that act early in embryological development?

Probable evolutionary relationships among the invertebrate groups are being established through anatomy, molecular biology, and genetics, even if they haven't been established through the fossil record. One tantalizing connection is between chordates, the group to which vertebrates belong (see below), and echinoderms, the group to which starfish and sea cucumbers belong. Based on embryology, RNA, and morphology, it appears as if the group to which humans and other vertebrates belong shared a common ancestor hundreds of millions of years ago with these primitive invertebrates. Although adult echinoderms don't look anything like chordates, their larval forms are intriguingly similar to primitive chordates. There are also biochemical similarities in the way they utilize phosphates – but read on to find out more about chordates!

Vertebrate evolution

Our species is among the vertebrates, creatures with a bony structure encircling our nerve cord. Vertebrates are included in a larger set of organisms called *chordates*. Although all vertebrates are chordates, not all chordates are vertebrates. The most primitive chordates look like stiff worms. A notochord or rod running along the back of the organism with a nerve cord running above it is characteristic of chordates. At some time in a chordate's life, it also has slits in the neck region (which become gills in many forms) and a tail. An example of a living chordate is a marine filter-feeding creature an inch or so long called amphioxus. To look at it, you wouldn't think it was very closely related to vertebrates, but it is. Amphioxus lacks vertebrae, but like vertebrates, it has a notochord, a dorsal nerve cord, a mouth, an anus, and a tail. Like vertebrates, it is the same on the right side of the body as it is on the left (that is, it is bilaterally symmetrical), and it has some other similarities in the circulatory system and muscle system that are structurally similar to vertebrates. It is probably fairly similar to an early chordate, but because it has been around the planet for a long time, it has evolved as well. Still, it preserves the diagnostic features of chordates in a relatively simple form.

Now that you have some idea what a primitive chordate was like, let's return to the fact that larval forms of echinoderms had similarities to primitive chordates. Unlike adult echinoderms, which are radially symmetrical like starfish, echinoderm larval forms are bilaterally symmetrical like chordates. Embryologically, echinoderms and chordates form the anus and mouth from the same precursors, and they have a number of other developmental similarities as well. One hypothesis for chordate origins is that the larval form of an early echinoderm could have become sexually mature without "growing up" – going through the full metamorphosis to an adult. This phenomenon is uncommon, but it is not unknown. It occurs in salamanders such as the axolotl, for example. [...]

In the Middle Cambrian is a small fossil called *Pikaia*, which is thought to be a primitive chordate because it looks rather like amphioxus. A new marine fossil discovered in the Late Cambrian Chengiang beds of China might even be a primitive vertebrate. Although *Haikouella* swam, it certainly didn't look much like a fish as we think of them today; it more resembled a glorified Amphioxus. From such primitive aquatic chordates as these eventually arose primitive jawless fish, then sharks and modern fish, and eventually the first land vertebrates: tetrapods (four-footed). These in turn became the ancestors of the other great groups of land animals, reptiles and mammals. Later, more detail will be provided about the evolution of many of these groups. But it is worthwhile to present four basic principles of biological evolution to keep in mind as you read the rest of the book. These are natural selection, adaptation, adaptive radiation, and speciation.

Principles of Biological Evolution

Natural selection and adaptation

Natural selection is the term given by Charles Darwin to what he – and almost all modern evolutionary biologists – considered to be the most powerful force of evolutionary change. In fact, the thesis that evolution is primarily driven by natural selection is sometimes called *Darwinism*. Unfortunately, many people misapply the term to refer to the concept of descent with modification itself, which is erroneous. Natural selection is not the same as *evolution*. As discussed in chapter 1, there is a conceptual difference between a phenomenon and the mechanisms or processes that bring it about.

When Darwin's friend T. H. Huxley learned of the concept of natural selection, he said, "How extremely stupid not to have thought of that!" (Huxley 1888), so obvious did the principle seem to him – after it was formulated. And indeed, it is a very basic, very powerful, idea. The philosopher Daniel Dennett has called natural selection "the single best idea anyone has ever had" (Dennett 1995: 21). Because of its generality, natural selection is widely found not only in nature but increasingly in engineering, computer programming, the design of new drugs, and other applications.

The principle is simple: generate a variety of possible solutions, and then pick the one that works best for the problem at hand. The first solution is not necessarily the optimal one – in fact, natural selection rarely results in a good solution to a problem in one pass. But repeated iterations of randomly generated solutions, combined with selection of the characteristics that meet the necessary criteria, result in a series of solutions that more closely approximate a good solution.

Engineers attempting to design new airplane wings have used natural selection approaches; molecular biologists trying to develop new drugs have also used the approach (Felton 2000). In living things, the "problem at hand," most broadly conceived, is survival and reproduction – passing on genes to the next generation: More narrowly, the "problem at hand" might be withstanding a parasite, finding a nesting site, being able to attract a mate, or being able to eat bigger seeds than usual when a drought reduces the number of small seeds. What is selected for depends on what, in the organism's particular circumstances, will be conducive to its survival and reproduction. The "variety of possible solutions" consists of genetically based variation which allow the organism to solve the problem.

Variation among members of a species is essential to natural selection, and fortunately it is common in sexually reproducing organisms. Some of these variations are obvious to us, such as differences in size, shape, or color. Other variations are invisible, such as genetically based biochemical and molecular differences that may be related to disease or parasite resistance, or the ability to digest certain foods. If the environment of a group of plants or animals presents a challenge – say heat, aridity, a shortage of hiding places, or a new predator – the individuals who just happen to have the genetic characteristics allowing them to survive longer and reproduce in that environment are going to be the ones most likely to pass on their genes to the next generation. The genes of these individual – including those that better suit them to the environment – will increase in proportion to those of other individuals as the population reproduces itself generation after generation. The environment *naturally selects* those individuals with the characteristics that provide for a higher probability of survival, and thus those characteristics will tend to increase over time.

So the essence of natural selection is genetic variation within a population, an environmental condition that favors some of these variations more than others, and differential reproduction of the individuals who happen to have the favored variations.

A classic example of natural selection followed the introduction of rabbits into Australia – an island continent where rabbits were not native. In 1859, an English, immigrant, Thomas Austin, released 12 pairs of rabbits so that he could go rabbit hunting. Unfortunately, except for the wedge-tailed eagle, a few large hawks, and dingoes (wild dogs) – and human hunters – rabbits have no natural enemies on Australia, and they reproduced like, well, rabbits. Within a few years, the rabbit population had expanded to such a large number that rabbits became a major pest, competing for grass with cattle, other domestic animals, and native Australian wildlife. Regions of the Australian outback that were infested with rabbits became virtual dustbowls as the little rodent herbivores nibbled down anything that was green. How could rabbit numbers be controlled?

Officials in Australia decided to import a virus from Great Britain that was fatal to rabbits but which was not known to be hazardous to native Australian mammals. The virus produced myxomatosis, or rabbit fever, which causes death fairly rapidly. It is spread from rabbit to rabbit by fleas or other blood-sucking insects. The virus first was applied to a test population of rabbits in 1950. Results were extremely gratifying: in some areas the count of rabbits decreased from 5,000 to 50 within six weeks. However, not all the rabbits were killed; some survived to reproduce. When the rabbit population rebounded, myxomatosis virus was reintroduced, but the positive effects of the first application were not repeated: many rabbits were killed, of course, but a larger percentage survived. Eventually, myxomatosis virus no longer proved effective in reducing the rabbit population. Subsequently, Australians have resorted to thousands of miles of rabbit-proof fencing to try to keep the rabbits out of at least some parts of the country.

How is this an example of natural selection? Consider how the three requirements outlined for natural selection were met:

1. Variation: The Australian rabbit population consisted of individuals that varied genetically in their ability to withstand the virus causing myxomatosis.
2. Environmental condition: Myxomatosis virus was introduced into the environment, making some of the variations naturally present in the population of rabbits more "valuable" than others.
3. Differential reproduction: Rabbits that happened to have variations allowing them to survive the viral disease reproduced more than others, leaving more copies of their genes in future generations. Eventually the population of Australian rabbits consisted of individuals that were more likely to have the beneficial variation. When myxomatosis

virus again was introduced into the environment, fewer rabbits were killed.

Natural selection involves *adaptation*: having characteristics that allow an organism to survive and reproduce in its environment. Which characteristics increase or decrease in the population through time depends on the value of the characteristic, and that depends on the particular environment – it's not "one size fits all." Because environments can change, it is difficult to precisely predict what characteristics will increase or decrease, though general predictions can be made. (No one would predict that natural selection would produce naked mole rats in the Arctic, for example.) As a result, natural selection is sometimes defined as *adaptive differential reproduction*. It is *differential reproduction* because some individuals reproduce more or less than others. It is *adaptive* because the reason for the differential in reproduction has to do with a value that a trait of set of traits has in a particular environment.

Natural selection and chance

The myxomatosis example illustrates two important aspects of natural selection: it is dependent on the genetic variation present in the population, and on the "value" of some of the genes in the population. Some individual rabbits just happened to have the genetically based resistance to myxomatosis virus even before the virus was introduced; the ability to tolerate the virus wasn't generated by the "need" to survive under tough circumstances. It is a matter of chance which particular rabbits were lucky enough to have the set of genes conferring resistance. So is it correct to say that natural selection is a chance process?

Quite the contrary. Natural selection is the opposite of chance. It is *adaptive* differential reproduction: the individuals that survive to pass on their genes do so because they have genes that are helpful (or at least not negative) in a particular environment. Indeed, there are chance aspects to the production of genetic variability in a population: Mendel's laws of genetic recombination are, after all, based on probability. However, natural selection itself is far from being a chance process. If indeed evolution is driven primarily by natural selection, then it is not the result of chance. Now, during the course of a species' evolution, unusual things may happen that are outside of anything genetics or adaptation can affect, such as a mass extinction caused by an asteroid striking Earth, but such events – though they may be dramatic – are exceedingly rare. Such contingencies do not make evolution a chance phenomenon any more than your life is governed by chance because there is a 1 in 2.8 million chance of your being struck by lighting.

Natural selection and perfection of adaptation

The first batch of Australian rabbits to be exposed to myxomatosis virus died in droves, though some survived to reproduce. Why weren't the offspring of these surviving rabbits completely resistant to the disease? A lot of *them* died, too, though a smaller proportion compared to those in the parent's generation. This is because natural selection usually does not result in perfectly adapted structures or individuals. There are several reasons for this, and one has to do with the genetic basis of heredity.

Genes are the elements that control the traits of an organism. They are located on chromosomes, in the cells of organisms. Because chromosomes are paired, genes also come in pairs, and for some traits, the two genes are identical. For mammals, genes that code for a four-chambered heart do not vary – or at least if there are any variants, the organisms that have them don't survive. But many genetic features do vary from individual to individual. Variation can be produced when the two genes of a pair differ, as they do for many traits. Some traits (perhaps most) are influenced by more than one gene, and similarly, one gene may have more than one effect. The nature of the genetic material and how it behaves is a major source of variation in each generation.

The rabbits that survived the first application of myxomatosis bred with one another, and because of genetic recombination, some offspring were produced that had myxomatosis resistance, and others were produced that lacked the adaptation. These latter were the ones who died in the second round when exposed to the virus. Back in Darwin's day, a contemporary of his invented a "sound bite" for natural selection: he called it "survival of the fittest," with "fit" meaning best adapted – not necessarily the biggest and strongest. Correctly understood, though, natural selection is survival of the fit *enough*. It is not, in fact, only the individuals who are most perfectly suited to the environment that survive; reproduction, after all, is a matter of degree, with some rabbits (or humans or spiders or oak trees) reproducing at higher than the average rate, and some at lower than the average rate. As long as an individual reproduces at all, though, it is *fit*, even if some are fitter than others.

Furthermore, just as there is selection within the rabbit population for resistance to the virus, so there is selection among the viruses! Viruses need to get copies of themselves into the next generation, and they can do this only in the body of a live rabbit. If the infected rabbit dies too quickly, the virus doesn't get a chance to spread. Viruses that are too virulent tend to be selected against, just as the rabbits that are too susceptible will also be selected against. The result is an evolutionary contest between host and pathogen – reducing the probability that the rabbit species will ever be fully free of the virus.

Another reason why natural selection doesn't result in perfection of adaptation is that once there has been any evolution at all (and there has been considerable animal evolution since the appearance of the first metazoa), there are constraints on the direction in which evolution can go. As discussed elsewhere, if a vertebrate forelimb is shaped for running, it would not be expected to become a wing at a later time; that is one kind of constraint. Another constraint is that natural selection has to work with structures and variations that are available, regardless of what sort of architecture could best do the job. If you needed a guitar but all you had was a toilet seat, you *could* make a sort of guitar by running strings across the opening, but it wouldn't be a perfect design. The process of natural selection works more like a tinkerer than an engineer (Monod 1971).

Evolution and tinkering

Some builders are engineers and some are tinkerers, and the way they go about constructing something differs quite a lot. An engineering approach to building a swing for little Charlie is to measure the distance from the tree branch to a few feet off the ground; go to the hardware store to buy some chain, hardware, and a piece of wood for the bench; and assemble the parts, using the appropriate tools: measuring devices, a drill, a screwdriver, screws, a saw, sandpaper, and paint. Charlie ends up with a really nice, sturdy swing that avoids the "down will come baby, cradle, and all" problem and that won't give slivers to his little backside when he sits on it. A tinkerer, on the other hand, building a swing for little Mary, might look around the garage for a piece of rope, throw it over the branch to see if it is long enough, and tie it around an old tire. Little Mary has a swing, but it isn't quite the same as Charlie's. It gets the job done, but it certainly isn't an optimal design: the rope may suspend little Mary too far off the ground for her to be able to use the swing without someone to help her get into it; the rope may be frayed and break; the swing may be suspended too close to the trunk, so Mary careens into it – you get the idea. The tinkering situation, where a structural problem is solved by taking something extant that can be bent, cut, hammered, twisted, or manipulated into something that more or less works, however, crudely, mirrors the process of evolution much more than do the precise procedures of an engineer. Nature is full of structures that work quite well – but it also is full of structures that just barely work or that, if one were to imagine designing from scratch, one would certainly not have chosen the particular modification that natural selection did.

Several articles by Stephen Jay Gould have discussed the seemingly peculiar ways some organisms get some particular job done. An angler fish has a clever "lure" resembling a wormlike creature that it waves at smaller fish to attract them close enough to eat. The lure springs from its forehead and is actually a modified dorsal fin spine (Gould 1980a). During embryological development, the panda's wrist bone is converted into a sixth digit, forming a grasping hand out of the normal five fingers of a bear paw, plus a "thumb" that is jury-rigged out of a modifiable bone (Gould 1980b). Like a tinkerer's project, it gets the job done, even if it isn't a great design. And after all, it's survival of the fit *enough*.

Natural selection is usually viewed as a mechanism that works on a population or sometimes a species to produce adaptations. Natural selection can also bring about adaptation on a very large scale through *adaptive radiation*.

Adaptive radiation

To be fruitful and multiply, all living things have to acquire energy (through photosynthesis, or from consuming other living things), they have to avoid predation and illness, and they have to reproduce. As is clear from the study of natural history, there are many different ways that organisms manage to perform these tasks, reflecting both the variety of environments on Earth and the variety of living things. Any environment – marine, terrestrial, arboreal, aerial, subterranean – contains many *ecological niches* that provide means that living things use to make a living. The principle of *adaptive radiation* helps to explain how niches get filled.

The geological record reveals many examples of the opening of a new environment and its subsequent occupation by living things. Island environments such as the Hawaiian Islands, the Galapagos Islands, Madagascar, and Australia show this especially well. The Hawaiian archipelago was formed as lava erupted from undersea volcanoes, and what we see as islands actually are the tips of volcanic mountains. Erosion produced soils, and land plants – their seeds or spores blown or washed in – subsequently colonized the islands. Eventually land animals reached the islands as well. Birds, insects, and a species of bat were blown to Hawaii or rafted there from other Pacific islands on chunks of land torn off by huge storms.

The Hawaiian honeycreepers are a group of approximately 23 species of brightly colored birds, ranging from four to eight inches long. They have been extensively studied by ornithologists and are shown to be very closely related. Even though they are closely related, honeycreeper species vary quite a bit from one another and occupy many different ecological niches. Some are insectivorous, some suck nectar from flowers, others are adapted to eating different kinds of seeds – one variety has even evolved to exploit a woodpecker-like niche. The best explanation for the similarity of honeycreepers in Hawaii is that they are all descended from a common ancestor. The best explanation for the diversity of these birds is that the descendants of this common ancestor diverged into many subgroups over time, as they became adapted to new, open ecological niches. Honeycreepers are, in fact, a good example of the principle of adaptive radiation, where one or a few individual animals arrive in a new environment that has empty ecological niches, and their descendants are selected to quickly evolve the characteristics needed to exploit these niches. Lemurs on Madagascar, finches on the Galapagos Islands, and the variety of marsupial mammals in Australia and prehistoric South America are other examples of adaptive radiation.

A major adaptive radiation occurred in the Ordovician period (about 430 million years ago), when plants developed protections against drying out, protections against ultraviolet radiation, vascular tissue to support erect stems, and other adaptations allowing life out of water (Richardson 1992). It was then that the dry land could be colonized by plants. The number of free niches allowed plants to radiate into a huge number of ways of life. The movement of plants from aquatic environments onto land was truly an Earth-changing event. Another major adaptive radiation occurred when vertebrates evolved adaptations (lungs and legs) permitting a similar movement onto land. These early tetrapods radiated into amphibians, reptiles, and mammals. During the late Cretaceous and early Cenozoic, about 65 million years ago, mammals began adaptively radiating after the demise of the dinosaurs opened up new ecological niches for them. Mammals moved into gnawing niches (rodents), a variety of grazing and browsing niches (hoofed quadrupeds, the artiodactyls and perissodactyls), insect-eating niches (insectivores and primates), and meat-eating niches (carnivores). Over time, subniches were occupied: some carnivores stalk their prey (lions, saber-toothed cats), others run it down (cheetahs, wolves); some (lions, wolves, hyenas) hunt large-bodied prey, some (foxes, bobcats) hunt small prey.

If a particular adaptive shift requires extensive changes, such as greatly increasing or reducing the size or number of parts of the body, the tendency is for that change to occur early in the evolution of the lineage rather than later. Though not a hard-and-fast rule, it follows logically from natural selection that the greatest potential for evolutionary change will occur before specializations of size or shape take place. Early in evolutionary history, the morphology of a major group tends to be more generalized, but as adaptive radiation takes place, structures are selected for to enable the organisms to adapt better to their environments. In most cases, these adaptations *constrain* or limit future evolution in some ways. The forelimbs of perch are committed to propelling them through the water and are specialized for this purpose; they will not become grasping hands.

We and all other land vertebrates have four limbs. Why? We tend to think of four limbs as being "normal," yet there are other ways to move bodies around. Insects have six legs and spiders have eight, and these groups of animals have been very successful in diversifying into many varieties and are represented in great numbers all over the world. So there's nothing especially superior about having four limbs, although apparently, since no organism has evolved wheels for locomotion, two or more legs apparently work better. But all land vertebrates have four limbs rather than six or eight because reptiles, birds, and mammals are descended from early four-legged creatures. These first land vertebrates had four legs because the swimming vertebrates that gave rise to them had two fins in front and two in back. The number of legs in land vertebrates was *constrained* because of the number of legs of their aquatic ancestors. Imagine what life on Earth would have looked like if the first aquatic vertebrates had had six fins! Might there have been more ecological niches for land creatures to move into? It certainly would have made sports more interesting if human beings had four feet to kick balls with – or four hands to swing bats or rackets.

We see many examples of constraints on evolution; mammalian evolution provides another example. After the demise of the dinosaurs, mammals began to radiate into niches that had previously been occupied by the varieties of dinosaurs. As suggested by the shape of their teeth, mammals of the late Cretaceous and early Paleocene were small, mostly undifferentiated creatures occupying a variety of insectivorous, gnawing, and seed-eating niches that dinosaurs were not exploiting. As new niches became available, these stem mammals quickly diverged into basic mammalian body plans: the two kinds of hoofed mammals, the carnivores, bats, insectivores, primates, rodents, sloths, and so on. Once a lineage developed (carnivores, for example), it radiated within the basic pattern to produce a variety of different forms (cats, dogs, bears, raccoons, etc.) in many sizes and shapes, all having inherited basic dental and skeletal traits from the early carnivore ancestor. Once a lineage is "committed" to a basic way of life, it is rare indeed for a major adaptive shift of the same degree to take place. Although both horses and bats are descended from generalized quadrupedal early mammalian ancestors, the bones in a horse's forelimbs have been modified for swift running: some bones have been greatly elongated, others have been lost completely, and others have been reshaped. A bat has the same basic bones in its forelimb, but they have been greatly modified in other ways: some bones have been elongated, others have been lost, and yet others have been reshaped for flight.

Humans belong in the primate group of mammals, and primates are characterized by relatively fewer skeletal changes than have occurred in other mammal lineages. A primate doesn't have the extensive remodeling of the forelimb and hand that resulted in a bat's wing or a horse's hoof. We primates have a relatively basic "four on the floor" quadruped limb pattern of one bone close to the body (the femur in the leg and the humerus in the arm), two bones next to this one (leg: tibia and fibula; arm: radius and ulna), a group of small bones after this (leg: tarsal or ankle bones; arm: carpal or wrist bones), and a fanlike spray of small bones at the end of the limb (leg: metatarsals and toe bones; arm: metacarpals and finger bones). Most primates locomote using four limbs; we human primates have taken this quadrupedal pattern and tipped it back so that all our weight is borne on the hind limbs (and not too successfully, as witnessed by hernias and the knee and lower back problems that plague our species). Being bipedal, though, meant we did not have to use our hands for locomotion, and they were thus freed for other purposes, like carrying things and making tools. Fortunately for human beings, our early primate ancestors did not evolve to have specialized appendages like those of horses or bats.

Which is better – to be generalized or specialized? It's impossible to say without knowing more about the environment or niche in which a species lives. Specialized organisms may do very well by being better able to exploit a resource than possible competitors; yet generalized organisms have an advantage in being able to adapt to a new environmental challenge.

Speciation

A *species* is composed of all the individuals that can exchange genes with one another. Some species are composed of very few individuals located in a restricted area, and others have millions of members spread out over large areas of the world. Some plant species are restricted to small areas of rain-forest habitat, while rats and humans live on literally every continent. It is more likely that an individual will mate with another individual living close by than farther away, and as a result, most species can be divided into smaller populations. Sometimes geographical factors, such as rivers or mountains or temperature gradients in different depths of water, will naturally carve species into populations.

Because of geographical differences among populations, natural selection tends to result in populations varying from one another. A typical widespread species may be divided into many differing populations. As long as they exchange genes at least at intervals, populations are likely to remain part of the same species. But how do new species form? New species form when members of a population or subdivision of a species no longer are able to exchange genes with the rest of the species. This is more likely to happen at the edges of the species range than in the center. We can say that *speciation* has occurred when a population becomes *reproductively isolated* from the rest of the species.

If a population at the end of the geographic range of a species is cut off from the rest of the species, through time it may become different from other populations. Perhaps natural selection is operating differently in its environment than it is in the rest of the species range, or perhaps the population has a somewhat different set of genes than other populations of the species. Just by the rules of probability, a small, peripheral population is not likely to have all the variants of genes that are present in the whole species, which might result in its future evolution taking a different turn.

No-longer exchanging genes with other populations of the species, and diverging genetically through time from them, members of a peripheral, isolated population might reach the stage where, were they to have the opportunity to mate with a member of the "parent" species, they would not be able to produce offspring. *Isolating mechanisms*, most of which are genetic but some of which are behavioral, can arise to prevent reproduction between organisms from different populations. Some isolating mechanisms prevent two individuals from mating at all: in some insects, for example, the sexual parts of males and females of related species are so different in shape or size that copulation can't take place. Other isolating mechanisms come into effect when sperm and egg cannot fuse for biochemical or structural reasons. An isolating mechanism could take the form of the prevention of implantation of the egg or of disruption of the growth of the embryo after a few divisions. Or the isolating mechanism could kick in later: mules, which result from crossing horses and donkeys, are healthy but sterile. Donkey genes thus are inhibited from entering into the horse species, and vice versa. When members of two groups aren't able to share genes because of isolating mechanisms, we can say that speciation between them has occurred. (Outside of the laboratory, it may be difficult to determine whether two species that no longer live in the same environment are reproductively isolated.)

The new species would of course be very similar to the old one – in fact, it might not be possible to tell them apart. Over time, though, if the new species manages successfully to adapt to its environment, it might also expand and bud off new species, which would be yet more different from the parent – now "grandparent" – species. This branching and splitting has, through time, given us the variety of species that we see today.

We can see this process of speciation operating today. Speciation in the wild usually takes place too slowly to be observed during the lifetime of any single individual, but there have been demonstrations of speciation under laboratory conditions. The geneticist Dobzhansky and his colleagues isolated a strain of Venezuelan fruit flies and bred it for several years. This strain of flies eventually reached a point of differentiation where it was no longer able to reproduce with other Venezuelan strains with which it had formerly been fertile. Speciation had occurred (Dobzhansky and Pavlovsky 1971).

Though not observed directly, good inferential evidence for speciation can be obtained

from environments that we know were only recently colonized. The Hawaiian and Galapagos islands have been formed within the last few million years from undersea volcanoes and have acquired their plants and animals from elsewhere. The Galapagos flora and fauna are derived from South America, whereas the native Hawaiian flora and fauna are more similar to those of the Pacific islands, which in turn are derived mostly from Asia. But Hawaiian species are reproductively isolated from their mainland counterparts.

One of the most dramatic examples of speciation took place among cichlid fish in the East African great lakes: Lake Victoria, Lake Malawi, and Lake Tanganyika. Geological evidence indicates that about 25,000 years ago, Lake Tanganyika underwent a drying spell that divided the lake into three separate basins. Perhaps as a result of this and similar episodes, the cichlid fish that had entered the lake from adjacent rivers and streams underwent explosive adaptive radiation. There are at least 175 species of cichlid fish found in Lake Tanganyika and nowhere else. Similar speciation events took place in Lake Victoria and Lake Malawi – only over shorter periods of time (Goldschmidt 1996). Large lakes like these can be watery versions of an island: interesting biological things can go on!

Occasionally speciation can take place very quickly. The London subway, or "Tube," was built during the 1880s. At that time, some mosquitoes found their way into the miles of tunnels, and they successfully bred in the warm air and intermittent puddles – probably several times per year. Because they were isolated from surface mosquitoes, differences that cropped up among them would not have been shared with their relatives above, and vice versa. In the late 1990s, it was discovered that the Tube mosquitoes were a different species from the surface species. One major, if unfortunate, difference is that the surface mosquitoes, *Culex pipiens*, bite birds, whereas the related Tube, species, called *Culex molestus*, has shifted its predation to people. What is surprising about this discovery is that it shows that at least among rapidly breeding insects like mosquitoes, speciation does not require thousands of years but can occur within a century (Bryne and Nichols 1999). Natural selection, adaptation, adaptive radiation, and speciation – these are the major principles that help us explain the pattern and understand the process of evolution.

Did Man Evolve from Monkeys?

So, to end with the question we began with: Did man evolve from monkeys? No. The concept of biological evolution, that living things shared common ancestry, implies that human beings did not descend from monkeys, but shared a common ancestor with them, and shared a common ancestor farther back in time with other mammals, and farther back in time with reptiles, and farther back in time with fish, and farther back in time with worms, and farther back in time with petunias. We are not descended from petunias, worms, fishes, or monkeys, but we shared common ancestors with all of these creatures, some more recently than others. The inference of common ancestry helps us make sense of biological variation. Humans are more similar to monkeys than we are to dogs because we shared a common ancestor with monkeys more recently than we shared a common ancestor with dogs. Humans, dogs, and monkeys are more similar to one another (they are all mammals) than they are to salamanders, because the species that provided the common ancestor of all mammals lived more recently than the species providing the common ancestors of salamanders and mammals. This historical branching relationship of species through time allows us to group species into categories such as "primates," "mammals," and "vertebrates," which allows us to hypothesize about other relationships. Indeed, the theory of evolution, as one famous geneticist put it, is what "makes sense" of biology. "Seen in the light of evolution, biology is, perhaps, the most satisfying science. Without that light it becomes a pile of sundry facts, some of them more or less interesting, but making no comprehensible whole" (Dobzhansky 1973: 129).

[. . .]

Evolution and Testing

What about the theory of evolution? Is it scientific? Some have claimed that since no one was there to see evolution occur, studying it cannot be scientific. Indeed, no paleontologist has ever observed one species evolving into another, but as we have seen, a theory can be scientific even if its phenomena are not directly observable. Evolutionary theory is built in the same way that theory is built in particle physics or any other field that uses indirect testing.

The essence of science is the testing of explanations against the natural world. I will argue that even though we cannot observe the evolution of, say, zebras and horses from a common ancestor, evolution is indeed a science. Such hypotheses about the patterns and descent of living things can still be tested.

The "big idea" of biological evolution is "descent with modification." Evolution is a statement about history and refers to something that happened, to the branching of species through time from common ancestors. The pattern that this branching takes and the mechanisms that bring it about are other components of evolution. We can therefore look at the testing of evolution in three senses: Can the "big idea" of evolution (descent with modification, common ancestry) be tested? Can the pattern of evolution be tested? Can the mechanisms of evolution be tested?

Testing the big idea

Hypotheses about evolutionary phenomena are tested just like hypotheses about other scientific topics: the trick (as in most science!) is to figure out how to formulate your question so it can be tested. The big idea of evolution – that living things have shared common ancestors – can be tested using the "if . . . then . . ." approach – testing by consequences – used by all scientists. The biologist John A. Moore suggested a number of these "if . . . then . . ." statements that could be used to test whether evolution occurred:

1. If living things descended with modification from common ancestors, then we would expect that "species that lived in the remote past must be different from the species alive today" (Moore 1984: 486). When we look at the geological record, this is indeed what we see. There are a few standout species that seem to have changed very little over hundreds of millions of years, but the rule is that the

farther back in time one looks, the more creatures differ from present forms.

2. If evolution occurred, we "would expect to find only the simplest organisms in the very oldest fossiliferous [fossil-containing] strata and the more complex ones to appear in more recent strata" (Moore 1984: 486). Again going to the fossil record, we find this is true. In the oldest strata, we find single-celled organisms, then simple multicelled organisms, and then simple versions of more complex invertebrate multicelled organisms (during the early Cambrian period). In later strata, we see the invasion of the land by simple plants, and then the evolution of complex seed-bearing plants, and then the development of the land vertebrates.

3. If evolution occurred, then "there should have been connecting forms between the major groups (phyla, classes, orders)" (Moore 1984: 489). To test this requires going again to the fossil record, but matters are complicated by the fact that not all connecting forms have the same probability of being preserved. For example, connecting forms between the very earliest invertebrate groups (which all are marine) are less likely to be found because of their soft bodies, which do not preserve as well as hard body parts such as shells and bones that can be fossilized. These early invertebrates also lived in shallow marine environments, where the probability of a creature's preservation is different depending on whether it lived under or on the surface of the seafloor: surface-living forms have a better record of fossilization due to surface sediments being glued together by bacteria. Fossilized burrowing forms haven't been found – although their burrows have. Connections between vertebrate groups might be expected to be found, because vertebrates are large animals with large calcium-rich bones and teeth that have a higher probability of fossilization than the soft body parts of the earliest invertebrates. There are, in fact, good transitions between fish and amphibians, and there are especially good transitions between reptiles and mammals. More and more fossils are being found that show structural transitions between reptiles (dinosaurs) and birds. Within a vertebrate

lineage, there are often fossils showing good transitional structures. We have good evidence of transitional structures showing the evolution of whales from land animals, and modem, large, single-hoofed horses from small, three-toed ancestors. Other examples can be found in reference books on vertebrate evolution such as Carroll (1998) or Prothero (1998).

In addition to the "if . . . then . . ." statements predicting what one would find if evolution occurred, one can also make predictions about what one would *not* find. If evolution occurred and living things have branched off the tree of life as lineages split from common ancestors, one would *not* find a major branch of the tree totally out of place. That is, if evolution occurred, paleontologists would not find mammals in the Devonian age of fishes, or seed-bearing plants back in the Cambrian. Geologists are daily examining strata around the world as they search for minerals, or oil, or other resources, and at no time has a major branch of the tree of life been found seriously out of place. Reports of "man tracks" being found with dinosaur footprints have been shown to be carvings, or eroded dinosaur tracks, or natural erosional features. If indeed there had not been an evolutionary, gradual emergence of branches of the tree of life, then there is no scientific reason why all strata would not show remains of living things all jumbled together.

In fact, one of the strongest sources of evidence for evolution is the consistency of the fossil record around the world. Similarly, the fact that when we look at the relationships among living things, we see that it is possible to group organisms in gradually broader classifications. There is a naturally occurring hierarchy of organisms that has been recognized since the seventeenth century: species can be grouped into genera, genera can be grouped into families, and on into higher categories. The splitting process of evolution generates hierarchy; the fact that animals and plants can be arranged in a "tree of life" is predicted by and explained by the inference of common descent.

Not only the "big idea" of evolution can be tested; so can more specific claims within that big idea. Such claims concern pattern and process, which require explanations of their own.

Pattern and process

Pattern

Consider that if evolution is fundamentally an aspect of history, then certain things happened and other things didn't. It is the job of evolutionary biologists and geologists to reconstruct the past as best they can, and try to ascertain what actually happened as the tree of life developed and branched. This is the *pattern* of evolution, and indeed, along with the general agreement about the gradual appearance of modern forms over the last 3.8 billion years, the scientific literature is replete with disputes among scientists about specific details of the tree of life, about which structures represent transitions between groups and how different groups are related. For instance, whales are known to be related to the group of hoofed mammals called artiodactyls, but are they more closely related to the hippopotamus branch of artiodactyls (suggested by molecular data) or the cattle branch (suggested by skeletal data)? Morphologically, most Neanderthal physical traits can be placed within the range of variation of living humans, but there are tests on fossil mitochondrial DNA that suggest modern humans and Neanderthals shared a common ancestor very, very long ago – no more recently than 300,000 years ago (Ovchinnikov *et al.* 2000). So are Neanderthals ancestral to modern humans, or not? There is plenty of room for argument about exactly what happened in evolution! But how do you test such statements!

Tests of hypotheses of relationship commonly use the fossil record, and it is the source of most of our conclusions about the relationships within a group. Unfortunately, sometimes one has to wait a long time before hypotheses can be tested. The fossil evidence has to exist (i.e., be capable of being preserved and actually *be* preserved), be discovered, and then painstakingly (and expensively) extracted. Only then can the analysis begin. Fortunately, we can test hypotheses about evolution – and the idea of descent with modification itself – using not only the fossil record but also anatomical, embryological, or biochemical evidence from living groups. One reason why evolution – the inference of common descent – is such a robust scientific idea is that so many different sources of information lead to the same conclusions.

We can use different sources of information to test a hypothesis about the evolution of the first primitive amphibians that colonized land. There are two main types of bony fish: the very large group of familiar ray-finned fish (fish such as trout, salmon, and sunfish) and the lobe-finned fish, represented today by only three species of lungfish and one species of coelacanth. In the Devonian, though, there were 19 families of lungfish and 3 families of coelacanths. Because of their many anatomical specializations, we know that ray-finned fish are not part of tetrapod (four-legged land vertebrate) ancestry; we and all other land vertebrates are descended from the lobefin line. Early tetrapods and lobefins both had teeth with wrinkly enamel, and shared characteristics of the shoulder girdle and jaws, plus a sac off the gut used for breathing (Prothero 1998: 358). But are we tetrapods more closely related to lungfish or coelacanths? . . .

It isn't that tetrapods evolved from lungfish, of course, but that lungfish and tetrapods shared a common ancestor, and shared a common ancestor with one another more recently than they shared a common ancestor with coelacanths. There is a large series of fossils filling the morphological gaps between ancestors of lungfish and tetrapods (Carroll 1998).

Another interesting puzzle about the pattern of evolution is ascertaining the relationships among the phyla, which are very large groupings of kinds of animals. All the many kinds of fish, amphibians, reptiles, birds, and mammals are lumped together in one phylum (Chordata) with some invertebrate animals such as sea squirts and the wormlike lancelet (amphioxus). Another phylum (Arthropoda) consists of a very diverse group of invertebrates that includes insects, crustaceans, spiders, millipedes, horseshoe crabs, and the extinct trilobites. So you can see that phyla contain a lot of diversity. Figuring out how such large groups might be related to one another is a challenging undertaking.

Phyla are diagnosed based on basic anatomical body plans – the presence of such features as segmentation, possession of shells, possession of jointed appendages, and so forth. Fossil evidence for most of these transitions is not presently available, so scientists have looked for other ways to ascertain relationships among these large groups. The recent explosions of knowledge in molecular biology and of developmental biology are opening up new avenues to test hypotheses of relationships – including those generated from anatomical and fossil data. Chordates for a long time have been related to echinoderms based on anatomical comparisons (larvae of some echinoderms are very similar to primitive chordates) and now are further linked through biochemical comparisons (ribosomal RNA) (Runnegar 1992). Ideas about the pattern of evolution can be and are being tested.

Process

Scientists studying evolution want to know not only the pattern of evolution but also the processes behind it: the mechanisms that cause cumulative biological change through time. The most important is natural selection (discussed in chapter 2), but there are other mechanisms (mostly operating in small populations, like genetic drift) that also are thought to bring about change. One interesting current debate, for example, is over the role of genetic factors operating early in embryological development. How important are they in determining differences among – and the evolution of – the basic body plans of living things? Are the similarities of early-acting developmental genes in annelid worms and in primitive chordates like amphioxus indicative of a shared common ancestry? Another debate has to do with the rate and pace of evolution: Do changes in most lineages proceed slowly and gradually, or do most lineages remain much the same for long periods that once in a while are "punctuated" with periods of rapid evolution? We know that individuals in a population compete with each other, and that populations of a species may outbreed each other, but can there be natural selection between lineages of species through time? Are there rules that govern the branching of a lineage through time? Members of many vertebrate lineages have tended to increase in size through time; is there a general rule governing size or other trends? All of these issues and many more constitute the process of evolution. Researchers are attempting to understand these processes by testing statements against the fossil and geological records as well as other sources of information.

Natural selection and other genetically based mechanisms are regularly tested and regularly prove robust. By now there are copious examples of natural selection operating in our modern world, and it is not unreasonable to extend their operation into the past. Farmers and agricultural experts are very aware of natural selection as insects, fungi, and other crop pests become resistant to chemical controls. Physicians similarly are very aware of natural selection as they try to counter antibiotic-resistant microbes. The operation of natural selection is not disputed in the creation/evolution controversy: both supporters and detractors of evolution accept that natural selection works. Creationists, however, claim that natural selection cannot bring about differences from one "kind" to another.

Pattern and process are both of interest in evolutionary biology, and each can be evaluated independently. Disputes about the pattern of evolutionary change are largely independent of disputes about the process. That is, arguments among specialists about how fast evolution can operate, or whether it is gradual or punctuated, are irrelevant to arguments over whether Neanderthals are ancestral to modern Europeans and vice versa. Similarly, arguments about either process or pattern are irrelevant to *whether* evolution took place (the "big idea" of descent with modification). This is relevant to the creation/evolution controversy because some of the arguments about pattern or process are erroneously used to support the claim that descent with modification did not occur. Such arguments confuse different levels of understanding.

Creationism and Testing

Creationism is a religious concept. Religion will be defined as a set of ideas concerning a nonmaterial reality; thus it would appear that – given science's concern for material explanations – science and creationism would have little in common. Yet the controversy that this book considers, the creationism/evolution controversy, includes the claim made by some that creationism is scientific, or can be made scientific, or has scientific elements. The question naturally arises, then, "Is creationism testable?"

As discussed, science operates by testing explanations of natural phenomena against the natural world. Explanations that are disproved are rejected; explanations that are not disproved – that are corroborated – are provisionally accepted (though at a later time they may be rejected or modified with new information). An important element of testing is being able to hold constant some of the conditions of the test, so that a causative effect can be correctly assigned.

The ultimate statement of creationism – that the present universe came about as the result of the action or actions of a divine Creator – is thus outside the abilities of science to test. If there is an omnipotent force in the universe, it would by definition be impossible to hold constant (to control) its effects. A scientist could control for the effects of temperature, light, humidity, or predators – but it would be impossible to control the actions of God!

The question of whether God created cannot be evaluated by science. Most believers conceive of God as omnipotent, so He can create everything all at once, a theological position known as "special creationism"; or He can create through the process of natural law, a theological position known as "theistic evolution." An omnipotent being could create the universe to appear as if it had evolved, but actually have created everything five minutes ago. The reason that the ultimate statement of creationism cannot be tested is simple: any action of an omnipotent Creator is compatible with any and all scientific explanations of the natural world. The methods of science cannot choose among the possible actions of an omnipotent Creator.

Science is thus powerless to test the ultimate claim of creationism, and must be agnostic about whether God did or did not create the material world. However, some types of creationism go beyond the basic statement "God created" to make claims of fact about the natural world. Many times these fact claims, such as those concerning the age of Earth, are greatly at variance with observations of science, and creationists sometimes evoke scientific support to support these fact claims. One creationist claim, for example, is that the Grand Canyon was laid down by the receding waters of Noah's Flood. In cases like this, scientific methods *can* be used to test creationist claims, because the claims are claims of fact. Of course, it is always possible to claim that the Creator performed miracles (that the Grand Canyon stratigraphy – which virtually

all geologists consider to be impossible to have been laid down during a year's time – was created through the special actions of an omnipotent Creator), but at this point one passes from science to some other way of knowing. If fact claims are made – assuming the claimer argues scientific support for such claims – then such claims can be tested by the methods of science; some scientific views are better supported than others, and some will be rejected as a result of comparing data and methodology. But such occasions leave the realm of science for that of religion if miracles are invoked.

Conclusion/Summary

Science is an especially good way of knowing about the natural world. It involves testing explanations against the natural world, discarding the ones that don't work and provisionally accepting the ones that do.

Theory-building is the goal of science. Theories explain natural phenomena and are logically constructed of facts, laws, and confirmed hyptheses. Knowledge in science, whether expressed in theories, laws, tested hypotheses, or facts, is provisional, though reliable. Although any scientific explanation may be modified, there are core ideas of science that have been tested so many times that we feel very confident about them and believe that there is an extremely low probability of their being discarded. The willingness of scientists to modify their explanations (theories) is one of the strengths of the method of science, and it is the major reason that knowledge of the natural world has increased exponentially over the last couple of hundred years.

Evolution, like other sciences, requires that natural explanations be tested against the natural world. Indirect observation and experimentation, involving "if . . . then . . ." structuring of questions and testing by consequence are the normal mode of testing in sciences such as particle physics and evolution, where phenomena cannot be directly observed.

The three elements of biological evolution – descent with modification, the pattern of evolution, and the process or mechanisms of evolution – can all be tested through the methods of science. The heart of creationism – that an omnipotent being created – is not testable by science, but fact claims about the natural world made by creationists can be.

References Cited

Bryne, Katharine, and Richard A. Nichols. 1999. *Culex pipiens* in London Underground Tunnels: Differentiation Between Surface and Subterranean Populations. *Heredity* 82: 7–15.

Carroll, Robert L. 1998. *Vertebrate Paleontology and Evolution*. New York: W. H. Freeman.

Dennett, Daniel C. 1995. *Darwin's Dangerous Idea*. New York: Simon and Schuster.

Dobzhansky, Theodosius. 1973. Nothing in Biology Makes Sense Except in the Light of Evolution. *American Biology Teacher* 25: 125–129.

Dobzhansky, Theodosius, and O. Pavlovsky. 1971. Experimentally Caused Incipient Species of *Drosophila*. *Nature* 230: 289–292.

Felton, Michael J. 2000. Survival of the Fittest in Drug Design. *Drug* 3 (9): 49–50, 53–54.

Goldschmidt, Tijs. 1996. *Darwin's Dreampond: Drama in Lake Victoria*, translated by S. Marx-MacDonald. Cambridge, MA: MIT Press.

Gould, Stephen Jay. 1980a. Double Trouble. In Gould's *The Panda's Thumb: More Reflections on Natural History*. New York: Norton.

Gould, Stephen Jay. 1980b. The Panda's Thumb. In Gould's *The Panda's Thumb: More Reflections on Natural History*. New York: Norton.

Gould, Stephen Jay. 1987. *Time's Arrow, Time's Cycle: Myth and Metaphor in the Discovery of Geological Time*. Cambridge: Harvard University Press.

Hazen, Robert M., Timothy R. Filley, Glenn A. Goodfriend. 2001. Selective adsorption of L- and D-amino acids on calcite: Implications for biochemical homochirality. *Proceedings of the National Academy of Sciences* 98 (10): 5487–5490.

Huxley, Thomas Henry. 1888. On the Reception of the "Origin of Species." In *The Life and Letters of Charles Darwin*, edited by F. Darwin. London: Thomas Murray.

Joyce, Gerald F. 1991. The Rise and Fall of the RNA World. *The New Biologist* 3: 399–407.

Lewis, Ricki. 1997. Scientists Debate RNA's Role at Beginning of Life on Earth. *The Scientist* 11 (7): 11.

Margulis, Lynn. 1993. *Symbiosis in Cell Evolution*, 2nd ed. New York: Freeman.

Miller, Stanley L. 1992. The Prebiotic Synthesis of Organic Compounds as a Step Towards the Origin of Life. In *Major Events in the History of Life*, edited by W. J. Schopf. Boston: Jones and Bartlett.

Monod, Jacques. 1971. *Chance and Necessity*. New York: Knopf.

Moore, John A. 1984. Science as a Way of Knowing – Evolutionary Biology. *American Zoologist* 24 (2): 467–534.

Nelson, Kevin E., Matthew Levy, and Stanley L. Miller. 2000. Peptide Nucleic Acids Rather Than RNA May Have Been the First Genetic Molecule. *Proceedings of the National Academy of Sciences* 97 (8): 3868–3871.

Ovchinnikov, I. V., A. Gotherstrom, G. P. Romanova, V. M. Kharitonov, K. Liden, and W. Goodwin. 2000. Molecular Analysis of Neanderthal DNA from the Northern Caucasus. *Nature* 404: 490–493.

Prothero, Donald R. 1998. *Bringing Fossils to Life: An Introduction to Paleontology*. Boston: WCB McGraw-Hill.

Richardson, John B. 1992. Origin and Evolution of the Earliest Land Plants. In *Major Events in the History of life*, edited by J. W. Schopf. Boston: Jones and Bartlett.

Rode, Bernd Michael. 1999. Peptides and the Origin of Life. *Peptides* 20: 773–786.

Runnegar, Bruce. 1992. Evolution of the Earliest Animals. In *Major Events in the History of Life*, edited by J. W. Schopf. Boston: Jones and Bartlett.

Schopf, J. William, ed. 1992. *Major Events in the History of Life*. Boston: Jones and Bartlett.

Silk, Joseph. 1994. *A Shore History of the Universe*. New York: Scientific American Library.

Strickberger, Monroe. 1996. *Evolution*, 2nd ed. Sudbury, MA: Jones and Bartlett.

PART II
EVOLUTION AND CHANCE

Introduction

The process of natural selection gives chance, probability, and randomness important roles in evolution. But understanding these roles is difficult and clarifying the concepts involved in the theory is essential for this understanding. Darwin employed the word 'chance' to describe variations, and one of the most puerile misunderstandings of the theory turns on the idea that natural selection is a process that leaves evolution to chance. In fact, natural selection tames and channels small chance events into large-scale adaptational trends. In doing so, it reveals that the appearance of design is the result of a process that involves no intentions or plans, but only the filtering and accumulation of random changes.

The first reading in this section (chapter 3) illustrates the crude mistake about the theory made by misunderstanding the role of chance in natural selection. The second, by Dawkins (chapter 4), gives a wonderfully clear explanation of how randomness really does operate in evolution while revealing the simple-minded error of Darwin's critics.

The last two papers focus on what is called *random* or *genetic drift*, which combines with natural selection to fix the actual course and rate of evolutionary change. The concept of drift has been a vexed one among biologists, and different understandings of its meaning have driven controversies about the importance of drift as a component of the neo-Darwinian theory of natural selection. Beatty's paper (chapter 5) explicates the concept of drift and shows how difficult it is to distinguish it empirically from adaptation. Brandon's article (chapter 6) argues that drift plays an absolutely coequal role with selection in nature, and that this fact must be recognized by any formulation of the theory. If Brandon's approach is correct, probabilities are indispensable in the process of natural selection and any description of its mechanism.

Further Reading

Beatty, J. (2008). Chance variation and evolutionary contingency: Darwin, Simpson, *The Simpsons*, and Gould. In M. Ruse (ed.), *The Oxford handbook of philosophy of biology* (pp. 189–210). Oxford: Oxford University Press.

Eble, G. (1999). On the dual nature of chance in evolutionary biology and paleobiology. *Paleobiology*, 25, 75–87.

Eells, E. (1991). *Probabilistic causality*. Cambridge: Cambridge University Press.

Gillies, D. (2000). *Philosophical theories of probability*. London: Routledge.

Hacking, I. (2006). *The emergence of probability: A philosophical study of early ideas about probability, induction and statistical inference*. Cambridge: Cambridge University Press.

Howie, D. (2002). *Interpreting probability: Controversies and developments in the early twentieth century*. Cambridge: Cambridge University Press.

Kimura, M. (1983). *The neutral theory of molecular evolution*. Cambridge: Cambridge University Press.

Kimura, M. (1992). Neutralism. In E. Keller & E. Lloyd (eds.), *Keywords in evolutionary biology* (pp. 225–230). Cambridge, MA: Harvard University Press.

Lennox, J., & Wilson, B. (1994). Natural selection and the struggle for existence. *Studies in History and Philosophy of Science*, 25, 65–80.

Matthen, M., & Ariew, A. (2002). Two ways of thinking about natural selection. *Journal of Philosophy*, 49, 55–83.

Millstein, R. (2002). Are random drift and natural selection conceptually distinct? *Biology and Philosophy*, 17, 33–53.

Rosenberg, A., & McShea, D. (2008). *Philosophy of biology: A contemporary introduction* (Chapters 2 and 3). London: Routledge.

Shanahan, T. (1991). Chance as an explanatory factor in evolutionary biology. *History and Philosophy of the Life Sciences*, 13, 249–269.

Von Plato, J. (1994). *Creating modern probability*. Cambridge: Cambridge University Press.

Wade, M., & Goodnight, C. (1998). Genetics and adaptation in metapopulations: When nature does many small experiments. *Evolution*, 52, 1537–1553.

3

Beyond the Reach of Chance

Michael Denton

He who believes that some ancient form was transformed suddenly . . . will further be compelled to believe that many structures beautifully adapted to all the other parts of the same creature and to the surrounduig conditions, have been suddenly produced . . . To admit all this is, as it seems in me, to enter into the realms of miracle, and to leave those of Science.

According to the central axiom of Darwinian theory, the initial elementary mutational changes upon which natural selection acts are entirely random, completely blind to whatever effect they may have on the function or structure of the organism in which they occur, "drawn," in Monod's words,[1] from "the realm of pure chance." It is only after an innovation has been disclosed by chance that it can be seen by natural selection and conserved.

Thus it follows that every adaptive advance, big or small, discovered during the course of evolution along every phylogenetic line must have been found as a result of what is in effect a purely random search strategy. The essential problem with this "gigantic lottery" conception of evolution is that all experience teaches that searching for solutions by purely random search procedures is hopelessly inefficient.

Consider first the difficulty of finding by chance English words within the infinite space of all possible combinations of letters. A section of this space would resemble the following block of letters:

```
FLNWCYTQONMCDFEUTYLDWPQXCZN
MIPQZXHGOTIRJSALXMZVTNCTDHEK
BUZRLHAJCFPTQOZPNOTJXDWHYGCB
ZUDKGTWIBMZGPGLAOTDJZKXUEMW
BCNXYTKGHSBQJVUCPDLWKSMYJVGX
UZIEMTJBYGLMPSJSKFURYEBWNQPCLX
KZUFMTYBUDISTABWNCPDORISMXKAL
QJAUWNSPDYSHXMCKFLQHAVCPDYRT
SIZSJRYFMAHZLVPRITMGYGBFMDLEPE
```

Within the total letter space would occur every single English word and every single English sentence and indeed every single English book that has been or will ever be written. But most of the space would consist of an infinity of pure gibberish.

Simple three letter English words would be relatively common. There are $26^3 = 17,000$

Michael Denton, *Evolution: A Theory in Crisis*, (Burnett, 1985), chapter 13, "Beyond the Reach of Chance." Reprinted by permission of Adler & Adler Publishers, Inc.

combinations three letters long, and, as there are about five hundred three letter words in English, then about one in thirty combinations will be a three letter word. All other three letter combinations are nonsense. To find by chance three letter words, eg "not," "bud," "hut," would be a relatively simple task necessitating a search through a string of only about thirty or so letters.

Because three letter words are so probable it is very easy to go from one three letter word to another by making random changes to the letter string. In the case of the word "hat" for example, by randomly substituting letters in the position occupied by h in the word we soon hit on a new three letter word:

> hat
>
> aat
>
> bat
>
> cat
>
> dat
>
> eat
>
> fat

Thus not only is it possible to find three letter English words by chance but because the probability gaps between them are small, it is easy to transform any word we find into a quite different word through a sequence of probable intermediates:

$$hat \rightarrow cat \rightarrow can \rightarrow tan \rightarrow tin$$

However, to find by chance longer words, say seven letters long such as "English" or "require," would necessitate a vastly increased search. There are 26^7 or 10^9, that is, one thousand million combinations of letters seven letters long. As there are certainly less than ten thousand English words seven letters long, then to find one by chance we would have to search through letter strings in the order of one hundred thousand units long. Twelve letter words such as "construction" or "unreasonable" are so rare that they occur only once by chance in strings of letters 10^{14} units long; as there are about 10^{14} minutes in one thousand million years one can imagine how long a monkey at a typewriter would take to type out by chance one English word twelve letters long. Intuitively it seems unbelievable that such

apparently simple entities as twelve letter English words could be so rare, so inaccessible to a random search.

The problem of finding words by chance arises essentially because the space of all possible letter combinations is immense and the overwhelming majority are complete nonsense; consequently meaningful sequences are very rare and the probability of hitting one by chance is exceedingly small.

Moreover, even if by some lucky fluke we were to find, say, one twelve letter word by chance, because each word is so utterly isolated in a vast ocean of nonsense strings it is very difficult to get another meaningful letter string by randomly substituting letters and testing each new string to see if it forms an English word. Take the word "unreasonable." There are a few closely related words such as "reasonable," "reason," "season," "treason," or "able," which can be reached by making changes to the letter sequence but the necessary letter changes are unfortunately highly specific and finding them by chance involves a far longer and more difficult search than was the case with three letter words.

Sentences, of course, even short ones, are even rarer and long sentences rare almost beyond imagination. Linguists have estimated a total of 10^{25} possible English sentences one hundred letters long, but as there are a total of 26^{100} or 10^{130} possible sequences one hundred letters long, then less than one in about 10^{100} will be an English sentence. The figure of 10^{100} is beyond comprehension – some idea of the immensity it represents can be grasped by recalling that there are only 10^{70} atoms in the entire observable universe.

Each English sentence is a complex system of letters which are integrated together in highly specific ways: firstly into words, then into word phrases, and finally into sentences. If the subsystems are all to be combined in such a way that they will form a grammatical English sentence then their integration must follow rigorously the *a priori* rules of English grammar. For example, one of the rules is that the letters in the sentence must be combined in such a way that they form words belonging to the lexicon of the English language.

However, random strings of English words, eg "horse," "cog," "blue," "fly," "extraordinary," do not form sentences because there exists a further set of rules – the rules of syntax which dictate, among other things, that a sentence must possess a subjective and a verbal clause.

On top of this there exists a further set of rules which governs the semantic relationship of the components of a sentence. Obviously, not all strings of English wards which are arranged correctly according to the rules of English syntax are meaningful. For example: "The raid (subject) ate (verb) the sky (object)." Each word is from the English lexicon and their arrangement satisfies the rules of syntax. However, the sequence disobeys the rules of semantics and is as nonsensical as a completely random string of letters.

The rules of English grammar are so stringent that only highly specialized letter combinations can form grammatical sentences and consequently, because of the immensity of the space of all possible letter combinations, such highly specialized strings are utterly lost within it, infinitely rare and isolated, absolutely beyond the reach of any sort of random search that could be conceivably carried out in a finite time even with the most advanced computers on earth. Moreover, because sentences are so rare and isolated, even if one was discovered by chance the probability gap between it and the nearest related sentence is so immeasurable that no conceivable sort of random change to the letter or word sequence will ever carry us across the gap.

Consider the sentence: "Because of the complexity of the rules of English grammar most English sentences are completely isolated." If we set out to reach another sentence by randomly substituting a new word in place of an existing word and then testing the newly created sequence of words to see if it made a grammatical sentence, we would find very few substitutions were grammatically acceptable and even to find *one* grammatical substitution would take an unbelievably long time if we searched by pure chance. Some of the few grammatical substitutions are shown below:

because
of
the
complexity → nature
of

the
rules → algorithms
of
English
grammar
most → all → some
english
sentences
are
completely → totally → invariably → always
isolated → immutable

There are about 10^5 words in the lexicon of the English language and, as there are sixteen words in the above sentence, we would have 1.6×10^5 possibilities to test. If there are, say, two hundred individual words out of the 1.6×100^6 which can be substituted grammatically, we would have to test about eight thousand words on average before we found a grammatical substitution.

Testing one new word per minute, it would take us five days working day and night to find by chance our first grammatical substitution, and to test all the possible words in every position in the sentence would take about three years, and after three years of searching all we would have achieved would be a handful of sentences closely related to the one with which we started.

Sentences are not the only complex systems which are beyond the reach of chance. The same principles apply, for example, to watches, which are also highly improbable, and where consequently each different functional watch is intensely isolated by immense probability gaps from its nearest neighbours.

To see why, we must begin by trying to envisage a universe of mechanical objects containing all possible combinations of watch components: springs, gears, levers, cogwheels, each of every conceivable size and shape. Such a universe would contain every functional watch that has ever existed on earth and every functional watch that could possibly exist at any time in the future. Although we cannot in this case calculate the rarity of functional combinations (watches that work) as we could in the case of words and sentences, common sense tells us that they would be exceedingly rare. Our imaginary universe would mostly consist of combinations of gears and cogwheels which would be entirely useless; each functional watch would, like a meaningful sentence, be an isolated island separated from all other islands of function by a surrounding infinity of junk composed only of incoherent and functionless combinations.

Again, as with sentences, because the total number of incoherent nonsense combinations of components vastly exceeds, by an almost inconceivable amount, the tiny fraction which can form coherent combinations, function is exceedingly rare. If we were to look by chance for a functional watch we would have to search for an eternity amid an infinity of combinations until we hit upon a functional watch.

The basic reason why functional watches are so exceedingly improbable is because, to be functional, a combination of watch components must satisfy a number of very stringent criteria (equivalent to the rules of grammar), and these can only be satisfied by highly specialized unique combinations of components which are coadapted to function together. One rule might be that all cogwheels must possess perfect regularly-shaped cogs; another rule might be that all the cogs must fit together to allow rotation of one wheel to be transmitted throughout the system.

It is obviously impossible to contemplate using a random search to find combinations which will satisfy the stringent criteria which govern functionality in watches. Yet, just as a speaker of a language cognizant with the rules of grammar can generate a functional sentence with great ease, so too a watchmaker has little trouble in assembling a watch by following the rules which govern functionality in combinations of watch components.

What is true of sentences and watches is also true of computer programs, airplane engines, and in fact of all known complex systems. Almost invariably, function is restricted to unique and fantastically improbable combinations of subsystems, tiny islands of meaning lost in an infinite sea of incoherence. Because the number of nonsense combinations of component subsystems vastly exceeds by unimaginable orders of magnitude the infinitesimal fraction of combinations in which the components are capable of undergoing coherent or meaningful interactions. Whether we are searching for a functional sentence or a functional watch or the best move in a game of chess, the goals of our search are in each case so far lost in an infinite space of possibilities that, unless we guide our search by

the use of algorithms which direct us to very specific regions of the space, there is no realistic possibility of success.

Discussing a well known checker-playing program. Professor Marvin Minsky of the Massachusetts Institute of Technology comments:[2]

> This game exemplifies the fact that many problems can in principle be solved by trying all possibilities – in this case exploring all possible moves, all the opponent's replies all the player's possible replies to the opponent's replies and so on. If this could be done, the player could see which move has the best chance of winning. In practice, however, this approach is out of the question, even for a computer; the tracking down of every possible line of play would involve some 10^{40} different board positions. A similar analysis for the game of chess would call for some 10^{120} positions. Most interesting problems present far too many possibilities for complete trial and error analysis.

Nevertheless, as he continues, a computer can play checkers if it is capable of making intelligent limited searches:

> Instead of tracking down every possible line of play the program uses a partial analysis (a "static evaluation") of a relatively small number of carefully selected features of a board position – how many men there are on each side, how advanced they are and certain other simple relations. This incomplete analysis is not in itself adequate for choosing the best move for a player in a current position. By combining the partial analysis with a limited search for some of the consequences of the possible moves from the current position, however, the program selects its move as if on the basis of a much deeper analysis. The program contains a collection of rules for deciding when to continue the search and when to stop. When it stops, it assesses the merits of the "terminal position" in terms of the static evaluation. If the computer finds by this search that a given move leads to an advantage for the player in all the likely positions that may occur a few moves later, whatever the opponent does, it can select this move with confidence.

The inability of unguided trial and error to reach anything but the most trivial of ends in almost every field of interest obviously raises doubts as to its validity in the biological realm. Such doubts were recently raised by a number of mathematicians and engineers at an international symposium entitled "Mathematical Challenges to the Neo-Darwinian Interpretation of Evolution,"[3] a meeting which also included many leading evolutionary biologists. The major argument presented was that Darwinian evolution by natural selection is merely a special case of the general procedure of problem solving by trial and error. Unfortunately, as the mathematicians present at the symposium such as Schutzenberger and Professor Eden from MIT pointed out, trial and error is totally inadequate as a problem solving technique without the guidance of specific algorithms, which has led to the consequent failure to simulate Darwinian evolution by computer analogues. For similar reasons, the biophysicist Pattee has voiced scepticism over natural selection at many leading symposia over the past two decades. At one meeting entitled "Natural Automata and Useful Simulations," he made the point:[4]

> Even some of the simplest artificial adaptive problems and learning games appear practically insolvable even by multistage evolutionary strategies.

Living organisms are complex systems, analogous in many ways to non-living complex systems. Their design is stored and specified in a linear sequence of symbols, analogous to coded information in a computer program. Like any other system, organisms consist of a number of subsystems which are all coadapted to interact together in a coherent manner: molecules are assembled into multimolecular systems, multimolecular assemblies are combined into cells, cells into organs and organ systems finally into the complete organism. It is hard to believe that the fraction of meaningless combinations of molecules, of cells, of organ systems, would not vastly exceed the tiny fraction that can be combined to form assemblages capable of exhibiting coherent interactions. Is it really possible that the criteria for function which must be satisfied in the design of living systems are at every level far less stringent than those which must be satisfied in the design of functional watches,

sentences or computer programs? Is it possible to design an automaton to construct an object like the human brain, laying down billions of specific connections, without having to satisfy criteria every bit as exacting and restricting as those which must be met in other areas of engineering?

Given the close analogy between living systems and machines, particularly at a molecular level, there cannot be any objective basis to the assumption that functional organic systems are likely to be less isolated or any easier to find by chance. Surely it is far more likely that functional combinations in the space of all organic possibilities are just as isolated, just as rare and improbable, just as inaccessible to a random search and just as functionally immutable by any sort of random process. The only warrant for believing that functional living systems are probable, capable of undergoing functional transformation by random mechanisms, is belief in evolution by the natural selection of purely random changes in the structure of living things. But this is precisely the question at issue.

If complex computer programs cannot be changed by random mechanisms, then surely the same must apply to the genetic programmes of living organisms. The fact that systems in every way analogous to living organisms cannot undergo evolution by pure trial and error and that their functional distribution invariably conforms to an improbable discontinuum comes, in my opinion, very close to a formal disproof of the whole Darwinian paradigm of nature. By what strange capacity do living organisms defy the laws of chance which are apparently obeyed by all analogous complex systems?

We now have machines which exhibit many properties of living systems. Work on artificial intelligence has advanced and the possibility of constructing a self-reproducing machine was discussed by the mathematician von Neumann in his now famous *Theory of Self-Reproducing Automata.*[5] Although some advanced machines can solve simple problems, none of them can undergo evolution by the selection of random changes in their structure without the guidance of already existing programs. The only sort of machine that might, at some future date, undergo some sort of evolution would be one exhibiting artificial intelligence. Such a machine would be capable of altering its own organization in an intelligent way. However evolution of this sort would be more akin to Lamarckian, but by no stretch of the imagination could it be considered Darwinian. The construction of a self-evolving intelligent machine would only serve to underline the insufficiency of unguided trial and error as a causal mechanism of evolution.

It was the close analogy between living systems and complex machines and the impossibility of envisaging how objects could have been assembled by chance that led the natural theologians of the eighteenth and early nineteenth centuries to reject as inconceivable the possibility that chance would have played any role in the origin of the complex adaptations of living things. William Paley, in his classic analogy between an organism and a watch, makes precisely this point:[6]

> Nor would any man hi his senses think the existence of the watch, with its various machinery, accounted for, by being told that it was one out of possible combinations of material forms; that whatever he had found in the place where he found the watch, must have contained some internal configuration or other; and that this configuration might be the structure now exhibited, viz. of the works of a watch, as well as a different structure.

It is true that some authorities have seen an analogy to evolution by natural selection in gradual technological advances. Jukes, for instance, in a recent letter to *Nature* drew an analogy between the evolution of the Boeing 747 from Bleriots' 1909 monoplane through the Boeing Clippers in the 1930s to the first Boeing jet airliner, the 707, which started in service in 1959 and which was the immediate predecessors of the 747s, and biological evolution. In his words:[7]

> The brief history of aircraft technology is filled with branching processes, phylogeny and extinctions that are a striking counterpart of three billion years of biological evolution.

Unfortunately, the analogy is false. At no stage during the history of the aviation industry was the design of any flying machine achieved by chance, but only by the most rigorous applications of all the rules which govern function in the field of aerodynamics. It is true, as Jukes states, that "wide-bodied jets evolved from small contraptions made in bicycle shops, or in junkyards," but they did not evolve by chance.

There is no way that a purely random search could ever have discovered the design of an aerodynamically feasible flying machine from a random assortment of mechanical components – again, the space of all possibilities is inconceivably large. All such analogies are false because in *all* such cases the search for function is intelligently guided. It cannot be stressed enough that evolution by natural selection is analogous to problem solving without any intelligent guidance, without any intelligent input whatsoever. No activity which involves as intelligent input can possibly be analogous to evolution by natural selection.

The above discussion highlights one of the fundamental flaws in many of the arguments put forward by defenders of the role of chance in evolution. Most of the classic arguments put forward by leading Darwinists, such as the geneticist H. J. Muller and many other authorities including G. G. Simpson, in defence of natural selection make the implicit assumption that islands of function are common, easily found by chance in the first place, and that it is easy to go from island to island through functional intermediates.

This is how Simpson, for example, envisages evolution by natural selection:[8]

> How natural selection works as a creative process can perhaps best be explained by a very much oversimplified analogy. Suppose that from a pool of all the letters of the alphabet in large, equal abundance you tried to draw simultaneously the letters c, a, and t, in order to achieve a purposeful combination of these into the word "cat." Drawing out three letters at a time and then discarding them if they did not form this useful combination, you obviously would have very little chance of achieving your purpose. You might spend days, weeks, or even years at your task before you finally succeeded. The possible number of combinations of three letters is very large and only one of these is suitable for your purpose. Indeed, you might well never succeed, because you might have drawn all the c's, a's, or t's in wrong combinations and have discarded them before you succeeded in drawing all three together. But now suppose that every time you draw a c, an a, or a t in a wrong

combination, you are allowed to put these desirable letters back in the pool and to discard the undesirable letters. Now you are sure of obtaining your result, and your chances of obtaining it quickly are much improved. In time there will be only *c*'s, *a*'s, and *t*'s in the pool, but you probably will have succeeded long before that. Now suppose that in addition to returning *c*'s, *a*'s, and *t*'s to the pool and discarding all other letters, you are allowed to clip together any two of the desirable letters when you happen to draw them at the same time. You will shortly have in the pool a large number of clipped *ca*, *ct*, and *at* combinations plus an also large number of the *t*'s, *a*'s and *c*'s needed to complete one of these if it is drawn again. Your chances of quickly obtaining the desired result are improved still more, and by these processes you have "generated a high degree of improbability" – you have made it probable that you will quickly achieve the combination cat, which was so improbable at the outset. Moreover, you have created something. You did not create the letters *c*, *a*, and *t*, but you have created the word "cat," which did not exist when you started

The obvious difficulty with the whole scheme is that Simpson assumes that finding islands of function in the first place (the individual letters *c*, *a*, and *t*) is highly probable and that the functional island "cat" is connected to the individual letter islands by intermediate functional islands *ca*, *ac*, *ct*, *at*, *ta*, *tc*, so that we can cross from letters to islands by natural selection in unit mutational steps. In other words, Simpson has assumed that islands of function are very probable, but this is the very assumption which must be proved to show that natural selection would work.

Obviously, if islands of function in the space of all organic possibilities are common, like three or four letter words, then of course functional biological systems will be within the reach of chance; and because the probability gaps will be small, random mutations will easily find a way across. However, as is evident from the above discussion, Simpson's scheme, and indeed the whole Darwinian framework, collapse completely if islands of function are like twelve letter words or English sentences.

These considerations of the likely rarity and isolation of functional systems within their respective total combinatorial spaces also reveal the fallacy of the current fashion of turning to saltational models of evolution to escape the impasse of gradualism. For as we have seen, in the case of every kind of complex functional system the total space of all combinatorial possibilities is so nearly infinite and the isolation of meaningful systems so intense, that it would truly be a miracle to find one by chance. Darwin's rejection of chance saltations as a route to new adaptive innovations is surely right. For the combinatorial space of all organic possibilities is bound to be so great, that the probability of a sudden macromutational event transforming some existing structure or converting *de novo* some redundant feature into a novel adaptation exhibiting, that "perfection of coadaptation" in all its component parts so obvious in systems like the feather, the eye or the genetic code and which is necessarily ubiquitous in the design of all complex functional systems biological or artifactual, is bound to be vanishingly small. Ironically, in any combinatorial space, it is the very same restrictive criteria of function which prevent gradual functional change which also isolate all functional systems to vastly improbable and inaccessible regions of the space.

To determine, finally, whether the distribution of islands of function in organic nature conforms to a probable continuum or an improbable discontinuum and to assess definitively the relevance of chance m evolution would be a colossal task. Just as in the case of the sentences and watches, we would have to begin by constructing a multi-dimensional universe filled with all possible combinations of organic chemicals. Within this space of all possibilities there would exist every conceivable functional biological system, including not only those which exist on earth, but all other functional biological systems which could possibly work elsewhere in the universe. The functional systems would range from simple protein molecules capable of particular catalytic functions right up to immensely complex systems such as the human brain. Within this universe of all possibilities we would find many strange biological systems, such as enzymes capable of transforming unique substrates not found on earth, and perhaps nervous systems resembling those found among vertebrates on earth, but far more advanced. We would also find many sorts of complex aggregates, the function of which may not be clear but which we could dimly conceive as being of some value on some alien planet.

Such a space would of course consist mainly of combinations which would have no conceivable biological function – merely junk aggregates ranging from functionless proteins to entirely disordered nervous systems reminiscent of Cuvier's incompatibilities. From the space we would be able to calculate exactly how probable functional biological systems are and how easy it is to go from one functional system to another, Darwinian fashion, in a series of unit mutational steps through functional intermediates. Of course if analogy is any guide then the space would in all probability conjure up a vision of nature more in harmony with the thinking of Cuvier and the early 19th-century typology than modern evolutionary thought in which each island of meaning is intensely isolated unlinked by transitional forms and quite beyond the reach of chance.

At present we are very far from being able to construct such a space of all organic possibilities and to calculate the probability of functional living systems. Nevertheless, for some of the lower order functional systems, such as individual proteins, their rarity in the space can be at least tentatively assessed.

A protein . . . is fundamentally a long chain-like molecule built up out of twenty different kinds of amino acids. After its assembly the long amino acid chain automatically folds into a specific stable 3D configuration. Particular protein functions depend on highly specific 3D shapes and, in the case of proteins which possess catalytic functions, depend on the protein possessing a particular active site, again of highly specific 3D configuration.

Although the exact degree of isolation and rarity of functional proteins is controversial it is now generally conceded by protein chemists that most functional proteins would be difficult to reach or to interconvert through a series of successive individual amino acid mutations. Zuckerkandl comments:[9]

Although, abstractly speaking, any polypeptide chain can be transformed into any other by successive amino acid substitutions and other mutational events, in concrete situations the pathways between a poorly and a highly adapted molecule will be mostly impracticable. Any such pathway, whether the theoretically shortest, or whether a longer one, will perforce include stages of favorable change as well as hurdles. Of the latter some will be surmountable and some not. Some of the latter will presumably be present along the pathways of adaptive change in a very large majority of ill adapted de novo polypeptide chains.

Consequently, when a protein molecule is selected for its weak enzymatic activity and in spite of limited substrate specificity, it will most often represent a dead end road.

The impossibility of gradual functional transformation is virtually self-evident in the case of proteins; mere casual observation reveals that a protein is an interacting whole, the function of every amino acid being more or less (like letters in a sentence or cogwheels in a watch) essential to the function of the entire system. To change, for example, the shape and function of the active site (like changing the verb in a sentence or an important cogwheel in a watch) in isolation would be bound to disrupt all the complex intramolecular bonds throughout the molecule, destabilizing the whole system and rendering it useless. Recent experimental studies of enzyme evolution largely support this view, revealing that proteins are indeed like sentences, and are only capable of undergoing limited degrees of functional change through a succession of individual amino acid replacements. The general consensus of opinion in this field is that significant functional modification of a protein would require several simultaneous amino acid replacements of a relatively improbable nature. The likely impossibility of major functional transformation through individual amino acid steps was raised by Brian Hartley, a specialist in this area, in an article in the journal *Nature* in 1974. From consideration of the atomic structure of a family of closely related proteins which, however, have different amino acid arrangements in the central region of the molecule, he concluded that their functional interconversion would be impossible:[10]

It is hard to see how these alternative arrangements could have evolved without going through an intermediate that could not fold correctly (i.e. would be non functional).

Here then, is at least one functional subset of the space of all organic possibilities which almost certainly conforms to the general discontinuous pattern observed in the case of other complex systems. But how discontinuous is the pattern of the distribution of proteins within the space of all organic possibilities? Might functional proteins be beyond the reach of chance?

In attempting to answer the question – how rare are functional proteins? – we must first decide what general restrictions must be imposed on a sequence of amino acids before it can form a biologically functional protein. In other words, what are the rules or criteria which govern functionality in an amino acid sequence?

First, a protein must be a stable structure so that it can hold a particular 3D shape for a sufficiently long period to allow it to undergo a specific interaction with some other entity in the cell. Second, a protein must be able to fold into its proper shape. Third, if a protein is to possess catalytic properties it must have an active site which necessitates a highly specific arrangement of atoms in some region of its surface to form this site.

From the tremendous advances that have been made over the past two decades in our knowledge of protein structure and function, there are compelling reasons for believing that these criteria for function would inevitably impose severe limitations on the choice of amino acids. It is very difficult to believe that the criteria for stability and for a folding algorithm would not require a relatively severe restriction of choice in at least twenty per cent of the amino acid chain. To get the precise atomic 3D shape or active sites may well require an absolute restriction in between one and five per cent of the amino acid sequence.

There is a considerable amount of empirical evidence for believing that the criteria for function must be relatively stringent. One line of evidence, for example, is the very strict conservation of overall shape and the exact preservation of the configuration of active sites in homologous proteins such as the cytochromes in very diverse species. Further evidence suggests that most mutations which cause changes in the amino acid sequence of proteins tend to damage function

to a greater or lesser degree. The effects of such mutations have been carefully documented in the case of haemoglobin, and some of them were described in an excellent article in *Nature* by Max Perutz,[11] who himself pioneered the X-ray crystallographic work which first revealed the detailed 3D structure of proteins. As Perutz shows, although many of the amino acids occupying positions on the surface of the molecule can be changed with little effect on function, most of the amino acids in the centre of the protein cannot be changed without having drastic deleterious effects on the stability and function of the molecule.

There are, in fact, both theoretical and empirical grounds for believing that the *a priori* rules which govern function in an amino acid sequence are relatively stringent. If this is the case, and all the evidence points to this direction, it would mean that functional proteins could well be exceedingly rare. The space of all possible amino acid sequences (as with letter sequences) is unimaginably large and consequently sequences which must obey particular restrictions which can be defined, like the rules of grammar, are bound to be fantastically rare. Even short unique sequences just ten amino acids long only occur once by chance in about 10^{13} average-sized proteins; unique sequences twenty amino acids long once in about 10^{26} proteins; and unique sequences thirty amino acids long once in about 10^{39} proteins!

As it can easily be shown that no more than 10^{40} possible proteins could have ever existed on earth since its formation, this means that, if proteins' functions reside in sequences any less probable than 10^{-40}, it becomes increasingly unlikely that any functional proteins could ever have been discovered by chance on earth.

We have seen . . . that envisaging how a living cell could have gradually evolved through a sequence of simple protocells seems to pose almost insuperable problems. If the estimates above are anywhere near the truth then this would undoubtedly mean that the alternative scenario – the possibility of life arising suddenly on earth by chance – is infinitely small. To get a cell by chance would require at least one hundred functional proteins to appear simultaneously in one place. That is one hundred simultaneous events each of an independent probability which could hardly be more than 10^{-20} giving a maximum combined probability of 10^{-2000}. Recently, Hoyle and Wickramasinghe in *Evolution from Space* provided a similar estimate of the chance of life originating, assuming functional proteins to have a probability of 10^{-20}:

> By itself, this small probability could be faced, because one must contemplate not just a single shot at obtaining the enzyme, but a very large number of trials such as are supposed to have occurred in an organic soup early in the history of the Earth. The trouble is that there are about two thousand enzymes, and the chance of obtaining them all in a random trial is only one part in $(10^{20})^{2000} = 10^{40,000}$ an outrageously small probability that could not be faced even if the whole universe consisted of organic soup.[12]

Although at present we still have insufficient knowledge of the rules which govern function in amino acid sequences to calculate with any degree of certainty the actual rarity of functional proteins, it may be that before long quite rigorous estimates may be possible. Over the next few decades advances in molecular biology are inevitably going to reveal in great detail many more of the principles and rules which govern the function and structure of protein molecules. In fact, by the end of the century, molecular engineers may be capable of specifying quite new types of functional proteins. From the first tentative steps in this direction it already seems that, in the design of new functional proteins, chance will play as peripheral a role as it does in any other area of engineering.[13]

The Darwinian claim that all the adaptive design of nature has resulted from a random search, a mechanism unable to find the best solution in a game of checkers, is one of the most daring claims in the history of science. But it is also one of the least substantiated. No evolutionary biologist has ever produced any quantitive proof that the designs of nature are in fact within the reach of chance. There is not the slightest justification for claiming, as did Richard Dawkins recently:[14]

> ... Charles Darwin showed how it is possible for blind physical forces to mimic the effects

of conscious design, and, by operating as a cumulative filter of chance variations, to lead eventually to organised and adaptive complexity, to mosquitoes and mammoths, to humans and therefore, indirectly, to books and computers.

Neither Darwin, Dawkins nor any other biologist has ever calculated the probability of a random search finding in the finite time available the sorts of complex systems which are so ubiquitous in nature. Even today we have no way of rigorously estimating the probability or degree of isolation of even one functional protein. It is surely a little premature to claim that random processes could have assembled mosquitoes and elephants when we still have to determine the actual probability of the discovery by chance of one single functional protein molecule!

Notes

1   Monod, J. (1972) *Chance and Necessity*, Collins, London, p. 114.

2   Minsky, M. (1966) "Artificial Intelligence," *Scientific American*, 215(3) September, pp. 246–60, see pp. 247–48.

3   Moorhead, P. S. and Kaplan, M. M., eds (1967) *Mathematical Challenges to the Darwinian Interpretation of Evolution*, Wistar Institute Symposium Monograph.

4   Pattee, H. H. (1966) "Introduction to Session One" in *Natural Automata and Useful Simulations*, eds H. H. Pattee *et al*, Spartan Books, Washington, pp. 1–2.

5   Von Neumann, J. (1966) *Theory of Self-Reproducing Automata*, University of Illinois Press, Urbana.

6   Paley, W. (1818) *Natural Theology on Evidence and Attributes of Deity*, 18th ed, Lackington, Allen and Co, and James Sawers, Edinburgh, p. 13.

7   Jukes, T. H. (1982) "Aircraft Evolution," *Nature*, 295, p. 548.

8   Simpson, G. G. (1947) "The Problem of Plan and Purpose in Nature," *Scientific Monthly*, 64: 481–495, see p. 493.

9   Zuckerkandl, E. (1975) "The Appearance of New Structures in Proteins During Evolution," *J. Mol. EvoL.* 7:1–57, see p. 21.

10 Rigby, P. W. J., Burleigh, B. D. Jnr, and Hartley, B. S. (1974) "Gene Duplication in Experimental Enzyme Evolution," *Nature*, 251:200–204, see p. 200.

11 Perutz, M. F. and Lehmann, H. (1968) "Molecular Pathology of Human Haemoglobin," *Nature*, 219:902–909.

12 Hoyle, F. and Wickramasinghe, C. (1981) *Evolution from Space*, J. M. Dent and Sons, London, p. 24.

13 Paba, C. (1983) "Designing Proteins and Peptides," *Nature*, 301:200.

14 Dawkins, R. (1982) "The Necessity of Darwinism," *New Scientist*, 94, (1301) 15 April, pp. 130–132, see p. 130.

Editor's note: The reference on p. 56 is to the earty-nineteenth century French biologist Georges Curvier, who denied the very possibility of evolution on the grounds that transitional organisms would simply be functionally inadequate. A reptile-bird intermediate, for instance, would supposedly fail both as a reptile and as a bird.

4

Accumulating Small Change

Richard Dawkins

We have seen that living things are too improbable and too beautifully "designed" to have come into existence by chance. How, then, did they come into existence? The answer, Darwin's answer, is by gradual, step-by-step transformations from simple beginnings, from primordial entities sufficiently simple to have come into existence by chance. Each successive change in the gradual evolutionary process was simple enough, *relative to its predecessor*, to have arisen by chance. But the whole sequence of cumulative steps constitutes anything but a chance process, when you consider the complexity of the final end-product relative to the original starting point. The cumulative process is directed by nonrandom survival. The purpose of this chapter is to demonstrate the power of this *cumulative selection* as a fundamentally nonrandom process.

If you walk up and down a pebbly beach, you will notice that the pebbles are not arranged at random. The smaller pebbles typically tend to be found in segregated zones running along the length of the beach, the larger ones in different zones or stripes. The pebbles have been sorted, arranged, selected. A tribe living near the shore might wonder at this evidence of sorting or arrangement in the world, and might develop a myth to account for it, perhaps attributing it to a Great Spirit in the sky with a tidy mind and a sense of order. We might give a superior smile at such a superstitious notion, and explain that the arranging was really done by the blind forces of physics, in this case the action of waves. The waves have no purposes and no intentions, no tidy mind, no mind at all. They just energetically throw the pebbles around, and big pebbles and small pebbles respond differently to this treatment so they end up at different levels of the beach. A small amount of order has come out of disorder, and no mind planned it.

The waves and the pebbles together constitute a simple example of a system that automatically generates non-randomness. The world is full of such systems. The simplest example I can think of is a hole. Only objects smaller than the hole can pass through it. This means that if you start with a random collection of objects above the hole, and some force shakes and jostles them about at random, after a while the objects above and below the hole will come to be nonrandomly sorted. The space below the hole will tend to contain objects smaller than the hole, and the space above will tend to contain objects larger than the hole. Mankind has, of course, long exploited this simple principle for generating non-randomness, in the useful device known as the sieve.

The Solar System is a stable arrangement of planets, comets and debris orbiting the sun, and it is presumably one of many such orbiting systems in the universe. The nearer a satellite is to its sun, the faster it has to travel if it is to counter the sun's gravity and remain in stable orbit. For any given orbit, there is only one speed at which a satellite can travel and remain in that orbit. If it were travelling at any other velocity, it would either move out into deep space, or crash into the Sun, or move into another orbit. And if we look at the planets of our solar system, lo and behold, every single one of them is travelling at exactly the right velocity to keep it in its stable orbit around the Sun. A blessed miracle of provident design? No, just another natural "sieve". Obviously all the planets that we see orbiting the sun must be travelling at exactly the right speed to keep them in their orbits, or we wouldn't see them there because they wouldn't be there! But equally obviously this is not evidence for conscious design. It is just another kind of sieve.

Sieving of this order of simplicity is not, on its own, enough to account for the massive amounts of nonrandom order that we see in living things. Nowhere near enough. [Think of] the analogy of the combination lock. The kind of nonrandomness that can be generated by simple sieving is roughly equivalent to opening a combination lock with one dial: it is easy to open it by sheer luck. The kind of non-randomness that we see in living systems, on the other hand, is equivalent to a gigantic combination lock with an almost uncountable number of dials. To generate a biological molecule like haemoglobin, the red pigment in blood, by simple sieving would be equivalent to taking all the amino-acid building blocks of haemoglobin, jumbling them up at random, and hoping that the haemoglobin molecule would reconstitute itself by sheer luck. The amount of luck that would be required for this feat is unthinkable, and has been used as a telling mind-boggler by Isaac Asimov and others.

A haemoglobin molecule consists of four chains of amino acids twisted together. Let us think about just one of these four chains. It consists of 146 amino acids. There are 20 different kinds of amino acids commonly found in living things. The number of possible ways of arranging 20 kinds of things in chains 146 links long is an inconceivably large number, which Asimov calls the "haemoglobin number." It is easy to calculate, but impossible to visualize the answer. The first link in the 146-long chain could be any one of the 20 possible amino acids. The second link could also be any one of the 20, so the number of possible 2-link chains is 20×20, or 400. The number of possible 3-link chains is $20 \times 20 \times 20$, or 8,000. The number of possible 146-link chains is 20 times itself 146 times. This is a staggeringly large number. A million is a 1 with 6 noughts after it. A billion (1,000 million) is a 1 with 9 noughts after it. The number we seek, the "haemoglobin number," is (near enough) a 1 with 190 noughts after it! This is the chance against happening to hit upon haemoglobin by luck. And a haemoglobin molecule has only a minute fraction of the complexity of a living body. Simple sieving, on its own, is obviously nowhere near capable of generating the amount of order in a living thing. Sieving is an essential ingredient in the generation of living order, but it is very far from being the whole story. Something else is needed. To explain the point, I shall need to make a distinction between "single-step" selection and "cumulative" selection. The simple sieves we have been considering so far in this chapter are all examples of single-step selection. Living organization is the product of cumulative selection.

The essential difference between single-step selection and cumulative selection is this. In single-step selection the entities selected or sorted, pebbles or whatever they are, are sorted once and for all. In cumulative selection, on the other hand, they "reproduce"; or in some other way the results of one sieving process are fed into a subsequent sieving, which is fed into . . . , and so on. The entities are subjected to selection of sorting over many "generations" in succession. The end-product of one generation of selection is the starting point for the next generation of selection, and so on for many generations. It is natural to borrow such words as "reproduce" and "generation," which have associations with living things, because living things are the main examples we know of things that participate in cumulative selection. They may in practice be the only things that do. But for the moment I don't want to beg that question by saying so outright.

Sometimes clouds, through the random kneading and carving of the winds, come to look like familiar objects. There is a much published photograph, taken by the pilot of a small aeroplane, of what looks a bit like the face of Jesus, staring out of the sky. We have all seen clouds that reminded us of something – a sea horse, say, or a smiling face. These resemblances come about by single-step selection, that is to say by a single coincidence. They are, consequently, not very impressive. The resemblance of the signs of the zodiac to the animals after which they are named, Scorpio, Leo, and so on, is as unimpressive as the predictions of astrologers. We don't feel overwhelmed by the resemblance, as we are by biological adaptations – the products of cumulative selection. We describe as weird, uncanny or spectacular, the resemblance of, say, a leaf insect to a leaf or a praying mantis to a cluster of pink flowers. The resemblance of a cloud to a weasel is only mildly diverting, barely worth calling to the attention of our companion. Moreover, we are quite likely to change our mind about exactly what the cloud most resembles.

Hamlet.	Do you see yonder cloud that's almost in shape of a camel?
Polonius.	By the mass, and 'tis like a camel, indeed.
Hamlet.	Methinks it is like a weasel.
Polonius.	It is backed like a weasel.
Hamlet.	Or like a whale?
Polonius.	Very like a whale.

I don't know who it was first pointed out that, given enough time, a monkey bashing away at random on a typewriter could produce all the works of Shakespeare. The operative phrase is, of course, given enough time. Let us limit the task facing our monkey somewhat. Suppose that he has to produce, not the complete works of Shakespeare but just the short sentence "Methinks it is like a weasel," and we shall make it relatively easy by giving him a typewriter with a restricted keyboard, one with just the 26 (capital) letters, and a space bar. How long will he take to write this one little sentence?

The sentence has 28 characters in it, so let us assume that the monkey has a series of discrete "tries," each consisting of 28 bashes at the keyboard. If he types the phrase correctly, that is the end of the experiment. If not, we allow him another "try" of 28 characters. I don't know any monkeys, but fortunately my 11-month old

daughter is an experienced randomizing device, and she proved only too eager to step into the role of monkey typist. Here is what she typed on the computer:

UMMK JK CDZZ F ZD DSDSKSM
S SS FMCV PU I DDRGLKDXBRDO
RDTE QDWFDVIOY UDSKZWDCCVYT
H CHVY NMGNBAYTDFCCVD D
RCDFYYYRM N DFSKD LD K WDWK
JJKAUIZMZI UXDKIDISFUMDKUDXI

She has other important calls on her time, so I was obliged to program the computer to simulate a randomly typing baby or monkey:

WDLDMNLT DTJBKWIRZREZLMQCO P
Y YVMQKZPGJXWVHGLAWFVCHQYOPY
MWR SWTNUXMLCDLEUBXTQHNZVJQF
FU OVAODVYKDGXDEKYVMOGGS VT
HZQZDSFZIHIVPHZPETPWVOVPMZGF
GEWRGZRPBCTPGQMCKHFDBGW ZCCF

And so on and on. It isn't difficult to calculate how long we should reasonably expect to wait for the random computer (or baby or monkey) to type METHINKS IT IS LIKE A WEASEL. Think about the total number of possible phrases of the right length that the monkey or baby or random computer *could* type. It is the same kind of calculation as we did for haemoglobin, and it produces a similarly large result. There are 27 possible letters (counting "space" as one letter) in the first position. The chance of the monkey happening to get the first letter – M – right is therefore 1 in 27. The chance of it getting the first two letters – ME – right is the chance of it getting the second letter – E – right (1 in 27) *given that* it has also got the first letter – M – right, therefore $1/27 \times 1/27$, which equals 1/729. The chance of it getting the first word – METHINKS – right is 1/27 for each of the 8 letters, therefore $(1/27) \times (1/27) \times (1/27) \ldots$, etc. 8 times, or (1/27) to the power 8. The chance of it getting the entire phrase of 28 characters right is (1/27) to the power 28, i.e. (1/27) multiplied by itself 28 times. These are very small odds, about 1 in 10,000 million million million million million. To put it mildly, the phrase we seek would be a long time coming, to say nothing of the complete works of Shakespeare.

So much for single-step selection of random variation. What about cumulative selection; how much more effective should this be? Very very much more effective, perhaps more so than we at first realize, although it is almost obvious when we reflect further. We again use our computer monkey, but with a crucial difference in its program. It again begins by choosing a random sequence of 28 letters, just as before:

WDLMNLT DTJBKWIRZREZLMQCO P

It now "breeds from" this random phrase. It duplicates it repeatedly, but with a certain chance of random error – "mutation" – in the copying. The computer examines the mutant nonsense phrases, the "progeny" of the original phrase, and chooses the one which, *however slightly*, most resembles the target phrase, METHINKS IT IS LIKE A WEASEL. In this instance the winning phrase of the next "generation" happened to be:

WDLTMNLT DTJBSWIRZREZLMQCO P

Not an obvious improvement! But the procedure is repeated, again mutant "progeny" are "bred from" the phrase, and a new "winner" is chosen. This goes on, generation after generation. After 10 generations, the phrase chosen for "breeding" was:

MDLDMNLS ITJISWHRZREZ MECS P

After 20 generations it was:

MELDINLS IT ISWPRKE Z WECSEL

By now, the eye of faith fancies that it can see a resemblance to the target phrase. By 30 generations there can be no doubt:

METHINGS IT ISWLIKE B WECSEL

Generation 40 takes us to within one letter of the target:

METHINKS IT IS LIKE AWEASEL

And the target was finally reached in generation 43. A second run of the computer began with the phrase:

Y YVMQKZPFJXWVHGLAWFVCHQXYOPY,

passed through (again reporting only every tenth generation):

Y YVMQKSPFTXWSHLIKEFV HQYSPY
YETHINKSPITXISHLIKEFA WQYSEY
METHINKS IT ISSLIKE A WEFSEY
METHINKS IT ISBLIKE A WEASES
METHINKS IT ISJLIKE A WEASEO
METHINKS IT IS LIKE A WEASEP

and reached the target phrase in generation 64. In a third run the computer started with:

GEWRGZRPBCTPGQMCKHFDBGW ZCCF

and reached METHINKS IT IS LIKE A WEASEL in 41 generations of selective "breeding."

The exact time taken by the computer to reach the target doesn't matter. If you want to know, it completed the whole exercise for me, the first time, while I was out to lunch. It took about half an hour. (Computer enthusiasts may think this unduly slow. The reason is that the program was written in BASIC, a sort of computer baby-talk. When I rewrote it in Pascal, it took 11 seconds.) Computers are a bit faster at this kind of thing than monkeys, but the difference really isn't significant. What matters is the difference between the time taken by *cumulative* selection, and the time which the same computer, working flat out at the same rate, would take to reach the target phrase if it were forced to use the other procedure of *single-step selection*: about a million million million million million years. This is more than a million million million times as long as the universe has so far existed. Actually it would be fairer just to say that, in comparison with the time it would take either a monkey or a randomly programmed computer to type our target phrase, the total age of the universe so far is a negligibly small quantity, so small as to be well within the margin of error for this sort of back-of-an-envelope calculation. Whereas the time taken for a computer working randomly but with the constraint of *cumulative selection* to perform the same task is of the same order as humans ordinarily can understand, between 11 seconds and the time it takes to have lunch.

There is a big difference, then, between cumu-lative selection (in which each improvement, however slight, is used as a basis for future building), and single-step selection (in which each new "try" is a fresh one). If evolutionary progress had had to rely on single-step selection, it would never have got anywhere. If, however, there was any way in which the necessary con-ditions for *cumulative* selection could have been set up by the blind forces of nature, strange and wonderful might have been the consequences. As a matter of fact that is exactly what happened on this planet, and we ourselves are among the most recent, if not the strangest and most won-derful, of those consequences.

It is amazing that you can still read calculations like my haemoglobin calculation, used as though they constituted arguments *against* Darwin's theory. The people who do this, often expert in their own field, astronomy or whatever it may be, seem sincerely to believe that Darwinism explains living organization in terms of chance – "single-step selection" – alone. This belief, that Darwinian evolution is "random," is not merely false. It is the exact opposite of the truth. Chance is a minor ingredient in the Darwinian recipe, but the most important ingredient is cumulative selection which is quintessentially *non*random.

Clouds are not capable of entering into cumu-lative selection. There is no mechanism whereby clouds of particular shapes can spawn daughter clouds resembling themselves. If there were such a mechanism, if a cloud resembling a weasel or a camel could give rise to a lineage of other clouds of roughly the same shape, cumulative selection would have the opportunity to get going. Of course, clouds do break up and form "daughter" clouds sometimes, but this isn't enough for cumulative selection. It is also necessary that the "progeny" of any given cloud should resemble its "parent" *more* than it resembles any old "parent" in the "population." This vitally important point is apparently misunderstood by some of the philosophers who have, in recent years, taken an interest in the theory of natural selection. It is further necessary that the chances of a given cloud's surviving and spawning copies should depend upon its shape. Maybe in some distant galaxy these conditions did arise, and the result, if enough millions of years have gone by, is an ethereal, wispy form of life. This might make a good science fiction story – *The White Cloud*, it could be called – but for our purposes a computer model like the monkey/Shakespeare model is easier to grasp.

5

Chance and Natural Selection

John Beatty

1. Introduction

Among the liveliest disputes in evolutionary biology today are disputes concerning the role of chance in evolution – more specifically, disputes concerning the relative evolutionary importance of natural selection vs. so-called "random drift". The following discussion is an attempt to sort out some of the broad issues involved in those disputes. In the first half of this paper, I try to explain the differences between evolution by natural selection and evolution by random drift. On some common construals of "natural selection", those two modes of evolution are completely indistinguishable. Even on a proper construal of "natural selection", it is difficult to distinguish between the "improbable results of natural selection" and evolution by random drift.

In the second half of this paper, I discuss the variety of positions taken by evolutionists with respect to the evolutionary importance of random drift vs. natural selection. I will then consider the variety of issues in question in terms of a conceptual distinction often used to describe the rise of probabilistic thinking in the sciences. I will argue, in particular, that what is going on here is not, as might appear at first sight, just another dispute about the desirability of "stochastic" vs. "deterministic" theories. Modern evolutionists do not argue so much about *whether* evolution is stochastic, but about *how* stochastic it is.

Charles Darwin's account of organic form appealed to chance in a way that did not settle well with his critics. As Darwin unhappily reported the opinion of the great scientist-philosopher, John Herschel, "I have heard, by a round-about channel, that Herschel says my book 'is the law of higgledy-piggledy.' What exactly this means I do not know, but it is evidently very contemptuous" (Darwin to Lyell, Dec. 12, 1859, in F. Darwin 1887, Vol. 2, p. 37). In time, though, Darwin was praised rather than scorned for his

John Beatty, "Chance and Natural Selection," *Philosophy of Science* (1984) 51, 183–211. Reprinted by permission of the publisher, the University of Chicago Press.

This article was written during the academic year 1982–1983, while I was a fellow at the Center for Interdisciplinary Research at the University of Bielefeld, in Bielefeld, West Germany. I was part of a group research project, organized by Lorenz Krüger, which studied the rise and role of probabilistic thinking in the sciences since 1800. I am very grateful to the staff of the Center, the faculty of the University of Bielefeld, and of course my fellow probabilists for their thoughts and for their friendship.

Robert Brandon, Lorenz Krüger, Elliott Sober, Kenneth Waters, and the referees of *Philosophy of Science* all helped me to clarify the issues discussed here. The residual unclarity distinguishes my contributions from theirs. Some of the residual unclarity must be attributed to Jonathan Hodge's and Alexander Rosenberg's critiques of the notion of "fitness" used here. Their thoughtful critiques have, I must admit, left me a *bit* confused about my position.

appeal to chance. For instance, looking back at the turn of the century, another great scientist-philosopher, C. S. Peirce, assessed Darwin's contribution in this regard more favorably:

> The Origin of Species was published toward the end of the year 1859. The preceding years since 1846 had been one of the most productive seasons – or if extended so as to cover the great book we are considering, the most productive period of equal length in the entire history of science from its beginnings until now. [For] the idea that chance begets order . . . was at that time put into its clearest light. (Peirce 1893, p. 183)

Since the turn of the century, however, and especially since the thirties, evolutionists have *further* appealed to chance in ways that Darwin himself might contemptuously have regarded as higgledy-piggledy views of nature. Indeed, proponents of one such appeal have coined the term "non-Darwinian evolution" to distance their views from his. Actually, the new appeals to chance have been matters of considerable dispute. And today those disputes are among the liveliest in the already lively field of evolutionary biology.

The following discussion is an attempt to sort out some of the broad issues involved in these disputes. The most general question at issue concerns the relative evolutionary importance of "random drift" vs. natural selection. But what does that mean? In the first half of this paper (Sections 2–3), I will try to make sense of that question. That will involve explaining the sense in which evolution by random drift is, properly speaking, an "alternative" to evolution by natural selection. Darwin did not conceive of chance as anything like an alternative to natural selection, but rather as "complimentary" to natural selection. Modern evolutionists, on the other hand, recognize alternative as well as complimentary roles of chance and natural selection. And yet, on some common construals of "natural selection", it is difficult (if not impossible) even to *distinguish* evolution by random drift from evolution by natural selection. Thus, in order to construe evolution by random drift and evolution by natural selection as proper alternatives, the concept of natural selection itself must first be properly interpreted. Unfortunately, though, even on a proper interpretation of "natural selection", it is difficult to

distinguish between what I call the "improbable results of natural selection" and evolution by random drift.

In the second half of this paper (Section 4–5), I discuss the variety of positions taken by evolutionists with respect to the evolutionary importance of random drift vs. natural selection. I will then consider the variety of issues in question in terms of a conceptual distinction often used to describe the rise of probabilistic thinking in the sciences. I will argue, in particular, that what is going on here is not, as might appear at first sight, just another dispute about the desirability of "stochastic" vs. "deterministic" theories. Modern evolutionists do not argue so much about *whether* evolution is stochastic, but about *how* stochastic it is.

2.   Chance in Darwinian Evolutionary Theory

Darwin usually invoked "chance" (or "accident") in the context of discussions about how variations, the materials of evolution, arise. His notion of "chance variation" was especially important for the purpose of distinguishing his own theory of evolution from the older "use and disuse" theory (Hodge and Kohn 1985). So a brief discussion of the differences between those theories might be helpful. The paradigm application of the theory of use and disuse is the account of the evolution of giraffes from shorter-necked, four-footed grazers. As the account goes, the ancestors of giraffes found themselves in an environment in which they had little upon which to graze, other than leaves of trees. They stretched their necks to reach more and more leaves, and were physically modified in the process. Their offspring inherited the modification – i.e., having longer necks as juveniles than their parents had as juveniles. The offspring also stretched to reach more and more leaves, lengthening their necks even beyond the length inherited. The third generation inherited the further modification. And so on, until all the descendants of the original group were quite long-necked. The important point here concerns how the variations that aid the survival of their possessors first arose – namely, in response to the survival needs of their possessors. That is, the chance of such a variation occurring was increased by the very fact that the organism needed it to survive.

On the Darwinian account, we are asked to consider what would have happened if, among the shorter-necked ancestors, some slightly longer-necked offspring happened "by chance" to be born. Those who happened to have the slightly longer necks would be able to reach slightly more food, and hence would slightly outsurvive and outreproduce the others. Assuming that neck length was inheritable, then, a slightly greater proportion of the next generation than of the previous generation would be longer-necked. This process alone would result in an ever-increasing frequency of the slightly longer-necked individuals. But now consider what would happen if, among those slightly longer-necked organisms, some slightly even longer-necked offspring happened by chance to be born. These in turn would slightly outsurvive and outreproduce the others. Thus, the proportions of longer- and longer-necked individuals would increase from generation to generation, until the present proportions were reached.

The important point for now concerns how, according to the Darwinian account, the variations that aid the survival and reproduction of their possessors first arise. They do not, as on the use-and-disuse account, arise in response to the survival and reproductive needs of their possessors. On the other hand, whether or not they further increase in frequency does depend on whether they serve those needs. But the probability of a particular variation occurring in an individual is not increased by the fact that that variation would promote the survival and reproduction of that individual. In that sense, it is a matter of "chance" that an organism would be born with a variation that promotes its survival and reproduction, though, again, not a matter of chance that that variation further increases in frequency in subsequent generations. As Darwin thus distinguished his account from the use-and-disuse account, the latter explains the evolution of adaptation in a way "analogous to a blacksmith having children with strong arms", while the former relies on "the other principle of those children which *chance* produced with strong arms, outliving the weaker ones" (Darwin 1839, passage 42). As Darwin elsewhere more eloquently explained the notion of chance variation,

[Evolution by natural selection] absolutely depends on what we in our ignorance call spontaneous or accidental variability. Let an architect be compelled to build an edifice with uncut stones, fallen from a precipice. The shape of each fragment may be called accidental. Yet the shape of each has been determined . . . by events and circumstances, all of which depend on natural laws; but there is no relation between these laws and the purpose for which each fragment is used by the builder. In the same manner the variations of each creature are determined by fixed and immutable laws; but these bear no relation to the living structure which is slowly built up through the power of selection. . . . (Darwin 1887, Vol. 2, p. 236)

There are several, more general notions of chance in terms of which one might try to understand the notion of chance variation. There is, for instance, the natural theological notion of chance vs. intelligently intended occurrences. There is also the old Aristotelian notion of chance as coincidence. Or we might invoke the Laplacean notion of chance, not in order to say anything about the event in question itself, but just as an admission of ignorance concerning the causal chain of events that resulted in that event. It seems to me possible to make a case for construing chance variation in each of these ways – Darwin sometimes seems to have favored one, and then at other times another more general meaning. Sometimes he spoke of chance variations in contrast with benevolently intended, useful variations. Sometimes he meant that it was a matter of coincidence that a useful variation should arise. And sometimes he seems only to have wanted to leave entirely open the question of how variations arise – thus making clear that his theory of evolution did not rely on the use and disuse account of that process. All three interpretations are suggested, for instance, by the quotation above! (See also Ghiselin 1969, Schweber 1982, and Sheynin 1980 with regard to these more general concepts of chance variation.) But perhaps the most important thing to consider is what is distinguishably Darwinian about the notion of chance variation. And that is, I think, best brought out simply in contrast to the use and disuse theories of evolution, from which Darwin wanted to distance himself.

3.   Chance in Modern Evolutionary Theory: Distinctions

3.1   The dispute concerning the relative evolutionary importance of natural selection vs. random drift raises genuinely new issues concerning the role of chance in evolution. What is *not* at issue is the role of chance in evolution as Darwin conceived it (except insofar as he *too narrowly* conceived it). Modern evolutionists assume, that is, that variations are indeed "random" in the sense intended by Darwin. As one contemporary evolutionist puts it, "Mutation is random in [the sense] that *the chance that a specific mutation will occur is not affected by how useful that mutation would be*" (Futuyma 1979, p. 249).

But while Darwin considered the *origin* of variations a matter of chance, he did not consider the possibility that their evolutionary *fates* might also be a matter of chance. In particular, he attributed their evolutionary fates to natural selection in or against their favor. Natural selection took over where chance variation left off: those organisms that were by chance better equipped to survive and reproduce actually outsurvived and outreproduced the organisms that were by chance less well equipped, and thus advantageous variations increased in frequency from generation to generation. In Darwin's scheme, in other words, chance and natural selection were "consecutive" rather than "alternative" stages of the evolutionary process. With the introduction of the concept of random drift, however, came the notion that the fate of a chance variation might itself be a matter of chance – i.e., the concept that chance and natural selection might be alternative rather than just consecutive stages of the evolutionary process.

But what does it mean to attribute the evolutionary fate of a variation to natural selection vs. random drift? The distinction will be easier to make if we confine our discussion to the fates of *genetic* variations. It will be useful, in other words, to think of evolutionary changes as changes in the gene and genotype frequencies of populations. (The Appendix consists of a review of genetic terminology that some readers might find useful at this point.) Thus, the kinds of evolutionary changes that I will be talking about are changes of the following sort. Of the alleles (genes) at a particular genetic locus of members of one generation of a particular population, 50% are of type *A* and 50% are of type *a*. Of the genotypes with respect to that locus, 25% are of the type *AA*, 50% are of the type *Aa*, and 25% are of the type *aa*. Of the alleles at that locus of members of a *later* generation of that population, 80% are of type *A* and 20% are of type *a*. And of the genotypes with respect to that locus of members of the later generation, 70% are of the type *AA*, 20% are of the type *Aa*, and 10% are of the type *aa*.

Though definitions of "evolution" in terms of such gene- and genotype-frequency changes are common (e.g., Wilson and Bossert 1971, p. 20), I recognize that we actually include more than just that kind of change under the term "evolution" (see Brandon 1978b). This restriction, however, simplifies the following discussion enormously.

3.2   I will rely, for the time being, on your intuitions about what it means to attribute gene- and genotype-frequency changes to natural selection. What does it mean to attribute the same to random drift? Since as early as 1932 (Dubinin and Romaschoff 1932; see also Dobzhansky 1937, p. 129), a popular approach to the exposition of random drift has been via a classic means of modelling chance processes – namely, the blind drawing of beads from an urn. The beads in this case are alleles – the different alleles are different colors, but they are otherwise indistinguishable by the blindfolded sampling agent. One urn of beads represents one generation of alleles – a finite number, characterized by particular allele frequencies. The next generation of alleles is determined by a blind drawing of beads from an urn. This second generation of alleles fills a new urn, blind drawings from which determine the frequencies of alleles in the third generation. And so on. The frequencies of alleles may differ from urn to urn – generation to generation – as a result of the fact that frequencies of otherwise indistinguishable beads sampled by blind drawings may not be representative of the frequencies in the urns from which the samples were drawn. The probability of drawing a representative sample from a population of a given finite size is easy to calculate – the smaller the population, the smaller that probability.

Such blind sampling could take at least two forms in nature. The first form is what might be called "indiscriminate parent sampling". By "parent sampling" I mean *the process of determining which organisms of one generation will be parents of the next, and how many offspring each parent will have.* This sort of sampling might be "indiscriminate" in the sense that any physical differences between the organisms of one generation might be irrelevant to differences in their offspring contributions. A forest fire, for instance, might so sample parents – killing some, sparing some – without regard to physical differences between them. Such sampling is indiscriminate in the same sense in which the usual model of blind drawing of beads from an urn is indiscriminate – that is, any physical differences (e.g., color) between the entities in question are irrelevant to whether or not they are sampled.

If a population is so maintained at a particular finite size – i.e., by sampling parents indiscriminately – its gene and genotype frequencies will "drift" from generation to generation. The reason is that the genotype frequencies of the parents, weighted according to their reproductive success, may by chance not be representative of the genotype frequencies of the parents' generation. There is no form of discrimination to ensure that the genotype frequencies *are* representative. To the extent that a parent sample is unrepresentative, and to the extent that the gene and genotype frequencies of the next generation reflect the appropriately weighted gene and genotype frequencies of their parents, the next generation's gene and genotype frequencies may diverge from those of the previous generation – i.e., an evolutionary change may occur.

Blind sampling in nature might also take the form of "indiscriminate gamete sampling". (This paragraph and the next can be skipped without much loss.) By "gamete sampling" I mean *the process of determining which of the two genetically different types of gametes produced by a heterozygotic parent is actually contributed to each of its offspring.* This sort of sampling might be indiscriminate in the sense that any physical difference between the two types of gametes produced by a heterozygote might be irrelevant to whether one or the other is actually contributed to any particular offspring. According to Mendel's Law (see the Appendix), there is no physical basis for a bias in the proportions of the two genetically different types of gametes *produced* by a

heterozygote. What we are now considering is that there is also no physical basis for a bias in the proportions of gametes that are actually *contributed* to a heterozygote's offspring.

If a population is so maintained at a particular finite size – i.e., by sampling the gametes of heterozygotes indiscriminately – the gene and genotype frequencies of the population will drift from generation to generation. For the gene frequencies of the gametes that are contributed to a generation of offspring may by chance not be representative of the gene frequencies of the parents' generation. Again, there is no form of discrimination to ensure that they *are* representative. Thus, indiscriminate gamete sampling, either together with indiscriminate parent sampling, or alone, can result in an evolutionary change, a so-called "random drift" of gene and genotype frequencies.

Some authors construe random drift entirely in terms of indiscriminate parent sampling (e.g., Sheppard 1967, pp. 126–127). More often, random drift is construed entirely in terms of indiscriminate gamete sampling (e.g., Wilson and Bossert 1971, p. 83). But these two processes are importantly similar agents of change – both involving elements of randomness, and "randomness" in the same sense in both cases. According to this sense of "randomness", sampling from a population is random when each member of the population has the same chance of being sampled. This is a common notion of random sampling – e.g., according to a philosophical introduction to probability theory, a sampling procedure gives random samples if:

1. Each member of the population has an equal probability of being selected as the first member of the sample.
2. Each member of the population, excluding those selected as previous members of the sample, has an equal chance of being selected as the nth member of the sample (Skyrms 1966, pp. 146–147; see also Hacking 1965, pp. 118–132).

Discussions of random drift, pedagogical or otherwise, generally do not include discussions of the concepts of "randomness" and "chance". Rather, evolutionists have, in general, simply relied on what is an *exemplary* model of random sampling in the literature of probability, statistics, and philosophy – i.e., the blind drawing of beads from an urn. Evolutionists are not, in general, inclined to defend any particular interpretation of "randomness". It is enough, for their purposes, to say of any process in nature that is sufficiently similar to their bead-drawing model of random drift, that evolution in this case is as much a matter of "random sampling" and "chance" as are the exemplary cases of random sampling upon which probabilists, statisticians, and philosophers rely.

In what follows, in comparing random drift to natural selection, I will be discussing only random drift via indiscriminate parent sampling, not via indiscriminate gamete sampling. So, in effect, I will be contrasting natural selection of parents and indiscriminate parent sampling. I could also contrast natural selection of gametes with indiscriminate gamete sampling, but that will not be necessary in order to make the kinds of points that I want to make.

3.3   Natural selection is not random sampling in the sense of "random" just discussed. Natural *selection* is, as its name suggests, a discriminate form of sampling – a sampling process that discriminates, in particular, on the basis of *fitness* differences. The general issue involved in the disputes concerning the relative evolutionary importance of random drift vs. natural selection, then, boils down to the question of the relative evolutionary importance of *sampling without regard to fitness differences vs. sampling with regard to fitness differences*. But I cannot unpack this issue without first discussing briefly the notion of fitness. For according to the most common conception of fitness, the distinction between random drift and natural selection of the fittest dissolves, and along with it dissolves the issue of the relative evolutionary importance of the two supposedly different sorts of processes.

According to the most common definition of "fitness", the fitness of an organism is a measure of its actual reproductive success. I. M. Lerner's definition of "fitness" is typical in this regard – according to Lerner, "the organisms who have more offspring are fitter in the Darwinian sense" (Lerner 1958, p. 10; see also Waddington 1968, p. 19; Mettler and Gregg 1969, p. 93; Crow and Kimura 1970, p. 5; Dobzhansky 1970, pp. 101–102; Wilson 1975, p. 585; Grant 1977, p. 66). Along

the same lines, the fitness of a genotype is the actual average offspring contribution of organisms of that type. For instance, to say that organisms of type *AA* are fitter than organisms of type *aa* is to say that organisms of the first type actually leave a higher average number of offspring.

But if fitness is so construed, then it is not clear what the supposed distinction between natural selection of the fittest (i.e., sampling on the basis of fitness differences), and sampling without regard to fitness differences, amounts to. How could parent sampling be *anything other* than fitness discriminating, when the organisms that leave the most offspring are by definition the fittest? On this conception of fitness, *all* parent sampling processes are fitness discriminating.

Consider, for instance, the proposed bead-drawing models of selection that are based on this notion of fitness – those models are indistinguishable from the bead-drawing models of random drift. Consider, in particular, the bead game called "Selection", described by Manfred Eigen and Ruth Schuster in their otherwise thoughtful book, *Laws of the Game* (Eigen and Schuster 1981, pp. 49–65). The game requires a checkerboard, the playing squares of which are identified by coordinates, and are occupied by colored beads, one bead per square. To begin, the beads that cover the board are of at least two colors – in whatever initial frequencies desired. The roll of two dice (each of which has the appropriate number of sides) picks out a square on the board. A bead of whatever color occupies that square is used to replace whatever color is on the square picked out by the next roll of the dice. This is followed by another color-determining roll, followed by another color-replacing roll. And so on, until eventually one color fills the entire board.

This is supposed to represent selection of the fittest color, where the fittest color is, as Eigen and Schuster put it, "the one that turns out to be the winner" (1981, p. 55). This is also, of course, just indiscriminate sampling. Indeed, Eigen and Schuster say of their model that "We are faced with the paradox that the selective process produces a winner in every game though competitors do not differ from each other at all" (1981, p. 55). Unfortunately, Eigen and Schuster are not just calling this game "Selection", they actually intend to explicate the concept of Darwinian selection in terms of the game. But if this is a proper explication of selection, then the problem of the relative evolutionary importance of random drift vs. natural selection is a pseudo-problem – there is no difference between them.

There is, however, another conception of fitness that keeps the issue alive. To motivate this conception briefly, consider the example of two genetically identical twins, one of whom is struck by lightning and killed, the other of whom is spared. As a result, say, the former leaves no offspring, while the latter goes on to be a parent. We balk at the consequences according to our conception of fitness. That is, we hesitate to attribute zero fitness to the former twin and relatively high fitness to his genetically identical, but seemingly luckier brother.

One way of accommodating these intuitions is to identify the fitness of an organism not with the actual number of offspring it contributes, but with the *number of offspring that it is physically disposed to contribute* (Brandon 1978a, Mills and Beatty 1979, Sober 1980, Burian 1983). Along the same lines, we can talk about the fitness of a genotype in terms of the average number of offspring that possessors of that genotype are physically disposed to contribute.

On this conception of fitness, the two physically identical twins, who must be physically disposed to contribute the same number of offspring, are equally fit. Hence we can say of the lightning that killed one and spared the other, that it did not sample the twins on the basis of their fitnesses – it was not an agent of natural selection. It was clearly an indiscriminate sampling agent – indiscriminate, that is, with regard to physical fitness differences between the organisms sampled.

3.4   That is perhaps as much about fitness as we really need to discuss in order to make meaningful the issue concerning the relative evolutionary importance of random drift vs. natural selection. But I am afraid that, as much as I would like the distinction between natural selection and random drift to be a clear-cut one, it is not as clear-cut as the preceding discussion suggests. (Elliott Sober and Kenneth Waters helped me considerably with the following discussion.) In order to complicate matters somewhat, let me elaborate a bit upon the notion of "fitness".

First, it is important to recognize that the fitness of an organism is the number of offspring it is physically disposed to contribute *in a particularly specified environment*. For instance, the number of offspring that a dark-colored moth is disposed to contribute is greater in a dark environment in which it is effectively camouflaged from its predators, than in a light environment in which it is more conspicuous. So all attributions of levels of fitness must be relative to particularly specified environments.

The specification of the environment, in turn, involves specifying a range of circumstances, each weighted according to the likelihood of its occurrence, or according to the likelihood of the organism(s) in question experiencing it. So one changes the environment, so to speak, by changing either the quality or the weighting of the component environmental circumstances. The ability of the dark moth to survive and reproduce in a uniformly dark environment is presumably different from its ability in a three-fourths dark environment, and that is presumably different from its ability in a half-dark environment, and so on.

The environmental circumstances relative to which fitness values are ascribed to members of a population are, ideally, all the environmental circumstances relevant to determining differences between the reproductive successes of those organisms. Some of these factors discriminate between the organisms on the basis of fitness differences between them. For instance, the combination of a dark background and color-sensitive birds favors the reproductive success of dark moths over light moths, *because of the difference in their color*. Some other factors among the specified environmental circumstances may be responsible for differences in reproductive success, but *not* in connection with any fitness differences between the organisms in question. Forest fires, for instance, might kill and spare moths without regard to any fitness differences between them. Not only the former factors, but also the latter belong in the environmental specifications relative to which fitness values are ascribed. Dark moths may be able to leave more offspring than light moths in environments in which the background is dark and the major predators are color sensitive, and in which forest fires are rare. But dark and light moths may have roughly the same

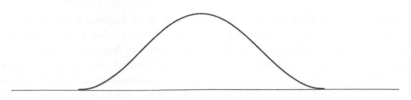

Figure 5.1

offspring-contribution dispositions in an environment in which forest fires are very frequent, even if the background is otherwise dark and the major predators are color sensitive. Similarly, the dark and light moths might have roughly equal fitnesses in an environment in which their predators are color *in*sensitive.

At any rate, relative to an environmental specification in terms of weighted environmental circumstances, there is a *range* of numbers of offspring that an organism is more or less disposed to contribute. For instance, in an environment made up of 60% dark-colored trees and 40% light-colored trees, a dark moth has a chance of visiting more or less than 60% dark trees, and more or less than 40% light trees. Given its physical structure, it may leave somewhat more offspring if it lands on more dark trees and somewhat fewer if it lands on fewer dark trees.

So for an organism in a particularly specified environment, there is a range of possible offspring contributions. And for each number within the range, there is a greater or lesser ability on the part of the organism to leave that number. Accordingly, we can talk about fitness *distributions* relative to specified environments. For instance, relative to a particularly specified environment, an organism might have a fitness distribution like the one in Figure 5.1. Here the *x*-axis represents the range of offspring contributions, and the *y*-axis represents the strength of the organism's ability to leave the corresponding number on the *x*-axis.

Somewhat alternatively, we can view this distribution as a probability distribution of offspring contributions, where the *x*-axis again represents the range of offspring contributions, and the *y*-axis the probability that the organism will contribute the corresponding number on the *x*-axis. The probabilities should, however, be interpreted as strengths of the organism's *physical ability* to contribute various numbers of

offspring, in line with *a propensity* interpretation of probability (Brandon 1978b, Mills and Beatty 1979). We can also construct a fitness distribution for a genotype. The range of this distribution would consist of the union of the ranges of possible offspring contributions of the individual members of the genotype. The height of the distribution at any point along the extended range would represent the group's average probability of leaving that many offspring.

Now the last thing I want to do in connection with this very general discussion of random drift, natural selection, and fitness, is to consider a possible scenario that requires that we give a bit more thought to the distinction between natural selection of the fittest and the indiscriminate sampling sources of random drift. How distinguishable are they, after all? Imagine again the case of the light and dark moths and their color-sensitive predators. Imagine too that they inhabit a forest in which 40% of the trees have light-colored bark and 60% have dark-colored bark. In this environment, we would say that the dark-colored moths are fitter, since the forest provides more camouflage for them than for the light moths. The fitness distributions of finite numbers of the two types might thus differ in the way shown in Figure 5.2.

But suppose that in one particular generation, we find, among remains of the moths killed by predators, a greater proportion of dark moths than was characteristic of the population as a whole, and a smaller proportion of light moths than was characteristic of the population. Say the actual average offspring contributions of the two types of moths in the previous generation differed as in Figure 5.3. And finally, let us say that we find the remains so distributed in the areas of dark and light trees that we have reason to believe that the dark moths were perched on light trees when attacked, and the light moths on dark trees. In other words, the dark moths

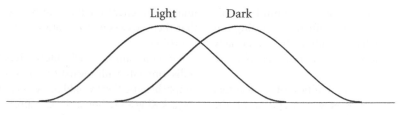

Figure 5.2

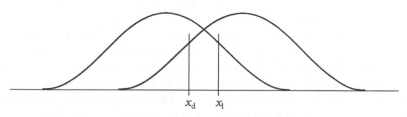

x_d = Actual average number of offspring of dark moths in E
x_l = Actual average number of offspring of light moths in E

Figure 5.3

chanced to land on light trees more frequently than on dark trees, even though the frequency of dark trees was greater.

That's the scenario. The question it raises is this. Is the change in frequency of genes and genotypes in question a matter of natural selection, or a matter of random drift? That is, is the change in question the result of sampling discriminately or indiscriminately with regard to fitness differences? It is not easy to maintain that the sampling was entirely indiscriminate with regard to differences in survival and reproductive ability. At least it is difficult to maintain that the death by predation of *conspicuously* dark moths in this environment is indiscriminate sampling, whereas the death of *conspicuously* light moths in the same environment is selection. On the other hand, it is also difficult to maintain that selection alone is the basis of the change. At least, it is difficult to maintain that the fittest were selected.

The problem here, I think, is that it is difficult to distinguish between random drift on the one hand, and the *improbable results of natural selection* on the other hand. Wherever there are fitness *distributions* associated with different types of organisms, there will be *ranges* of outcomes of natural selection – some of the outcomes within those ranges will be more probable than others, but all of the outcomes within the ranges are possible outcomes of natural selection. And yet some outcomes within a fitness distribution (the outer-lying outcomes of a bell-shaped fitness distribution, for instance), are in a sense "less representative" of the offspring-contribution abilities of the organisms in question. Consider a cointossing analogy. "Fifty heads : fifty tails" and "ninety heads : ten tails" are both possible outcomes of one hundred flips of a fair coin, where by "fair coin" I mean a coin equally disposed to land heads up and tails up. And yet "ninety heads : ten tails" is in a sense less representative of the fairness of the coin. So too in the case of the light and dark moths just discussed, the actual average offspring contributions shown in Figure 5.3 are possible results of natural selection, but not very "representative" of the fitnesses of the light and dark moths. To the extent that those outcomes are *less* representative of the physical abilities of those organisms to survive and reproduce in the environment in question, any evolutionary change that results will be *more* a matter of random drift. In other words, it seems that we must say of some evolutionary changes that they are to some extent, or in some sense, a matter of natural selection *and* to some extent, or in some sense, a matter of random drift. And the reason (one of the reasons) we must say this is that it is conceptually difficult to distinguish natural selection from random drift, especially where *improbable results of natural selection* are concerned.

Even given these difficulties of conceptual analysis, though, the new conception of the role of chance in evolution *clearly* goes beyond Darwin's conceptions, and raises questions that did not arise for him. In Darwin's scheme of things, recall, chance events and natural selection were consecutive rather than alternative stages of the evolutionary process. There was no question as to which was more important at a particular stage. But now that we have the concept of random drift taking over where random variation leaves off, we are faced with just such a question. That is, given chance variations, are further changes in the frequencies of those variations more a matter of chance or more a matter of natural selection?

4.   Chance in Modern Evolutionary Theory: Issues

4.1   As I suggested earlier, there are at least two versions of this general question. At issue in the first version is the extent of so-called selectively "neutral" mutations. The most likely candidates for this title were discovered during the molecular-biology boom of the fifties and sixties. It was discovered then that different permutations of the sequences of bases that make up DNA code for different amino acids. But it was also discovered that the sequence of bases that codes for a particular amino acid can sometimes be changed in a certain way such that the same amino acid is produced. These were called "synonymous" changes, in the information language of the field. It is certainly a plausible enough suggestion that synonymous changes have no effect on the numbers of offspring which organisms are physically disposed to contribute. Claims for neutrality have, however, also been extended to grosser phenotypic differences, from single amino acid differences between proteins of the same functional family, to differences between blood types (Wright 1940), to differences between banding patterns on snail shells (Diver 1940). Many such changes have not, upon first thought, had any

imaginable effect on the numbers of offspring that their possessors were physically disposed to contribute.

Sampling among such alternatives would be indiscriminate with regard to fitness differences, simply because there would be no such differences. As Jack King and Thomas Jukes expressed the basic idea in their classic position paper, "Natural selection is the editor, rather than the composer, of the genetic message. One thing the editor does *not* do is to remove changes which it is unable to perceive" (King and Jukes 1969, p. 788). As a result of this sort of indiscriminate sampling, the frequencies of neutral genetic alternatives may drift. In the words of the self-styled "neutralists", the "fates" of these genetic alternatives are matters of "chance" (e.g., Nei 1975, p. 165).

The neutralist theory of evolution is mathematically very sophisticated, though some of its features can be discussed informally. The theory is "stochastic" rather than "deterministic", in the sense that, given the gene and genotype frequencies of one generation, and values for all the other variables of the theory (like the size of the population), one can at best calculate the *probabilities* of possible gene and genotype frequencies in successive generations. It is not possible to predict *one specific gene or genotype frequency* for each successive generation.

Figure 5.4 is an example of the sorts of calculations that the theory allows (from Roughgarden 1979, p. 62). In this case, we are talking about an infinite number of populations, each of which has a constant size of 8 individuals, and each of which starts off with gene frequencies of 0.5 A and 0.5 a – i.e., 8 A alleles and 8 a alleles. This is represented by the top distribution, a bar that signifies that 100% of the populations have 8 A genes at time $t = 0$. According to the theory, over the course of generations, the distribution of populations spreads out with respect to the frequency of A alleles in each, until finally all of the populations have either no A alleles, or 100% A alleles (i.e., either 0 or 16 A alleles). We cannot predict whether or not a particular population will have, say, 50% A and 50% a alleles after one generation, but we can predict the *probability* that it will be 50 : 50, in terms of the proportion of populations that will still be 50 : 50 after one generation. The probability is about 0.2. Figure 5.5 is a somewhat more general way of representing

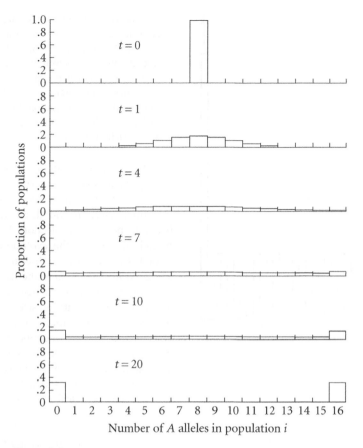

Figure 5.4

what Figure 5.4 represents. This is the first of such diagrams, from Sewall Wright's pioneering 1931 work on drift. According to the diagram, the frequencies of populations with all or none of a particular allele increase faster, the smaller the population size N.

That is, of course, just a glimpse of the general theory of the random drift of frequencies of selectively neutral genetic alternatives. But even that glimpse is important for our purposes. It is important, too, that we consider the sorts of issues that divide investigators in this area. There are more specific as well as more general issues. The more specific issues concern whether or not a change in frequency of a *specific set of genetic alternatives* can be ascribed to their selective neutrality and the consequent random drift of their frequencies. Complementarily, one might expect that the more general issues concern whether *all or no* evolutionary changes – or whether all of a

certain kind or none of a certain kind of evolutionary change – are due to random drift alone. It is important to recognize, however, that even the most general issues surrounding this version of the importance of random drift do not boil down to questions of all or none, but to questions of *more or less*. I cannot improve on Douglas Futuyma's assessment of this situation, so I will simply quote it. (I especially like this quotation because it is from a textbook in evolutionary biology, and communicates to students of evolution a valuable lesson concerning the nature of evolutionary theorizing, and of the nature of disputes in that field.)

The answer to the question, Is enzyme polymorphism due to selection or drift? will, of course, be: Both. But this is really no answer at all. No neutralist would deny that some few enzyme polymorphisms are maintained by selection, and

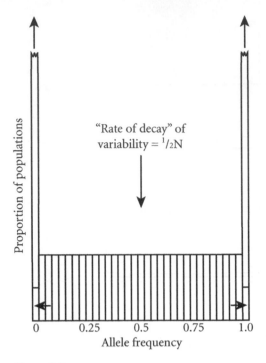

Figure 5.5

mutations may not be as rare as previously considered." However, I also added a note of caution: "It is important to note that probably not all synonymous mutations are neutral, even if most of them are nearly so." Perhaps this was a pertinent statement in light of the recent finding that synonymous codons are often used in "non-random" or unequal fashion (Kimura 1981). (Kimura 1982, p. 8)

4.2 The proponents of the second version of the importance of random drift discourage dividing evolutionary changes into those due entirely to natural selection and those due to random drift with no role for selection. They encourage, instead, analysis of the evolutionary effects of natural selection and indiscriminate sampling acting *concurrently*. That is, they rather encourage dividing evolutionary changes into those due *predominantly* to natural selection, those due *predominantly* to random drift, and those in which natural selection and random drift *both* play important roles. Consider, for instance, Wright's criticism of the evolutionary perspective of R. A. Fisher and E. B. Ford – that they misconstrued the importance-of-drift problem:

even the staunchest selectionist would not deny that some amino acid substitutions must have such trivial effects that they have vanishingly small effects on fitness and so fluctuate in gene frequency by drift alone. The real question is, What fraction of the variation is attributable to each factor . . . ? (Futuyma 1979, p. 340)

Similarly, Masatoshi Nei, a principle figure in the neutralist disputes, characterizes the strongest neutralist position as only "a majority rule" (1975, p. 165). As he explains, even the strongest neutralist position does not *rule out* selection with regard to some sets of genetic alternatives. It asserts instead that the majority of gene-frequency changes involve selectively neutral alternatives. Motoo Kimura, perhaps the principal neutralist, has emphasized over and over again the slack in his relatively very strong position. As he recently reported, along these lines,

. . . in one of my papers on the neutral theory (Kimura 1968), I wrote: "the recent findings of 'degeneracy' of DNA code, that is, existence of two or more base triplets coding for the same amino acid, seem to suggest that neutral

They hold that fluctuations of gene frequencies of evolutionary significance must be supposed to be due either wholly to variations in selection (which they accept) or to accidents of sampling. The antithesis is to be rejected. The fluctuations of some genes are undoubtedly governed *largely* by violently shifting conditions of selection. But for others in the same populations, accidents of sampling should be much *more* important and for still others *both* may play significant roles. It is a question of the relative values of certain coefficients. (Wright 1948, p. 291, emphases added)

Wright's criticism was twofold. First, he was warning against reducing the importance-of-drift question to an all-or-none neutralism issue – i.e., to an issue concerning whether all or no evolutionary changes are due to random drift with no role for selection. This is an issue that I just claimed was not really an issue with regard to the neutralist version of the importance of drift. And Fisher and Ford also denied being concerned with it (Fisher and Ford 1950, p. 175). But Wright was also warning against reducing the importance-of-drift question to even a more-or-less neutralism issue – i.e., to an issue concerning whether more or fewer evolutionary changes are due to random drift with no role for selection. Wright considered that the proper version of the importance of random drift relative to natural selection was one that concerned the *concurrent* relative roles of indiscriminate and selective sampling in accounting for evolutionary change with regard to each set of genetic alternatives. As Theodosius Dobzhansky, following Wright, also expressed this version,

> An evolutionary change need not be due either to directed or to random processes [for example, either to natural selection or to indiscriminate sampling]; quite probably it is the result of a combination of both types. The theoretically desirable and rarely achieved aim of investigation is to quantify the respective contributions of the different factors of gene frequency change, as well as their interactions. (Dobzhansky 1970, p. 231)

As for the various ways in which selective and indiscriminate sampling can act concurrently, several possibilities come to mind. The change in frequencies of A and a genes, and AA, Aa, and aa genotypes over the course of generations may be due predominantly to natural selection during some generations, and to indiscriminate sam-

pling during others. Indiscriminate and selective sampling are only loosely "concurrent" in this case. Or the changes in frequencies of those genes and genotypes may be due to a combination of discriminate parent sampling and indiscriminate gamete sampling, or vice versa. Or the changes may be due to sampling in some geographical subpopulations by entirely indiscriminate sampling agents, like forest fires, while the rest of the population is sampled selectively. Or the changes may be due to the sort of more-or-less selective and more-or-less indiscriminate samplings that I discussed in Section 3.4.

The general theory that describes the concurrent roles of random drift and natural selection is, like the neutralist theory, stochastic rather than deterministic – again in the sense that, given the gene and genotype frequencies of one generation, and values for all the other variables of the theory (including, again, size of the population, and in this case fitness values as well), one can at best calculate the *probabilities* of possible gene and genotype frequencies in successive generations.

The general theory includes, among many other things, variations on the U-shaped curves that describe the effects of random drift without selection. For example, figures 5.6(a) and 5.6(b), again from Wright's pioneering 1931 work, show the effects of the same variety of selection

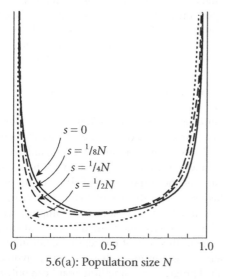

5.6(a): Population size N

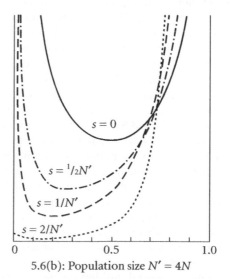

5.6(b): Population size $N' = 4N$

Figure 5.6

pressures in small (6a), and then in large (6b) populations. As in figures 5.4 and 5.5, we are dealing with *proportions* of populations and their gene frequencies. So all we can predict about a population with the help of these diagrams is the *probability* that it will have a particular combination of gene frequencies. Also, as in figure 5.4, all the populations in question had 50 : 50 gene frequencies at time $t = 0$. figures 5.6(a) and 5.6(b) show the distributions at some much later time. When there is no selection – represented by the case where the selection coefficient, s, equals zero – the characteristic U-shaped distribution results, although it "squeezes out" more populations with intermediate gene frequencies in the case of the smaller populations. For both the large and small populations, greater selection results in a greater skewing of the curve – that is, more and more of those populations have greater and greater frequencies of the gene whose frequency is represented on the x-axis. But this shows up more clearly in the case of the larger population, where in the case of strong selection for the allele in question (i.e., where $s = 2/N$), almost all of the populations have between 70% and 100% of that allele; whereas in the case of weaker selection (i.e., where $s = {}^1\!/_2N$), many more of the populations have less than 70% of the allele in question.

There again you have only a glimpse of a theory, but hopefully a glimpse that will be of some use. Again, somewhat aside from the theory, there are more or less specific and more or less general issues that divide investigators in this area. Specific issues with regard to this version of the importance of drift are those that concern the relative, concurrent importances of selective and indiscriminate sampling in accounting for changes in frequency of a *specific* set of genetic or genotypic frequencies (e.g., Kerr and Wright 1954, Dobzhansky and Pavlovsky 1957, Lamotte 1959). The general issues are, as in the case of the neutralism version of the importance of random drift, *more or less* issues rather than *all or none* issues. In other words, the issues do not simply boil down to whether indiscriminate sampling has a certain relative importance in *every* evolutionary change, but whether it has a certain relative importance in *many* evolutionary changes. For instance, Wright, who comes as close as anyone to defending a general position on the relative importance of random drift, nevertheless warns,

"There is no theoretical necessity for supposing that evolution has proceeded in the same way in all groups. In some it may proceed largely under direct selection pressure following change of conditions, in other cases it may be determined by random differentiation . . . , with or without selection" (Wright 1940, p. 181). And as Dobzhansky so nicely summed up this version of the importance of random drift, and the general issues associated with it,

Since evolution as a biogenic process obviously involves an interaction of all of the [agents of evolutionary change, including natural selection and indiscriminate sampling], the problem of the relative importance of the different agents presents itself. For years this problem has been the subject of discussion. The results of this discussion so far are notoriously inconclusive. . . . One of the possible sources of the situation may be that a theory which would fit the entire living world is in general unattainable, since the evolution of the different groups may be guided by different agents. To a certain extent this possibility is undoubtedly correct. In species that occur in great abundance in a fairly continuous area of distribution, the population size factor is bound to be less important than in species that are subdivided into numerous local colonies each having a small effective breeding population. [For] organisms whose environment is in the throes of a cataclysmic change, natural selection and mutation pressure are more important than [for] organisms living in a relatively constant environment. Nevertheless, one can hardly eschew trying to sketch some sort of a general picture of evolution. (Dobzhansky 1937, p. 186)

I have tried, so far, to distinguish two versions, and two sets of issues, concerning the evolutionary importance of random drift. But, of course, the distinction can be collapsed. As one investigates smaller and smaller fitness differences, one moves from an investigation of the second version of the importance of drift, to the first. Whether – or to the extent that – research programs into the two versions of the importance of random drift have maintained separate identities in spite of this sort of conceptual continuity, or whether this continuity has, at times, favored a unified program, is something that I can only say that I would like to know more about.

5.   Chance in Nature

At any rate, given a suitable interpretation of fitness and natural selection, there are two more or less distinguishable versions of the evolutionary importance of indiscriminate vs. selective sampling. Now if we step back from these positions and do not concentrate so much on the *variety* of issues involved, we see in the development of evolutionary biology what we see in the development of so many scientific disciplines in the nineteenth and twentieth centuries, namely the rise of stochastic theories. We have already discussed, very briefly, the respects in which evolutionary theories that encompass genetic drift are stochastic vs. deterministic. It might also be helpful to discuss the stochasticity of evolutionary theory by comparing the stochastic "version" of evolutionary theory with the effectively deterministic version that is presented in the opening chapters of most textbooks (prior to the introduction of the stochastic version).

The deterministic version, like the stochastic version, is based on the so-called "Hardy–Weinberg Law". According to that principle, in an infinitely large population in which individuals mate randomly with respect to their geneotypes, and in which there is no selection, mutation, or migration, the gene and genotype frequencies remain at equilibrium. More specifically, if the only alleles at a locus are A and a, and if we represent the proportion of A genes by p and the proportion of a genes by q, the genotypic frequencies of the population will remain in the equilibrium frequencies,

p²*AA* : 2pq*Aa* : q²*aa*.

Now assume for the moment that population size is not a variable of the theory – that the theory only applies to populations of infinite size. In an infinite population, any "local" misrepresentations of gene and genotype frequencies due to indiscriminate sampling would balance out. So by insisting that the population be infinitely large, evolution by genetic drift is effectively ruled out.

Given this version of the Hardy–Weinberg Law, given values for p and q in one generation, and given 1) magnitudes of selection for or against the various genotypes, 2) rates of mutation with respect to the various alleles, 3) magnitudes of preferential mating between the different genotypes, and 4) magnitudes of migration of the different genotypes, we can calculate the exact gene and genotype frequencies in the succeeding generations. For instance, if the frequencies of A and a are p and q respectively, and if the relative average fitnesses of the genotypes *AA*, *Aa*, and *aa* are 1, 1, and 1 − s respectively, then the change in frequency of A (Δp) in one generation is,

$$\Delta p = (spq^2)/(1 - sq^2).$$

The theory of evolution, thus construed, would be neatly deterministic, but would be grossly incomplete with respect to the many finite populations that supposedly also undergo evolutionary change! To cover these, population size must be considered as a variable. We can then formulate the sorts of stochastic extensions of the theory that we have already discussed. Without them, the theory of evolution would be awfully incomplete.

As in the cases of thermodynamics and quantum mechanics, stochastic theories play a very important role in evolutionary biology. In spite of all the disputes concerning the evolutionary importance of random drift, no one denies that evolutionary theory should be stochastic. But just to say that evolutionary theory is stochastic, and that is that, is to overlook an important respect in which stochasticity continues to be an issue in evolutionary biology. I do not mean that the issue has been transformed into the sort of stochasticity vs. *in-principle* stochasticity issue characteristic of quantum mechanics. I have in mind yet another respect in which the stochasticity issue in evolutionary biology is far from being settled. It is kept alive by the variety of issues and disputes concerning the importance of random drift. In order to see the connection, it will be useful to consider the disputes concerning the importance of random drift in the context of other, somewhat similar disputes in evolutionary biology.

Evolutionists recognize a variety of modes of evolution – a variety of ways in which the gene and genotype frequencies of a population may be changed (basically, any of the ways in which Hardy–Weinberg equilibrium might be upset). There is evolution via the genetic mutation of one allele to another, evolution via the migration of genetically different types of organisms into or out of the population, evolution via the preferential mating of organisms with particular genotypes, and of course evolution via random drift, and evolution via natural selection. Evolutionary theory describes the separate and concurrent outcomes of these various processes, depending on the values of the relevant variables. So, for instance, evolutionary theory describes what happens in a large population when there is no migration, no mutation, and moderate selection; what happens in a small population when there is moderate migration, moderate mutation, and no selection; etc.

Agreement with regard to the theory as described still leaves open some very important questions – namely, questions concerning the relative evolutionary importances of the various modes of evolution *in nature* – hence, the magnitudes of the relevant variables *in nature*. It is remarkable how many modern evolutionists have been of the opinion that the work that "remains to be done" in evolutionary biology is just that of measuring these magnitudes. Timofeef-Ressovsky's 1940 assessment of the work-to-be-done is representative of this opinion, and has been echoed often since then. As Timofeef explained the situation, evolutionary theorists of the likes of R. A. Fisher, J. B. S. Haldane, Sergei Tschetverikov, and Sewall Wright had succeeded in

> ... showing us the relative efficacy of various evolutionary factors under the different conditions possible within the populations (Wright 1932). It does not, however, tell us anything about the real conditions in nature, or the actual empirical values of the coefficients of mutation, selection, or isolation [limitation of population size]. It is the task of the immediate future to discover the order of magnitude of the coefficients in free-living populations of different plants and animals; this should form the aim and content of an empirical population genetics (Buzzati-Traverso, Jucci, and Timofeef-Ressovsky 1938). (Timofeef-Ressovsky 1940, p. 104)

And as A. J. Cain expressed the same position much more recently,

> The researches of R. A. Fisher, J. B. S. Haldane, and Sewall Wright laid the foundations of the mathematics of population genetics. ... What they did was to provide a mathematical theory covering all possible contingencies, from which quantitative predictions of both deterministic and stochastic processes could be made. That was a great gain, but it does not tell us what of all these possibilities are actually exemplified in the wild – what, in short, are relevant to the actual process of evolution. (Cain 1979, pp. 599–600)

So evolutionists can agree on *theory* – on what are the possible modes of evolution, and how evolution occurs given various possible magnitudes of the relevant variables – and at the same time disagree vehemently as to the actual magnitudes of these variables and thus the relative importances of these modes of evolution *in nature*. Since Darwin, evolutionists have continually haggled over such matters. Neo-Lamarckians concerned themselves for a long time with the relative evolutionary importance of an evolutionary factor that is no longer included in evolutionary theory – namely, the inheritance of acquired characters. Darwin and Moritz Wagner argued about the importance of migration relative to selection (see Sulloway 1979). In the early twentieth century, William Bateson and W. F. R. Weldon argued over the importance of mutation size and pressure relative to selection (see Provine 1971). There are even disputes as to which is the most important kind of selection. For instance, there was a long, complicated, and even bitter controversy between Dobzhansky and H. J. Muller as to the relative evolutionary importance of selection in favor of heterozygotes vs. selection in favor of homozygotes (see Lewontin 1974). Representative of this interest of evolutionary biologists in ranking the modes of evolution according to their importance is the title of a book by A. L. and A. C. Hagedoorn, published in 1921: *The Relative Value of the Processes Causing Evolution*.

These are rarely all-or-none issues. The disputants defend the importance of their favorite modes of evolution without ruling the others entirely out of the question. As Stephen Gould and Richard Lewontin characterize such issues, "In natural history, all possible things happen sometimes; you generally do not support your favoured phenomenon by declaring rivals impossible in theory" (Gould and Lewontin 1979,

p. 151). The neo-Lamarckians recognized selection, though their neo-Darwinian opponents did not reciprocate. Darwin did not altogether deny the evolutionary effects of migration, nor did Wagner completely rule out selection. Bateson certainly did not ignore selection, nor did Weldon ignore mutation. Dobzhansky always admitted cases of selection in favor of homozygotes, and Muller always admitted cases of selection in favor of heterozygotes.

The structure of these disputes thus has the important effect of not rendering entirely fruitless those parts of evolutionary theory that deal with the supposedly less important modes of evolution. So those parts of the theory are rarely threatened by elimination. These disputes can take place, in other words, without bringing any part of evolutionary theory into question, either for reasons of incorrectness or lack of fruitfulness.

The issues concerning the relative evolutionary importance of random drift are very similar: What is more important in accounting for evolutionary change, indiscriminate sampling in finite populations, or discriminate sampling? As has been discussed, these issues are not all-or-none, but more-or-less issues. As Gould characterizes the random drift issues, "The question, as with so many issues in the complex sciences of natural history, becomes one of relative frequency" (Gould 1980, p. 122). Thus, no neutralist *denies* selection. No selectionist *denies* neutral mutations. No one who investigates the concurrent action of natural selection and random drift believes that either of those factors is *always* overwhelming. The stochastic theory that describes the evolution of neutral alternatives, and that describes the concurrent effects of different degrees of indiscriminate sampling and natural selection, is thus not at issue.

But precisely because the actual relative magnitudes of population size and indiscriminate sampling on the one hand, and selection pressure on the other hand, *are* at issue, stochasticity is still an issue in evolutionary biology. Evolutionary biologists are content to have a stochastic theory of evolution, but are not at all in agreement concerning how important a role to attribute to chance in accounting for actual evolutionary change. The stochasticity issue in evolutionary biology is decidedly not *whether* chance plays a role in evolutionary biology, but *to what extent*.

I will close with reference, once more, to the long-standing dispute between Wright and Fisher concerning the relative evolutionary importance of random drift vs. natural selection. In his biography of Wright, William Provine thoughtfully analyzes the development of their agreements and disagreements. As Provine points out, Wright and Fisher were ultimately able, in published work and correspondence, to work out differences concerning the mathematical theory involved. For instance, Wright succeeded in convincing Fisher that the rate of decay of neutral variability was $1/2N$ (figure 5.5) rather than $1/4N$ as Fisher originally thought.

As Wright acknowledged in his review of Fisher's landmark, *The Genetical Theory of Natural Selection* (Fisher 1930), "our mathematical results on the distribution of gene frequencies are now in complete agreement, as far as comparable" (Wright 1930, p. 352). But as Wright also made clear, agreement with regard to mathematical theory still left room for considerable disagreement. And the most important source of disagreement was Fisher's assumption of large population sizes in nature, which effectively ruled out the actual importance of evolution by random drift. Again, this issue was somewhat detachable from the mathematical theory. As for the real disagreement, as Wright later put it, "It is a question of the relative values of certain coefficients" (1948, p. 291). Thus, evolutionists like Wright and Fisher could agree that small population size and chance are important in theory, all the while disagreeing considerably as to their importance in nature. It is in this sense that the stochasticity issues in evolutionary biology are still far from being resolved.

References

Brandon, R. (1978a), "Adaptation and Evolutionary Theory", *Studies in History and Philosophy of Science* 9: 181–200.

Brandon, R. (1978b), "Evolution", *Philosophy of Science* 45: 96–109.

Burian, R. (1983), "Adaptation", in M. Grene (ed.), *Dimensions of Darwinism*. Cambridge: Cambridge University Press.

Buzzati-Traverso, A., Jucci, C. and Timofeef-Ressovsky, N. W. (1938), "Genetics of Populations", *Ricerca Scientifica*, Year 9, Vol. 1.

Cain, A. J. (1979), "Introduction to General Discussion", *Proceedings of the Royal Society of London* B205: 599–604.

Crow, J. F. and Kimura, M. (1970), *An Introduction to Population Genetics Theory*. New York: Harper and Row.

Darwin, C. (1839), "N Notebook", in H. E. Gruber and P. H. Barrett, *Darwin on Man*. New York: Dutton, 1974.

Darwin, C. (1859), *On the Origin of Species*. London: Murray. Cambridge: Harvard University Press (facsimile edition), 1959.

Darwin, C. (1887), *The Variation of Animals and Plants under Domestication*, 2nd ed. (2 vols.). New York: Appleton.

Darwin, F. (ed.) (1887), *The Life and Letters of Charles Darwin* (3 vols.). London: Murray.

Diver, C. (1940), "The Problem of Closely Related Snails Living in the Same Area", to J. S. Huxley (ed.), *The New Systematics*. Oxford: Clarendon.

Dobzhansky, T. (1937), *Genetics and the Origin of Species*. New York: Columbia University Press.

Dobzhansky, T. (1970), *Genetics of the Evolutionary Process*. New York: Columbia University Press.

Dobzhansky, T. and Pavlovsky, O. (1957), "An Experimental Study of Interaction Between Genetic Drift and Natural Selection", *Evolution* 11: 311–319.

Dubinin, N. P. and Romaschoff, D. D. (1932), "The Genetic Structure of Species and their Evolution" (in Russian). *Biologichesky Zhurnal* 1: 52–95.

Eigen, M. and Winkler, R. (1981), *Laws of the Game*. New York: Knopf.

Fisher, R. A. (1930), *The Genetical Theory of Natural Selection*. Oxford: Oxford University Press.

Fisher, R. A. and Ford, E. B. (1950), "The 'Sewall Wright Effect'", *Heredity* 4: 117–119.

Futuyma, D. J. (1979), *Evolutionary Biology*. Sunderland, Massachusetts: Sinaucr.

Ghiselin, M. T. (1969), *The Triumph of the Darwinian Method*. Berkeley: University of California Press.

Gould, S. J. (1980), "Is a New and General Theory of Evolution Emerging?" *Paleobiology* 6: 119–130.

Gould, S. J. and Lewontin, R. C. (1979), "The Spandrels of San Marco and the Panglossian Paradigm: A Critique of the Adaptationist Programme", *Proceedings of the Royal Society of London* B205: 581–598.

Hacking, I. (1965), *Logic of Statistical Inference*. Cambridge: Cambridge University Press.

Hagedoorn, A. L. and A. C. (1921), *The Relative Value of the Processes Causing Evolution*. The Hague: Nijoff.

Hodge, J. and Kohn, D. (1985), "The Immediate Origins of the Theory of Natural Selection", in D. Kohn (ed.), *The Darwinian Heritage: A Centennial Retrospect*. Princeton: Princeton University Press.

Kerr, W. E. and Wright, S. (1954), "Experimental Studies of the Distribution of Gene Frequencies in Very Small Populations of *Drosophila melanogaster*", *Evolution* 8: 172–177.

Kimura, M. (1981), "Possibility of Extensive Neutral Evolution under Stabilizing Selection with Special Reference to Non-Random Usage of Synonymous Codons", *Journal of Molecular Evolution* 17: 121–122.

Kimura, M. (1982), "The Neutral Theory as a Basis for Understanding the Mechanism of Evolution and Variation at the Molecular Level", in M. Kimura (ed.), *Molecular Evolution, Protein Polymorphism, and the Neutral Theory*. New York: Springer.

King, J. L. and Jukes, T. H. (1969), "Non-Darwinian Evolution", *Science* 164: 788–798.

Lamotte, M. (1959), "Polymorphism of Natural Populations of *Cepaea nemoralis*", *Cold Spring Harbor Symposia on Quantitative Biology* 24: 65–84.

Lerner, I. M. (1958), *The Genetic Basis of Selection*. New York: Wiley.

Lewontin, R. C. (1974), *The Genetic Basis of Evolutionary Change*. New York: Columbia University Press.

Mettler, L. E. and Gregg, T. G. (1969), *Population Genetics and Evolution*. Englewood Cliffs, New Jersey: Prentice Hall.

Mills, S. K. and Beatty, J. (1979), "A Propensity Interpretation of Fitness", *Philosophy of Science* 46: 263–286.

Nei, M. (1975), *Molecular Population Genetics and Evolution*. New York: Elsevier.

Peirce, C. (1893), "Evolutionary Love", *Monist 3*: 176–200.

Provine, W. B. (1971), *The Origins of Theoretical Population Genetics*. Chicago: University of Chicago Press.

Provine, W. B. (1989), *Sewall Wright: Geneticist and Evolutionist*. Chicago: University of Chicago Press.

Roughgarden, J. (1979), *Theory of Population Genetics and Evolutionary Ecology: An Introduction*. New York: Macmillan.

Schweber, S. (1982), "Aspects of Probabilistic Thought in Great Britain During the Nineteenth Century: Darwin and Maxwell", in A. Shimony and H. Feshbach (eds.), *Physics and Natural Philosophy*. Cambridge: MIT Press.

Sheppard, P. M. (1967), *Natural Selection and Heredity*. London: Hutchinson.

Sheynin, O. B. (1980), "On the History of the Statistical Method in Biology", *Archive for History of the Exact Sciences* 22: 323–371.

Skyrms, B. (1966), *Choice and Chance*. Belmont, California: Dickenson.

Sober, E. (1980), "Evolutionary Theory and the Ontological Status of Properties", *Philosophical Studies 40*: 147–161.

Sulloway, F. J. (1979), "Geographical Isolation in Darwin's Thinking: The Vicissitudes of a Crucial Idea", *Studies in the History of Biology 3*: 23–65.

Timofeef-Ressovsky, N. W. (1940), "Mutations and Geographical Variation", in J. S. Huxley (ed.), *The New Systematics*. Oxford: Clarendon.

Waddington, C. H. (1968), "The Basic Ideas of Biology", *Towards a Theoretical Biology 1*. Chicago: Aldine Publishing Company, pp. 1–41.

Wilson, E. O. (1975), *Sociobiology*. Cambridge: Harvard University Press.

Wilson, E. O. and Bossert, W. H. (1971), A *Primer of Population Biology*. Sunderland, Massachusetts: Sinauer.

Wright, S. (1930), "The Genetical Theory of Natural Selection: A Review", *Journal of Heredity 21*: 349–356.

Wright, S. (1931), "Evolution in Mendelian Populations", *Genetics 16*: 97–159.

Wright, S. (1932), "The Roles of Mutation, Inbreeding, Crossbreeding, and Selection in Evolution", *Proceedings of the Sixth International Congress of Genetics 1*: 356–366.

Wright, S. (1940), "The Statistical Consequences of Mendelian Heredity in Relation to Speciation", in J. S. Huxley (ed.), *The New Systematics*. Oxford: Clarendon.

Wright, S. (1948), "On the Roles of Directed and Random Changes in Gene Frequency in the Genetics of Populations", *Evolution 2*: 279–294.

Appendix

Recall that the hereditary material of sexual organisms comes in pairs of morphologically similar, or homologous, chromosomes. The genes reside linearly along the chromosomes; the two genes that lie opposite one another on homologous chromosomes are said to occupy the same chro-

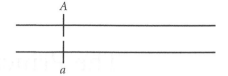

mosomal location, or "locus", and are said to be "alleles" with respect to that locus. In the diagram, A and a are alleles at a particular locus. The two alleles at a locus may be materially different, as in the case of A and a (as the different symbols are supposed to represent), or they may be materially identical. Different combinations of alleles may result in alternative states of the same general character, like blue eye color vs. brown eye color.

The particular combination of alleles that an organism has is called its genotype. We can speak of the genotype at a particular locus – for instance, we can speak of the Aa genotype in the case at hand – or we can speak of the entire genotype. If the two alleles at a locus are different, the organism is said to have a "heterozygotic" genotype with respect to that locus. Otherwise, it would be said to have a "homozygotic" genotype with respect to that locus.

Finally, the gametes that a sexual organism produces (i.e., the sperm/pollen or egg cells that it produces) each appropriately contain one allele from each locus. According to "Mendel's Law", there is no physical basis for a bias in the number of gametes containing one or the other allele of a heterozygotic organism. In other words, in the long run, a heterozygotic organism should produce gametes, half of which contain one allele, and the other half of which contain the other allele.

6

The Principle of Drift: Biology's First Law

Robert N. Brandon

Drift is to evolution as inertia is to Newtonian mechanics. Both are the "natural" or default states of the systems to which they apply. Both are governed by zero-force laws. The zero-force law in biology is stated here for the first time.

The Newtonian analogy to evolutionary theory is fairly common in presentations of population genetics, especially in reference to the Hardy–Weinberg Law. In the philosophy of biology, Elliott Sober's[1] presentation is canonical. Recent literature in the philosophy of biology has both criticized[2] and defended[3] the Newtonian analogy to evolutionary theory. Here I show exactly how the Newtonian analogy is helpful. More interestingly, I show how the analogue of Newton's First Law, what I term The Principle of Drift, is, in a very important way, non-Newtonian. Newtonian or not, it is one of the two fundamental laws of biology (along with the Principle of Natural Selection).

I.   Newton's Principle of Inertia

The focus of this paper is on what is, and what is not, an appropriate zero-force law in Evolutionary Theory. Before turning to that, I want to emphasize a few well-known points about the paradigm of zero-state laws – Newton's First Law, also known as the Principle of Inertia. After reviewing these points I will explore what sort of explanations Newton's First Law supports.

Newton expressed the First Law in two parts, since he thought there was an absolute difference between objects at rest and objects moving at a constant velocity. Of course relativity theory rejects this distinction and therefore we often see simpler one-clause statements of the principle. But for our purposes, the Newtonian version is best. It states that:

1. An object at rest will stay at rest unless acted on by a net external force.
2. An object in motion with velocity v will remain in motion with velocity v unless acted on by a net external force.

(Of course, relativity aside, the first clause can be seen as a special case of the second, where $v = 0$.) Newton's First Law represented a major break from the tradition of Aristotelian physics. While Aristotle would have agreed with the first

Robert Brandon, "The Principle of Drift: Biology's First Law," *Journal of Philosophy* (2006), 102, 319–335. Reprinted by permission of the *Journal of Philosophy* and by Robert Brandon.

My thanks to Richard Lewontin, Dan McShea, Fred Nijhout, Grant, Ramsey, Mark Rausher, and Alex Rosenberg for helpful comments on earlier drafts of this paper. I also want to thank Greg Wary for help with references and Leonore Fleming for editorial help.

clause stated above, he would have disagreed with the second. For Aristotle the maintenance of constant velocity requires the constant action of a force. (Put anachronistically, for Newton $F = ma$, while for Aristotle $F = mv$.) Thus Newton's First Law changes what calls for an explanation in terms of force. Or to put the point in more Aristotelian terms, Newton fundamentally altered the conception of what constituted "natural" motion.

All this is familiar. But let us now turn our attention to Newtonian explanations of inertial motion. Suppose an object is at rest at time t. Over the time interval t-t' it remains at rest. How do we explain this behavior? The acceleration, a, during this time interval is 0. Applying Newton's Second Law, $F = ma$, we see that $F = 0$. Thus, according to the First Law the object remains at rest. This is elementary physics. We will see if it is elementary philosophy.

What sort of explanation is this? According to the covering-law model of explanation,[4] to explain a phenomenon is to subsume it under some one or more general laws. That is exactly what we did above, and so, based on that model we would say that our explanation is a standard scientific explanation. But if one considers this model of explanation to be inadequate,[5] as I do, then how do we think about this explanation? That, not surprisingly, depends on the alternative model one adopts. It is well beyond the scope of this paper to argue for one theory of explanation over others. But elsewhere I have argued for a causal-mechanical model of explanation in evolutionary biology.[6] Since our ultimate concern is with evolutionary biology, let us ask the question of how that model treats the above bit of Newtonian physics.

If one thinks of Newtonian forces as Newtonian causes – and that certainly seems reasonable – then one might think that the explanation of inertial phenomena is noncausal. Our explanation above did not cite a force; therefore it did not cite a cause. Rather it cited the absence of causes. So how can it be a causal explanation? One response to this question seems so unattractive that I will simply mention it without laboring to refute it. One might say that Newton's First Law tells us what does, and what does not, require explanation. Inertial phenomena do not. Although there is something right

about this response, and we will turn to that shortly, saying a whole class of well understood phenomena cannot be explained seems just wrong. A second response is to say that the above reasoning is correct, that the explanation is non-causal, and that shows the limitations of the causal-mechanical model of explanation. Although a number of philosophers of science would be sympathetic to this response, I am not. Salmon (*op. cit.*) argues that the goal of explanatory science is to uncover the causal structure of the world. The Principle of Inertia (or rather its Relativistic counterpart) seems to be a fundamental part of the causal structure of our world. Because of that, I would consider our inertial explanation to be a perfectly good causal-mechanical explanation.

But, returning to the point made above, there is something different about this sort of Newtonian explanation versus one that explains some noninertial phenomenon in terms of some net force acting on the object. The latter sort of phenomenon requires "special" explanations. That is the lesson of Newton's First Law. Thus we could mark this distinction by saying that noninertial phenomena require "special-causal-explanations," while inertial phenomena require what I will call "default-causal-explanations."

In Newtonian mechanics, the Principle of Inertia clearly delimits what can, and cannot, be explained by default explanations. That is, it delimits the class of objects that fall under the zero-force law. We will see if there is any analogous law in biology.

II. The Hardy–Weinberg Law as a Zero-Force Law

The existence of a zero-force law presumes the existence of forces. Are there forces of evolution? John A. Endler[7] argued that the analogy to Newtonian physics was more misleading than helpful. More recently, Dennis Walsh, Tim Lewens, and André Ariew (*op. cit*), and Mohan Matthen and Ariew separately (*op. cit.*) have argued against the force analogy. I will not try to respond to these arguments in a point by point fashion; in part because I have offered detailed criticisms of the "emergentist statistical interpretation" of evolutionary theory defended in the two last mentioned papers,[8] but also because I think a direct positive case can be made for considering certain evolutionary processes as forces.

Evolutionary forces

The processes I have in mind are selection, mutation, migration, and nonrandom mating. In physics a force is a vector quantity. That means it has both a magnitude and a direction. Also various forces in Newtonian physics (for example, gravity, electrical force, friction) can all be measured in a standard quantitative unit (*à la* Newton). Is there a common currency to measure the processes mentioned above? From a population genetics point of view all can be measured in terms of their effects on gene frequencies (see caveat below). From the point of view of molecular evolution they would be measured on a finer grain – the level of gene sequences. And one could take a purely phenotypic point of view and measure in terms of phenotypic distributions. Let us ignore this embarrassment of riches and focus on the population genetics perspective.

Consider mutation and migration. Both are similar in that they are already net forces that could be broken down into their components. Migration consists of immigration into a population and emigration from it. Both can have effects on gene frequencies, which can then be added together to get the force of migration; similarly for mutation. In a simple system with one locus and two alleles, A_1 and A_2, there is an actual number of mutations from A_1 to A_2, and an actual number from A_2 to A_1. The change in A_1 is simply the second number minus the first. There is no problem in conceiving of these as forces. (Although some might insist on considering them as net forces rather than as component forces: that would not militate against the view that evolutionary theory is a theory of forces.)

Nonrandom mating is different. I mention it now, because we are about to consider the Hardy–Weinberg Law, and nonrandom mating is one of the processes that can cause a population to deviate from H–W equilibrium. Nonrandom mating has two basic forms: positive assortative mating and negative assortative mating. Let me say explicitly that here we are solely concerned

with forms of nonrandom mating that do not involve differential reproduction among different genotypes. (So, for example, we are not here talking about positive assortative mating where rare genotypes have lower reproductive success because of difficulty in finding mates.) Nonrandom mating has an immediate effect on genotype frequencies, not gene frequencies. (Consider a population with two alleles, A_1 and A_2, at frequencies p and q. If the homozygotes preferentially mate with each other, p and q will remain unchanged but there will be a decrease in the number of heterozygotes.) Nonrandom mating can have selective effects (for example, sexual selection or selection against inbreeding), but from the point of view of the H–W law it is its effect on genotypic frequencies that matters. This contradicts the statement made above that all of these processes can be measured in terms of effects on gene frequencies. But we can achieve consistency by measuring mutation and migration in terms of effects on genotypic frequencies. From a population genetic point of view that is inconvenient, but possible. So I consider this a minor technical point and will henceforth ignore it.

Selection presents a new challenge to the view that these evolutionary processes are forces. People tend to think of directional selection when they think of selection (this point will be explored and defended below). The other two patterns of selection are stabilizing and disruptive. These terms come from quantitative genetics so are usually defined in terms of phenotypic distributions. Thus directional selection is when one extreme in the distribution is favored (for example, the taller the fitter). Stabilizing selection is when a single point in the distribution is favored (for example, 5m tall is the fittest). Disruptive selection occurs when 2 (or more) points in the distribution are favored (for example, 2m tall and 6m tall are the fittest). But these terms have come to be used more generally and can be translated into genetic terms. For instance in a two-allele model, if A_1 is rare and A_1A_1 is fitter than A_1A_2, which is fittter than A_2A_2, then we would call that directional selection. But remember that a force is supposed to have a magnitude and a direction. This seems totally unproblematic for directional selection. But how about the other two forms? What is the direction of stabilizing selection? If a population, or some subpart of it, finds itself to the left of the selected point, then the direction of the force is to the right. But if the population is on the right of that point then the direction is to the left. Is our force analogy in trouble? No, or rather, only if it is problematic to think of gravity as a force. In our solar system two objects of equal mass and equal distance from the Sun, but on opposite sides of the Sun, will experience a force of equal magnitude but opposite direction. That is exactly analogous to two subpopulations equidistant from the selected point in stabilizing selection but on opposite sides (assuming the selection gradient is the same on both sides). That gravitational force is relative to position is obvious in Newtonian mechanics. So that a similar relativity exists with respect to stabilizing and disruptive selection is not problematic.

Thus, although it has not been exactly straightforward, we have made a positive case for treating certain evolutionary processes as forces. Notice that drift has yet to be mentioned.

The Hardy–Weinberg Law. There are two importantly different ways of stating the H–W law. One statement of it lumps drift together with mutation, migration, and the rest. The other does not. Here are the two versions:

H–W$_1$: If a population exists with two alleles, A_1 and A_2, with frequencies p and q respectively, then in a single generation the population will settle into genic and genotypic equilibrium with gene frequencies p and q, and genotypic frequencies of $A_1A_1 = p^2$; $A_1A_2 = 2pq$; and $A_2A_2 = q^2$ – provided that there is no selection, mutation, migration, nonrandom mating, or *drift*.

H–W$_2$: If an *infinite* population exists with two alleles, A_1 and A_2, with frequencies p and q respectively, then in a single generation the population will settle into genic and genotypic equilibrium with gene frequencies p and q, and genotypic frequencies of $A_1A_1 = p^2$; $A_1A_2 = 2pq$; and $A_2 A_2 = q^2$ – provided that there is no selection, mutation, migration, or non-random mating.

If H–W$_1$ is to be considered a zero-force law then drift needs to be a force just like the other evolutionary processes. However, I think that there are overwhelming reasons for not considering it a force.

Consider first the idea that a force has both a magnitude and direction. Drift has a magnitude that can be probabilistically predicted prior to the fact, and can be quantitatively accessed after the fact.[9] But drift definitely does not have a direction. (That, of course, is why it is called *drift*.) Given a population consisting of two selectively neutral alleles, A_1 and A_2, at frequencies $p = q = .5$, we can predict that one of the two alleles will go to fixation (in the absence of evolutionary forces), but we cannot predict which will. That means drift has no direction. Christopher Stevens (*op. cit.*) argues that drift does have a direction at the genotypic level. He suggests that it predictably leads to loss of heterozygosity and increase of homozygosity. That is a prediction, but a prediction without a direction. It does not say which of A_1A_1 or A_2A_2 will increase in frequency. In physics that would be like saying that a 20-Newton force is acting on object A. Such a statement either makes no sense (the magnitude, but not direction has been specified) or is incomplete (oops, I meant a 20-Newton downward force). Notice that we were able to specify the direction of the other evolutionary processes mentioned in H–W$_1$.

Second, and perhaps more importantly, drift is not a "special" force in evolution; it is the default position. By that I mean that it is part and parcel of a constitutive process of any evolutionary system – namely, the sampling process. Sampling takes place in a number of ways. I will not try to exhaustively name all of them, because I cannot, and I suspect no one can. That is, I suspect there are as of yet undiscovered sampling processes going on in evolutionary systems. But here is a partial list of sampling processes. Evolution requires reproducing entities that form lineages, parent-offspring lineages. So in many cases there is sampling of parents (not all potential parents actually reproduce). In sexual diploid organisms there is gametic sampling in the formation of a new diploid cell (fertilized zygote). There is also chromosomal sampling in the formation of gametes. There is further sampling of parts of chromosomes in crossover events during meiosis. At the ecological level there is habitat sampling when organisms are dispersed over heterogeneous selective environments.[10] The process that leads to drift, sampling, is a necessary part of any evolutionary system. It is not some new process added to the basic evolutionary process. In contrast, mutation, migration and nonrandom mating are clearly separable processes, which may, or may not, be a part of some particular evolutionary scenario.

(Asking if drift is a force is just like asking if inertia is a force in Newtonian mechanics. The lesson of Newton's First law is that it is not.)

Selection, I have argued[11] is both conceptually and empirically distinguishable from drift. When nonequal probabilities govern the sampled entities, selection becomes possible. However the possibility of drift does not go away just because the equiprobable distribution is no longer in force. Drift becomes impossible only in the highly implausible condition that I have labeled a state of *Maximal Probability Difference* – that is, when all of the fitnesses of the competing entities equal either 1 or 0, with both values being present. (See Figure 6.1.)

Selection, like drift, depends on the sampling process. However, unlike drift, it has a direction and is separable from drift. Thus I think it is reasonable to treat selection as a "special" evolutionary force (on a par with mutation, and so on), while it is clearly unreasonable to think of drift that way. I will offer more support for this position

shortly. For now, let us tentatively adopt the position that drift is not a force. It follows that H–W$_1$ is not a zero-force law. For it to be true it needs to mix legitimate evolutionary forces with one non-force (drift). H–W$_1$ is useful in many ways. It is true. It just is not a zero-force law.

H–W$_2$ does not run afoul of the above problem. It mentions only legitimate evolutionary forces in its proviso-clause. However, it is unclear how it applies to real populations, all of which are finite. I am not criticizing H–W$_2$ on the basis of its employing an idealization. That is normal and useful in science. But when we talk about frictionless planes in physics and get results that apply in that idealized situation, we ultimately want to be able to apply it to real world situations. So how do we apply H–W$_2$ to finite populations? One answer is that we do not. I presume that most would be unsatisfied with this. A second answer is that in applying H–W$_2$ to finite populations we transform it back into H–W$_1$, that is, we stick drift back into the proviso-clause. But then, as we have seen, we no longer have a zero-force law.

A final answer is that we do not revert to H–W$_1$, we leave the provisos alone, and we assert H–W$_2$ is true of finite populations. Then H–W$_2$ basically says of finite populations: If no

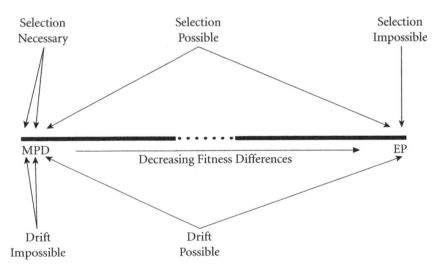

Figure 6.1 The heavy horizontal line, with dotted center section, represents the infinite number of possible fitness distributions from Maximal Probability Differences (MPD – all fitness = 0 or = 1, with some of both) on the left to the Equiprobable Distribution (EP – all fitness the same) on the right. The arrows emanating from the different descriptions of the modalities of selection and drift indicate the areas of the distribution falling under these descriptions.

evolutionary forces act on a population then its gene and genotypic frequencies remain unchanged. That is logically equivalent to the following: If a population changes in gene or genotypic frequencies then some evolutionary force is acting on it. Assuming, as we are for the moment, that I am right in not classifying drift as an evolutionary force, then these statements are false. A population with no evolutionary forces acting on it will drift. (And so, a population that changes over generational time may or may not be experiencing evolutionary forces.) This truth is exactly what guides contemporary work on trying to distinguish selection from drift. For instance, at the molecular level we compare synonymous substitutions (substitutions in the third place in a codon that end up coding for the same amino acid), with nonsynonymous substitutions to see where, and with what strength, selection is acting.[12]

H–W$_2$ applied in a straightforward manner to finite populations is not only false in what it says, it is also misleading in what it suggests. The first gloss on H–W$_2$, stated in the paragraph above, basically says: If no forces then no change. That does not strictly imply the following: If no change then no force; but it does suggest it. It suggests that lack of change is a sign of lack of force. (This suggestion is clear in Sober's Newtonian analogy (*op. cit.*).) But in fact, the opposite is true: If no change then a set of forces must be acting. Again this truth is important in that it licenses the sorts of inferences biologists need to make in understanding equilibrium situations (more on this shortly).

Thus H–W$_2$ is either: (a) A true zero-force law inapplicable to any real population; or (b) Applicable, but then equivalent to H–W$_1$, and so not a zero-force law; or (c) Applicable, but false.

And so the Hardy–Weinberg Law fails to be a zero-force law in evolutionary biology. Is there a zero-force law in evolutionary theory?

III. The Principle of Drift as a Zero-Force Law

The answer is "yes." But the law in question has heretofore been unstated. I will state it first, and then defend its truth, its usefulness, and its status as a zero-force law.

The evolutionary principle of drift

(A) A population at equilibrium will tend to drift from that equilibrium unless acted on by an evolutionary force.

(A population at rest will tend to start moving unless acted on by an external force.)

(B) A population on evolutionary trajectory *t*, caused by some net evolutionary force F, will tend to depart from the extrapolated path predicted based on F alone (in either direction or magnitude or both) even if no other evolutionary force intervenes, unless F continues to act.

(A population in motion will tend to stay in motion, but change its trajectory, unless continually acted on by an external force.)

The Principle of Drift (PD) is stated in two clauses, just like Newton's version of his Principle of Inertia. I have stated each clause first in more precise biological language and then, in parentheses, in a more physics-like language that facilitates a comparison to Newton's First Law. Clause (B) is clearly non-Newtonian. In fact, it is Aristotelian in that it requires the constant action of a force to keep a population in the same state of motion. But Aristotle would not be happy with Clause (A), which is both non-Newtonian and non-Aristotelian. So in this way the PD is quite unlike Newton's First Law.

However, in another way, there is a strict parallel to Newton's First Law. Recall our discussion above where we saw how Newton's law sets out inertia as the default state of an object. I argued that Newtonian explanations of inertial states are importantly different from Newtonian explanations of non-inertial states. The latter require the citation of "special" Newtonian causes, while the former are what I called default-causal-explanations. The PD serves exactly the same purpose. It clearly sets out the default state of a population, or in other words, it is a principle of default evolutionary action. In the absence of evolutionary forces drift occurs, or, more precisely drift will tend to occur (see figure 6.1 above). Do not misunderstand this, drift can certainly occur in the presence of evolutionary forces as well. So, for instance, in clause (B) above, a population continually acted on by net evolutionary force, F, will quite frequently depart from the trajectory predicted from F alone. F needs to be very strong relative to the effective population size for it to dominate drift.[13] Drift is the default state of evolutionary systems.

We can explain evolutionary change due to drift. The drift producing potential of the sampling processes that are constitutive of the evolutionary process is a fundamental part of the causal structure of our world. Thus such explanations are causal but differ from evolutionary explanations that cite specific evolutionary forces. Just as in the Newtonian case, the first sort of explanation is a default-causal-explanation and the second a special-causal-explanation. It is the PD that enables us to make this distinction.

Is the PD true? I think its truth is patently obvious once one potential misunderstanding of it is removed. One might think that the PD takes a stand on an empirical issue, the truth of which is in doubt. That is, one might think that the PD is somehow committed to some particular claim about the relative importance of drift versus selection. But this is a misunderstanding. Newton's First Law says nothing about the relative frequency of inertial objects versus objects experiencing a net force in the universe. Indeed, think of the paradigm example of a system to which Newton's Laws apply – our solar system. All of the major objects in that system are experiencing constant acceleration due to gravitational forces. Likewise, the PD is perfectly compatible with the claim that selection and other evolutionary forces are the primary movers of the vast majority of populations.

Consider clause (A). It says that in the absence of evolutionary forces (and in particular, selection) a population previously at equilibrium will start to drift. The empirical debate over the relative importance of drift versus selection is irrelevant to this claim. By hypothesis clause (A) removes selection. Then, by hypothesis, the existing variants are selectively neutral. From that it follows that drift will occur.

Consider clause (B). It says that in the absence of evolutionary forces (and in particular, selection) a population previously moving in state space with trajectory t will depart from that trajectory. Again, the truth of this is fairly obvious. If the trajectory were due to one or more of the special evolutionary forces, then removing those forces would tend to change the trajectory.

So the truth of the Principle of Drift is not hostage to any particular resolution of the empirical selectionist/neutralist debate. However, some philosophers might think its truth is dependent on a questionable metaphysical assumption – namely, that the evolutionary process is fundamentally indeterministic. First, the question of determinism versus indeterminism for the whole world, or for some part of it (the evolutionary process), is an empirical question, not a metaphysical question. I believe, and have defended the view, that the evolutionary process is indeed indeterministic,[14] but I think that the only assumption that has been at work in describing the Principle of Drift is that drift is a real phenomenon. Drift happens. Drift is distinguishable from other sorts of evolutionary change. No biologists doubt this. And that is all I have assumed. The (misguided) metaphysical determinist needs to explain how drift can exist in a fundamentally deterministic world. (Since that is a fantasy world I leave that exercise to him or her.)

It is also a misunderstanding to think that I have defended the Principle of Drift at the expense of the H–W Law. I have argued that the H–W Law is not a zero-force law, but I have not denied its utility. In fact its strength relative to the Principle of Drift flows from its limitations. The H–W Law is explicitly couched in the terms of the population genetics of diploid sexual organisms. That allows it, for instance, to be used to show that assortative mating is occurring in a population when genotypic frequencies depart from the H–W prediction. The PD is not couched in this specific language and could not be used for this particular inference. But, as John Beatty[15] pointed out some time ago, this very specificity shows that the H–W Law is no law at all. It depends on conditions that are evolutionarily derived, and not at all necessary for the evolutionary process. The evolutionary forces that brought the H–W into play could change and render H–W totally inapplicable to life on Earth (for example, if all lineages were to go asexual). The Principle of Drift is different. It is not predicated on diploidy or sex. It is based solely on the fact that evolution consists of one or more sampling processes. In that way it is perfectly general and is a law of nature.

One way to think about the lawfulness of biological generalizations is to ask if they would be exportable to life elsewhere in the universe. Were I asked to take Evolutionary Theory to some sentient beings elsewhere in the universe I would pack the Principle of Natural Selection. Where there is life there is natural selection. I would also pack the Principle of Drift. Along with a few personal items I would also pack my math books. And that is all. I would not bother with the H–W Law because I would not necessarily expect to find diploid sexual organisms. If I did find them I would realize that $(p + q)^2 = p^2 + 2pq + q^2$ and would thereby recreate H–W.

So although my point is not to criticize the H–W Law, it is clear that it is not fundamental, unlike the PD. Likewise, and perhaps more importantly, it does not provide appropriate null hypotheses in evolution. It suggests evolutionary stasis as a null hypothesis. But this is just wrong. In the absence of evolutionary forces drift occurs. So drift is the appropriate null hypothesis.

Finally, and relatedly, the H–W law is not a zero-force law. The PD is. It is in virtue of this fact that it does provide the appropriate null hypothesis.

IV. Consequences

Exactly how Evolutionary Theory relates to Newtonian mechanics may seem an esoteric topic in philosophy of biology. It is not. It has real consequences. I have explored the biological consequences of the PD elsewhere.[16] Here the focus will be on the more purely philosophical.

First, the very concept of evolution suggests the view that stasis is the default condition of evolutionary systems and that change is what is fundamental to evolution. Thus evolution is often defined as *change* in gene frequencies. Those objecting to a narrow population genetics perspective have suggested other definitions, such as, evolution is *change* in phenotypic distributions, or evolution is *change* in developmental programs. But from the perspective of this paper, all such definitions are entirely misleading. Change is a heterogeneous category in evolution. It may be driven by special evolutionary forces, or it may result from no net forces at all. Stasis, on the other hand, seems to be quite homogeneous in its cause. There is no way to produce

long-term stasis other than strong stabilizing selection.

Second, this privileging of change, the failure to see the fundamentally different causes of change (default versus special), and the failure to see stasis as a strong signal of selection, together have led to empirically inadequate and explanatorily empty conceptions of fitness. Thus, starting with R. A. Fisher,[17] many have sought to *define* fitness in terms of evolutionary *change*. (For Fisher, fitness is the per capita rate of increase of a type.) Recently Matthen and Ariew (*op. cit.*) have argued that evolutionary biologists *must* measure fitness in terms of evolutionary change. But if one of the primary ways selection acts is in stabilizing selection, then these views are entirely wrong-headed. Strong stabilizing selection (or what is called *purifying selection* at the molecular level) results in no change (in gene frequencies, or at the molecular level, in DNA sequence). And indeed, at the molecular level the strongest evidence of the past action of selection is evolutionary conservation of DNA sequence. (As, for example, in the PAX6 gene found in virtually identical forms in fruit fly, zebrafish, mice, and humans.[18]) But the Fisherian is conceptually precluded from seeing this as strong selection, since nothing is changing. Further, it should be noted against the argument that biologists must/should measure fitness in terms of evolutionary change, that in fact, biologists measure fitness in that way only very rarely and only as a last resort.[19]

Finally, the views put forward in this paper have implications for what is perhaps the most widely debated question in philosophy of biology – namely, whether or not there are laws in biology. I used to think that biology had exactly one law – the Principle of Natural Selection.[20] I now see it has two.

Perhaps more interesting than counting the number of fundamental laws in evolutionary theory is recognizing the thoroughly probabilistic nature of those laws. I have argued extensively elsewhere[21] that the Principle of Natural Selection is an instantiation of (a specially structured version of) what is called the Principle of Direct Inference in Probability Theory.[22] This is the principle that allows one to infer from statistical probabilities (or chances) to actual frequencies.[23] For example, if coin A has a probability of heads

= 0.7 (in a certain chance set-up) while coin B has a probability of heads = 0.3 (in the same chance set-up), then if we toss both coins 10 times (in the same set-up) then (probably) coin A will land on heads more than coin B. That is an application of the Principle of Direct Inference. Consider the PNS: If organism A is fitter than organism B in environment E, then (probably) A will have more offspring in E than B. Clearly, this is a perfectly analogous application of the same principle. In this paper we have seen how the Principle of Drift is purely the result of the finite sampling processes that are part and parcel of the evolutionary process. Thus, for those who are reductionistically minded (I am not), these two laws point to probability theory as the reductive foundation of evolution, not to physics or chemistry.

V.   Biology's First Law

Newton's First Law is so called not because it was the first law of physics discovered or articulated. Here "First" does not relate to temporal priority. Rather it describes a sort of logical, or better, nomological priority. Newton's Law of Inertia describes the default, or "natural," state of Newtonian systems. It describes what happens when no net Newtonian forces are acting on such a system. It is, as we have seen, a zero-force law. I have shown that The Principle of Drift is analogous to Newton's First Law in these ways. That is why I call it biology's first law.[24]

But from a temporal point of view the Principle of Drift is new, or at least relatively new. However, in the paper that introduced the Hardy–Weinberg law, G. H. Hardy talks about a drift phenomenon, without using that word.[25] This is probably the first discussion of evolutionary drift.[26] It would be an interesting historical project to see just how and when drift as the default state of evolutionary systems became embedded into evolutionary thinking. My point is that, unlike the Principle of Natural Selection, which one might reasonably trace back to Darwin, the Principle of Drift is relatively modern. It has been first named and articulated in this paper. However, contemporary evolutionary biologists regularly make the inferences it licenses. So I do not claim to have made a new biological discov-

ery here. What I have done is to systematize these important inferences into a single law, a law that is fundamental to contemporary evolutionary biology.

VI.   Conclusions

How similar is modern evolutionary theory to Newtonian mechanics? Taken in isolation, that question is at best silly, at worst meaningless. Is the answer 17? Taken in the context of recent literature in the philosophy of biology we can say that it is not as similar to Newtonian mechanics as Christopher Stevens thinks, but much more so than either Walsh, Lewens, and Ariew, or Matthen and Ariew would allow. What I have found surprising is how useful is the comparison. Newton's First Law provides the model for a zero-force law. The Principle of Drift, which is stated here for the first time, is in one important way non-Newtonian. For a population to maintain its trajectory through state-space the constant action of evolutionary forces is required. But, the Principle of Drift is like Newton's Principle of Inertia in that it is a zero-force law, and so it does provide appropriate null-hypotheses. The Hardy–Weinberg law fails to be a zero-force law. It fails not just because it is not a law, but also because it mixes evolutionary forces (selection, migration, mutation, nonrandom mating) with a nonforce (drift). And thus it fails to provide the appropriate null-hypothesis for evolutionary scenarios.

Notes

1   Sober, *Conceptual Issues in Evolutionary Biology* (Cambridge: MIT, 1984, second edition).

2   See Denis M. Walsh, Tim Lewens, and André Ariew, "Trials of Life: Natural Selection and Random Drift," *Philosophy of Science*, LXIX (2002): 452–73; Mohan Matthen and André Ariew, "Two Ways of Thinking about Fitness and Natural Selection," this JOURNAL, XCLX, 2 (February 2002): 55–83.

3   See Christopher Stephens, "Selection, Drift, and the 'Forces' of Evolution," *Philosophy of Science*, LXXI (2001): 550–70.

4   See, for example, Carl Hempel, *Aspects of Scientific Explanation* (New York: Free Press, 1965).

5 See Wesley Salmon, *Statistical Explanation and Statistical Relevance* (Pittsburgh: University Press, 1971), and his *Scientific Explanation and the Causal Structure of the World* (Princeton: University Press, 1984).

6 See Brandon, *Adaptation and Environment* (Princeton: University Press, 1990), where I was inspired by Salmon's *Scientific Explanation and the Causal Structure of the World*.

7 Endler, *Natural Selection in the Wild* (Princeton: University Press, 1986).

8 See Brandon and Grant Ramsey, "What's Wrong with the Emergentist Statistical Interpretation of Natural Selection and Random Drift," in Michael Ruse and David L. Hull, eds., *The Cambridge Companion to Philosophy of Biology* (New York: Cambridge, 2007).

9 See Brandon, "The Difference between Selection and Drift: A Reply to Millstein," *Biology and Philosophy*, xx (2005): 153–70.

10 See Brandon, *Adaptation and Environment*, especially chapter 2, "The Concept of Environment in the Theory of Natural Selection."

11 See Brandon, "The Difference between Selection and Drift."

12 See, for example, Martin Kreitman, "Methods to Detect Selection in Populations with Applications to the Human," *Annual Review of Genomics and Human Genetics*, I (2002): 539–59; Ziheng Yang and Joseph P. Bielawski, "Statistical Methods for Detecting Molecular Adaptation," *Trends in Ecology and Evolution*, xv (2000): 496–503; Michael Bamshad and Stephen P. Wooding, "Signatures of Natural Selection in the Human Genome," *Nature Reviews Genetics*, IV (2003): 99–111.

13 See Brandon and H. Fred Nijhout, "The Empirical Non-equivalence of Genic and Genotypic Models of Selection: A (Decisive) Refutation of Genic Selectionism and Pluralistic Genic Selectionism," forthcoming in *Philosophy of Science*, LXXIII (2006): 277–97.

14 See Brandon and Scott Carson, "The Indeterministic Character of Evolutionary Theory: No 'No Hidden Variables Proof' but No Room for Determinism Either," *Philosophy of Science*, LXIII (1996): 315–37.

15 Beatty, "What's Wrong with the Received View of Evolutionary Theory?" in P. D. Asquith and R. N. Giere, eds., *PSA 1980*, Volume 2 (East Lansing, MI: Philosophy of Science Association, 1980), pp. 397–426.

16 See Brandon and Nijhout; and see D. McShea and Brandon, "Diversity and Complexity: the Zero-Force Expectation for Biology" (in preparation).

17 Fisher, *The Genetical Theory of Natural Selection* (New York: Oxford, 1930).

18 See, for example, R. Quiring, U. Walldorf, U. Kloter, and W. J. Gehring, "Homology of the Eyeless Gene of *Drosophila* to the *Small Eye* Gene in Mice and *Aniridia* in Humans," *Science*, CCLXV (1994): 785–89.

19 See Endler; Brandon and Ramsey; and J. G. Kingsolver, H. E. Hoekstra, J. M. Hoekstra, D. Berrigan, S. N. Vignieri, C. E. Hill, A. Hoang, P. Gibert, and P. Beerli, "The Strength of Phenotypic Selection in Natural Populations," *The American Naturalist*, CLVII (2001): 245–61.

20 See Brandon and Alex Rosenberg, "Philosophy of Biology," in Peter Clark and Katherine Hawley, eds., *Philosophy of Science Today* (New York: Oxford, 2003), pp. 147–80.

21 See Brandon, "Adaptation and Evolutionary Theory," *Studies in History and Philosophy of Science*, IX (1978): 181–206, and *Adaptation and Environment*.

22 See Ian Hacking, *Logic of Statistical Inference* (New York: Cambridge, 1965), pp. 1–12.

23 There are other interpretations of this principle, which instead of connecting objective probabilities with outcomes, connect them with subjective probabilities; see Isaac Levi, "Direct Inference," this JOURNAL, IXXIVL, 1 (January 1977): 5–29. My view is that an objectivist theory of probability requires principles to connect observation of relative frequencies to objective probabilities and to connect objective probabilities with observable events. On this view probabilities are no different from any other sort of theoretical entity, for example, black holes. If we are meaningfully to posit black holes we need principles to go from observables to black holes and vice versa; similarly for objective probabilities. Subjective probabilities are superfluous on this view. Given these principles we can reconstruct subjective probabilities, but they do no work. My view of evolutionary theory's requirement of objective probabilities is similar to Karl Popper's view of quantum mechanics. See Popper, "The Propensity Interpretation of Probability," *The British Journal for the Philosophy of Science*, x (1959): 25–42. My view of objective probability finds precedent in the works of Hans Reichenbach, Salmon, and Hacking, among others. In particular, I follow Salmon's idea of objectively homogeneous reference classes as being the solution to the reference class problem. I have specifically developed these ideas in the context of evolutionary biology and have argued that my notion of "selective environmental homogeneity" solves the reference

class problem in evolutionary biology. For the most recent development of this argument, see Brandon, "The Difference between Selection and Drift: A Reply to Millstein," *Biology and Philosophy*, xx (2005): 153–70. These issues involving the interpretation of probability and the principle of direct inference are enormously complex, important, and well beyond the scope of this paper. I would only add that my view of them seems to be just what evolutionary theory requires.

24 In calling this biology's first law I do not mean to suggest that evolutionary biology is the whole of biology. However, I do mean to suggest something that virtually all biologists would agree with, namely, that evolutionary biology lies at the center of biology.

25 G. H. Hardy, "Letter to the Editor," *Science*, xxviii (1908): 49–50.

26 Richard Lewontin, personal communication, 2005.

PART III
THE TAUTOLOGY PROBLEM

Introduction

Ever since Darwin adopted a slogan of Herbert Spencer's to describe natural selection – "the survival of the fittest" – Darwin's theory has been beset with the problem of defining fitness. The trouble is that almost any definition that seems faithful to his theory appears to make the theory into a tautology – a proposition true by definition instead of true in virtue of facts about the world. This would be a serious problem for the theory since definitions and their consequences explain nothing, let alone the tree of life on this planet. Consider whether "all bachelors are unmarried males" together with the fact that Joe is a bachelor explains why he is one. Evidently it does not. His being an unmarried male merely re-describes the fact that he is a bachelor. If "x is fitter than y" is defined as meaning "x survives to leave more offspring than y," it will turn out that the principle of natural selection tells us that those organisms that survive to leave more offspring – the fitter ones – are the ones that survive to leave more off-spring. This statement has no more explanatory content than "all bachelors are unmarried males."

Opponents of Darwin's theory have seized on this problem since its inception in the nineteenth century to try to overturn it on logical grounds alone. Gould's article (chapter 7) both expounds some of these arguments and shows how they misunderstand the theory. One confusion people make is to treat the units in which fitness is measured – number of offspring, or descendants more generally – as giving the meaning of fitness, a mistake rather like treating inches or centimeters as giving the meaning of "space."

Nevertheless, many biologists and philosophers think that there is a deeper and more serious problem here than superficial critics allege. For biology requires that fitness be quantitatively measurable and yet no units biologists have hit on can do this job in all contexts. Brandon's paper (chapter 8) shows that providing a definition for fitness is no easy matter, but he also shows that intelligent attempts to do so reveal important features of the theory.

Further Reading

Abrams, M. (1999). Propensities in the propensity interpretation of fitness. *Southwest Philosophy Review*, 15, 27–35.

Ariew, A. (2008). Population thinking. In M. Ruse (ed.), *The Oxford handbook of philosophy of biology* (pp. 64–86). Oxford: Oxford University Press.

Beatty, J., & Finsen, S. (1989). Rethinking the propensity interpretation: A peek inside Pandora's Box. In M. Ruse (ed.), *What the philosophy of biology is: Essays for David Hull* (pp. 17–31). Dordrecht: Kluwer.

Bethel, T. (1976). Darwin's mistake. *Harper's Magazine*, 252, 70–75.

Brandon, R., & Beatty, J. (1984). Discussion: The propensity interpretation of 'fitness' – no interpretation is no substitute. *Philosophy of Science*, 51, 342–347.

Campbell, R., & Robert, J. (2005). The structure of evolution by natural selection. *Biology and Philosophy*, 20, 673–696.

Hull, D. (1969). What philosophy of biology is not. *Synthese*, 20, 157–184.

Lennox, J., & Wilson, B. (1994). Natural selection and the struggle for existence. *Studies in History and Philosophy of Science*, 25, 65–80.

Mills, S., & Beatty, J. (1979). A propensity interpretation of fitness. *Philosophy of Science*, 46, 263–286.

Popper, K. (1978). Natural selection and the emergence of mind. *Dialectica*, 32, 339–355.

Rosenberg, A. (1982). On the propensity definition of fitness. *Philosophy of Science*, 49, 268–273.

Rosenberg, A., & McShea, D. (2008). *Philosophy of biology: A contemporary introduction* (Chapters 2 and 3). London: Routledge.

Rosenberg, A., & Williams, M. (1986). Fitness as primitive and propensity. *Philosophy of Science*, 53, 412–418.

Sober, E. (1984). *The nature of selection: Evolutionary theory in philosophical focus*. Cambridge, MA: MIT Press.

Sober, E. (2001). Two faces of fitness. In R. Singh, D. Paul, C. Krimbas, & J. Beatty (eds.), *Thinking about evolution: Historical, philosophical, and political perspectives* (pp. 309–321). Cambridge: Cambridge University Press.

Waters, C. (1986). Natural selection without survival of the fittest. *Biology and Philosophy*, 1, 207–225.

Weber, M. (1996). Fitness made physical: The supervenience of biological concepts revisited. *Philosophy of Science*, 63, 411–431.

7

Darwin's Untimely Burial

Stephen Jay Gould

In one of the numerous movie versions of *A Christmas Carol*, Ebenezer Scrooge encounters a dignified gentleman sitting on a landing, as he mounts the steps to visit his dying partner, Jacob Marley, "Are you the doctor?" Scrooge inquires. "No," replies the man, "I'm the undertaker; ours is a very competitive business." The cutthroat world of intellectuals must rank a close second, and few events attract more notice than a proclamation that popular ideas have died. Darwin's theory of natural selection has been a perennial candidate for burial. Tom Bethell held the most recent wake in a piece called "Darwin's Mistake" (*Harper's*, February 1976): "Darwin's theory, I believe, is on the verge of collapse. . . . Natural selection was quietly abandoned, even by his most ardent supporters, some years ago." News to me, and I, although I wear the Darwinian label with some pride, am not among the most ardent defenders of natural selection. I recall Mark Twain's famous response to a premature obituary: "The reports of my death are greatly exaggerated."

Bethell's argument has a curious ring for most practicing scientists. We are always ready to watch a theory fall under the impact of new data, but we do not expect a great and influential theory to collapse from a logical error in its formulation. Virtually every empirical scientist has a touch of the Philistine. Scientists tend to ignore academic philosophy as an empty pursuit. Surely, any intelligent person can think straight by intuition. Yet Bethell cites no data at all in sealing the coffin of natural selection, only an error in Darwin's reasoning: "Darwin made a mistake sufficiently serious to undermine his theory. And that mistake has only recently been recognized as such. . . . At one point in his argument, Darwin was misled."

Although I will try to refute Bethell, I also deplore the unwillingness of scientists to explore seriously the logical structure of arguments. Much of what passes for evolutionary theory is as vacuous as Bethell claims. Many great theories are held together by chains of dubious metaphor and analogy. Bethell has correctly identified the hogwash surrounding evolutionary theory. But we differ in one fundamental way: for Bethell, Darwinian theory is rotten to the core; I find a pearl of great price at the center.

Natural selection is the central concept of Darwinian theory – the fittest survive and spread their favored traits through populations. Natural selection is defined by Spencer's phrase "survival of the fittest," but what does this famous bit of jargon really mean? Who are the fittest? And how is "fitness" defined? We often read that fitness involves no more than "differential reproductive success" – the production of more surviving offspring than other competing members

Stephen Jay Gould, "This View of Life: Darwin's Untimely Burial," *Natural History* (1976) 85, 24–30.

of the population. Whoa! cries Bethell, as many others have before him. This formulation defines fitness in terms of survival only. The crucial phrase of natural selection means no more than "the survival of those who survive" – a vacuous tautology. (A tautology is a phrase – like "my father is a man" – containing no information in the predicate ("a man") not inherent in the subject ("my father"). Tautologies are fine as definitions, but not as testable scientific statements – there can be nothing to test in a statement true by definition.)

But how could Darwin have made such a monumental, two-bit mistake? Even his severest critics have never accused him of crass stupidity. Obviously, Darwin must have tried to define fitness differently – to find a criterion for fitness independent of mere survival. Darwin did propose an independent criterion, but Bethell argues quite correctly that he relied upon analogy to establish it, a dangerous and slippery strategy. One might think that the first chapter of such a revolutionary book as *Origin of Species* would deal with cosmic questions and general concerns. It doesn't. It's about pigeons. Darwin devotes most of his first forty pages to "artificial selection" of favored traits by animal breeders. For here an independent criterion surely operates. The pigeon fancier knows what he wants. The fittest are not defined by their survival. They are, rather, allowed to survive because they possess desired traits.

The principle of natural selection depends upon the validity of an analogy with artificial selection. We must be able, like the pigeon fancier, to identify the fittest beforehand, not only by their subsequent survival. But nature is not an animal breeder; no preordained purpose regulates the history of life. In nature, any traits possessed by survivors must be counted as "more evolved"; in artificial selection, "superior" traits are defined before breeding even begins. Later evolutionists, Bethell argues, recognized the failure of Darwin's analogy and redefined "fitness" as mere survival. But they did not realize that they had undermined the logical structure of Darwin's central postulate. Nature provides no independent criterion of fitness; thus, natural selection is tautological.

Bethel then moves to two important corollaries of his major argument. First, if fitness only means survival, then how can natural selection be a "creative" force, as Darwinians insist. Natural selection can only tell us how "a given type of animal became more numerous"; it cannot explain "how one type of animal gradually changed into another." Secondly, why were Darwin and other eminent Victorians so sure that mindless nature could be compared with conscious selection by breeders. Bethell argues that the cultural climate of triumphant industrial capitalism had defined any change as inherently progressive. Mere survival in nature could only be for the good: "It is beginning to look as though what Darwin really discovered was nothing more than the Victorian propensity to believe in progress."

I believe that Darwin was right and that Bethell and his colleagues are mistaken: criteria of fitness independent of survival can be applied to nature and have been used consistently by evolutionists. But let me first admit that Bethell's criticism applies to much of the technical literature in evolutionary theory, especially to the abstract mathematical treatments that consider evolution only as an alteration in numbers, not as a change in quality. These studies do assess fitness only in terms of differential survival. What else can be done with abstract models that trace the relative successes of hypothetical genes A and B in populations that exist only on computer tape? Nature, however, is not limited by the calculations of theoretical geneticists. In nature, A's "superiority" over B will be *expressed* as differential survival, but it is not *defined* by it – or, at least, it better not be so defined, lest Bethell *et al.* triumph and Darwin surrender.

My defense of Darwin is neither startling, novel, nor profound. I merely assert that Darwin was justified in analogizing natural selection with animal breeding. In artificial selection, a breeder's desire represents a "change of environment" for a population. In this new environment, certain traits are superior a priori; (they survive and spread by our breeder's choice, but this is a *result* of their fitness, not a definition of it). In nature, Darwinian evolution is also a response to changing environments. Now, the key point: certain morphological, physiological, and behavioral traits should be superior a priori as designs for living in new environments. These traits confer fitness by an engineer's criterion of good design, not by the empirical fact of their survival and spread. It got colder before the woolly mammoth evolved its shaggy coat.

Why does this issue agitate evolutionists so much? OK, Darwin was right: superior design in changed environments is an independent criterion of fitness. So what? Did anyone ever seriously propose that the poorly designed shall triumph? Yes, in fact, many did. In Darwin's day, many rival evolutionary theories asserted that the fittest (best designed) must perish. One popular notion – the theory of racial life cycles – was championed by a former inhabitant of the office I now occupy, the great American paleontologist Alpheus Hyatt. Hyatt claimed that evolutionary lineages, like individuals, had cycles of youth, maturity, old age, and death (extinction). Decline and extinction are programmed into the history of species. As maturity yields to old age, the best-designed individuals die and the hobbled, deformed creatures of phyletic senility take over. Another anti-Darwinian notion, the theory of orthogenesis, held that certain trends, once initiated, could not be halted, even though they must lead to extinction caused by increasingly inferior design. Many nineteenth-century evolutionists (perhaps a majority) held that Irish elks became extinct because they could not halt their evolutionary increase in antler size, thus, they died – caught in trees or bowed (literally) in the mire. Likewise, the demise of saber-toothed "tigers" was often attributed to canine teeth grown so long that the poor cats couldn't open their jaws wide enough to use them.

Thus, it is not true, as Bethell claims, that any traits possessed by survivors must be designated as fitter. "Survival of the fittest" is not a tautology. It is also not the only imaginable or reasonable reading of the evolutionary record. It is testable. It had rivals that failed under the weight of contrary evidence and changing attitudes about the nature of life. It has rivals that may succeed, at least in limiting its scope.

If I am right, how can Bethell claim, "Darwin, I suggest, is in the process of being discarded, but perhaps in deference to the venerable old gentleman, resting comfortably in Westminster Abbey next to Sir Isaac Newton, it is being done as discreetly and gently as possible with a minimum of publicity." I'm afraid I must say that Bethell has not been quite fair in his report of prevailing opinion. He cites the gadflies C. H. Waddington and H. J. Muller as though they epitomized a consensus. He never mentions the leading selectionists of our present generation – E. O. Wilson or D. Janzen, for example. And he quotes the architects of neo-Darwinism – Dobzhansky, Simpson, Mayr, and J. Huxley – only to ridicule their metaphors on the "creativity" of natural selection. (I am not claiming that Darwinism should be cherished because it is still popular; I am enough of a gadfly to believe that uncriticized consensus is a sure sign of impending trouble. I merely report that, for better or for worse, Darwinism is alive and thriving, despite Bethell's obituary.)

But why was natural selection compared to a composer by Dobzhansky; to a poet by Simpson; to a sculptor by Mayr; and to, of all people, Mr. Shakespeare by Julian Huxley? I won't defend the choice of metaphors, but I will uphold the intent, namely, to illustrate the essence of Darwinism – the creativity of natural selection. Natural selection has a place in all anti-Darwinian theories that I know. It is cast in a negative role as an executioner, a headsman for the unfit (while the fit arise by such non-Darwinian mechanisms as the inheritance of acquired characters or direct induction of favorable variation by the environment). The essence of Darwinism lies in its claim that natural selection creates the fit. Variation is ubiquitous and random in direction. It supplies the raw material only. Natural selection directs the course of evolutionary change. It preserves favorable variants and builds fitness gradually. In fact, since artists fashion their creations from the raw material of notes, words, and stone, the metaphors do not strike me as inappropriate. Since Bethell does not accept a criterion of fitness independent of mere survival, he can hardly grant a creative role to natural selection.

According to Bethell, Darwin's concept of natural selection as a creative force can be no more than an illusion encouraged by the social and political climate of his times. In the throes of Victorian optimism in imperial Britain, change seemed to be inherently progressive; why not equate survival in nature with increasing fitness in the nontautological sense of improved design.

I am a strong advocate of the general argument that "truth" as preached by scientists often turns out to be no more than prejudice inspired by prevailing social and political beliefs. I have devoted several essays to this theme because I believe that it helps to "demystify" the practice of science by showing its similarity to all creative human activity. But the truth of a general argument does not validate any specific application, and I maintain that Bethell's application is badly misinformed.

Darwin did two very separate things: he convinced the scientific world that evolution had occurred and he proposed the theory of natural selection as its mechanism. I am quite willing to admit that the common equation of evolution with progress made Darwin's first claim more palatable to his contemporaries. But Darwin failed in his second quest during his own lifetime. The theory of natural selection did not triumph until the 1940s. Its Victorian unpopularity, in my view, lay primarily in its denial of general progress as inherent in the workings of evolution. Natural selection is a theory of *local* adaptation to changing environments. It proposes no perfecting principles, no guarantee of general improvement; in short, no reason for general approbation in a political climate favoring innate progress in nature.

Darwin's independent criterion of fitness is, indeed, "improved design," but not "improved" in the cosmic sense that contemporary Britain favored. To Darwin, improved meant only "better designed for an immediate, local environment." Local environments change constantly: they get colder or hotter, wetter or drier, more grassy or more forested. Evolution by natural selection is no more than a tracking of these changing environments by differential preservation of organisms better designed to live in them: hair on a mammoth is not progressive in any cosmic sense. Natural selection can produce a trend that tempts us to think of more general progress – increase in brain size does characterize the evolution of group after group of mammals. But big brains have their uses in local environments; they do not mark intrinsic trends to higher states. And Darwin delighted in showing that local adaptation often produced "degeneration" in design – anatomical simplification in parasites, for example.

If natural selection is not a doctrine of progress, then its popularity cannot reflect the politics that Bethell invokes. If the theory of natural selection contains an independent criterion of fitness, then it is not tautological. I maintain, perhaps naïvely, that its current, unabated popularity must have something to do with its success in explaining the admittedly imperfect information we now possess about evolution. I rather suspect that we'll have Charles Darwin to kick around for some time.

8

Adaptation and Evolutionary Theory

Robert N. Brandon

There is virtually universal disagreement among students of evolution as to the meaning of adaptation.

(Lewontin, 1957)

Much of past and current disagreement on adaptation centers about the definition of the concept and its application to particular examples: these arguments would lessen greatly if precise definitions for adaptations were available.
(Bock and von Wahlert, 1965)

The development of a predictive theory [of evolution] depends on being able to specify when a population is in better or worse evolutionary state. For this purpose an objective definition of adaptedness is necessary.

(Slobodkin, 1968)

The conception of adaptation was not introduced into biology in 1859. Rather what Darwin did was to offer a radically new type of explanation of adaptations and in so doing he altered the conception. As the above quotes indicate we have not in the last century sufficiently delimited this conception and it is important to do so.

In this paper we will analyse and, I hope, clarify one aspect of the conception of adaptation. One of the aims of this paper is a theoretically adequate definition of relative adaptedness. As we will see such analysis cannot be divorced from an analysis of the structure of evolutionary theory. The other major aim of this paper is to expose this structure, to show how it differs from the standard philosophical models of scientific theories, and to defend this differentiating feature (and hence to show the inadequacy of certain views about the structure of scientific theories which purport to be complete).

Reprinted from Robert Brandon, *Studies in the History and Philosophy of Science*, Vol. 9, *Adaptation and Evolutionary Theory*, pp. 9, 181–206. Copyright 1978, with permission from Elsevier.

I owe a debt of gratitude to all those who read earlier versions of this paper and helped me improve it. Where possible I have tried to footnote contributions. Here I want to give special thanks to Ernst Mayr and Paul Ziff whose comments and criticisms have had pervasive effects on the evolution of this paper.

A note on defining is needed. Definitions are often thought to be of two kinds, descriptive and stipulative. (See, for example, Hempel (1966), chapter 7.) Descriptive definitions simply describe the meaning of terms already in use; stipulative definitions assign, by stipulation, special meaning to a term (either newly coined or previously existing). According to this view the project of defining a term is either purely descriptive or purely stipulative. This view is mistaken. The project at hand calls for neither pure linguistic analysis nor pure stipulation; it is much more complex. Briefly, we examine the conceptual network of evolutionary biology. We find that according to evolutionary theory there is a biological property, adaptedness, which some organisms have more of than others. Those having more of it, or those better adapted, tend to leave more offspring. And this is the mechanism of evolution. The project calls for conceptual analysis but such analysis is sterile unless it is coupled with an examination of the biological property which is the object of the conception. Any definition which fails to fit the conceptual network must be rejected, as must any which fails to apply to the property. The project calls for an element of stipulation but our stipulatory freedom is constrained both by theoretical and conceptual requirements and, one hopes, by the real world.

A note on the restricted scope of this paper is also needed. Biologists talk about the adaptedness of individual organisms and of populations. Selection occurs at the level of individuals and, presumably, at higher levels. That is, there is intrapopulational selection and interpopulational selection. It is vital that we keep these levels separate and that we see the relation between selection and adaptation.[1] Selection at the level of individual organisms has as its cause differences in individual adaptedness and its effect is adaptions for individual organisms. We will follow standard practice in calling selection at this level natural selection. Any benefit to the population from natural selection is purely fortuitous. One must distinguish between a group of adapted organisms and an adapted group of organisms. For instance, a herd of fleet gazelles is not necessarily a fleet herd of gazelles. Similarly group selection will have as its cause differences in group adaptedness and as its effect group adaptations. The theory of group selection is quite clear; its

occurrence in nature is controversial. One could speak of an abstract theory of evolution which covers natural selection, group selection and even the selection of tin cans in junk yards. But most of the interesting problems don't arise at this level of generality. In this paper we will be primarily concerned with natural selection, *i.e.* with intraspecific intraenvironmental selection. Thus we will be concerned with the adaptedness of individual organisms, not with the adaptedness of populations.

Let me illustrate the confusion that results from the failure to relate adaptedness to the proper level of selection. One of the more prominent definitions of relative adaptedness is due to Thoday.[2] Basically it says: a is better adapted than b if and only if a is more likely than b to have offspring surviving 10^8 (or some other large number) years from now. Either the long-range probability of offspring corresponds to the short-range probability of offspring or it does not. (Corresponds means: a's long-range probability of offspring is greater than b's long-range probability of offspring if and only if a's short-range probability of offspring is greater than b's short-range probability of offspring.) If it does correspond then we should stick to the more easily measurable short-range probability. If not, then since natural selection is not foresighted, *i.e.* it operates only on the differential adaptedness of present organisms to present environments, the long-range probability of offspring is irrelevant to natural selection.

Why has Thoday's definition been so favorably received? Because the long-range probability of descendants is important to selection at or above the species level. For instance, one plausible explanation of the predominance of sexual reproduction over asexual modes of reproduction is that the long-range chances of survival are greater for populations having sex (see Maynard Smith, 1975, pp. 185ff). But if one is interested in selection at the population level then the relevant notion of adaptedness would be that which applies to populations. Until recently even biologists have failed to distinguish intra- and inter-populational selection. Thoday's definition, not being selection relative, lends itself to this confusion. To keep matters as clear as possible we will only be concerned with natural selection and with that notion of adaptation which properly relates to it.

1. The Role of the Concept of Relative Adaptedness in Evolutionary Theory

The following three statements are crucial components of the Darwinian (or neo-Darwinian) theory of evolution:[3]

1. Variation: There is (significant) variation in morphological, physiological and behavioral traits among members of a species.
2. Heredity: Some traits are heritable so that individuals resemble their relations more than they resemble unrelated individuals and, in particular, offspring resemble their parents.
3. Differential Fitness: Different variants (or different types of organisms) leave different numbers of offspring in immediate or remote generations.

When the conditions described above are satisfied organic evolution occurs. A thorough examination into the history of our awareness of these conditions would be interesting and worthwhile but will not be attempted here (see Mayr, 1977). Suffice it to say that in Darwin's time each was a non-trivial statement. In what follows we will examine them predominantly from our own point of view.

Ignoring the parenthetical 'significant' (1) could not help but be true. The uniqueness of complex material systems is now taken for granted; and so we expect variation among individuals of a species. Their similarity needs explaining not their variation. (1) becomes less empty from our point of view when 'significant' is added. What sort of variation is significant? That which can lead to adaptive evolutionary changes. Though the world is such that individuals must be unique the recognition of this fact is of fairly recent origins and is necessary for an evolutionary world view.

Unlike (1), (2) is not at all trivial. There is no metaphysical necessity in offspring resembling their parents. (2) can now be derived from our modern theories of genetics; in Darwin's time it was an observation common to naturalists and animal breeders. Darwin's theories of heredity were notoriously muddled but fortunately a correct theory of genetics is not a prerequisite for a Darwinian theory of evolution (see Mayr, 1977, p. 325). What is important to note is that given

that there is variation, (1), and that some of the traits which vary are heritable, (2), it follows that the variation within a species tends to be preserved. (Of course this tendency can be counterbalanced by other factors.)

When (3) holds, when there are differences in reproductive rates, it follows from (1) and (2) that the variation status quo is disrupted, that is, that there are changes in the patterns of variation within the species. For our purposes we can count such changes as evolution. (For a fuller explication of the concept of evolution see Brandon, 1978.) Thus when (1)–(3) hold evolution occurs.

We have seen that (1) is in a sense trivial and requires no explanation. We have also seen that (2) is non-trivial and is to be explained by modern theories of genetics, but that this explanation is not essential to Darwinian theory. In contrast, the distinguishing feature of a Darwinian theory of evolution is its explanation of (3).[4] The focus of this paper is the conception used for such explanations.

The distinguishing feature of a Darwinian theory of evolution is explaining evolutionary change by a theory of natural selection. Of course, that is not the only possible sort of explanation of evolution. In his own time Darwin convinced the majority of the scientific community that evolution has and does occur but hardly anyone bought his natural-selection-explanation of it. (For an excellent source book on the reception of Darwin's theory see Hull, 1973.) The alternatives of Darwin's day, e.g. divine intervention and the unfolding of some predetermined plan, are no longer scientifically acceptable. But there is one present day alternative we should consider.

It is not surprising that in finite populations of unique individuals some variants leave more offspring than others. We would expect such differences in reproductive success simply from chance. And if there are chance differences in reproductive success between two types of organisms (or similarity classes of organisms) we expect one type ultimately to predominate by what statisticians call random walk. If we can explain (3) and so the occurrence of evolution in terms of chance is the hypothesis of natural selection necessary?

It is becoming the received view in the philosophy of science that hypotheses are not evaluated in isolation but rather in comparison with rival hypotheses. This view is, I think, for the most part correct but not entirely; some hypotheses we reject as unacceptable without comparison with specific alternatives. Unacceptable hypotheses are those that violate deeper-seated beliefs, theories or metaphysics. Similarly some forms of explanation are unacceptable in that no investigation into the particular phenomenon is required to reject them. We reject them without considering any particular alternative explanation simply because we believe there must be a better alternative. For example, accepting Darwinian theory we reject the explanation that bees make honey in order to provide food for bears without examining bees, bears or honey. (An acceptable form of explanation is not one which is necessarily correct or even accepted; it simply is one which is not unacceptable.)

The theory of evolution by chance or by random walk has been developed in recent years and is often called the theory of non-Darwinian evolution, or better, the neutrality theory of evolution (see King and Jukes, 1969). We cannot give it the discussion it deserves but it is worth pointing out that explanation in terms of chance is an acceptable form of explaining short term evolutionary change but not of any interesting sort of long term evolutionary change. (The truth of this hinges on what counts as interesting. I will not try to delimit interesting long term evolutionary change; suffice it to say that any seemingly directed change is interesting.)

The neutrality theory supposes that certain alternative alleles (and so certain protein molecules coded by them) are functionally equivalent, *i.e.* are selectively *neutral*. Given this supposition the neutrality theory predicts (and so is able to explain) the sorts of changes in frequencies of these alleles expected by a process of random sampling in different situations. As Ayala (1974) points out these predictions differ both qualitatively and quantitatively from those given by the selectionist theory. (Ayala presents data on different species of *Drosophila* which tend to corroborate the natural selectionist hypothesis and refute the neutrality hypothesis.) Whether evolution by random walk is a common or rare phenomenon we cannot reject *a priori* a chance-explanation of short term evolutionary change.

The situation is different for interesting long term evolutionary phenomena. Of course we do not directly observe long term evolutionary change. What we observe and try to explain are the products of such change. Presumably any complex feature of an organism is the product of long term evolutionary change. On the one hand some complex features of organisms, such as the eye of a human, are so obviously useful to their possessor that we cannot believe that this usefulness plays no part in explaining their existence. That is, given Darwinian theory and the obvious usefulness of sight we have a better alternative to the chance-explanation. On the other hand there are features whose usefulness is unclear for which we still reject chance-explanations because of their high degree of complexity and constancy. Complexity and constancy are not made likely on the hypothesis of evolution by random sampling. A good example is lateral lines in fish. This organ is structurally complex and shows a structural constancy within taxa, yet until recently it was not known how the lateral line was useful to its possessor. In this case the rejection of a chance-explanation was good policy; studies eventually showed that the lateral line is a sense organ of audition. (This example is taken from G. C. Williams, 1966, pp. 10–11.)

One can contrast the lateral line in fish with the tailless condition of Manx cats. This feature is not even constant within the species and a non-existent tail is hardly complex. (Actually what is relevant concerning complexity is that the historical process leading from tailed to tailless is most probably not complex.) Furthermore legend has it that Manx cats originated on the Isle of Man in what would be a small isolated population; thus increasing the probable role of chance. The tailless condition of Manx cats may have evolved by natural selection but for all we know the best explanation of it is the explanation in terms of chance.

It is important to keep in mind the possibility of evolution by random walk for it is important that Darwinian explanations be testably different (at least in principle) from chance-explanations. What is the Darwinian explanation of (3)? The conventional wisdom is that Darwin explained (3) by his postulate of the 'struggle for existence' (or in Spencer's words, which Darwin later used, 'the survival of the fittest') and that this explanation, or this discovery of the mechanism of evolution, was Darwin's greatest contribution.

How does 'the struggle for existence' or 'the survival of the fittest' explain (3)? Following current practice let us define the *reproductive success* or the *Darwinian fitness* of an organism in terms of its actual genetic contribution to the next generation. I will not try to make this definition precise and complete. The genetic contribution to the next generation can usefully be identified with the number of sufficiently similar offspring when 'sufficiently similar' is sufficiently explicated. This would disallow, for example, sterile offspring from counting towards Darwinian fitness. There are two options: either let the Darwinian fitness of an individual equal its actual number of sufficiently similar offspring or let the Darwinian fitness of an individual equal the mean number of sufficiently similar offspring of members of the similarity class to which it belongs. In either case Darwinian fitness is defined in terms of numbers of *actual* offspring. I should point out that most biologists use the words 'fitness' and 'adaptedness' interchangeably. In this paper 'fitness' will only be used to refer to Darwinian fitness. Adaptedness, as we will see, cannot be identified with Darwinian fitness. (3) says that Darwinian fitness is correlated with certain morphological, physiological or behavioral traits. Why is there this correlation? Why is there differential fitness? Darwin's answer, which he arrived at after reading Malthus' *Essay on Population*,[5] was that since in each generation more individuals are produced than can survive to reproduce there is a struggle for existence. In this 'struggle' (which in its broadest sense is a struggle of the organism with its environment not just with other individuals, see Darwin, 1859, p. 62) certain traits will render an organism *better adapted* to its environment than conspecifics with certain other traits. The better adapted individuals will tend to be fitter (*i.e.* produce more offspring) than the less well adapted. Why are those who happen to be the fittest in fact the fittest? The Darwinian answer is: They are (for the most part) better adapted to their environment.

What does this explanation presuppose? It seems to presuppose the following as a law of nature:

(D)   If *a* is better adapted than *b* in environment *E* then (probably) *a* will have more (sufficiently similar) offspring than *b* in *E*.

Certainly if (D) is a true law then the Darwinian explanation is acceptable. Darwin seems to presuppose (D) but it is not to be found stated explicitly in the *Origin*. Nor is it to be found in modern evolutionary works. But if one examines work in modern evolutionary biology – the theorizing done, the inferences made, the explanations offered – one finds that (D) or something like (D) is required as the foundation of evolutionary theory. I take it that this conclusion will be so uncontroversial that it need not be further supported by examining examples of evolutionary reasoning. But later in this paper we will give some examples to show how (D) is to be employed.

Philosophers of science talk about laws more often than they display actual examples of them. In particular many people have discussed whether or not 'the survival of the fittest' is a tautology without displaying something other than that phrase which might be a tautology. (As for example J. J. C. Smart, 1963, p. 59.) The phrase itself, not being a declarative sentence, could not be a tautology. An exception is Mary Williams.[6] She has attempted to give a 'precise, concise and testable' version of that phrase, and so has attempted to give a precise, concise and testable version of the fundamental law of evolutionary theory.

William defines the clan of a set β as the members of β plus all their descendants. On a phylogenetic tree the clan of β would be those nodes which are in β plus all nodes after them which are on a branch which passes through one of the original nodes. A subclan is either a whole clan or a clan with one or more branches removed. A Darwinian subclan is a subclan which is held together by cohesive forces so that it acts as a unit with respect to selection (this crucial concept is not defined by Williams; she takes it as primitive). Informally Williams' version of the fundamental law of evolutionary theory states that for any subclan D_1 of any Darwinian subclan D,

> If D_1 is superior in fitness to the rest of D for sufficiently many generations . . . then the proportion of D_1 in D will increase during these generations. (1970, p. 362)

(D) is a 'law'[7] about properties of individual organisms; Williams' version is a law about properties of sets of organisms. Which is fundamental? Some properties of sets (notable exceptions being set-theoretic properties such as cardinality) are a function of the properties of the sets' members. In particular, as Williams herself points out (1973, p. 528), the fitness of a clan is to be identified with the average fitness of the members of the clan. Thus the property of individuals (or more precisely the property of individuals in some environment) – what we will call adaptedness, what Williams calls fitness – is fundamental. Likewise (D) is fundamental in that Williams' law can be derived from it and the laws of population genetics but not *vice versa*. Perhaps the only way of testing (D) is to apply it to fairly large populations and so to test something like Williams' law, but this does not change our conclusion. (D) is required as the foundation of evolutionary theory.

2.   Four Desiderata of Definitions of Relative Adaptedness

We have seen the role the relational concept of adaptedness is to play in a Darwinian theory of evolution: It is the explanatory concept in what I have called the fundamental law of evolutionary theory. Philosophers have not been able to come up with a set of necessary and jointly sufficient conditions for scientific lawhood, but there is wide agreement on some necessary conditions. In particular laws of the empirical sciences are to be empirically testable universal statements. It is also highly desirable, whether or not definitionally necessary, that laws be empirically correct or at least nearly true. One cannot just look at the surface logic of a statement in order to determine whether or not it is a scientific law (as done in Ruse, 1975). To determine whether (D) is a scientific law we will have to look deeply into the conception of adaptation. My strategy is to try to construct a definition of relative adaptedness that makes (D) a respectable scientific law. In this section I will argue that from any definition (construction, explication) of this concept we would want the following: (a) independence from actual reproductive values; (b) generality; (c) epistemological applicability; and (d) empirical correctness. After arguing for the above desiderata I will show how current definitions fail to satisfy all four and then I will produce a general argument showing that no explication

of the concept will satisfy all four desiderata. In the final section I will attempt to draw the ramifications of this result.

(a)   Independence

The relational concept of adaptation is to explain differential fitness. To do so (D) must not be a tautology. Clearly if (D) is to be a scientific law rather than a tautology the relational concept of adaptation cannot be defined in terms of actual reproductive values. That is, we cannot define it as follows:

> *a* is better adapted than *b* in *E* iff *a* has more offspring than *b* in *E*

('iff' is shorthand for 'if and only if'.) Most biologists treat 'fitness' and 'adaptedness' as synonymous and many define relational fitness in just this way. (See Stern, 1970, p. 47 where he quotes Simpson, Waddington, Lerner and Mayr[8] to this effect. Stern approves of this definition.) They thus deprive evolutionary theory of its explanatory power.

To avoid turning (D) into a tautology it seems we must also avoid defining relative adaptedness in terms of probable reproductive values. That is, the following definition also seems to render (D) a tautology:

> *a* is better adapted than *b* in *E* iff *a* will probably have more offspring than *b* in *E*

(See Munson, 1971, p. 211 for a definition of this form; but he substitutes survival for reproductive values.) Actually things are not as simple as they seem to be. Whether or not the above definition makes (D) tautologous depends on the interpretation of probability being used. More will be said about this, but for the moment we may conclude the obvious: If the relational concept of adaptation is to play its explanatory role in evolutionary theory it must be defined so that (D) does not become a tautology. We will call this requirement the condition of independence from actual reproductive values.

(b)   Generality

As stated earlier we are primarily interested in intraspecific selection and so for the set of ordered triple $\langle x, y, z \rangle$ which satisfy '*x* is better adapted than y in z' the first two members of those triples will be members of the same species. In other, less formal, words we are interested in what it is for one alligator to be better adapted than another alligator to their particular environment but not in what it is for one elephant to be better adapted than one swallow to their environment (since they are not in direct reproductive competition with each other, see Ghiselin, 1974). But we do expect one and the same explication or definition of relative adaptedness to apply to ants, birds and elephants. That is, we want (D) to be a general law that applies to the whole biosphere.

Suppose for some precursors of modern giraffes it was true that one was better adapted than another to their environment if and only if it was taller than the other. (Suppose this only for the sake of this discussion. Even within a given species it is doubtful that any single-dimensional analysis of adaptedness will be adequate.) It won't do to define relative adaptedness in terms of relative height because even though such a definition may truly apply to some giraffe precursors it will not apply to most other plants and animals. Such a definition would make (D) a true law of giraffe precursors but make it false or inapplicable to other plants and animals. If (D) is to be a general law our definition of relative adaptedness must meet what we will call the condition of generality; that is, it must apply to all plants and animals.

(c) Epistemological applicability

One way of stating this requirement is to say that our definition of relative adaptedness must render (D) testable. However, I prefer to stress another side of what is perhaps the same coin and say that our definition of relative adaptedness should tell us something about how (D) is to be applied to particular cases. I choose this stress because I think testing (D) is a pipe dream, whereas applying it to explain certain phenomena should not be. (Such thoughts are in consonance with Scriven, 1959, and Mayr, 1961.)

One sometimes hears talk of adaptedness as a 'close correlation with the environment'. We could define relative adaptedness as follows:

a is better adapted than b in E iff a is more closely correlated than b to E

This is a good example of a definition which fails epistemological applicability. Without further information we have no idea how it applies to particular organisms, simply because we have no idea what it means. Consider the following:

a is better adapted than b in E iff God prefers a to b in E

At least for those theistically inclined there is no problem of meaning here. But this definition is clearly useless since we have no way of knowing which organisms God favors.

The definition discussed above in terms of relative height is a good example of a definition which meets the requirement of epistemological applicability. We know what it is for one organism to be taller than another. Unfortunately this definition lacks generality (or if general then it is empirically incorrect).

To say a definition is epistemologically applicable does not imply that there is an easy mechanical test for its application. Perhaps a paradigm for an epistemologically applicable definition of relative adaptedness is the definition in terms of actual reproductive values (which explains its popularity). But if we try to apply it to two female Pacific salmon in the sea we are faced with real difficulties. We would have to try to follow them up river to their spawning ground. And if we managed to do that and if they both managed to make it we would be faced with the task of counting numerous eggs dispersed in the water. And then we would have to follow each egg's progress to sexual maturity or to death.

But these practical difficulties need not matter. What matters is that theoretically we know what it is for a to be better adapted than b in E and that for at least some cases we can apply it and so test (D) and in those cases where we cannot test (D) we have a good explanation of why we cannot. Thus by requiring epistemological applicability I do not mean to require an operational definition, theoretical applicability is enough.

(d) Empirical correctness

I hardly need to argue that we want our definition of relative adaptedness to be empirically correct but I do need to say something about what it is for our definition to be empirically correct and how we go about determining its correctness.

There may be many features of organisms, such as strength, beauty or even longevity, which we will be disappointed to find out are not invariably selected. In fact quite often there is no selection for higher fecundity.[9] The best adapted may not always be the strongest or the most beautiful or even the most prolific. But natural selection, rather than personal or collective taste, must be the ultimate criterion against which we test our explication of adaptedness.[10] If we define natural selection in terms of relative adaptedness (as we will, see below p. 111) then those selected will by definition be the better adapted. Yet it does not follow that those organisms with higher reproductive values will by definition be better adapted. (If it did then (D) would be tautologous.) We must allow that some instances of differential reproduction are not instances of natural selection.

If natural selection is to be defined in terms of relative adaptedness how can we use it to test the empirical correctness of our definition of relative adaptedness? Suppose for a certain species of organisms we pick out 2 similarity classes of members of this species, A and B. (For our purposes these classes should be formed on the basis of the functional or epigenetical similarity of the genotypes of the members, see Brandon, 1978.) Suppose further that by our definition of relative adaptedness all members of A are better adapted than any member of B to their mutual environment. Our theory of natural selection, of which (D) is a major component, tells us that in statistically large populations (where chance differences in fitness are cancelled out) A's will have a higher average reproductive rate than B's. If repeated observations (either in the lab or in the field) show that A's do in fact outreproduce B's then our definition of relative adaptedness fits these facts of natural selection and so is corroborated; if not then it is on its way to being falsified (of course no one observation would falsify it).

It should be clear that any definition that fails to satisfy the condition of independence from actual reproductive values will fail to be testable in the way described above. Yet it is important to note that once we accept some theory of adaptedness, that is, some theory of what it is for an organism to be adapted to its environment, we can criticize a definition failing (a) as empirically incorrect. In fact, as we will see, on any decent theory of adaptedness any definition failing (a)

will also fail (d).[11] We want our definition of relative adaptedness to fit the facts of natural selection. We cannot accept a definition which renders (D) false.

To summarize, our strategy is to construct a definition of relative adaptedness that makes (D) a respectable scientific law (from the received point of view of philosophy of science). Requirement (a) is that (D) cannot be a tautology. Requirement (b) is that (D) must be general, *i.e.* universally applicable throughout the biosphere. Requirement (c) is that (D) not be so vague or so obscure that we have no idea how to apply it to particular cases (or that (D) be testable). And requirement (d) is that (D) must not be false (or more precisely, that (D) must be nontautologously true).

3. Current Definitions and the Possibility of Satisfying the Four Desiderata

Let us now examine current approaches to the problem of defining relative adaptedness in the light of the four desiderata discussed above. As I said earlier the simplest approach is perhaps the most popular: *a* is better adapted than *b* in *E* iff *a* has more offspring than *b* in *E*. Besides making (D) a tautology and so stripping the concept of its explanatory power this approach totally ignores the fact that natural selection is a statistical phenomenon. Differential fitness may be correlated with certain differences in traits but the correlation is not expected to be perfect. For example, in a certain population of moths darker winged individuals may on average produce more offspring than lighter winged individuals but this certainly does not imply that for every pair of moths the darker winged one will have a greater number of offspring than the lighter winged one. Appreciating that natural selection is a statistical rather than a deterministic process has led some theorists to suggest a more sophisticated approach to our problem (see Mayr, 1963, pp. 182–184).

This more sophisticated approach would define relative adaptedness in terms of the statistical probability of reproductive success. How is this probability to be determined? Suppose we separate the members of a population (of moths, for example) into similarity classes formed on the basis of the functional or epigenetical similarity of their genotypes. To fix ideas let us say that we form two such classes and that the members of one are all darker winged than any of the members of the other (this difference being the result of genetic differences between members of the two classes). Further suppose that these classes are epistemically homogeneous with respect to reproduction; *i.e.* no other division of this class of moths that we can make (based on our knowledge) will be statistically more relevant to reproduction, except divisions based tautologously on actual reproduction. We can now determine the probability of reproductive success of any individual as a simple function of the average reproductive success of the members of the similarity class to which it belongs. And so the reproductive success of the individual is statistically determined by the functional properties of its genotype.

This approach, which we will call the statistical approach, fits some existing paradigms of statistical explanation (see Salmon, 1970), but, as I will show, it fails not only desideratum (a) but also (d). The statistical approach is most closely related to the frequentist interpretation of probability which identifies the probability of an event with its relative frequency 'in the long run'. The leading proponents of this interpretation have been Richard von Mises and Hans Reichenbach. In what follows I am only criticizing the application of this interpretation of probability to defining relative adaptedness. This, of course, does not constitute a general criticism of that interpretation. In the next section I will suggest a definition using a rival conception of probability.

Since the statistical approach uses actual reproductive values its empirical correctness cannot be tested by prediction and observation. It can only be tested against certain general theoretical principles. Consider the following case. Four dogs are on an island; two German shepherds one of each sex and two basset hounds one of each sex. Both bitches go into heat, basset mounts basset and German shepherd mounts German shepherd. While copulating the shepherds are fatally struck by lightning. The bassets, on the other hand, raise a nice family. Are the bassets therefore better adapted to the island environment

than the shepherds? To put the question another way, do we count this differential reproduction as natural selection?

Biologists usually define natural selection simply as differential reproduction (of genes, genotypes or phenotypes). But this is due to carelessness not lack of understanding. Most biologists would agree that the above case is not an instance of natural selection but rather a case of chance differences in fitness. (Not that it could not be natural selection, but nothing in the story indicates that it is. We can elaborate the story in ways that make it clear that it is not a case of natural selection. For instance, the only food source for dogs on our island might be animals whose size and ferocity would make it relatively easier for the larger shepherds to eat than the bassets. Furthermore lightning might be a rare phenomenon and indifferent between bassets and shepherds.) How then shall we characterize natural selection? The concept must be defined in terms of the as yet undefined notion of adaptedness. Natural selection is not just differential reproduction but rather is differential reproduction which is due to the adaptive superiority of those who leave more offspring.

Even without a definition of relative adaptedness we can be confident that cases like the basset–shepherd case are not instances of natural selection. Given that natural selection is a statistical phenomenon it should not be surprising that in small populations Darwinian fitness is not always correlated with adaptedness. Yet the statistical approach to defining relative adaptedness cannot recognize this. According to our story the basset–shepherd case is unique; no such population of dogs has ever been nor will ever be on this island nor on any sufficiently similar island. Thus our four dogs exhaust the data available for the statistical approach. So according to the statistical approach the bassets are better adapted to the island environment than the shepherds. Yet by ecological analysis, in which we determine what it takes for a dog to survive and reproduce on our island, we conclude that the shepherds are better adapted to the island than the bassets. This conflict raises questions concerning the empirical correctness of the statistical approach.

If the basset–shepherd case were just an *ad hoc* counter-example dreamed up to refute the statistical approach then perhaps we should ignore it. But statistically small populations are not uncommon in nature and they are of considerable evolutionary significance (especially for speciation by what Mayr calls the *founder principle*, see Mettler and Gregg, 1969, pp. 130–135; and Mayr, 1963). When applied to small populations the statistical approach will quite predictably conflict with our best analyses of the organism-environment relation, and so we are led to conclude that this approach which renders (D) a tautology is also empirically incorrect. (It should be clear that defining the relative adaptedness of an individual in terms of *its* actual reproductive success is likewise empirically incorrect.)

Let me criticize the statistical approach in a slightly different way to show the connection between its empirical incorrectness and its explanatory failure. The role in evolutionary theory of the relational concept of adaptation is to explain differential fitness. The question is: Why are those features which happen to be highly correlated with reproductive success in fact highly correlated with reproductive success? The Darwinian answer is: Organisms having these features are (for the most part) better adapted to their environment than their conspecifics lacking them. This higher degree of adaptedness causes the fitter organisms to be fitter and is the explanation of their higher fitness. The idea behind the statistical approach to defining relative adaptedness is that high statistical correlations between certain features and Darwinian fitness will indeed be causal connections and so will explain differential fitness. Yet we have seen that there are conceptually clear-cut types of cases (involving small populations) where the high statistical correlation is not a causal connection (in any interesting sense) and so cannot be used to explain differential fitness. In our basset–shepherd case certain distinctively basset features (such as shortness and color of coat) are perfectly correlated with fitness. Yet our bassets are fitter than our shepherds not because they are shorter or are a certain color, but rather because the shepherds were in the wrong place at the wrong time. In our case it's not that shepherds are characteristically in the wrong place at the wrong time but just that they happened to be once. Due to small population size once is enough and so an essentially random process has radically altered our island

population of dogs. Here differential fitness is explained (some might worry over how this is an explanation – I can't concern myself with that here) in terms of a chance process and small population size. Thus if evolutionists are to explain what they want to explain, if they are to have the sort of explanatory theory they want, some other approach to the problem of adaptation is needed.

Early in this paper we were led to distinguish adaptedness from Darwinian fitness. As we have seen from the basset–shepherd example, in small populations the two do not always coincide. Are there other types of cases where the two do not coincide? I can think of only three candidates for such cases: cases of artificial selection, cases of domestication such as in modern man where selection seems to have been relaxed and cases of sexual selection. But none of these types of cases are ones where the correlation of fitness and adaptedness should not be expected and it is important to see why this is so. I will focus my attention on artificial selection; what is said about it can easily be applied to the other two types of cases by analogy.

Artificial selection quite often results in organisms which could not survive in their 'natural' habitat. Organisms which under 'natural' conditions would be the fittest are prevented from breeding while other organisms, less fit under 'natural' conditions, are allowed to breed. By such a process we end up with chickens without feathers, dogs so small they can fit in your hand and fruit flies with legs where they should have antennae. Such cases, it could be claimed, are clear cases where Darwinian fitness does not coincide with adaptedness. But how could one argue for this claim?

Suppose we are following the relative frequency of a segregating genetic entity, say a chromosome inversion in a population of fruit flies. We divide this population into two genetically identical subpopulations, leave one sub-population in its original habitat and move the other to some new and different habitat. After a few generations we observe that the frequency of this chromosome inversion has changed in the moved population (while remaining the same in the control population). Are we to conclude that this change in frequency is the result of some divergence between fitness and adaptedness, since some flies which would have been less fit in the original environment have had a higher relative fitness in the new environment? Obviously not. Whatever adaptedness is it has something to do with the organism-environment relation. With a change in environment a change in relative adaptedness is not unexpected. Man is often thought of as the zenith of evolution yet he can hardly get by in his fishy ancestors' environment.

Artificial selection is just a human induced change in environment. I presume that it is true that a fly with leg-like antennae would not be as well adapted to his ancestral home as many of his more normal relatives. But is he not much better adapted than his normal relatives to the laboratory where the experimenter is selecting for an extra set of legs? In this environment he is much better able to survive and reproduce than his more normal colleagues. The flies are living and breeding in the laboratory; what would be their relative adaptedness in the wild is irrelevant to an assessment of their relative adaptedness in the lab.

To argue that in cases of artificial selection fitness and adaptedness do not coincide is clearly to ignore the environment in which the selection is taking place; in particular it is to ignore the experimenter's or breeder's part in this environment. But that is no more justified than ignoring the part of predators in the prey's environment and is a bit of anthropocentricism. To objective biologists experimenters and breeders are no different than those English birds who for hundreds of years have steadfastly selected against (i.e. eaten) moths not cryptically colored.

Thus artificial selection is just a type of natural selection. This point will have a crucial role to play in an argument later in this paper so I should make it clear that it is not a quibble over words. How would we reply to one who says that by 'natural selection' he means all cases of selection excluding those involving man? To this we should reply that the concept he has defined is not as useful for theoretical purposes as the more inclusive concept we have defined. He can try to use words however he wants, but he can't justify an anthropocentric point of view towards the concept of adaptedness.

We have seen that the simplest approach to defining relative adaptedness, which does so in terms of actual reproductive values, and the

more sophisticated statistical approach fail both desiderata (a) and (d). This failure, especially the failure to meet (a), is fairly apparent and is presumably due to the neglect of theorists to formulate desiderata concerning the concept of relative adaptation. However there is the novel approach by Walter Bock and Gerd von Wahlert (1965) which might be taken as an attempt to meet (a)–(d); at least it does not obviously fail them.

Bock and von Wahlert argue that a measure of adaptedness should be expressed in terms of energy requirements. First they point out that the energy available to an organism at any given time (from both internal and external sources) is limited and that there is interindividual variation in the amount of energy available to organisms (as well as intraindividual variation over the life-span of an individual). Next they point out that for an organism to maintain the proper relation to its environment (i.e. to stay alive) it must expend energy. The amount of energy expended will vary depending, for example, on whether the organism is resting or escaping predation. Since an organism must expend energy to live and reproduce and since its available energy is limited it is advantageous, they argue, for the organism to minimize the amount of energy required to maintain successfully its ecological niche (p. 287). Thus the following definition is suggested by their work:

> a is better adapted than b in E iff a requires less energy to maintain successfully its niche in E than does b.

There are a number of problems with this definition. First we must ask whether it really meets requirement (a). Stern (1970, p. 48) suggests that it does not. He asks what it means to successfully maintain a niche. He quotes Bock and von Wahlert as follows: 'The relative factor of survival or the relative number of progeny left which is usual when comparing the adaptedness of individuals is accounted for by the relative nature of the term "successful"' (Bock and von Wahlert, p. 287). This, according to Stern, 'is tantamount to admitting that their criterion is really subservient to reproduction, and that success in adaptation is still to be measured by more conventional means. That a niche will be maintained more successfully if less energy is required is clearly only an unsupported conclusion, not a matter of definition.' (Stern, p. 48). But here Stern misses the point. Bock and von Wahlert clearly assert that 'unsupported conclusion'. They say, 'The less energy used, the more successfully . . . the niche will be maintained.' (p. 287). If they are right then differences in fitness can and will be explained in terms of differences in energy requirements. It remains for us to ask whether they are right.

We may not be able to answer this question. Although their definition of relative adaptedness seems to be applicable (*i.e.* it seems to satisfy desideratum (c)) it may not be. We can turn to Bock and von Wahlert for suggestions on how their definition is to be applied to particular cases. Unfortunately they do not discuss intraspecific comparisons; but from their discussion of comparing the energy requirements of sparrows *vs* woodpeckers for clinging to vertical surfaces we can reconstruct how they would make such a comparison (see Bock and von Wahlert, pp. 287 ff.). They would determine the amount of energy expended in clinging to a vertical surface by measuring the amount of oxygen consumed. Thus for two woodpeckers they would determine which is better adapted to clinging to vertical surfaces by measuring their oxygen consumption while clinging to some surface. One would be better adapted than the other if it used less oxygen than the other. Recall that we want to explain differential reproductive success. One could test the hypothesis that if one woodpecker requires less energy to cling to a vertical surface than another then it (probably) will have more offspring than the other. But it is not likely to be true. Even for woodpeckers there is more to life than hanging on trees. What seems to be needed is a determination of all the activities necessary for survival and reproduction in a particular environment. We would then compare the relative adaptedness of two organisms by comparing their energy requirements for these activities. But would not these activities have to be weighted according to their importance? How would they be weighted? And isn't it possible, and even fairly frequent, that one organism can bypass some 'necessary activity' because of some difference from his conspecifics in morphology, physiology or behavior? These questions lead me to believe that the Bock and von Wahlert definition is in fact not epistemologically applicable (*i.e.* it fails (c)) but I will not pursue this further.

Rather let us grant for the sake of argument that it is applicable and ask whether or not it is empirically correct.

I have already outlined how to test the empirical correctness of a definition of relative adaptedness (see above, pp. 110–111). In brief, we take paradigmatic cases of natural selection and see if the definition fits the case. In the well known case of melanism in English moths we would check to see if darker winged moths required on average less energy than lighter winged moths. I have raised doubts whether the Bock and von Wahlert definition is so testable and since I can't overcome the problems raised for its testability I can't subject it to this case-study type of test. But if it is testable (or epistemologically applicable) it can, I will argue, be shown to be empirically incorrect.

Suppose we have in our laboratory a population of genetically diverse individuals whose diversity is phenotypically expressed in an easily recognizable manner. By Bock's and von Wahlert's definition some variants are better adapted than others. I, as a perverse Popperian, prevent the so-called 'better adapted' from breeding while allowing the so-called 'less well adapted' to breed. I do this in a large population over a number of generations. Since artificial selection is just a type of natural selection we have here a case of natural selection which does not fit Bock's and von Wahlert's definition. If more falsifying cases are wanted we can produce them. And so, it seems, if Bock's and von Wahlert's definition is epistemologically applicable it is not empirically correct. Clearly this argument applies not only to the Bock and von Wahlert definition but to all definitions which meet desiderata (a)–(c).

This argument is not conclusive. When we begin to select for the so-called 'less well adapted' we change the environment of the organisms. It is open for the theorist whose definition we are criticizing to claim that our change of environment has reversed his estimations of adaptedness, adaptedness being environment relative. This doesn't deter us; again we try to refute the implications of the definition. But what if our most perverse efforts fail to contradict the proposed definition? Here I think we must conclude that empirical correctness has been purchased at the price of epistemological applicability. (Consider how one would try to defend the Bock and von Wahlert definition against such countercases.) That is, the definition has become so vague and malleable as to make (D) unfalsifiable. My claim is that for any proposed definition of relative adaptedness satisfying desiderata (a)–(c) I can produce cases showing that it fails (d) (*i.e.* is empirically incorrect) and that to resist falsification by artificial selection is to give up (c) (*i.e.* is to cease being epistemologically applicable or testable). To exhaustively prove this would be to take every possible definition of relative adaptedness and produce the relevant countercases. It is not surprising that I can't do this. But I do hope my argument is convincing.

I'm sure some will feel that this argument from artificial selection is a cheap victory. If we could find a definition of relative adaptedness that truly applied to all organisms in 'natural' environments wouldn't we be justified in ignoring counterexamples produced by artificial selection? That is a difficult theoretical question but we can say this: Such a definition would represent a tremendous advance in our knowledge of ecology and would be welcomed. But artificial selection is as much a natural phenomenon as predation, starvation, mate selection, *etc.* The argument from artificial selection should, if nothing else, decrease the plausibility of the possibility of such a definition. Naturalists are well aware that natural selection is an opportunistic process, often leading to evolutionary dead ends and extinction. Are not some 'natural' cases of selection just as bizarre as our concocted cases?

The point emphasized in the argument from artificial selection is this: The environments in which organisms find themselves competing are radically different from each other, and at least practically speaking there is no way to specify all possible environments. Thus there is conflict between desiderata (c) and (d). To make (D) testable is to expose (D) to falsification from some radically new ecological situation. And to protect (D) from such falsification is to make it so general that it ceases to be applicable. This point should be accepted even by those who fail to subsume artificial selection under natural selection. Having given good reasons to doubt that any definition of relative adaptedness will satisfy (a)–(d) the question should be: Is there any reason to suppose such a definition possible? I've found none.

4. A Suggested Definition

The attempt has been to construct a definition of relative adaptedness that renders (D) an explanatory law. Accepting the received view of philosophy of science I pointed out that for (D) to be an explanatory law it must be nontautologous, general, testable and true. I argued that for (D) to be such the definition of relative adaptedness must satisfy desiderata (a)–(d). Finally I showed that no definition of relative adaptedness can satisfy (a)–(d). In the light of these conclusions I will now suggest what I take to be the best possible definition of relative adaptedness.

Recall our desiderata. Apparently we will have to give up at least one of them. We should retain (a) and (d); tautologies and false statements explain little (one should note that giving up (a) would also entail giving up (d)). As we will see there is a trade-off between desiderata (b) and (c) and my suggested definition will, in a sense, preserve both.

First I will suggest a non-technical definition of relative adaptedness and then a more technical version. The non-technical version follows:

(RA) *a* is better adapted than *b* in *E* iff *a* is better able to survive and reproduce in *E* than is *b*.

This definition avoids tautology, that is, it is independent of actual reproductive values. (We can confidently assert that a particular Mercedes-Benz 450 SEL is *able* to do 150 mph while knowing that it never has and never will go that fast.) It is also a general definition and it is empirically correct (insofar as this makes sense, at least it is not empirically incorrect). But how are we to apply it to particular situations? I think it is clear that as it stands (RA) is not epistemologically applicable. So this suggested definition has the effect of preserving (a), (b) and (d) at the expense of (c), and given that we cannot have all four, (RA)'s obvious failure of (c) is a virtue. It is an unpretentious definition; it wears its epistemological inapplicability on its sleeve.

We can construct a more technical (and more pretentious) definition. Earlier I criticized what I called the statistical approach to defining relative adaptedness. This approach identified adaptedness with the statistical mean of observed reproductive rates. As pointed out then it is not too distorting to call the interpretation of probability used in this approach the frequentist interpretation. There are other interpretations of probability. Some, for instance the logical and subjective interpretations (associated with Carnap and de Finetti respectively), are here irrelevant. But the approach, best expounded by Hacking (1965) (also see Popper, 1959), on which probabilities are deduced from theory rather than identified with observed frequencies is relevant.

In discussing the basset–shepherd case I said that observed reproductive rates can conflict with estimations of adaptedness based on ecological analysis. Suppose our ecological theories to be so well developed that for any given environment and organism we could deduce the distribution of probabilities of the number of offspring left by that organism (in the next generation). That is, from our theories we deduce for each organism O and environment E a range of possible numbers of sufficiently similar offspring, $Q_1^{OE}, Q_2^{OE}, \dots,$ Q_n^{OE} and for each Q_i^{OE} our theory associates a number $P(Q_i^{OE})$ which is the probability (or chance or propensity) of O leaving Q_i sufficiently similar offspring in E. Given all this we define the adaptedness O in E (symbolized as $A(O, E)$) as follows:

$$A(O,E) = \Sigma\, P(Q_i^{OE})Q_i^{OE}$$

That is, the adaptedness of O in E equals the expected value of its genetic contribution to the next generation. (The units of value are arbitrary. All that matters here are the ordinal relations among the numbers associated with each pair $<o, E>$. Outside of this context the numbers have no significance.) Our new more exacting definition of relative adaptedness, (RA'), is as follows:[12]

(RA') a is better adapted than b in E iff $A(a, E) > A(b, E)$.

Two things should be clear: First, (RA') only makes sense for intraspecific intra-environmental comparisons. Second, (RA') is a step in the right direction only on the proper interpretation of probability.

Before evaluating (RA') I should say something about its basic presupposition: *viz.* that from detailed ecological analysis we can give good estimates of the probabilities of reproductive

success of organisms in environments independent of observations of their actual reproductive success. For example, given the characteristics of a certain island environment and the particular characteristics of some basset hounds and German shepherds such theories should be able to predict the relative reproductive success of each even without any relevant statistics. Clearly such predictions are falsifiable (as falsifiable as any statistical hypothesis), but do we have any reason to expect them to be successful? There are few, if any, outstanding examples of such success in the corpus of biological science. On the other hand, there seems to be no theoretical obstacle to successful predictions of this sort.

The informal definition of relative adaptedness suggested above, (RA), satisfied desiderata (a), (b), and (d) but not (c). How does (RA') fare on our desiderata? Given the proper interpretation of probability it satisfies (a). On this interpretation the probability of reproductive success (or expected genetic contribution to future generations) is some biological property of the organism and its environment (just as the probability of heads for a coin is a physical property of the coin and the tossing device). The organism in its environment has this property even if it is struck by lightning prior to leaving any offspring (just as the chance of heads may be $^1/_2$ for a coin even if it is unique and is melted before it is ever tossed). Thus (RA') is independent of actual reproductive values. The occurence of 'probably' in (D) may be confusing but (RA') does not turn (D) into a tautology.[13] (RA') clearly satisfies (b); that is, it is general. Like (RA), (RA') is not empirically incorrect and so we will say it satisfies (d), *i.e.* that it is empirically correct. Although (RA')'s failure to satisfy (c) may not be as apparent as (RA)'s it also fails to be epistemologically applicable. If there were a single all encompassing theory of adaptedness from which we could derive the adaptedness (as defined above) of any organism in any environment then (RA)'s would be epistemologically applicable. But, as I've argued, no such theory is possible. (I presented Bock and von Wahlert's theory as an attempt at such completeness.)

How is the suggested definition useful? It is useful as what we might call a schematic definition. It is neither applicable nor testable but particular instances of it are. What do I mean by an instance of (RA′)? Formally, in an instance of (RA′) we fix the value of the environmental parameter 'E' and limit the range of the individual variables 'a' and 'b' to a particular population of organisms living in E. Such an instantiation would represent a hypothesis concerning what it takes for certain types of organisms to survive and reproduce in a certain type of environment. Good hypotheses of this kind can only result from detailed ecological analysis. (Where 'ecological' is used in a broad, perhaps too broad, sense. I would include in such analysis the study of the sorts of genetic variation that occur and are likely to occur in the relevant organisms and the study of the phenotypic effects of this variation.)

For a simplified example suppose that the only variation in a certain population of moths is in wing color. These moths all rest on dark colored tree trunks during the day. Birds prey on the moths by sight in daytime. We analyse this simplified situation as follows: The darker the wing color the closer it is to the color of the tree trunks. Moths whose wings are colored most like the tree trunks are least likely to be eaten by birds. Moths less likely to be eaten are more likely to leave offspring. Thus we instantiate (RA′) as follows:

Moth a is better adapted than b in (our specified) E iff a's wings are darker colored than b's (in E).

(I am here primarily interested in illustrating certain logical points, but I don't want to appear to take an overly naive and sanguine view towards the sort of ecological analysis necessary for complex organisms in complex environments.) Lewontin (1977) discusses some of the problems involved. Suffice it to say that although successful ecological analysis is difficult it does not seem to be impossible.

With a schematic definition of relative adaptedness (D) becomes a schematic law, and with an instantiation of (RA′) we get an instantiation of (D). For our moths (D) says:

If a is darker winged than b (in E) then (probably) a will have more offspring than b (in E).

Such an instantiation of (D) is clearly testable (in fact it has been tested, see Kettlewell, 1955 and 1956). Moreover it does what we want it to do, it explains differential reproduction and so

explains evolution by natural selection (as in this instance we explain the evolution of industrial melanism in certain species of English moths).

To summarise; I have suggested that we give up epistemological applicability and adopt a schematic definition of relative adaptedness, (RA′). This correlatively makes (D) schematic and so not testable. When we instantiate (RA′) we give up generality for applicability. Likewise instances of (D) becomes testable and explanatory but not general.

5. The Structure of Evolutionary Theory

(D) is the fundamental law of evolutionary theory. What sort of foundation is (D) for a scientific theory? Critics have often maintained that evolutionary theory rests on a tautology. As I hope I have made clear, (D) is not a tautology. But I have shown that no definition of relative adaptedness can render (D) non-tautologous, general, testable and true. (D) as a schematic law is not testable, instantiations of (D) are not general. This may not be so bad. If disconfirming an instantiation of (D) disconfirms (D) then (D) may be a respectable law. But this relation between (D) and its instances does not hold. That is, no amount of falsification of instances of (D) even begins to falsify (or disconfirm) (D).

Consider the instantiation of (D) concerning moths. If through experiments and observations it proved to be false then our response would be and should be that we have incorrectly analysed the ecological situation. Perhaps the birds prey on these moths using heat-sensing devices, making color variation irrelevant (unless that variation is correlated with variation in heat irradiation). We reanalyse the situation and test our new hypothesis. If the falsification of one instance of (D) doesn't even begin to cast doubt on (D) will large numbers of falsifications change matters? If, as is the case, some instances of (D) have proved successful then even large numbers of falsifications of instances of (D) will not cast doubt on (D). If no instance of (D) ever succeeded then we would doubt the usefulness of (D) but even this would not lead us to say (D) is false. In our world (where some instances of (D) have successfully explained and predicted certain phenomena) no set of test results could falsify (D). Thus (D) is unfalsifiable.

With this in mind and given that through informative instantiations of (RA') we get testable and explanatory instances of (D) one might question the status of the schematic (RA') and (D). Neither meets our philosophical expectations so why should they be granted any status in our expurgated science? To answer this question we must consider some of the aims of scientific inquiry and some of the criteria by which theories are judged. Perhaps the distinguishing feature between science and myth is that science, unlike myth, aims at testable explanations. So theories and laws are judged according to their (in-principle-) testability. Instantiations of (D) fare well on this criterion, (D) itself does not. But scientific inquiry also aims at the systematic unification of broad bodies of diverse phenomena. Without (D) there is no theory of evolution, there are only low level theories about the evolution of certain organisms in certain environments. (And at present there are very few of those.) With (D) Darwinian theory is possible.

I have not simply presented a case where philosophy of science is at variance with actual science. Rather I have presented a case where two philosophical principles conflict. There is, as I have shown, a trade-off between desiderata (b) and (c), and so a conflict between testability and systematic unification. I have suggested adopting (RA') and so treating (D) as a schematic law as the best possible solution to this dilemma.

6. Summary

The conception of adaptation has been one of the most troublesome and yet one of the most important concepts in the biological sciences. I hope that this paper has cleared up much of that trouble. We have constructed an adequate definition of relative adaptedness. Our analysis of the conception of relative adaptedness went hand in hand, as it had to, with an analysis of the structure of evolutionary theory. We found that Darwinian evolutionary theory has as its foundation what I called a schematic law; thus its structure does not fit any existing philosophical paradigms for scientific theories. Heretofore schematic definitions and schematic laws have not been recognized or investigated by philosophers of science.

In constructing a definition of relative adaptedness we posited the biological property of adaptedness. In this paper I said much about what this property is and what it is not. But its particular ontological status has not been discussed and remains somewhat mysterious.

Notes

1 'G. C. Williams (1966) does an excellent job of clarifying these matters. Also see Lewontin (1970).

2 Thoday (1953) and (1958). Actually he uses the word 'fitness' not 'adaptedness' but I think he is like most biologists in using the words interchangeably.

3 This characterization of evolutionary theory is adopted from Lewontin (1977). For less satisfactory versions see Lewontin (1968) and (1970). For a more historical and fuller sketch of the major components of the theory see Mayr (1977).

4 Perhaps one should not speak of *the* distinguishing feature of Darwinian theory. One should recognize that evolutionary theory is not a monolithic whole. For instance, theories of speciation are quite distinct from the part of Darwinian theory on which we are focusing; *viz.* the theory of evolution within a species by natural selection. *A propos* the history of the subject it is useful to distinguish four subtheories or four parts of Darwin's theory (pointed out to me by Ernst Mayr): (a) Evolution at all; (b) Gradual evolution; (c) Evolution by common descent; and (d) Evolution by natural selection. Nevertheless both from a historical and contemporary perspective the most salient feature of a Darwinian theory of evolution is its explanation of evolution by natural selection.

5 Malthus (1798). It seems that Malthus was more of a coagulant than a catalyst for Darwin's ideas on this matter. See Hull (1973), pp. 344, 345, and Mayr (1977).

6 Another exception is Michael Ruse (1971). He has attacked the problem from a historical perspective and has tried to show that what Darwin said on natural selection was not tautologous.

7 In speaking of (D) as a 'law' I could continue to put 'law' in scare-quotes in order not to prejudge its status, but I will not. We will, in due course carefully evaluate its status.

8 Mayr, it seems, was quoted out of context. See Mayr (1963), pp. 182–184.

9 As shown by Lack (1954). This must be quite surprising to those with only a superficial understanding of evolution. For example Popper (1972, p. 271) thinks it is 'one of the countless difficulties of Darwin's theory' that natural selection should do anything other than increase fecundity. The explanation is really quite simple: Increased fecundity often results in a decreased number of offspring surviving in the next generation. See Williams (1966, chp. 6) for discussion.

10 To some unfamiliar with the problem of adaptation this may not be obvious. Rather than reargue the generally accepted I refer the reader to Stern (1970) which is a good introduction into the relevant literature.

11 See below pp. 110–111. Of course one might wonder how a definition could fail both (a) and (d), or how a tautology could be empirically incorrect. It can be in just this sense: *given* an adequate theory of adaptedness we have a notion of adaptedness which differs from any notion failing (a) (*i.e.* any notion which identifies adaptedness with actual reproductive success). These two notions will not be extensionally equivalent. So, from the standpoint of our theory, the definition which fails (a) will also fail (d).

12 The move to this sort of definition was suggested to me by Hilary Putnam.

13 (D) becomes something like an instance of what Hacking calls the Law of Likelihood and is analogous to the following: If the chance of heads for coin a is $1/2$ and the chance of heads for b is $1/4$ then (probably) when both coins are tossed a small number of times a will land on heads more than b will.

Bibliography

F. J. Ayala, 'Biological Evolution: Natural Selection or Random Walk?,' *Am. Scient.* **62** (1974), 692–701.

W. J. Bock and G. von Wahlert, 'Adaptation and the Form-Function Complex,' *Evolution* **19** (1965), 269–299.

R. N. Brandon, 'Evolution,' *Philosophy of Science* **45** (1978), 96–109.

C. Darwin, *On the Origin of the Species* (London: John Murray, 1859).

T. Dobhansky, 'Adaptedness and Fitness,' *Population Biology and Evolution*, R. C. Lewontin (ed.) (Syracuse: Syracuse University Press), pp. 109–121.

M. T. Ghiselin, 'A Radical Solution to the Species Problem,' *Systematic Zoology* **23** (1974), 536–544.

I. Hacking, *Logic of Statistical Inference* (Cambridge: Cambridge University Press, 1965).

C. G. Hempel, *Philosophy of Natural Science,* (Englewood Cliffs: Prentice-Hall, 1966).

D. L. Hull, *Darwin and his Critics* (Cambridge: Harvard University Press, 1973).

H. B. D. Kettlewell, 'Selection Experiments on Industrial Melanism in the Lepidoptera,' *Heredity* **9** (1955), 323–342.

H. B. D. Kettlewell, 'Further Selection Experiments on Industrial Melanism in the Lepidoptera,' *Heredity* **10** (1956), 287–301.

J. L. King and T. H. Jukes, 'Non-Darwinian Evolution,' *Science* **164**, 788–798.

D. Lack, 'The Evolution of Reproductive Rates,' *Evolution as a Process*, J. S. Huxley, A. C. Hardy and E. B. Ford (eds.) (London: Allen & Unwin, 1954), pp. 143–156.

R. C. Lewontin, 'The Adaptations of Populations to Varying Environments,' *Symposium of Quantitative Biology* **22** (1957), 395–408.

R. C. Lewontin, 'The Concept of Evolution,' *International Encyclopedia of the Social Sciences* (New York: Macmillan, 1968), pp. 202–210.

R. C. Lewontin, 'The Units of Selection,' *Annual Reivew of Systematics and Ecology* **1** (1970), 1–18.

R. C. Lewontin, 'Adattamento,' *Enciclopedia Eiaudi* (Torino, Italy, 1977).

T. R. Malthus, *An Essay on the Principle of Population.* London (1798).

J. Maynard Smith, *The Theory of Evolution* 3rd edn. (Middlesex: Penguin Books Ltd, 1975).

E. Mayr, 'Cause and Effect in Biology,' *Science* **134** (1961), 1501–1506.

E. Mayr, *Animal Species and Evolution* (Cambridge: Harvard University Press, 1963).

E. Mayr, 'Darwin and Natural Selection,' *Am. Scient.* **65** (1977), 321–327.

L. E. Mettler and T. G. Gregg, *Population Genetics and Evolution* (Englewood Cliffs: Prentice-Hall, 1969).

R. Munson, 'Biological Adaptation,' *Philosophy of Science* **38** (1971), 200–215.

K. R. Popper, 'The Propensity Interpretation of Probability,' *Br. J. Phil.* **10** (1959), 25–42.

K. R. Popper, *Objective Knowledge* (Oxford: Oxford University Press, 1972).

M. Ruse, 'Natural Selection in *The Origin of Species,' Stud. Hist. Phil. Sci.* I (1971), 311–351.

M. Ruse, 'Charles Darwin's Theory of Evolution: An Analysis,' *J. Hist. Biol.* **8** (1975), 219–241.

W. Salmon, 'Statistical Explanation,' R. G. Colodny (ed.) *The Nature and Function of Scientific Theories* (Pittsburgh Studies in the Philosophy of Science IV, pp. 173–321, 1970).

M. Scriven, 'Explanation and Prediction in Evolutionary Theory,' *Science* **130** (1959), 477–482.

L. B. Slobodkin, 'Toward a Predictive Theory of Evolution,' *Population Biology and Evolution,*

(Syracuse: Syracuse University Press, 1968), pp. 187–205.

J. J. C. Smart, *Philosophy and Scientific Realism*, R. C. Lewontin (ed.) (London: Routledge & Kegan Paul, 1963).

J. T. Stern, 'The Meaning of "Adaptation" and its Relation to the Phenomenon of Natural Selection,' *Evolutionary Biology* **4** (1970), 39–66.

J. M. Thoday, 'Components of Fitness,' *Symposium of the Society for Experimental Biology* **7** (1953), 96–113.

J. M. Thoday, 'Natural Selection and Biological Process, *A Century of Darwin*, S. A. Barnett (ed.) (London: Heinemann, 1958), pp. 313–333.

G. C. Williams, *Adaptation and Natural Selection* (Princeton: Princeton University Press, 1966).

M. B. Williams, 'Deducing the Consequences of Evolution,' *Journal of Theoretical Biology* **29** (1970), 343–385.

M. B. Williams, 'Falsifiable Predictions of Evolutionary Theory,' *Philosophy of Science* **40** (1973), 518–537.

PART IV
ADAPTATIONISM

Introduction

Darwin's theory explains adaptation (and through it, complexity and diversity) so powerfully that there is a temptation to identify all traits and characters of biological systems – genes, organelles, cells, tissues, organs, individuals, families, lines of descent, and species – as adaptations. The result has on the one hand vastly extended the potential reach of the theory of natural selection to explain traits as genetically encoded adaptations – including human traits, dispositions, capacities, abilities. On the other hand, it has made some biologists, and Darwinian theorists in the human sciences, incautious about the evidence needed to ground a Darwinian explanation of traits as adaptations, and about the alternative explanations for traits as non-adaptational that biology, and even Darwin's theory, allows for.

Lewontin and Gould provide the most powerful and influential critique of what they consider adaptationist excesses. Significantly, their paper (chapter 9 of this volume) focuses on the role of drift, which is treated in the previous section, and non-adaptational (physical, chemical, biophysical) constraints, which both help shape biological traits and may even swamp natural selection in determining their form.

Mayr, one of the founders of the combination of Darwin's theory and Mendelian genetics that gave rise to the Neo-Darwinian "evolutionary synthesis," defends the adaptationist perspective against these criticisms (chapter 10).

It is important for all biologists, biologically-inspired students of human affairs, and philosophers of biology to be sensitive to the terms of this debate and to resolve it in their own research programs.

Further Reading

Brandon, R., & Rausher, M. (1996). Testing adaptationism: A comment on Orzack and Sober. *American Naturalist*, 148, 189–201.

Dawkins, R. (1976). *The selfish gene*. Oxford: Oxford University Press.

Dawkins, R. (1982). *The extended phenotype*. Oxford: Oxford University Press.

Dawkins, R. (1986). *The blind watchmaker: Why the evidence of evolution reveals a universe without design*. New York: Norton.

Dennett, D. (1995). *Darwin's dangerous idea: Evolution and the meanings of life*. New York: Simon & Schuster.

Dupré, J. (ed.). (1987). *The latest on the best: Essays on evolution and optimality*. Cambridge, MA: MIT Press.

Gould, S., & Vrba, E. (1982). Exaptation: A missing term in the science of form. *Paleobiology*, 8, 4–15.

Kimura, M. (1968). Evolutionary rate at the molecular level. *Nature*, 217, 624–626.

Kimura, M. (1983). The neutral theory of molecular evolution. Cambridge: Cambridge University Press.

Kimura, M. (1986). DNA and the Neutral Theory. *Philosophical Transactions of the Royal Society of London. Series B, Biological Sciences*, 312, 343–354.

Lewontin, R. (1993). *Biology as ideology: The doctrine of DNA*. New York: Harper Collins.

Maynard Smith, J. (1988). *Did Darwin get it Right?: Essays on games, sex and evolution*. London: Chapman & Hall.

Orzack, S. (2008). Testing adaptive hypotheses, optimality models, and adaptationism. In M. Ruse (ed.), *The Oxford handbook of philosophy of biology* (pp. 87–112). Oxford: Oxford University Press.

Orzack, S., & Sober, E. (1994). Optimality models and the test of adaptationism. *American Naturalist, 143*, 361–380.

Orzack, S., & Sober, E. (eds.). (2001). *Adaptationism and optimality*. Cambridge: Cambridge University Press.

Prum, R., & Brush, A. (2002). The evolutionary origin and diversification of feathers. *The Quarterly Review of Biology, 77*, 261–295.

Rosenberg, A., & McShea, D. (2008). *Philosophy of biology: A contemporary introduction* (Chapter 3). London: Routledge.

Sober, E. (1998). Six sayings about adaptationism. In D. Hull & M. Ruse (eds.), *The Philosophy of Biology* (pp. 72–86). Oxford: Oxford University Press.

Williams, G. (1966). *Adaptation and natural selection*. Princeton, NJ: Princeton University Press.

Wimsatt, W. (1987). False models as means to truer theories. In M. Nitecki & H. Hoffman (eds.), *Neutral models in biology* (pp. 23–55). Oxford: Oxford University Press.

9

The Spandrels of San Marco and the Panglossian Paradigm: A Critique of the Adaptationist Programme

Stephen Jay Gould and Richard C. Lewontin

An adaptationist program has dominated evolutionary thought in England and the United States during the past forty years. It is based on faith in the power of natural selection as an optimizing agent. It proceeds by breaking an organism into unitary "traits" and proposing an adaptive story for each considered separately. Trade-offs among competing selective demands exert the only brake upon perfection; non-optimality is thereby rendered as a result of adaptation as well. We criticize this approach and attempt to reassert a competing notion (long popular in continental Europe) that organisms must be analyzed as integrated wholes, with *Baupläne* so constrained by phyletic heritage, pathways of development, and general architecture that the constraints themselves become more interesting and more important in delimiting pathways of change than the selective force that may mediate change when it occurs. We fault the adaptationist program for its failure to distinguish current utility from reasons for origin (male tyrannosaurs may have used their diminutive front legs to titillate female partners, but this will not explain why they got so small); for its unwillingness to consider alternatives to adaptive stories; for its reliance upon plausibility alone as a criterion for accepting speculative tales; and for its failure to consider adequately such competing themes as random fixation of alleles, production of non-adaptive structures by developmental correlation with selected features (allometry, pleiotropy, material compensation, mechanically forced correlation), the separability of adaptation and selection, multiple adaptive peaks, and current utility as an epiphenomenon of nonadaptive structures. We support Darwin's own pluralistic approach to identifying the agents of evolutionary change.

Introduction

The great central dome of St. Mark's Cathedral in Venice presents in its mosaic design a detailed iconography expressing the mainstays of Christian faith. Three circles of figures radiate out from a central image of Christ: angels, disciples, and virtues. Each circle is divided into quadrants, even though the dome itself is radially symmetrical in structure. Each quadrant meets one of the four spandrels in the arches below the dome. Spandrels – the tapering triangular spaces formed by the intersection of two rounded arches at right angles – are necessary architectural by-products of mounting a dome on rounded arches. Each spandrel contains a design admirably

Stephen Gould and Richard Lewontin. "The Spandrels of San Marco and the Panglossian Paradigm: A Critique of the Adaptationist Programme," *Proceedings of the Royal Society of London* (1978) 205, pp. 581–598. Reprinted by permission of the Royal Society.

fitted into its tapering space. An evangelist sits in the upper part flanked by the heavenly cities. Below, a man representing one of the four biblical rivers (Tigris, Euphrates, Indus, and Nile) pours water from a pitcher in the narrowing space below his feet.

The design is so elaborate, harmonious, and purposeful that we are tempted to view it as the starting point of any analysis, as the cause in some sense of the surrounding architecture. But this would invert the proper path of analysis. The system begins with an architectural constraint: the necessary four spandrels and their tapering triangular form. They provide a space in which the mosaicists worked; they set the quadripartite symmetry of the dome above.

Such architectural constraints abound, and we find them easy to understand because we do not impose our biological biases upon them. Every fan-vaulted ceiling must have a series of open spaces along the midline of the vault, where the sides of the fans intersect between the pillars. Since the spaces must exist, they are often used for ingenious ornamental effect. In King's College Chapel in Cambridge, for example, the spaces contain bosses alternately embellished with the Tudor rose and portcullis. In a sense, this design represents an "adaptation," but the architectural constraint is clearly primary.

The spaces arise as a necessary by-product of fan vaulting; their appropriate use is a secondary effect. Anyone who tried to argue that the structure exists because the alternation of rose and portcullis makes so much sense in a Tudor chapel would be inviting the same ridicule that Voltaire heaped on Dr. Pangloss: "Things cannot be other than they are. . . . Everything is made for the best purpose. Our noses were made to carry spectacles, so we have spectacles. Legs were clearly intended for breeches, and we wear them." Yet evolutionary biologists, in their tendency to focus exclusively on immediate adaptation to local conditions, do tend to ignore architectural constraints and perform just such an inversion of explanation.

As a closer example, recently featured in some important biological literature on adaptation, anthropologist Michael Harner has proposed (1977) that Aztec human sacrifice arose as a solution to chronic shortage of meat (limbs of victims were often consumed, but only by people

of high status). E. O. Wilson (1978) has used this explanation as a primary illustration of an adaptive, genetic predisposition for carnivory in humans. Harner and Wilson ask us to view an elaborate social system and a complex set of explicit justifications involving myth, symbol, and tradition as mere epiphenomena generated by the Aztecs as an unconscious rationalization masking the "real" reason for it all: need for protein. But Sahlins (1978) has argued that human sacrifice represented just one part of an elaborate cultural fabric that, in its entirety, not only represented the material expression of Aztec cosmology, but also performed such utilitarian functions as the maintenance of social ranks and systems of tribute among cities.

We strongly suspect that Aztec cannibalism was an "adaptation" much like evangelists and rivers in spandrels, or ornamented bosses in ceiling spaces: a secondary epiphnomenon representing a fruitful use of available parts, not a cause of the entire system. To put it crudely: a system developed for other reasons generated an increasing number of fresh bodies; use might as well be made of them. Why invert the whole system in such a curious fashion and view an entire culture as the epiphenomenon of an unusual way to beef up the meat supply? Spandrels do not exist to house the evangelists. Moreover, as Sahlins argues, it is not even clear that human sacrifice was an adaptation at all. Human cultural practices can be orthogenetic and drive toward extinction in ways that Darwinian processes, based on genetic selection, cannot. Since each new monarch had to outdo his predecessor in even more elaborate and copious sacrifice, the practice was beginning to stretch resources to the breaking point. It would not have been the first time that a human culture did itself in. And, finally, many experts doubt Harner's premise in the first place (Ortiz de Montellano 1978). They argue that other sources of protein were not in short supply, and that a practice awarding meat only to privileged people who had enough anyway, and who used bodies so inefficiently (only the limbs were consumed, and partially at that), represents a mighty poor way to run a butchery.

We deliberately chose nonbiological examples in a sequence running from remote to more familiar: architecture to anthropology. We did this because the primacy of architectural constraint and the epiphenomenal nature of adaptation are not obscured by our biological prejudices in these examples. But we trust that the message for biologists will not go unheeded: if these had been biological systems, would we not, by force of habit, have regarded the epiphenomenal adaptation as primary and tried to build the whole structural system from it?

The Adaptationist Program

We wish to question a deeply engrained habit of thinking among students of evolution. We call it the adaptationist program, or the Panglossian paradigm. It is rooted in a notion popularized by A. R. Wallace and A. Weismann, (but not, as we shall see, by Darwin) toward the end of the nineteenth century: the near omnipotence of natural selection in forging organic design and fashioning the best among possible worlds. This program regards natural selection as so powerful and the constraints upon it so few that direct production of adaptation through its operation becomes the primary cause of nearly all organic form, function, and behavior, Constraints upon the pervasive power of natural selection are recognized, of course (phyletic inertia primarily among them, although immediate architectural constraints, as discussed in the last section, are rarely acknowledged). But they are usually dismissed as unimportant or else, and more frustratingly, simply acknowledged and then not taken to heart and invoked.

Studies under the adaptationist program generally proceed in two steps:

1. An organism is atomized into "traits" and these traits are explained as structures optimally designed by natural selection for their functions. For lack of space, we must omit an extended discussion of the vital issue, "What is a trait?" Some evolutionists may regard this as a trivial, or merely a semantic problem, it is not. Organisms are integrated entities, not collections of discrete objects. Evolutionists have often been led astray by inappropriate atomization, as D'Arcy Thompson (1942) loved to point out. Our favorite example involves the human chin (Gould 1977, pp. 381–382; Lewontin 1978).

If we regard the chin as a "thing," rather than as a product of interaction between two growth fields (alveolar and mandibular), then we are led to an interpretation of its origin (recapitulatory) exactly opposite to the one now generally favored (neotenic).

2. After the failure of part-by-part optimization, interaction is acknowledged via the dictum that an organism cannot optimize each part without imposing expenses on others. The notion of "trade-off" is introduced, and organisms are interpreted as best compromises among competing demands. Thus interaction among parts is retained completely within the adaptationist program. Any suboptimality of a part is explained as its contribution to the best possible design for the whole. The notion that suboptimality might represent anything other than the immediate work of natural selection is usually not entertained. As Dr. Pangloss said in explaining to Candide why he suffered from venereal disease: "It is indispensable in this best of worlds. For if Columbus, when visiting the West Indies, had not caught this disease, which poisons the source of generation, which frequently even hinders generation, and is clearly opposed to the great end of Nature, we should have neither chocolate nor cochineal." The adaptationist program is truly Panglossian. Our world may not be good in an abstract sense, but it is the very best we could have. Each trait plays its part and must be as it is.

At this point, some evolutionists will protest that we are caricaturing their view of adaptation. After all, do they not admit genetic drift, allometry, and a variety of reasons for nonadaptive evolution? They do, to be sure, but we make a different point. In natural history, all possible things happen sometimes; you generally do not support your favored phenomenon by declaring rivals impossible in theory. Rather, you acknowledge the rival but circumscribe its domain of action so narrowly that it cannot have any importance in the affairs of nature. Then, you often congratulate yourself for being such an undogmatic and ecumenical chap. We maintain that alternatives to selection for best overall design have generally been relegated to unimportance by this mode of argument. Have we not all heard the catechism about genetic drift: it can only be important in populations so small that they are likely to become extinct before playing any sustained evolutionary role (but see Lande 1976).

The admission of alternatives in principle does not imply their serious consideration in daily practice. We all say that not everything is adaptive; yet, faced with an organism, we tend to break it into parts and tell adaptive stories as if trade-offs among competing, well-designed parts were the only constraint upon perfection for each trait. It is an old habit. As Romanes complained about A. R. Wallace in 1900: "Mr. Wallace does not expressly maintain the abstract impossibility of laws and causes other than those of utility and natural selection. . . . Nevertheless, as he nowhere recognizes any other law or cause . . . he practically concludes that, on inductive or empirical grounds, there is *no* such other law or cause to be entertained."

The adaptationist program can be traced through common styles of argument. We illustrate just a few; we trust they will be recognized by all:

1. If one adaptive argument fails, try another. Zig-zag commissures of clams and brachiopods, once widely regarded as devices for strengthening the shell, become sieves for restricting particles above a given size (Rudwick 1964). A suite of external structures (horns, antlers, tusks), once viewed as weapons against predators, become symbols of intraspecific competition among males (Davitashvili 1961). The Eskimo face, once depicted as "cold engineered" (Coon *et al.* 1950), becomes an adaptation to generate and withstand large masticatory forces (Shea 1977). We do not attack these newer interpretations; they may all be right. We do wonder, though, whether the failure of one adaptive explanation should always simply inspire a search for another of the same general form, rather than a consideration of alternatives to the proposition that each part is "for" some specific purpose.

2. If one adaptive argument fails, assume that another must exist; a weaker version of the first argument. Costa and Bisol (1978), for example, hoped to find a correlation between genetic polymorphism and stability of environment in the deep sea, but they failed. They conclude (1978, pp. 132, 133): "The degree of genetic polymorphism found would seem to indicate absence of correlation with the particular environmental factors which characterize the sampled area. The results suggest that the adaptive strategies of organisms belonging to different phyla are different."

3. In the absence of a good adaptive argument in the first place, attribute failure to imperfect understanding of where an organism lives and what it does. This is again an old argument. Consider Wallace on why all details of color and form in land snails must be adaptive, even if different animals seem to inhabit the same environment (1899, p. 148): "The exact proportions of the various species of plants, the numbers of each kind of insect or of bird, the peculiarities of more or less exposure to sunshine or to wind at certain critical epochs, and other slight differences which to us are absolutely immaterial and unrecognizable, may be of the highest significance to these humble creatures, and be quite sufficient to require some slight adjustments of size, form, or color, which natural selection will bring about."

4. Emphasize immediate utility and exclude other attributes of form. Fully half the explanatory information accompanying the full-scale Fiberglass *Tyrannosaurus* at Boston's Museum of Science reads: "Front legs a puzzle: how *Tyrannosaurus* used its tiny front legs is a scientific puzzle; they were too short even to reach the mount. They may have been used to help the animal rise from a lying position," (We purposely choose an example based on public impact of science to show how widely habits of the adaptationist program extend. We are not using glass beasts as straw men; similar arguments and relative emphases, framed in different words, appear regularly in the professional literature.) We don't doubt that *Tyrannosaurus* used its diminutive front legs for something. If they had arisen *de novo*, we would encourage the search for some immediate adaptive reason. But they are, after all, the reduced product of conventionally functional homologues in ancestors (longer limbs of allosaurs,

for example). As such, we do not need an explicitly adaptive explanation for the reduction itself. It is likely to be a developmental correlate of allometric fields for relative increase in head and hindlimb size. This nonadaptive hypothesis can be tested by conventional allometric methods (Gould 1974, in general; Lande 1978, on limb reduction) and seems to us both more interesting and fruitful than untestable speculations based on secondary utility in the best of possible worlds. One must not confuse the fact that a structure is used in some way (consider again the spandrels, ceiling spaces, and Aztec bodies) with the primary evolutionary reason for its existence and conformation.

Telling Stories

All this is a manifestation of the rightness of things, since if there is a volcano at Lisbon it could not be anywhere else. For it is impossible for things not to be where they are, because everything is for the best.

> Dr. Pangloss on the great Lisbon earthquake of 1755, in which up to 50,000 people lost their lives

We would not object so strenuously to the adaptationist program if its invocation, in any particular case, could lead in principle to its rejection for want of evidence. We might still view it as restrictive and object to its status as an argument of first choice. But if it could be dismissed after failing some explicit test, then alternatives would get their chance. Unfortunately, a common procedure among evolutionists does not allow such definable rejection for two reasons. First, the rejection of one adaptive story usually leads to its replacement by another, rather than to a suspicion that a different kind of explanation might be required. Since the range of adaptive stories is as wide as our minds are fertile, new stories can always be postulated. And if a story is not immediately available, one can always plead temporary ignorance and trust that it will be forthcoming, as did Costa and Bisol (1978), cited above. Second, the criteria for acceptance of a story are so loose that many pass without proper confirmation. Often, evolutionists use *consistency*

with natural selection as the sole criterion and consider their work done when they concoct a plausible story. But plausible stories can always be told. The key to historical research lies in devising criteria to identify proper explanations among the substantial set of plausible pathways to any modern result.

We have, for example (Gould 1978) criticized Barash's (1976) work on aggression in mountain bluebirds for this reason. Barash mounted a stuffed male near the nests of two pairs of bluebirds while the male was out foraging. He did this at the same nests on three occasions at ten-day intervals: the first before eggs were laid, the last two afterward. He then counted aggressive approaches of the returning male toward both the model and the female. At time one, aggression was high toward the model and lower toward females but substantial in both nests. Aggression toward the model declined steadily for times two and three and plummeted to near zero toward females. Barash reasoned that this made evolutionary sense, since males would be more sensitive to intruders before eggs were laid than afterward (when they can have some confidence that their genes are inside). Having devised this plausible story, he considered his work as completed (1976, pp. 1099, 1100):

> The results are consistent with the expectations of evolutionary theory. Thus aggression toward an intruding male (the model) would clearly be especially advantageous early in the breeding season, when territories and nests are normally defended. . . . The initial aggressive response to the mated female is also adaptive in that, given a situation suggesting a high probability of adultery (i.e., the presence of the model near the female) and assuming that replacement females are available, obtaining a new mate would enhance the fitness of males. . . . The decline in male-female aggressiveness during incubation and fledgling stages could be attributed to the impossibility of being cuckolded after the eggs have been laid. . . . The results are consistent with an evolutionary interpretation.

They are indeed consistent, but what about an obvious alternative, dismissed without test by Barash? Male returns at times two and three, approaches the model, tests it a bit, recognizes it as the same phoney he saw before, and doesn't bother his female. Why not at least perform the obvious test for this alternative to a conventional adaptive story: expose a male to the model for the *first* time after the eggs are laid?

After we criticized Barash's work, Morton *et al.* (1978) repeated it, with some variations (including the introduction of a female model), in the closely related eastern bluebird *Sialia sialis.* "We hoped to confirm," they wrote, that Barash's conclusions represent "a widespread evolutionary reality, at least within the genus *Sialia.* Unfortunately, we were unable to do so." They found no "anticuckoldry" behavior at all: males never approached their females aggressively after testing the model at any nesting stage. Instead, females often approached the male model and, in any case, attacked female models more than males attacked male models. "This violent response resulted in the near destruction of the female model after presentations and its complete demise on the third, as a female flew off with the model's head early in the experiment to lose it for us in the brush" (1978, p. 969). Yet, instead of calling Barash's selected story into question, they merely devise one of their own to render both results in the adaptationist mode. Perhaps, they conjecture, replacement females are scarce in their species and abundant in Barash's. Since Barash's males can replace a potentially "unfaithful" female, they can afford to be choosy and possessive. Eastern bluebird males are stuck with uncommon mates and had best be respectful. They conclude: "If we did not support Barash's suggestion that male bluebirds show anticuckoldry adaptations, we suggest that both studies still had 'results that are consistent with the expectations of evolutionary theory' (Barash 1976, p. 1099), as we presume any careful study would." But what good is a theory that cannot fail in careful study (since by "evolutionary theory," they clearly mean the action of natural selection applied to particular cases, rather than the fact of transmutation itself)?

The Master's Voice Reexamined

Since Darwin has attained sainthood (if not divinity) among evolutionary biologists, and since all sides invoke God's allegiance, Darwin has often been depicted as a radical selectionist at heart who invoked other mechanisms only in retreat, and only as a result of his age's own lamented ignorance about the mechanisms of heredity. This view is false. Although Darwin regarded selection as the most important of evolutionary mechanisms (as do we), no argument from opponents angered him more than the common attempt to caricature and trivialize his theory by stating that it relied exclusively upon natural selection. In the last edition of the *Origin,* he wrote (1872, p. 395):

> As my conclusions have lately been much misrepresented, and it has been stated that I attribute the modification of species exclusively to natural selection, I may be permitted to remark that in the first edition of this work, and subsequently, I placed in a most conspicuous position – namely at the close of the Introduction – the following words: "I am convinced that natural selection has been the main, but not the exclusive means of modification." This has been of no avail. Great is the power of steady misinterpretation.

Romanes, whose once famous essay (1900) on Darwin's pluralism versus the panselectionism of Wallace and Weismann deserves a resurrection, noted of this passage (1900, p. 5): "In the whole range of Darwin's writings there cannot be found a passage so strongly worded as this: it presents the only note of bitterness in all the thousands of pages which he has published." Apparently, Romanes did not know the letter Darwin wrote to *Nature* in 1880, in which he castigated Sir Wyville Thomson for caricaturing his theory as panselectionist (1880, p. 32):

> I am sorry to find that Sir Wyville Thomson does not understand the principle of natural selection. . . . If he had done so, he could not have written the following sentence in the Introduction to the Voyage of the Challenger: "The character of the abyssal fauna refuses to give the least support to the theory which refers the evolution of species to extreme variation guided only by natural selection." This is a standard of criticism not uncommonly reached by theologians and metaphysicians when they write on scientific subjects, but is something new as coming from a naturalist. . . . Can Sir Wyville Thomson name any one who has said that the evolution of species depends only on natural selection? As far

as concerns myself, I believe that no one has brought forward so many observations on the effects of the use and disuse of parts, as I have done in my "Variation of Animals and Plants under Domestication"; and these observations were made for this special object. I have likewise there adduced a considerable body of facts, showing the direct action of external conditions on organisms.

We do not now regard all of Darwin's subsidiary mechanisms as significant or even valid, though many, including direct modification and correlation of growth, are very important. But we should cherish his consistent attitude of pluralism in attempting to explain Nature's complexity.

A Partial Typology of Alternatives to the Adaptationist Program

In Darwin's pluralistic spirit, we present an incomplete hierarchy of alternatives to immediate adaptation for the explanation of form, function, and behavior.

1. No adaptation and no selection at all. At present, population geneticists are sharply divided on the question of how much genetic polymorphism within populations and how much of the genetic differences between species is, in fact, the result of natural selection as opposed to purely random factors. Populations are finite in size, and the isolated populations that form the first step in the speciation process are often founded by a very small number of individuals. As a result of this restriction in population size, frequencies of alleles change by *genetic drift*, a kind of random genetic sampling error. The stochastic process of change in gene frequency by random genetic drift, including the very strong sampling process that goes on when a new isolated population is formed from a few immigrants, has several important consequences. First, populations and species will become genetically differentiated, and even fixed for different alleles at a locus in the complete absence of any selective force at all.

Second, alleles can become fixed in a population *in spite of natural selection*. Even if an allele is favored by natural selection, some proportion of populations, depending upon the product

of population size N and selection intensity s, will become homozygous for the less fit allele because of genetic drift. If Ns is large, this random fixation for unfavorable alleles is a rare phenomenon, but if selection coefficients are on the order of the reciprocal of population size ($Ns = 1$) or smaller, fixation for deleterious alleles is common. If many genes are involved in influencing a metric character like shape, metabolism, or behavior, then the intensity of selection on each locus will be small and Ns per locus may be small. As a result, many of the loci may be fixed for nonoptimal alleles.

Third, new mutations have a small chance of being incorporated into a population, even when selectively favored. Genetic drift causes the immediate loss of most new mutations after their introduction. With a selection intensity s, a new favorable mutation has a probability of only $2s$ of ever being incorporated. Thus one cannot claim that, eventually, a new mutation of just the right sort for some adaptive argument will occur and spread. "Eventually" becomes a very long time if only one in 1,000 or one in 10,000 of the "right" mutations that do occur ever get incorporated in a population.

2. No adaptation and no selection on the part at issue; form of the part is a correlated consequence of selection directed elsewhere. Under this important category, Darwin ranked his "mysterious" laws of the "correlaton of growth." Today, we speak of pleiotropy, allometry, "material compensation" (Rensch 1959, pp. 179–187) and mechanically forced correlations in D'Arcy Thompson's sense (1942; Gould 1971). Here we come face to face with organisms as integrated wholes, fundamentally not decomposable into independent and separately optimized parts.

Although allometric patterns are as subject to selection as static morphology itself (Gould 1966), some regularities in relative growth are probably not under immediate adaptive control. For example, we do not doubt that the famous 0.66 interspecific allometry of brain size in all major vertebrate groups represents a selected "design criterion," though its significance remains elusive (Jerison 1973). It is too repeatable across too wide a taxonomic range to represent much else than a series of creatures similarly well designed for their different sizes. But another common allometry, the 0.2 to 0.4 intraspecific scaling among homeothermic adults differing in body size, or among races within a species, probably does not require a selectionist story, though many, including one of us, have tried to provide one (Gould 1974). R. Lande (personal communication) has used the experiments of Falconer (1973) to show that selection upon *body size alone* yields a brain-body slope across generations of 0.35 in mice.

More compelling examples abound in the literature on selection for altering the timing of maturation (Gould 1977). At least three times in the evolution of arthropods (mites, flies, and beetles), the same complex adaptation has evolved, apparently for rapid turnover of generations in strongly r-selected feeders on superabundant but ephemeral fungal resources: females reproduce as larvae and grow the next generation within their bodies. Offspring eat their mother from inside and emerge from her hollow shell, only to be devoured a few days later by their own progeny. It would be foolish to seek adaptive significance in paedomorphic morphology per se; it is primarily a by-product of selection for rapid cycling of generations. In more interesting cases, selection for small size (as in animals of the interstitial fauna) or rapid maturation (dwarf males of many crustaceans) has occurred by progenesis (Gould 1977, pp. 324–336), and descendant adults contain a mixture of ancestral juvenile and adult features. Many biologists have been tempted to find primary adaptive meaning for the mixture, but it probably arises as a by-product of truncated maturation, leaving some features "behind" in the larval state, while allowing others, more strongly correlated with sexual maturation, to retain the adult configuration of ancestors.

3. The decoupling of selection and adaptation:

(i) Selection without adaptation. Lewontin (1979) has presented the following hypothetical example: "A mutation which doubles the fecundity of individuals will sweep through a population rapidly. If there has been no change in efficiency of resource utilization, the individuals will leave no more offspring than before, but simply lay twice as many eggs, the excess dying because of resource limitation. In what sense are the individuals or the population as a whole

better adapted than before? Indeed, if a predator on immature stages is led to switch to the species now that immatures are more plentiful, the population size may actually decrease as a consequence, yet natural selection at all times will favour individuals with higher fecundity."

(ii) Adaptation without selection. Many sedentary marine organisms, sponges and corals in particular, are well adapted to the flow regimes in which they live. A wide spectrum of "good design" may be purely phenotypic in origin, largely induced by the current itself. (We may be sure of this in numerous cases, when genetically identical individuals of a colony assume different shapes in different microhabitats.) Larger patterns of geographic variation are often adaptive and purely phenotypic as well. Sweeney and Vannote (1978), for example, showed that many hemimetabolous aquatic insects reach smaller adult size with reduced fecundity when they grow at temperatures above and below their optima. Coherent, climatically correlated patterns in geographic distribution for these insects – so often taken as a priori signs of genetic adaptation – may simply reflect this phenotypic plasticity.

"Adaptation" – the good fit of organisms to their environment – can occur at three hierarchical levels with different causes. It is unfortunate that our language has focused on the common result and called all three phenomena "adaptation": the differences in process have been obscured, and evolutionists have often been misled to extend the Darwinian mode to the other two levels as well. First, we have what physiologists call "adaptation": the phenotypic plasticity that permits organisms to mold their form to prevailing circumstances during ontogeny. Human "adaptations" to high altitude fall into this category (while others, like resistance of sickling heterozygotes to malaria, are genetic, and Darwinian). Physiological adaptations are not heritable, though the capacity to develop them presumably is. Second, we have a "heritable" form of non-Darwinian adaptation in humans (and, in rudimentary ways, in a few other advanced social species): cultural adaptation (with heritability imposed by learning). Much confused thinking in human sociobiology arises from a failure to distinguish this mode from Darwinian adaptation based on genetic variation. Finally, we have adaptation arising from the conventional Darwinian mechanism of selection upon genetic variation. The mere existence of a good fit between organism and environment is insufficient for inferring the action of natural selection.

4.   Adaptation and selection but no selective basis for differences among adaptations. Species of related organisms, or subpopulations within a species, often develop different adaptations as solutions to the same problem. When "multiple adaptive peaks" are occupied, we usually have no basis for asserting that one solution is better than another. The solution followed in any spot is a result of history; the first steps went in one direction, though others would have led to adequate prosperity as well. Every naturalist has his favorite illustration. In the West Indian land snail *Cerion*, for example, populations living on rocky and windy coasts almost always develop white, thick, and relatively squat shells for conventional adaptive reasons. We can identify at least two different developmental pathways to whiteness from the mottling of early whorls in all *Cerion*, two paths of thickened shells and three styles of allometry leading to squat shells. All twelve combinations can be identified in Bahamian populations, but would it be fruitful to ask why – in the sense of optimal design rather than historical contingency – *Cerion* from eastern Long Island evolved one solution, and *Cerion* from Acklins Island another?

5.   Adaptation and selection, but the adaptation is a secondary utilization of parts present for reasons of architecture, development, or history. We have already discussed this neglected subject in the first section on spandrels, spaces, and cannibalism. If blushing turns out to be an adaptation affected by sexual selection in humans, it will not help us to understand why blood is red. The immediate utility of an organic structure often says nothing at all about the reason for its being.

Another, and Unfairly Maligned, Approach to Evolution

In continental Europe, evolutionists have never been much attracted to the Anglo-American penchant for atomizing organisms into parts and trying to explain each as a direct adaptation. Their general alternative exists in both a strong and a weak form. In the strong form, as advocated by such major theorists as Schindewolf (1950), Remane (1971), and Grassé (1977), natural selection under the adaptationist program can explain superficial modifications of the *Bauplan* that fit structure to environment: why moles are blind, giraffes have long necks, and ducks webbed feet, for example. But the important steps of evolution, the construction of the *Bauplan* itself and the transition between *Baupläne*, must involve some other unknown, and perhaps "internal," mechanism. We believe that English biologists have been right in rejecting this strong form as close to an appeal to mysticism.

But the argument has a weaker – and paradoxically powerful – form that has not been appreciated, but deserves to be. It also acknowledges conventional selection for superficial modifications of the *Bauplan*. It also denies that the adaptationist program (atomization plus optimizing selection on parts) can do much to explain *Baupläne* and the transitions between them. But it does not therefore resort to a fundamentally unknown process. It holds instead that the basic body plans of organisms are so integrated and so replete with constraints upon adaptation (categories 2 and 5 of our typology) that conventional styles of selective arguments can explain little of interest about them. It does not deny that change, when it occurs, may be mediated by natural selection, but it holds that constraints restrict possible paths and modes of change so strongly that the constraints themselves become much the most interesting aspect of evolution.

Rupert Riedl, the Austrian zoologist who has tried to develop this thesis for English audiences (1977 and 1975, translated into English by R. Jeffries in 1978) writes:

> The living world happens to be crowded by universal patterns of organization which, most obviously, find no direct explanation through environmental conditions or adaptive radiation, but exist primarily through universal requirements which can only be expected under the systems conditions of complex organization itself. . . . This is not self-evident, for the whole of the

huge and profound thought collected in the field of morphology, from Goethe to Remane, has virtually been cut off from modern biology. It is not taught in most American universities. Even the teachers who could teach it have disappeared.

Constraints upon evolutionary change may be ordered into at least two categories. All evolutionists are familiar with *phyletic* constraints, as embodied in Gregory's classic distinction (1936) between habitus and heritage. We acknowledge a kind of phyletic inertia in recognizing, for example, that humans are not optimally designed for upright posture because so much of our *Bauplan* evolved for quadrupedal life. We also invoke phyletic constraint in explaining why no molluscs fly in air and no insects are as large as elephants.

Developmental constraints, a subcategory of phyletic restrictions, may hold the most powerful rein of all over possible evolutionary pathways. In complex organisms, early stages of ontogeny are remarkably refractory to evolutionary change, presumably because the differentiation of organ systems and their integration into a functioning body is such a delicate process so easily derailed by early errors with accumulating effects. Von Baer's fundamental embryological laws (1828) represent little more than a recognition that early stages are both highly conservative and strongly restrictive of later development. Haeckel's biology law, the primary subject of late nineteenth-century evolutionary biology, rested upon a misreading of the same data (Gould 1977). If development occurs in integrated packages and cannot be pulled apart piece by piece in evolution, then the adaptationist program cannot explain the alteration of developmental programs underlying nearly all changes of *Bauplan*.

The German palaeontologist A. Seilacher, whose work deserves far more attention than it has received, has emphasized what he calls "*bautechnischer*, or *architectural*, constraints" (Seilacher 1970). These arise not from former adaptations retained in a new ecological setting (phyletic constraints as usually understood), but as architectural restrictions that never were adaptations but rather were the necessary consequences of materials and designs selected to build basic *Baupläne*. We devoted the first section of this chapter to nonbiological examples in this category. Spandrels must exist once a blueprint specifies that a dome shall rest on rounded arches. Architectural constraints can exert a far-ranging influence upon organisms as well. The subject is full of potential insight because it has rarely been acknowledged at all.

In a fascinating example, Seilacher (1972) has shown that the divaricate form of architecture occurs again and again in all groups of molluscs, and in brachiopods as well. This basic form expresses itself in a wide variety of structures: raised ornamental lines (not growth lines because they do not conform to the mantle margin at any time), patterns of coloration, internal structures in the mineralization of calcite and incised grooves. He does not know what generates this pattern and feels that traditional and nearly exclusive focus on the adaptive value of each manifestation has diverted attention from questions of its genesis in growth and also prevented its recognition as a general phenomenon. It must arise from some characteristic pattern of inhomogeneity in the growing mantle, probably from the generation of interference patterns around regularly spaced centers; simple computer simulations can generate the form in this manner (Waddington and Cowe 1969). The general pattern may not be a direct adaptation at all.

Seilacher then argues that most manifestations of the pattern are probably nonadaptive. His reasons vary but seem generally sound to us. Some are based on field observations: color patterns that remain invisible because clams possessing them either live buried in sediments or remain covered with a periostracum so thick that the colors cannot be seen. Others rely on more general principles: presence only in odd and pathological individuals, rarity as a developmental anomaly, excessive variability compared with much reduced variability when the same general structure assumes a form judged functional on engineering grounds.

In a distinct minority of cases, the divaricate pattern becomes functional in each of the four categories. Divaricate ribs may act as scoops and anchors in burrowing (Stanley 1970), but they are not properly arranged for such function in most clams. The color chevrons are mimetic in one species (*Pteria zebra*) that lives on hydrozoan branches; here the variability is strongly reduced.

The mineralization chevrons are probably adaptive in only one remarkable creature, the peculiar bivalve *Corculum cardissa* (in other species they either appear in odd specimens or only as post-mortem products of shell erosion). This clam is uniquely flattened in an anterioposterior direction. It lies on the substrate, posterior up. Distributed over its rear end are divaricate triangles of mineralization. They are translucent, while the rest of the shell is opaque. Under these windows dwell endosymbiotic algae!

All previous literature on divaricate structure has focused on its adaptive significance (and failed to find any in most cases). But Seilacher is probably right in representing this case as the spandrels, ceiling holes, and sacrificed bodies of our first section. The divaricate pattern is a fundamental architectural constraint. Occasionally, since it is there, it is used to beneficial effect. But we cannot understand the pattern or its evolutionary meaning by viewing these infrequent and secondary adaptations as a reason for the pattern itself.

Galton (1909, p. 257) contrasted the adaptationist program with a focus on constraints and modes of development by citing a telling anecdote about Herbert Spencer's fingerprints:

Much has been written, but the last word has not been said, on the rationale of these curious papillary ridges; why in one man and in one finger they form whorls and in another loops. I may mention a characteristic anecdote of Herbert Spencer in connection with this. He asked me to show him my Laboratory and to take his prints, which I did. Then I spoke of the failure to discover the origin of these patterns, and how the fingers of unborn children had been dissected to ascertain their earliest stages, and so forth. Spencer remarked that this was beginning in the wrong way; that I ought to consider the purpose the ridges had to fulfil, and to work backwards. Here, he said, it was obvious that the delicate mouths of the sudorific glands required the protection given to them by the ridges on either side of them, and therefrom he elaborated a consistent and ingenious hypothesis at great length, I replied that his arguments were beautiful and deserved to be true, but it happened that the mouths of the ducts did not run in the valleys between the crests, but along the crests of the ridges themselves.

We feel that the potential rewards of abandoning exclusive focus on the adaptationist program are very great indeed. We do not offer a counsel of despair, as adaptationists have charged; for nonadaptive does not mean non-intelligible. We welcome the richness that a pluralistic approach, so akin to Darwin's spirit, can provide. Under the adaptationist program, the great historic themes of developmental morphology and *Bauplan* were largely abandoned; for if selection can break any correlation and optimize parts separately, then an organism's integration counts for little. Too often, the adaptationist program gave us an evolutionary biology of parts and genes, but not of organisms. It assumed that all transitions could occur step by step and underrated the importance of integrated developmental blocks and pervasive constraints of history and architecture. A pluralistic view could put organisms, with all their recalcitrant yet intelligible complexity, back into evolutionary theory.

References

Baer, K. E. von. 1828. *Entwicklungsgeschichte der Tiere*, Königsberg: Bornträger.

Barash, D. P. 1976. Male response to apparent female adultery in the mountain-bluebird: an evolutionary interpretation, *Am. Nat.*, 110: 1097–1101.

Coon, C. S., Carn, S. M., and Birdsell, J. B. 1950. *Races*, Springfield Ohio, C. Thomas.

Costa, R., and Bisol, P. M. 1978. Genetic variability in deep sea organisms, *Biol. Bull.*, 155: 125–133.

Darwin, C. 1872. *The origin of species*, London, John Murray.

Darwin, C. 1880. Sir Wyville Thomson and natural selection, *Nature*, London, 23: 32.

Davitashvili, L. S. 1961. *Teoriya polovogo othora* (Theory of sexual selection), Moscow, Akademil Nauk.

Falconer, D. S. 1973. Replicated selection for body weight in mice, *Genet. Res.*, 22: 291–321.

Galton, F. 1909. *Memories of my life*, London, Methuen.

Gould, S. J. 1966. Allometry and size in ontogeny and phylogeny, *Biol. Rev.*, 41: 587–640.

Gould, S. J. 1971. D'Arcy Thompson and the science of form, *New Literary Hisl.*, 2, No. 2, 229–258.

Gould, S. J. 1974. Allometry in primates, with emphasis on scaling and the evolution of the brain. In *Approaches to primate paleobiology*, *Contrib. Primatol.*, 5: 244–292.

Gould, S. J. 1977. *Ontogeny and phylogeny*, Cambridge, Ma., Belknap Press.

Gould, S. J. 1978. Sociobiology: the art of storytelling, *New Scient.*, 80: 530–533.

Grassé, P. P. 1977. *Evolution of living organisms*, New York, Academic Press.

Gregory, W. K. 1936. Habitus factors in the skeleton fossil and recent mammals, *Proc. Am. phil. Soc.*, 76: 429–444.

Harner, M. 1977. The ecological basis for Aztec sacrifice. *Am. Ethnologist*, 4: 117–135.

Jerison, H. J. 1973. *Evolution of the brain and intelligence*, New York, Academic Press.

Lande, R. 1976. Natural selection and random genetic drift in phenotypic evolution, *Evolution*, 30: 314–334.

Lande, R. 1978. Evolutionary mechanisms of limb loss in tetrapods, *Evolution*, 32: 73–92.

Lewontin, R. C. 1978. Adaptation, *Scient. Am.*, 239 (3): 156–169.

Lewontin, R. C. 1979. Sociobiology as an adaptationist program, *Behav. Sci.*, 24: 5–14.

Morton, E. S., Geitgey, M. S., and McGrath, S. 1978. On bluebird "responses to apparent female adultery." *Am. Nat.*, 112: 968–971.

Ortiz de Montellano, B. R. 1978. Aztec cannibalism: an ecological necessity? *Science*, 200: 611–617.

Remane, A. 1971. Die *Grundlagen des natürlichen Systems der vergleichenden Anatomie und der Phylogenetik.* Königstein-Taunus: Koeltz.

Rensch, B. 1959. *Evolution above the species level*, New York, Columbia University Press.

Riedel, R. 1975. *Die Ordnung des Lebendigen*, Hamburg, Paul Parey, tr. R. P. S. Jefferies, *Order in living systems: A systems analysis of evolution*, New York, Wiley, 1978.

Riedel, R. 1977. A systems-analytical approach to macro-evolutionary phenomena, *Q. Rev. Biol.*, 52: 351–370.

Romanes, G. J. 1900. The Darwinism of Darwin and of the post-Darwinian schools. In *Darwin, and after Darwin*, vol. 2, new ed., London, Longmans, Green and Co.

Rudwick, M. J. S. 1964. The function of zig-zag deflections In the commissures of fossil brachlopods, *Palaeontology*, 7: 135–171.

Sahlins, M. 1978. Culture as protein and profit, *New York Review of Books*, 23: Nov., pp. 45–53.

Schindewolf, O. H. 1950. *Grundfragen der Paläontologie*, Stuttgart. Schweizerbart.

Seilacher, A. 1970. Arbeitskonzept zur Konstruktionsmorphologie, *Lethaia*, 3: 393–396.

Seilacher, A. 1972. Divaricate patterns in pelecypod shells, *Lethaia*, 5: 325–343.

Shea, B. T. 1977. Eskimo craniofacial morphology, cold stress and the maxillary sinus, *Am. J. Phys. Anthrop.*, 47: 289–300.

Stanley, S. M. 1970. Relation of shell form to life habits in the Bivalvia (Mollusca). *Mem. Geol. Soc. Am.*, no. 125, 296 pp.

Sweeney, B. W., and Vannote, R. L. 1978. Size variation and the distribution of hemimetabolous aquatic insects: two thermal equilibrium hypotheses. *Science*, 200: 444–446.

Thompson, D. W. 1942. *Growth and form*, New York, Macmillan.

Waddington, C. H., and Cowe, J. R. 1969. Computer simulation of a molluscan pigmentation pattern, *J. Theor. Biol.*, 25: 219–225.

Wallace, A. R. 1899. *Darwinism*, London, Macmillan.

Wilson, E. O. 1978. *On human nature*, Cambridge, Ma., Harvard University Press.

10

How to Carry Out the Adaptationist Program?

Ernst Mayr

To have been able to provide a scientific explanation of adaptation was perhaps the greatest triumph of the Darwinian theory of natural selection. After 1859 it was no longer necessary to invoke design, a supernatural agency, to explain the adaptation of organisms to their environment. It was the daily, indeed hourly, scrutiny of natural selection, as Darwin had said, that inevitably led to ever greater perfection. Ever since then it has been considered one of the major tasks of the evolutionist to demonstrate that organisms are indeed reasonably well adapted, and that this adaptation could be caused by no other agency than natural selection. Nevertheless, beginning with Darwin himself (remember his comments on the evolution of the eye), evolutionists have continued to worry about how valid this explanation is. The more generally natural selection was accepted after the 1930s, and the more clearly the complexity of the genotype was recognized, particularly after the 1960s, the more often the question was raised as to the meaning of the word *adaptation*. The difficulty of the concept adaptation is best documented by the incessant efforts of authors to analyze it, describe it, and define it. Since I can do no better myself, I refer to a sample of such efforts (Bock and von Wahlert 1965; Bock 1980; Brandon 1978; Dobzhansky 1956, 1968; Lewontin 1978, 1979; Muller 1949; Munson 1971; Stern 1970; Williams 1966; Wright 1949). The one thing about which modern authors are unanimous is that adaptation is not teleological, but refers to something produced in the past by natural selection. However, since various forms of selfish selection (e.g., meiotic drive, many aspects of sexual selection) may produce changes in the phenotype that could hardly be classified as "adaptations," the definition of adaptation must include some reference to the selection forces effected by the inanimate and living environment. It surely cannot have been anything but a lapse when Gould wanted to deny the designation "adaptation" to certain evolutionary innovations in clams, with this justification; "The first clam that fused its mantle margins or retained its byssus to adulthood may have gained a conventional adaptive benefit in its local environment. But it surely didn't know that its invention would set the stage for future increases in diversity" (Gould and Calloway 1980, p. 395). Considering the strictly a posteriori nature of an adaptation, its potential for the future is completely irrelevant, as far as the definition of the term adaptation is concerned.

Ernst Mayr, "How to Carry Out the Adaptationist Program?" *The American Naturalist* (1983) 121, 324–334. Reprinted by permission of the publisher, the University of Chicago Press.

An early draft of the manuscript was read by W. Bock, S. J. Gould, and R. Lewontin. I am indebted to them for numerous suggestions for changes, not all of which I was able to adopt.

A program of research devoted to demonstrate the adaptedness of individuals and their characteristics is referred to by Gould and Lewontin (1979) as an "adaptationist program." A far more extreme definition of this term was suggested by Lewontin (1979, p. 6) to whom the adaptationist program "assumes without further proof that all aspects of the morphology, physiology and behavior of organisms are adaptive optimal solutions to problems." Needless to say, in the ensuing discussion I am not defending such a sweeping ideological proposition.

When asking whether or not the adaptationist program is a legitimate scientific approach, one must realize that the method of evolutionary biology is in some ways quite different from that of the physical sciences. Although evolutionary phenomena are subject to universal laws, as are most phenomena in the physical sciences, the explanation of the history of a particular evolutionary phenomenon can be given only as a "historical narrative." Consequently, when one attempts to explain the features of something that is the product of evolution, one must attempt to reconstruct the evolutionary history of this feature. This can be done only by inference. The most helpful procedure in an analysis of historical narratives is to ask "why" questions; that is, questions (to translate this into modern evolutionary language) which ask what is or might have been the selective advantage that is responsible for the presence of a particular feature.

The adaptationist program has recently been vigorously attacked by Gould and Lewontin (1979) in an analysis which in many ways greatly pleases me, not only because they attack the same things that I questioned in my "bean bag genetics" paper (Mayr 1959), but also because they emphasize the holistic aspects of the genotype as I did repeatedly in discussions of the unity of the genotype (Mayr 1970, chap. 10; 1975). Yet I consider their analysis incomplete because they fail to make a clear distinction between the pitfalls of the adaptationist program as such and those resulting from a reductionist or atomistic approach in its implementation. I will try to show that basically there is nothing wrong with the adaptationist program, if properly executed, and that the weaknesses and deficiencies quite rightly pointed out by Gould and Lewontin are the result of atomistic and deterministic approaches.

In the period after 1859 only five major factors were seriously considered as the causes of evolutionary change, or, as they are sometimes called, the agents of evolution. By the time of the evolutionary synthesis (by the 1940s), three of these factors had been so thoroughly discredited and falsified that they are now no longer considered seriously by evolutionists. These three factors are: inheritance of acquired characters, intrinsic directive forces (orthogenesis, etc.), and saltational evolution (de Vriesian mutations, hopeful monsters, etc.). This left only two evolutionary mechanisms as possible causes of evolutionary change (including adaptation), chance, and selection forces. The identification of these two factors as the principal causes of evolutionary change by no means completed the task of the evolutionist. As is the case with most scientific problems, this initial solution represented only the first orientation. For completion it requires a second stage, a fine-grained analysis of these two factors: What are the respective roles of chance and of natural selection, and how can this be analyzed?

Let me begin with chance. Evolutionary change in every generation is a two-step process, the production of genetically unique new individuals and the selection of the progenitors of the next generation. The important role of chance at the first step, the production of variability, is universally admitted (Mayr 1962), but the second step, natural selection, is on the whole viewed rather deterministically: Selection is a non-chance process. What is usually forgotten is what an important role chance plays even during the process of selection. In a group of sibs it is by no means necessarily only those with the most superior genotypes that will reproduce. Predators mostly take weak or sick prey individuals but not exclusively so, nor do localized natural catastrophes (storms, avalanches, floods, etc.) kill only inferior individuals. Every founder population is largely a chance aggregate of individuals and the outcome of genetic revolutions, initiating new evolutionary departures, may depend on chance constellations of genetic factors. There is a large element of chance in every successful colonization. When multiple pathways toward the acquisition of a new adaptive trait are possible, it is often a matter of a momentary constellation of chance factors as to which one will be taken (Bock 1959).

When one attempts to determine for a given trait whether it is the result of natural selection or of chance (the incidental byproduct of stochastic processes), one is faced by an epistemological dilemma. Almost any change in the course of evolution might have resulted by chance. Can one ever prove this? Probably never. By contrast, can one deduce the probability of causation by selection? Yes, by showing that possession of the respective feature would be favored by selection. It is this consideration which determines the approach of the evolutionist. He must first attempt to explain biological phenomena and processes as the product of natural selection. Only after all attempts to do so have failed, is he justified in designating the unexplained residue tentatively as a product of chance.

The evaluation of the impact of selection is a very difficult task. It has been demonstrated by numerous selection experiments that selection is not a phantom. That it also operates in nature is a conclusion that is usually based only on inference, but that is increasingly often experimentally confirmed. Very convincing was Bates' demonstration that the geographic variation of mimics parallels exactly that of their distasteful or poisonous models. The agreement of desert animals with the variously colored substrate also strongly supports the power of selection. In other cases the adaptive value of a trait is by no means immediately apparent.

As a consequence of the adaptationist dilemma, when one selectionist explanation of a feature has been discredited, the evolutionist must test other possible adaptationist solutions before he can resign and say: This phenomenon must be a product of chance. Gould and Lewontin ridicule the research strategy: "If one adaptive argument fails, try another one." Yet the strategy to try another hypothesis when the first fails is a traditional methodology in all branches of science. It is the standard in physics, chemistry, physiology, and archeology. Let me merely mention the field of avian orientation in which sun compass, sun map, star navigation, Coriolis force, magnetism, olfactory clues, and several other factors were investigated sequentially in order to explain as yet unexplained aspects of orientation and homing. What is wrong in using the same methodology in evolution research?

At this point it may be useful to look at the concept of adaptation from a historical point of view. When Darwin introduced natural selection as the agent of adaptation, he did so as a replacement for supernatural design. Design, as conceived by the natural theologians, had to be perfect, for it was unthinkable that God would make something that was less than perfect. It was on the basis of this tradition that the concept of natural selection originated. Darwin gave up this perfectionist concept of natural selection long before he wrote the *Origin*. Here he wrote, "Natural selection tends only to make each organic being as perfect as, or slightly more perfect than, the other inhabitants of the same country with which it has to struggle for existence. And we see that this is the degree of perfection attained under nature" (1859, p. 201). He illustrated this with the biota of New Zealand, the members of which "are perfect . . . compared with another" (p. 201), but "rapidly yielding" (p. 201) to recent colonists and invaders. After Darwin, some evolutionists forgot the modesty of Darwin's claims, but other evolutionists remained fully aware that selection cannot give perfection, by observing the ubiquity of extinction and of physiological and morphological insufficiencies. However, the existence of some perfectionists has served Gould and Lewontin as the reason for making the adaptationist program the butt of their ridicule and for calling it a Panglossian paradigm. Here I dissent vigorously. To imply that the adaptationist program is one and the same as the argument from design (satirized by Voltaire in *Candide*) is highly misleading. When *Candide* was written (in 1759), a concept of evolution did not yet exist and those who believed in a benign creator had no choice but to believe that everything "had to be for the best." This is the Panglossian paradigm, the invalidity of which has been evident ever since the demise of natural theology. The adaptationist program, a direct consequence of the theory of natural selection, is something fundamentally different. Parenthetically one might add that Voltaire misrepresented Leibniz rather viciously. Leibniz had not claimed that this is the best possible world, but only that it is the best of the possible worlds. Curiously one can place an equivalent limitation on selection (see below). Selection does not produce perfect genotypes, but it favors the best which the numerous constraints upon it allow. That such constraints exist was ignored by those evolutionists who interpreted every trait of an organism as an ad hoc adaptation.

The attack directed by Gould and Lewontin against unsupported adaptationist explanations in the literature is fully justified. But the most absurd among these claims were made several generations ago, not by modern evolutionists. Gould and Lewontin rightly point out that some traits, for instance the gill arches of mammalian embryos, had been acquired as adaptations of remote ancestors but, even though they no longer serve their original function, they are not eliminated because they have become integral components of a developmental system. Most so-called vestigial organs are in this category. Finally, it would indeed be absurd to atomize an organism into smaller and smaller traits and to continue to search for the ad hoc adaptation of each smallest component. But I do not think that this is the research program of the majority of evolutionists. Dobzhansky well expressed the proper attitude when saying: "It cannot be stressed too often that natural selection does not operate with separate 'traits.' Selection favors genotypes ... The reproductive success of a genotype is determined by the totality of the traits and qualities which it produces in a given environment" (1956, p. 340). What Dobzhansky described reflects what I consider to be the concept of the adaptationist program accepted by most evolutionists, and I doubt that the characterization assigned to the adaptations program by Gould and Lewontin, "An organism is atomized into traits and these traits are explained as structures optimally designed by natural selection for their functions" (p. 585), represents the thinking of the average evolutionist.

By choosing this atomistic definition of the adaptationist program and by their additional insistence that the adaptive control of every trait must be "immediate," Gould and Lewontin present a picture of the adaptations program that is indeed easy to ridicule. The objections cited by them are all based on their reductionist definition. Of course, it is highly probable that not all secondary byproducts of relative growth are "under immediate adaptive control." In the case of multiple pathways it is, of course, not necessary that every morphological detail in a

convergently acquired adaptation be an ad hoc adaptation. This is true, for instance, in the case cited by them, of the adaptive complex for a rapid turnover of generations that evolved at least three times independently in the evolution of the arthropods. Evolution is opportunistic and natural selection makes use of whatever variation it encounters. As Jacob (1977) has said so rightly: "Natural selection does not work like an engineer. It works like a tinkerer."

Considering the evident dangers of applying the adaptationist program incorrectly, why are the Darwinians nevertheless so intent on applying it? The principal reason for this is its great heuristic value. The adaptationist question, "What is the function of a given structure or organ?" has been for centuries the basis for every advance in physiology. If it had not been for the adaptationist program, we probably would still not yet know the functions of thymus, spleen, pituitary, and pineal. Harvey's question "Why are there valves in the veins?" was a major stepping stone in his discovery of the circulation of blood. If one answer turned out to be wrong, the adaptationist program demanded another answer until the true meaning of the structure was established or until it could be shown that this feature was merely an incidental byproduct of the total genotype. It would seem to me that there is nothing wrong with the adaptationist program, provided it is properly applied.

Consistent with the modern theory of science, adaptationist hypotheses allow a falsification in most cases. For instance, there are numerous ways to test the thesis that the differences in beak dimensions of a pair of species of Darwin's finches on a given island in the Galapagos is the result of competition (Darwin's character divergence). One can correlate size of preferred seeds with bill size and study how competition among different assortments of sympatric species of finches affects bill size. Finally, one can correlate available food resources on different islands with population size (Boag and Grant 1981). As a result of such studies the adaptationist program leads in this case to a far better understanding of the ecosystem.

The case of the beak differences of competing species of finches is one of many examples in which it is possible, indeed necessary, to investigate the adaptive significance of individual traits. I emphasize this because someone might conclude from the preceding discussion that a dissection of the phenotype into individual characters is inappropriate in principle. To think so would be a mistake. A more holistic approach is appropriate only when the analysis of individual traits fails to reveal an adaptive significance.

What has been rather neglected in the existing literature is the elaboration of an appropriate methodology to establish adaptive significance. In this respect a recent analysis by Traub (1980) on adaptive modifications in fleas is exemplary. Fleas are adorned with a rich equipment of hairs, bristles, and spines, some of which are modified into highly specialized organs. What Traub (and various authors before him) found is that unrelated genera and species of fleas often acquire convergent specializations on the same mammalian or avian hosts. The stiffness, length, and other qualities of the mammalian hair are species specific and evidently require special adaptations that are independently acquired by unrelated lineages of fleas. "The overall association [between bristles and host hair] is so profound that it is now possible to merely glance at a new genus or species of flea and make correct statements about some characteristic attributes of its host" (Traub 1980, p. 64). Basically, the methodology consists in establishing a tentative correlation between a trait and a feature of the environment, and then to analyze in a comparative study, other organisms exposed to the same feature of the environment and see whether they have acquired the same specialization. There are two possible explanations for a failure of confirmation of the correlation. Either the studied feature is not the result of a selection force or there are multiple pathways for achieving adaptedness.

When the expanded comparative study results in a falsification of the tentative hypothesis, and when other hypotheses lead to ambiguous results, it is time to think of experimental tests. Such tests are not only often possible, but indeed are now being made increasingly often, as the current literature reveals (Clarke 1979). Only when all such specific analyses to determine the possible adaptive value of the respective trait have failed, is it time to adopt a more holistic approach and to start thinking about the possible adaptive significance of a larger portion of the phenotype, indeed possibly of the Bauplan as a whole.

Thus, the student of adaptation has to sail a perilous course between a pseudoexplanatory reductionist atomism and stultifying nonexplanatory holism. When we study the literature, we find almost invariably that those who were opposed to nonexplanatory holism went too far in adopting atomism of the kind so rightly stigmatized by Gould and Lewontin, while those who were appalled by the simplistic and often glaringly invalid pseudoexplanations of the atomists usually took refuge in an agnostic holism and abandoned all further effort at explanation by invoking best possible compromise, or integral component of Bauplan, or incidental byproduct of the genotype. Obviously neither approach, if exclusively adopted, is an appropriate solution. How do Gould and Lewontin propose to escape from this dilemma?

While castigating the adaptationist program as a Panglossian paradigm, Gould and Lewontin exhort the evolutionists to follow Darwin's example by adopting a pluralism of explanations. As much as I have favored pluralism all my life, I cannot follow Darwin in this case and, as a matter of fact, neither do Gould and Lewontin themselves. For Darwin's pluralism, as is well known to the historians of science, consisted of accepting several mechanisms of evolution as alternatives to natural selection, in particular the effects of use and disuse and the direct action of external conditions on organisms. Since both of these subsidiary mechanisms of Darwin's are now thoroughly refuted, we have no choice but to fall back on the selectionist explanation.

Indeed, when we look at Gould and Lewontin's "alternatives to immediate adaptation," we find that all of them are ultimately based on natural selection, properly conceived. It is thus evident that the target of their criticism should have been neither natural selection nor the adaptationist program as such, but rather a faulty interpretation of natural selection and an improperly conducted adaptationist program. Gould and Lewontin's proposals (1979, pp. 590–593) are not "alternatives to the adaptationist program," but simply legitimate forms of it. Such an improved adaptationist program has long been the favored methodology of most evolutionists. There is a middle course available between a pseudoexplanatory reductionist atomism and an agnostic nonexplanatory holism. Dobzhansky

(1956) in his stress on the total developmental system and adjustment to a variable environment and my own emphasis on the holistic nature of the genotype (1963, chap. 10; 1970, chap. 10 [considerably revised]; 1975) have been attempts to steer such a middle course, to mention only two of numerous authors who adopted this approach. They all chose an adaptationist program, but not an extreme atomistic one.

Much of the recent work in evolutionary morphology is based on such a middle-course adaptationist program, for instance Bock's (1959) analysis of multiple pathways and my own work on the origin of evolutionary novelties (1960). A semiholistic adaptationist program often permits the explanation of seemingly counter-intuitive results of selection. For instance, the large species of albatrosses (*Diomedea*) have only a single young every second year and do not start breeding until they are 6 to 8 yr old. How could natural selection have led to such an extraordinarily low fertility for a bird? However, it could be shown that in the stormy waters of the south temperate and subantarctic zones only the most experienced birds have reproductive success and this in turn affects all other aspects of the life cycle. Under the circumstances the extraordinary reduction of fertility is favored by selection forces and hence is an adaptation (Lack 1968).

The critique of Gould and Lewontin would be legitimate to its full extent if one were to adopt (1) their narrow reductionist definition of the adaptationist program as exclusively "breaking an organism into unitary traits and proposing an adaptive story for each considered separately" (p. 581) and (2) their characterization of natural selection, in the spirit of natural theology, as a mechanism that must produce perfection.

Since only a few of today's evolutionists subscribe to such a narrow concept of the adaptationist program, Gould and Lewontin are breaking in open doors. To be sure, it is probable that many evolutionists have a far too simplistic concept of natural selection: They are neither fully aware of the numerous constraints to which natural selection is subjected, nor do they necessarily understand what the target of selection realty is, nor, and this is perhaps the most important point, do they appreciate the importance of stochastic processes, as is rightly emphasized by Gould and Lewontin.

Darwin, as mentioned above, was aware of the fact that the perfecting of adaptations needs to be brought only to the point where an individual is "as perfect as, or slightly more perfect than" any of its competitors. And this point might be far from potentially possible perfection. What could not be seen as clearly in Darwin's day as it is by the modern evolutionist, is that there are numerous factors in the genetics, developmental physiology, demography, and ecology of an organism that makes the achievement of a more perfect adaptation simply impossible. Gould and Lewontin (1979) and Lewontin (1979) have enumerated such constraints and so have I (Mayr 1982) based in part on independent analysis.

Among such constraints, the following seem most important.

1. *A capacity for nongenetic modification.* The greater the developmental flexibility of the phenotype, the better a species can cope with a selection pressure without genetic reconstruction. This is important for organisms that are exposed to highly unpredictable environmental conditions. When the phenotype can vary sufficiently to cope with varying environmental challenges, selection cannot improve the genotype.

2. *Multiple pathways.* Several alternative responses are usually possible for every environmental challenge. Which is chosen depends on a constellation of circumstances. The adoption of a particular solution may greatly restrict the possibilities of future evolution. When the ancestor of the arthropods acquired an external skeleton, his descendants henceforth had to to cope with frequent molts and with a definite limitation on body size. Yet, to judge from the abundance and diversity of arthropods in the water and on land, it was apparently a fortunate choice in other respects.

3. *Stochastic processes.* An individual with a particular genotype has only a greater probability of reproductive success than other members of its population, but no certainty. There are far too many unpredictable chance factors in the environment to permit a deterministic outcome of the selection process. With the benefit of hindsight, one might come to the conclusion that selection has sometimes permitted a less perfect solution than would have seemed available. Virtually all evolutionists have underestimated the

ubiquity and importance of stochastic processes. The kind of constraints to which natural selection is subjected, becomes even more apparent when we look at the process of selection more closely.

4. *The target of selection* is always a whole individual, rather than a single gene or an atomized trait, and an individual is a developmentally integrated whole, "fundamentally not decomposable into independent and separately optimized parts" (Gould and Lewontin, p. 591). For this reason, adaptation is by necessity always a compromise between the selective advantages of different organs, different sexes, different portions of the life cycle, and different environments. Even if the human chin is not the direct product of an ad hoc selection pressure, it is indirectly so as the compromise between two growth fields each of which is under the influence of selection forces.

A pleiotropic gene or gene complex may be selected for a particularly advantageous contribution to the phenotype even if other effects of this gene complex are slightly deleterious. To uncouple the opposing effects if apparently not always easy.

Since it is sufficient when an individual is competitively superior to most other individuals of its population, it may achieve this by particular features, indeed sometimes by a single trait. In that case natural selection "tolerates" the remainder of the genotype even when some of its components are more or less neutral or even slightly inferior.

5. *Cohesion of the genotype.* Development is controlled by a complex regulatory system, the components of which are often so tightly interconnected with each other, that any change of an individual part, a gene, could be deleterious. For instance, it is apparently less expensive in the development of a mammal to go through a gill arch stage than to eliminate this circuitous path and to approach the adult mammalian stage more directly. Allometry is another manifestation of regulatory systems. A selectively favored increase (or decrease) of body size may result in a slightly deleterious change in the proportions of certain appendages. Selection will determine the appropriate compromise between the advantages of a changed body size and the disadvantages of correlated changes in the proportion of appendages. The capacity of natural selection

to achieve deviations from allometry has been established by numerous investigations. It was realized by students of morphology as far back as Étienne Geoffroy St. Hilaire, that there is competition among organs and structures. Geoffrey expressed this in his loi de balancement. The whole is a single interacting system. Organisms are compromises among competing demands. Wilhelm Roux, almost 100 years ago, referred to the competitive developmental interactions as the *struggle of parts* in organisms. The attributes of every organism show to what an extent it is the result of a compromise. Every shift of adaptive zones leaves a residue of morphological features that are actually an impediment. Reductionists have asked, Why has selection not been able to eliminate these weaknesses? The answer would seem to be that these are inseparable parts of a whole which, as a whole, is successful.

There are chance components in all these processes, but it must be stated emphatically that selection and chance are not two mutually exclusive alternatives, as was maintained by many authors from the days of Darwin to the earlier writings of Sewall Wright and to the arguments of some anti-Darwinians of today. Actually there are stochastic perturbations ("chance events") during every stage of the selection process.

The question whether or not the adaptationist program ought to be abandoned because of presumptive faults can now be answered. It would seem obvious that little is wrong with the adaptationist program as such, contrary to what is claimed by Gould and Lewontin, but that it should not be applied in an exclusively atomistic manner. There is no better evidence for this conclusion than that which Gould and Lewontin themselves have presented. Aristotelian "why" questions are quite legitimate in the study of adaptations, provided one has a realistic conception of natural selection and understands that the individual-as-a-whole is a complex genetic and developmental system and that it will lead to ludicrous answers if one smashes this system and analyzes the pieces of the wreckage one by one.

A partially holistic approach that asks appropriate questions about integrated components of the system needs to be neither stultifying nor agnostic. Such an approach may be able to avoid the Scylla and Charybdis of an extreme atomistic or an extreme holistic approach.

Summary

1. The adaptationist program attempts to determine what selective advantages have contributed to the shaping of the phenotype.

2. Evolutionary change falls far short of being a perfect optimization process. Stochastic processes and other constraints upon selection prevent the achievement of perfect adaptedness. Evolutionists must pay more attention to these constraints than they have in the past. However, as already stressed by Darwin (1859, p. 201) there is no selective premium on perfect adaptation.

3. Even though the adaptationist program has been occasionally misapplied, particularly in an uncontrolled reductionist manner, its heuristic power justifies its continued adoption under appropriate safeguards. The application of the adaptationist program has led to important discoveries in many branches of biology.

Literature cited

Boag, P. T., and P. R. Grant. 1981. Intense natural selection in a population of Darwin's finches (Geospizinae) in the Galapagos. Science 214:82–85.

Bock, W. J. 1959. Preadaptation and multiple evolutionary pathways. Evolution 13:194–211.

Bock, W. J. 1980. The definition and recognition of biological adaptation. Am. Zool. 20:217–227.

Bock, W., and G. von Wahlert. 1965. Adaptation and the form-function complex. Evolution 19:269–299.

Brandon, R. N. 1978. Adaptation and evolutionary theory. Stud. Hist. Philos. Scl. 9:191–206.

Clarke, B. C. 1979. The evolution of genetic diversity. Proc. R. Soc. Lond., B. Biol. Sci. 205:453–474.

Darwin, C. 1859. On the origin of species by means of natural selection or the preservation of favored races in the struggle for life. John Murray, London.

Dobzhansky, Th. 1956. What is an adaptive trait? Am. Nat. 90:337–347.

Dobzhansky, Th. 1968. Adaptedness and fitness. Pages 109–121 in R. Lewontin, ed. Population biology and evolution. Syracuse University Press, Syracuse, N.Y.

Gould, S. J., and C. B. Calloway. 1980. Clams and brachiopods – ships that pass in the night. Paleobiology 6(4):383–396,

Gould, S. J., and R. Lewontin. 1979. The spandrels of San Marco and the Pangloesian paradigm: a critique of the adaptationist programme. Proc. R. Soc. Lond., B. Biol. Sci, 205:581–598.

Jacob, F. 1977. Evolution and tinkering. Science 196:1161–1166.

Lack, D. 1968. Ecological adaptations for breeding in birds. Methuen, London.

Lewontin, R. C. 1978. Adaptation. Sci. Am. 239:156–169.

Lewontin, R. C. 1979. Sociobiology as an adaptationist program. Behav. Sci. 24:5–14.

Mayr, E. 1959. Where are we? Cold Spring Harbor Symp. Quant, Biol. 24:409–440.

Mayr, E. 1960. The emergence of evolutionary novelties. Pages 349–380 in S. Tax, ed. The evolution of life. University of Chicago Press, Chicago.

Mayr, E. 1962. Accident or design, the paradox of evolution. Pages 1–14 in The evolution of living organisms: a symposium of the Royal Society of Victoria held in Melbourne, December, 1959. Melbourne University Press, Melbourne.

Mayr, E. 1963. Animal species and evolution. Harvard University Press, Cambridge, Mass.

Mayr, E. 1970. Population, species, and evolution. Harvard University Press, Cambridge, Mass.

Mayr, E. 1975. The unity of the genotype. Biol. Zentralbl. 94:377–388.

Mayr, E. 1982. Adaptation and selection. Biol. Zentralbl. 101:66–77.

Muller, H. J. et al. 1949. Natural selection and adaptation. Proc. Am. Philos. Soc. 93:459–519.

Munson, R. 1971. Biological adaptation. Philos. Sci. 38:200–215.

Roux, W. 1881. Kampf der Thelle im Organismus. Jena, Leipzig.

Stern, J. T. 1970. The meaning of "adaptation" and its relation to the phenomenon of natural selection. Evol. Biol. 4:39–66.

Traub, R. 1980. Some adaptive modifications in fleas. Pages 33–67 in R. Traub and H. Starcke, eds. Fleas. A. A. Baldema, Rotterdam.

Williams, G. C. 1966. Adaptation and natural selection. Princeton University Press, Princeton, N.J.

Wright, S. 1949. Adaptation and selection. Pages 365–389 in G. Jepsen, E. Mayr, and G. G. Simpson, eds. Genetics, paleontology, and evolution. Princeton University Press, Princeton, N.J.

PART V

BIOLOGICAL FUNCTION AND TELEOLOGY

Introduction

Biologists have been explaining events, processes, and other biological traits in terms of their function as far back as Aristotle. Harvey's discovery that the function of the heart is to pump the blood has remained an unrevised and established biological fact, one which explains why the heart beats – in order to pump the blood – since the mid-seventeenth century. Not even the eclipse of appeals to the all-powerful designer-god has reduced the role which such claims have in biology. Indeed, molecular biology is as replete with functional attributions as any other part of the discipline.

But the demand that we identify the function of biological structures, and the continued acceptance of explanations of their operation in terms of their function raises the problem of teleology: how can appeal to goals, purposes, ends – all more or less implicated in attributions of function – be justified in a world from which such things have been banished first by physical science and then by the biologist's insistence that all things biological be given a causal, and not a purposive or teleological, explanation?

In the first selection in this section (chapter 11), Mark Perlman traces out what he calls the "Modern Philosophical Resurrection of Teleology," complete with a categorization of the various positions about the role of function analysis in biology and its relationship to physical causality. His paper identifies two of the most prominent accounts of the nature of function in contemporary philosophy of science: (1) a "present and future" focus on function as nothing but a causal role in a larger system that includes the component to which function is ascribed; (2) a backward-looking, historical, or etiological analysis of function due to Larry Wright that accords it a Darwinian provenance.

The next paper, by Robert Cummins (chapter 12), gives the original canonical treatment of the causal role conception of function. The final paper, by Peter Godfrey-Smith (chapter 13), reprises the etiological account, but with an emphasis on recent history of natural selection as giving the content of a functional claim.

Further Reading

Achinstein, P. (1977). Function statements. *Philosophy of Science*, 44, 341–367.

Allen, C., Bekoff, M., & Lauder, G. (eds.) (1998). *Nature's purposes: Analyses of function and design in biology*. Cambridge, MA: MIT Press.

Ariew, A., Cummins, R., & Perlman, M. (eds.) (2002). *Functions: New essays in the philosophy of psychology and biology*. Oxford: Oxford University Press.

Arnhart, L. (1998). *Darwinian natural right*. Binghamton, NY: SUNY Press.

Arp, R. (2007). Evolution and two popular proposals for the definition of function. *Journal for General Philosophy of Science*, 37, 2–12.

Ayala, F. (1970). Teleological explanations in evolutionary biology. *Philosophy of Science*, 37, 1–15.

Bechtel, W. (1989). Functional analyses and their justification. *Biology and Philosophy*, 4, 159–162.

Bogdan, R. (1994). *Grounds for cognition: How goal-guided behavior shapes the mind*. Hillsdale, NJ: Lawrence Erlbaum Associates.

Buller, D. (ed.) (1999). *Function, selection, and design*. Binghamton, NY: SUNY Press.

Cummins, R. (1975). Functional analyses. *Journal of Philosophy*, 72, 741–765.

FitzPatrick, W. (2000). *Teleology and the norms of nature*. New York: Garland.

Godfrey-Smith, P. (1993). Functions: Consensus without unity. *Pacific Philosophical Quarterly*, 74, 196–208.

Godfrey-Smith, P. (1994). A modern history theory of functions. *Nôus*, 28, 344–362.

Godfrey-Smith, P. (1996). *Complexity and the function of mind in nature*. Cambridge: Cambridge University Press.

Griffiths, P. (1993). Functional analysis and proper function. *British Journal for the Philosophy of Science*, 44, 409–422.

Griffiths, P. (1996). The historical turn in the study of adaptation. *British Journal for the Philosophy of Science*, 47, 51–532.

Lennox, J. (1993). Darwin was a teleologist. *Biology and Philosophy*, 8, 409–421.

Millikan, R. (1984). *Language, thought, and other biological categories*. Cambridge, MA: MIT Press.

Millikan, R. (1989). In defense of proper functions. *Philosophy of Science*, 56, 288–302.

Neander, K. (1991). Functions as selected effects: The conceptual analyst's defense. *Philosophy of Science*, 58, 168–184.

Rosenberg, A., & McShea, D. (2008). *Philosophy of biology: A contemporary introduction* (Chapters 1 and 3). London: Routledge.

Trivers, R. (1985). *Social evolution*. Menlo Park, CA: Benjamin Cummings.

Walsh, D. (2008). Teleology. In M. Ruse (ed.), *The Oxford handbook of philosophy of biology* (pp. 113–137). Oxford: Oxford University Press.

11

The Modern Philosophical Resurrection of Teleology

Mark Perlman

"Why is there air?"
"To blow up volleyballs."
Bill Cosby

I. Introduction

Many objects in the world have functions. Typewriters are for typing. Can-openers are for opening cans. Lawnmowers are for cutting grass. That is what these things are for. Every day around the world people attribute functions to objects. Some of the objects with functions are organs or parts of living organisms. Hearts are for pumping blood. Eyes are for seeing. Countless works in biology explain the "Form, Function, and Evolution of . . ." everything from bee dances to elephant tusks to pandas' 'thumbs'. Many scientific explanations, in areas as diverse as psychology, sociology, economics, medical research, and neuroscience, rest on appeals to the function and/or malfunction of things or systems. They talk of how humans and other organisms or their parts work, what their functions are, why they are

present, and how different situations will affect them and how they will react. Philosophers, going back to Aristotle, used to make generous use of functions in describing objects, organisms, their interactions, and even as the basis of ethics and metaphysics. And yet, since the Enlightenment, talk of the function of natural objects, teleological function, began to be viewed with suspicion, as the mechanical model of the world replaced the old Aristotelian model. From a religious standpoint, it used to be easy to see how objects in the natural world could have natural functions, for God was said to instill functions by design throughout Creation. But philosophers became increasingly (and wisely) reluctant to invoke God to solve every difficult philosophical problem, and became unwilling to indulge in such religious explanations of teleology. It is easy to see how artifacts produced by

Mark Perlman, "The Modern Philosophical Resurrection of Teleology," *The Monist* (2004) 87, 3–51. Copyright © 2004, *The Monist: An International Quarterly Journal of General Philosophical Inquiry*, Peru, Illinois, 61354. Reprinted by permission.
Thanks to Randy Dipert and Anthonie Meijers for their suggestions and comments, and to Katherine Munn for comments and editorial suggestions. Special thanks also go to André Ariew for his useful comments and criticisms (especially for the sections on artifacts), and for getting me interested in teleology in the first place.

humans would have functions, derived from the intentions of the human designers, but without God, it seemed impossible to believe that teleology has a place in Nature.

By the twentieth century, analytic philosophers were positively allergic to any mention of teleology or teleological function. It was seen as an insidious metaphysical notion that was to be tossed out with the rest of metaphysics. The Logical Positivists insisted that, to be meaningful, each individual sentence had to be reducible to a set of confirming observations. This made it almost impossible to countenance attribution of functions to things, since to distinguish something's function from coincidental behavior requires a pattern of action beyond an individual instance, and integration into a larger body of theory. With Quine's rejection of this reductionist dogma in "Two Dogmas of Empiricism" (1951), it might have seemed that teleological functions could be allowable as parts of an overall theory that matched well with observations. Quine wrote: "To call a posit a posit is not to patronize it", and this could have been specifically applied to positing teleological functions. But it was not, and so, despite this possible remedy, teleology retained its bad reputation, and philosophers remained suspicious of teleological functions. Talk of teleological function was still taken by philosophers as a *façon de parler*, legitimate only if reducible to well-behaved physicalistic and non-teleological phenomena. The "traditional" view of functional explanation, as promoted by Hempel and Nagel, was that functional analyses were "incomplete" deductive-nomological explanations of the presence of the analyzed item. Another common view was to see attributions of functions as part of a Wittgensteinian language game, without any ontological significance.

Yet, while it is perhaps not so surprising that philosophers would go against common sense in rejecting teleology (indeed some take it as an essential part of philosophy to oppose common sense), it is surprising that analytic philosophers, with their strong focus on science, would reject a notion that is so central to some areas of science, most notably, biology and engineering sciences. Of course, the Positivists would have said that biology's reliance on teleology jeopardized its standing as a science at all, and certainly prevented it from being a basic natural science. They were similarly suspicious of sociology and psychology. The Logical Behaviorists even sought to sanitize psychology of all such metaphysically polluted terms. Indeed, Logical Positivists had ambitions to sanitize all of science from metaphysics. But these projects failed to find anything to adequately fill the important role teleological functions play, particularly in biology and psychology. So began the modern philosophical movement to legitimize teleology.

The first shots came in the late 1960s and early 1970s, with papers by Ernst Nagel (1961, 1972), Wright (1973, 1976), Boorse (1976, 1977), Wimsatt (1972), Mayr (1974), Woodfield (1976), and others. In 1975, Robert Cummins's paper "Functional Analysis" broke new ground in giving a naturalistic and ahistorical analysis of functions based on causal role, an approach that had little trace of the metaphysical, and proposed an account that science could accept. The historical, or etiological, side to naturalizing teleology was brought to prominence with Ruth Garrett Millikan's landmark 1984 book, *Language, Thought, and Other Biological Categories*. Following that there was a flood of different accounts of functions, how they would or would not serve as the basis of this and that, and the functions literature took off.

The rather unfortunate thing is that, with teleology involved so widely in so many fields, it is difficult for anyone to really master all the relevant literature, and for this reason many theories about teleology are incomplete. Those who focus on the biology side often fail to take all the relevant philosophical discussions into account, and even among the philosophers, there seem to be various disparate strands of discussion. Things have recently improved in this regard, but there are still mistakes (or at least exaggerations) being made in discussing teleology, because of incomplete knowledge of all the relevant literature. (We will discuss this problem more in section III.B.4.)

So it may be useful to develop a taxonomy of teleological theories, and assess them from a larger perspective. Of course, any taxonomy is likely to oversimplify certain issues, and various writers may object to the place in the scheme I have put them, but this in itself may help clarify how the various parts fit together. One helpful step was made by Walsh and Ariew (1996), but they were focused on functions in the biological context (and really only on the naturalistic reductionist theories), and I have made the present taxonomy broader. Walsh and Ariew (1996: 493) divide the naturalistic reductionist theories into the causal side (Cummins) and evolutionary side, and they further divide the evolutionary side into "historical" (the etiological theories of Millikan and company) and "current" (represented by Bigelow and Pargetter). My taxonomy deviates from their's in some ways, and I have added more subdivisions. In what follows I will seek to organize various theories of functions into categories (listing names of the main proponents of each), and summarize the major distinctions made in the literature.

How far this taxonomy extends (or could extend) is a subject of debate. The entire modern discussion of teleology and function has been and remains focused on natural, biological functions. This is partly because modern science seems to have purged physics and chemistry of any teleology, whereas the same cannot be said for biological sciences. Biology cannot, or at least in practice does not, eliminate functions and purposes. Despite the Positivists' desire, it is difficult to see how we could eliminate from biological explanations any mention of the functions and purposes of organs, mechanisms and systems in organisms. Teleology also seems central to many explanations in physiology, medicine, psychology, sociology, and anthropology.

We also ought to consider the question of human-created artifacts, since many of these obviously have purpose and function. So we may want to extend the theories of functions in the taxonomy to explanations in engineering and design theory, as well as the parts of sociology and anthropology dealing with artifacts. But just because we want to extend one of these theories does not guarantee that that particular theory, or general approach, will be suitable for applying to artifacts. Some of the writers listed above, particularly Cummins, Millikan, and Neander, do indeed have ambitions to extend their accounts of functions to artifactual objects. But there are others who challenge such application to artifacts (such as Vermaas and Houkes (2003)). We will consider the issue of artifact function in section VI below. We will also examine whether these are all theories trying to explain the same type of functions, or whether different theories are in fact accounting for different types of function.

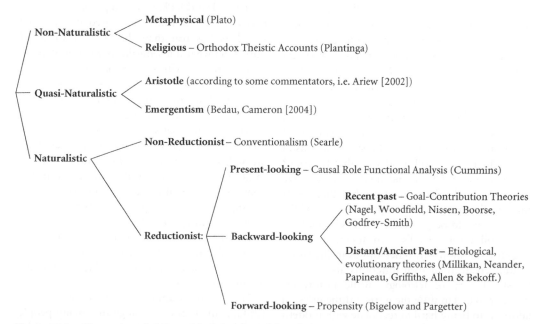

Figure 11.1   Categories of philosophical theories of functions.

II. Non-Naturalistic and Quasi-Naturalistic Views

The religious (theistic) view goes back at least to St. Thomas Aquinas, and is based on an Aristotelian metaphysical framework. It emerges from an analogy with human creation of common objects with functions. Human intelligent agents endow artifacts with purpose, design, and function. Natural objects and organisms are similarly directed toward ends, so Aristotle concludes that there is purpose behind the objects in nature – they have functions. Understanding the natural world is to see how objects fit into an overarching teleological hierarchy. Going even farther into teleology, theists like Aquinas argue there must be a Being, an intelligent Grand Designer, God, who is the source of design and function throughout the natural world, including the organisms in it.

There are well-known difficulties with this view which I will not rehearse here, but I will simply point out that theistic views of the source of functions have the same sort of circularity problems that threaten causal argument for God's existence. Consider Aquinas' Aristotelian causal argument for God: Everything must have a cause, there cannot be an infinite regress of causes, and nothing comes from nothing, so there must be a first cause, which is God. But because everything must have a cause, this argument is threatened with vicious circularity. The argument then requires desperate maneuvers to prevent the circularity – God is self-caused, or needs no cause because God is a necessary Being, and so on. If we make God the ultimate source of the functions of items in the natural world, we get the same kind of problems: If God has a function, to create things and instill functions in them, then who instills God with that function? A Higher God? That would invite a regress. We could say God has no function, but that seems unappealing to the theist – the creation of the universe would be purposeless. Or we might say that God creates God's own function, which begs the question why other things in the world cannot create their own functions also. Thus a theistic theory of teleology, while historically popular, has daunting theoretical difficulties, in addition to being hard to reconcile with modern science. Moreover, the Creationist model brings

huge epistemological problems with it – to know or discover the function of an organ or mechanism we would have to discover what the Creator had in mind.

Yet the religious, specifically theistic, non-naturalistic side of the teleology debate is still being defended, for example, by Plantinga (1993a, 1993b, 2000). On his view we cannot naturalize functions, and yet we believe that there really are "(for natural organisms) such things as proper function, damage, design, dysfunction, and all the rest," and this gives us a "powerful argument against naturalism" (1993: 214). Plantinga's work on God and Christian philosophy makes no secret of his choice for the real source of teleology.

We can also have non-theistic non-naturalistic theories, and I have simply called these "metaphysical". The most well-known example of this is Plato's view of the reality-giving transcendent Forms. If a material object has the Form of Heart, then it would seem that part of its participating in that Form is to have a certain function, to be *for* pumping blood. In performing its function the heart is contributing to the Good of the organism. Thus we could say that all correct functioning of an item is ultimately linked to the Form of the Good. This may be the ultimate sign of a non-naturalistic view – investing value into the natural world. Plato had no problem with this because, for one thing, he viewed the material world as less real than the abstract world of Forms. So what if a semi-illusory realm of existence is invested with value properties? Yet some contemporary philosophers would argue that teleological functions are essentially value-laden, and thus not naturalizable. How can we have function at all, what something is *good for*, if we take the Good out of function? Of course, there are well-known difficulties for Plato's view – a multitude of metaphysical puzzles about the relation of material objects and the Forms they are supposed to participate in, potential infinite regresses, and epistemological issues of how we could know the Forms, as well as the lack of any empirical evidence that such Forms exist at all. Thus the metaphysical route is viable only for fans of epistemological mysteries.

The category I am calling "quasi-naturalistic" contains theories of teleology that many people would assume were simply more non-naturalistic theories. The common interpretation of Aristotle is non-naturalistic – seeing him as proposing fishy (and non-explanatory) vital forces and bizarre backwards causation in nature. But there are good reasons to see Aristotle as engaging in a much more scientifically respectable enterprise. André Ariew (2002: 18) argues that "Aristotle's final aitia are not *causes* but *explanations* or *reasons*," and that Aristotle is proposing teleology through a very respectable argument scheme – inference to the best explanation. Though functions are not reducible to material causes (which is why so many have called them non-naturalistic), they are part of what Aristotle would take to be a scientific explanation of why things happen as they do. But is this kind of explanation naturalistic or not?

The teleologist says (according to Ariew 2002: 17) that "orderly patterns occur *for the sake of* the form while monstrosities do not." So form is the end toward which change is normally directed. But as such, form can still be value-neutral – a result that we could consider good, bad, or toward which we might be indifferent. What would make such an account fully non-naturalistic would be to go beyond something that merely "acts for an end" to acting for an end that "achieves the best result" or "achieves the Good". If we assume that value is a non-natural property, then stating the teleology as "achieving the Good" or "achieving the best result" makes the value-centered theory of teleological function non-naturalistic. If Aristotle had actually done this, instill value into his theory of the natural world (as many commentators believe he does), then his teleology would indeed belong to the fully non-naturalistic category. Plato does indeed make ends value-laden, whereas Aristotle does not. For Aristotle leaves functions at the level of being "for the sake of" some end, without adding value to that end. Yet Aristotle's medieval followers, like Aquinas, did indeed take Aristotle's view to license this move, and this has lead many people to mistake Aristotle for a non-naturalist.

Some might even take this argument against non-naturalism in Aristotle to make Aristotle's view count as fully naturalistic, but I think there is enough teleology in it to make that problematic. To preserve the difference between Aristotle's teleological science and modern science, I put him in the category of "quasi-naturalistic".

Another view in the "quasi-naturalistic" category takes functions to be irreducible emergent properties, along the lines of emergentism of mental properties. Emergent properties are said to depend for their existence on the interactions of underlying component properties (usually material properties), but which cannot be reduced to those underlying properties, and which can have causal powers that cannot be identified with the causal powers or relations of the underlying components. Thus it is said that mental states depend on physical particles (atoms composing neurons) for their existence, yet the interactions and causal powers of mental states could be irreducible to any molecular or neural events. Mark Bedau (1990, 1991, 1992a,b, 1993, 1997) has defended an emergentist view of teleology along the same lines – functions depend on lower-level physical properties, but they themselves are irreducible and have properties and causal powers beyond any in the physical level. The article in this volume by Richard Cameron defends such an emergentist line on teleology.

Bedau thinks that teleology is fundamentally value-laden, and thus cannot be reducible to any value-free natural properties. In this sense the view of functions as emergent is a non-naturalistic view, but only, Bedau (1991) says, on a "narrow" sense of naturalistism. On a "broader" view of naturalism, which would be accepting of Aristotelian teleology as in some sense still natural, emergentist teleology could also be considered naturalistic. So as to not beg the question against this "broader" notion, and still to allow that adherents of naturalism would not consider these theories truly naturalistic, I include both emergentist teleology and Aristotle's teleology in the "quasi-naturalistic" category.

III. Naturalistic Theories – To Reduce or not to Reduce?

Among naturalistic theories of teleology and function, we see a division into those that reduce teleology and function to some more basic natural property, and those that take it as natural but irreducible. One might shy away from reduction because one believes it is either impossible or undesirable. But science is generally in the business of reduction, and so we should expect the urge to achieve the reduction of teleology, if it can be done. The reduction issue links us to a large body of rich and complex thought on scientific reduction – so large that I will not detour us into it here. Suffice it to say that it is at least plausible that functions could be natural and yet not reducible to a more fundamental science, just as geology might not be reducible to physics.[1]

III. A. Non-reductive naturalism

John Searle (1990: 14) argues that functions "are never intrinsic but are always observer relative". Searle contends that functions are not in nature itself, but always attributed by human observers. We do "discover" functions in nature, he says, "only within a set of prior *assignments* of value" (including purposes, teleology, and other functions). Naturalism excludes values from the natural world, and our vocabulary of "functions" adds a set of values to the vocabulary of "causes". To say the heart's function is to pump blood (and not to make a thumping noise) is based on our previous value judgment that survival is valuable, and, because pumping blood is necessary for survival, the heart's function is to pump blood. If our system of thought put greater value on production of thumping noises, then that would steer our function attribution toward the heart's function being to make a 'heartbeat'. This conventional approach to functions is non-reductive just in case (as seems to be the case) value cannot be reduced to natural properties or objects. For Searle, functions are external to the object and its history – they are ontologically dependent on human intentionality. But the theory is still naturalistic because functions attain no supernatural status – they're just properties attributed to objects by natural human creatures. Functions derived from intentionality remain natural because Searle considers intentionality itself to be a natural property.

III. B. Reductive naturalism – which way to look?

Borrowing some of the terminology from Bigelow and Pargetter (1987), we can divide reductive naturalistic theories of function into three main categories: Present-looking, Backward-looking, and Forward-looking. Even this is contentious because various people may object to their placement in the scheme, or the evaluative slant they take such a placement to involve. My placement of Cummins as having a Present-looking view of functions clashes with Bigelow & Pargetter, who categorize Cummins' view as eliminativism of functions. But I don't think Cummins intends to be eliminating functions – he means to be analyzing them.[2] The real eliminativists (or perhaps "fictionalists") would be Dennett (1981, 1983, 1987, 1995) and the Churchlands (1979, 1981, 1984, 1986, 1989), who do not reduce teleology to any mechanisms, but see teleology as simply an illusion. Some would also want to consider Searle an eliminativist about functions, but he doesn't consider our value systems (from which functions are assigned) fictions. Even if they are not part of the natural world, they are neither supernatural nor non-existent. Searle views intentionality as a natural property, though at a higher level of complexity than the most basic level, analogous with the fluidity of water, which is not a property of individual water molecules, but a higher level property of collections of water molecules. Thus I categorize Searle as a non-reductive naturalist about functions.

In the Present-looking approach, represented by its main advocate, Cummins (1975, 2002), causal role functional analysis is an attempt to ground functions in the current interrelation of structural parts to the whole of a system and its capacities. The Backward-looking approach has two subgroups, depending on how far back they look. Some theories refer only to recent history in their analyses, such as the goal-contribution theories of Woodfield (1976), Nagel (1961), Nissen (1997), and Boorse (1976, 1977, 2002), and the modern history account of Godfrey-Smith (1994). Others emphasize ancient history, following Millikan (1984, 1986, 1989a,b,c, 1993) in basing functions and teleology on etiological factors – evolution by natural selection (Neander (1991a, 1991b, 2002), Papineau (1987), Griffiths (1993), Kitcher (1993), Allen and Bekoff (1995)). The Forward-looking approach of Bigelow and Pargetter (1987) sees functions as propensities toward future survival, and is thus biological and formulated in the context of evolutionary theory, but not historical/etiological. We will proceed to discuss each approach in detail.

III. B. 1. Looking at the present: causal
roles and functional analysis
Various philosophical disputes surround naturalistic teleology as it applied to biology and psychology. The most significant one is whether or
not accounts of functions should rest on history.
Cummins's (1975) functional analysis account
has functions determined by causal role, the
contribution of the capacities of subsystems to
the capacities of larger systems. The functional
explanations explain how a complex system has
the sophisticated capacities it has by virtue of the
capacities of its parts to do a series of simpler tasks.
This is typically formulated by a flow-chart
which links together these sub-capacities, which
Cummins calls the "functions" of the parts.
Thus an item only has a function in this sense
relative to a particular supersystem under consideration. Parts of a tree may have different
functions relative to the forest than they do
relative to a single tree that they compose. This
kind of analysis allows a single part to have a wide
variety of functions, relative to different sized
supersystems and different contingencies. The
functions that such a Cummins-style functional
analysis yields have come to be called by many
(including Millikan) "Cummins functions".
The implication is that Cummins-functions are
less than full functions (and certainly different
from Millikan's "proper functions"). These
Cummins functions are independent of the
history of how the items and their capacities
developed or evolved.

Cummins argues not only that the function
of a structure *need* not have played a role in the
structure's etiology, but the more extreme position that a structure's function *cannot* play a role
in its etiology. Cummins writes: "to 'explain' the
presence of the heart in vertebrates by appeal to
what the heart *does* is to 'explain' its presence
by appeal to factors which are causally irrelevant
to its presence . . ." (1975: 390). To perform a
function, the structure must already be there.

> To explain the presence of a naturally occurring
> structure or physical process – to explain why
> it is there, why such a thing exists in the place
> (system, context) it does – this *does* require
> specifying factors which causally determine the
> appearance of that structure or processes.
> (1975: 390)

So explaining the presence of a structure
requires explaining the antecedent causes, and
because its function cannot precede its presence,
that function cannot be required, or even relevant,
to explaining the presence of the structure.
This is different from explaining the structure of
an artifact, which did have a function planned
before its structure existed – it was designed so
as to perform that function. For biological structures, a naturalistic theory cannot make such an
assumption, and the explanation of a structure
must rest solely on the causes of the structure itself.

The functional analysis route to teleology
Cummins advocates is to answer the question of
"why-is-it-there" by analyzing the contribution
that parts of a system make to the operation of
the whole, that is, by explaining "how-does-
it-work" (Cummins 2002: 158). He thinks that
teleology should focus on the issue of "how-
does-it-work" instead of "what-is-it-for" to
explain the presence of a structure. Thus he sees
the historical evolutionary approach as intent on
answering the wrong question and missing the
important issue of functional explanation in
terms of the operation of complex structure.

Cummins specifically rejects natural selection
as a way of providing function and structure.
He discusses the evolutionary option that "by
performing their respective functions [the struc-
tures] help species incorporating them to survive,
and thereby contribute to their own continued
presence in organisms of those species, and this
might seem to explain the presence of those
structures in the organisms incorporating them"
(1975: 393–394). But he rejects this as resting
on a fundamental misunderstanding of evolutionary theory. The question of whether an
organism comes to have a certain kind of organ
or mechanism depends on its genetic plan, and
"alterations in the plan are not the effects of the
presence or exercise of the structures the plan
specifies." Alterations of the plan typically involve
random mutation, and nature selects which
organisms, with which alterations, will survive and
reproduce.

> If a plan is altered so that it specifies s' rather
> than s, then the organisms inheriting this plan
> will incorporate s' regardless of the function or
> survival value of s' in those organisms. If the altera-
> tion is advantageous, the number or organisms

inheriting that plan may increase, and, if it is disadvantageous, their number may decrease. But this typically has no effect on the plan, and therefore no effect on the occurrence of *s'* in the organisms in question. (Cummins 1975: 394)

"Natural selection cannot alter a plan, but it can trim the set. Thus we may be able to explain why a given plan is not a failure by appeal to the functions of the structures it specifies . . ." (Cummins 1975: 394). Cummins continues this attack on basing teleology on natural selection in his recent paper of 2002: "Selection cannot produce traits *in the sense relevant to neo-teleology*" (2002: 163). Selection does not work by selecting winged organisms over those without wings or with proto-wings that as yet cannot produce flight. Selection favors organisms with better wings over those with poorer wings. Selection is not driven by the having of a function, but rather by functioning better, and that depends on the entire system, not just the alleged function of a particular part.

> Better wing designs need not improve flight, but may simply make it more efficient, or make development less error prone, or make the structure less fragile. Hence selected changes in wing design that accumulate to yield the current design we seek to explain need not be related to the wing's function. Indeed, they may even compromise flight in the interest of other factors. (Cummins 2002: 168)

Thus it is analysis of the interrelations of parts of the structure and their relation to the environment that explains the presence of a structure. Natural selection does not.

The common criticism is that this type of functional analysis gives us the wrong answer about what the function of certain structures is, and in some cases gives functions to things that we think do not have them. As Boorse (2002) explains,

> It implies that the function of mists is to make rainbows (Bigelow and Pargetter 1987: 184), the function of rocks in a river is to widen the river delta (Kitcher 1993: 390), the function of clouds [is] to make rain with which to fill the streams and rivers (Millikan 1989a: 294), and the func-

tion of a piece of dirt stuck in a pipe is to regulate the water flow (Griffiths 1993: 411).

All these kinds of objections are intended to show that Cummins functions are not sufficiently teleological *in the right way*.

III. B. 2. a.   Looking to the distant past – historical, evolutionary, etiological functions
In opposition to the causal role approach to functions is the historical approach, basing functions on causal history. The modern origination of such an approach came from Wright (1973):

> The function of X is Z **means:**
> 1. X is there because it does Z.
> 2. Z is a consequence (or result) of X's being there.

The function comes about and is reinforced in the same way that the function of artifacts is created and reinforced by the human designers of the artifacts: the function of a screwdriver is to turn screws – the screwdriver was made in order to turn screws, and the turning of screws is a result of the screwdrivers being there to turn them. In the case of a natural item, it is nature, not people, that put the function there, and we can see this by examining the history of production of the item. The historical factors Wright appeals to are diverse, and do not focus solely on specifically evolutionary history as determined by Darwinian natural selection, yet he is often misunderstood to have devised the evolutionary approach.

The real focus on evolution and natural selection came with Millikan's landmark 1984 book, *Language, Thought, and Other Biological Categories: New Foundations for Realism*. She sees organs or mechanisms of an organism having their functions due specifically to their evolutionary history. Having those functions has contributed to the reproduction of organisms of that kind, and this evolutionary contribution gives the organs (or mechanisms) the functions they have. More exactly, Millikan develops a technical theoretical concept she calls "proper function" – a *proper function* is a property of a mechanism in an organism needed for survival in *Normal* conditions, that is, in conditions which the majority of organisms of that kind have lived and evolved. *X* performs its proper function when it does the sort of thing that has been historically responsible for replication of *X*s. Proper functions can fail to be fulfilled fairly often and still be to an adaptational advantage: they have to be fulfilled enough times for survival. So a Normal *X* need not fulfill its proper function most of the time – just often enough to keep the organism out of danger. A mechanism that indicates the presence of predators in an organism's immediate environment need not be reliable in the sense of only being tokened when a predator is around – it can give a lot of false positives and still be a good mechanism. (But one false negative may be too many, for just one will be almost a certain chance of being eaten.)

It is common for things to seldom perform their proper function, but a few performances may be enough. Millikan's example of this is human reproduction: the function of sperm is to reach and fertilize an egg, and that is the function of all sperm even if few actually make it. Bee dances are another example she cites: the orientation of the bee's dance tells the other bees where food is. If the dance is mis-oriented, its content is still *direction of food*; it just fails to fulfill its proper function, and is a misrepresentation of the direction of food.

Millikan actually has defended two versions of her view that teleology can be based on natural selection. While both versions are ultimately intended to provide a theory of representational content, neither of them assigns proper functions to the individual representations themselves. In *Language, Thought, and Other Biological Categories* (1984) her view is that it is not the representations themselves but the producers and consumers of the representations that have the Proper Functions, to produce and process representations which represent the environment correctly. In "Thoughts Without Laws: Cognitive Science Without Content" (1986) (and also Millikan 1989a and 1989b) she emphasizes the proper function of the consumer side of the representations, the mechanisms that use the representations produced by perception.

One aspect that Millikan consistently makes clear is that her view is the proposal of a technical theoretical concept, "proper function", not an analysis of the common sense notion or philosophical notion of function (though proper functions are a subset of functions). Yet many people ignore or miss or forget that point (even after she explicitly corrects them), and take her theory to be a definition or analysis of our concept of teleological 'function' in general. (Perhaps if she had named it 'punction' people would have been less able, likely, or willing to miss this point.) However, there have been writers, most notably Karen Neander (1991a, 1991b, 2002), while acknowledging Millikan's avoidance of conceptual analysis, who want to erect a Millikan-like theory as an analysis of the general teleological concept of 'function'. Peter Schwartz, in his paper in this volume of *The Monist* opposes this tendency to have the teleology debate be an exercise of conceptual analysis, and favors the technique (devised by Carnap (1950) and Quine (1960)) of "philosophical explication", in concert with Millikan. Other recent defenders of "neo-teleology" include Griffiths (1993), Kitcher (1993), and Allen and Bekoff (1995).

Papineau's (1987) etiological theory of function centers on the related contents of beliefs and desires. Belief content is dependent on the objects of desire. The biologically advantageous function of representational mechanisms is to bring about satisfaction of desires. The satisfaction condition of a desire is whichever condition it is that desire's biological purpose to cause. The truth condition of a belief is whichever condition guarantees that the actions generated by that belief will fulfill the belief's biological purpose, namely, of satisfying desires. So the content of belief is derived from beliefs having the proper function of producing desire-satisfying behavior. This makes it seem that all desires have the same purpose – survival and reproduction, or merely replication of the organism's genes. But Papineau argues that "different desires have been designed to foster this end in determinately different ways" (1987: 67n8). So even after one recognizes that eating chocolate isn't healthy, and won't help one pass on one's genes, one's desire for chocolate persists, because the connection between eating and reproduction is mediated by several intervening layers of beliefs and desires.

Such etiological theories are also criticized for allegedly yielding the wrong answer for certain biological functions. Bedau (1991) tells a story (attributed to Robert Van Gulick) of a stick in a stream that creates a backwash so that it is pinned to the side due to its backwash, and being so pinned reinforces and perpetuates the backwash so that it remains pinned, and being pinned continues the backwash, generating a self-perpetuating cycle that satisfies the requirements of etiological theories. The stick and its backwash are reproducing the property that keeps it pinned. So is the natural function of the stick and its backwash to keep it pinned? Surely not, yet that is what the etiological theory is alleged to dictate.

Another proposed counterexample to etiological accounts is the possibility of a spontaneously created organism (resulting from a freak lightning strike in an organic environment), molecularly identical to organisms with an evolutionary history and organs with functions based on that history. (Such a spontaneous organism has been nicknamed 'Swampman'.) On the etiological accounts, such an organism would have a heart, lungs, eyes, etc. without functions, because it has no evolutionary history. Millikan accepts this consequence, and does not see it as a problem for the multitude of real organisms whose organs' functions her theory does explain. But the debate over this continues, and what to say about all these proposed counterexamples is yet unsettled.

III. B. 2. b.   Looking to the recent past – goal-contribution and modern history accounts

Some theories which include past history only go back a short while, unlike the evolutionary selectionist theories of Millikan and company. The recent history approach has various names, but centers around goals and their fulfillment. To some this may not seem historical at all, with the goal being a state present in the organism, but to analyze attainment of goals requires that the goals precede the situations that involve the goals and in which the goals may or may not be accomplished. Boorse (2002) explains the details of the development of this view, citing its origins in 'cybernetic' accounts which use ideas from systems theory, by such writers as Rosenbleuth, Wiener, Bigelow (1943), and Beckner (1959). It is applied to functions by Nagel (1961), and later by Boorse (1976, 1977), and Nissen (1997). The basic idea is that a system can be "directively organized" or "goal-directed" toward some result when the system will vary its behavior to deal with

its environment in order to reach that result. Such a system shows flexibility to alter its initial plan to go around obstacles which arise. Nissen (1997) points out that "function" and "purpose" are not the only teleological notions regularly attributed to organisms – animal behaviors such as "fleeing", "protecting", "hiding", "migrating" are also teleological. The source of the teleology in all these notions is goal-directedness. Nissen also adds additional requirements to a goal-directed account, such as negative feedback and internal representation of the goal-state. This last requirement puts a lower bound on function attributions to animals – only those with sufficient mental capacity to represent goals can have them, so dogs, cats, and birds have them, but insects may not. Nissen (1997: 228) seems willing to accept this result for lower animals, whereas Boorse (2002: 108n) says: "Surely, however, moths can flee birds just as obviously as mice can; the difference is only whether they know they are doing so." Boorse's description of the goal-contribution approach to functions is as follows:

> X performs the function Z in the G-ing of S at t if and only if at t, the Z-ing of X is a causal contribution to G (2002: 70).

> A trait X's contribution to a goal, then, if made sufficiently often, becomes the function of X, or X's function, if it is X's only regular contribution; and is a function of X, among X's functions, otherwise (2002: 72).

These goal-directed accounts differ from evolutionary accounts in that, though the goals of organisms are, at the highest level, survival and reproduction, not all functions reduce to just these evolutionary goals. Individual systems can have many lower-level goals that would not be correctly characterized as depending on adaptation, natural selection, or evolution generally. They need reach back into history only as far as the establishment of the goal, not all the way back to ancient history of the development of the species and its organs and mechanisms. Boorse (2002: 73–74) is careful to correct what is, in his view, a mistaken view of goal-contribution accounts as selectionist (i.e. a version of the backward-looking evolutionary theories described above). He points out that "a trait can causally promote its bearer's survival in the absence of any selection process favoring the trait or its bearer." Since "a trait may have no current rivals in the actual population", such a trait will not be selected for (over alternative traits actually found in the population), it simply persists due to contribution to the survival of the organisms. Thus I have placed goal-contribution accounts in the recent-history category, apart from Millikan-type evolutionary selectionist theories.

One nice feature of goal-directed accounts is that, being based in some sense on intentional states (goals, plans, desires, etc.), they can apply to artifacts as well as organisms. The artifact side seems almost a given – it is the biological side that defenders see as the contentious one in need of defense. Boorse (2002) develops what he calls the "general goal contribution" (GGC) account of functions, and responds to a wide variety of objections to goal-directed accounts.

Godfrey-Smith's account also restricts historical determination of function to modern history, rather than the distant past.

> Functions are dispositions and powers which explain the recent maintenance of a trait in a selective context. . . . [Millikan] explicitly allows powers which were important in ancient history, but not in modern history, to be functions (1984 p. 32). In her 1989b treatment her emphasis is different, and she claims the relevant past is "*usually*" only yesterday. But perhaps, as far as functions go, it *must* be only yesterday. (1994: 468)

His reason for favoring modern history over ancient history is that many traits develop and are adapted for one purpose, and later become important for another. Feathers are one example Godfrey-Smith cites (1994: 471). The first birds had only rudimentary wings, but were covered with feathers. The evolutionary origin of these feathers was as insulation, and only later, with improved wing design allowing more effective flight, did feathers come to have flight involved with their function. Though they still offer insulation, feathers are now maintained because they facilitate flight. So it is their recent history that forms the function of organs, not their evolutionary origins. "Recent" may cover a period of a week or a thousand years, depending on relevance to the current assignment of function.

III. B. 3.   Looking forward: functions as propensities

Another view of functions defines them in a forward-looking way, connecting them to their role in future events, as natural propensities. According to Bigelow and Pargetter, "something has a (biological) function just when it confers a survival enhancing propensity on a creature that possesses it" (1987: 192). They equate "survival enhancing propensity" with "propensity for selection" (1987: 194). They view this as an advantage over historical theories because it is "forward-looking", whereas historical theories are obviously "backward-looking". They think functions should be "characterized by reference to possibly nonexistent future events" (1987). The way to make functions forward-looking is "to construe them in the manner of dispositions" (1987), and this is related to the evolutionary account of fitness. This account, like fitness, is relativized to environment, so that "a character may confer propensities which are survival-enhancing in the creature's usual habitat, but which would be lethal elsewhere. When we speak of the function of a character, therefore, we mean that the character generates propensities that are survival-enhancing in the creature's natural habitat" (1987). Such propensities are dispositional, specified subjunctively – "a character *would* give a survival-enhancing propensity to a creature . . . in the creature's natural habitat", even if the creature dies or is not in that habitat. So a character can have a function even if it does not perform its function, and that is one of the crucial features of function generally. Since it is the actual structural form of the creature that causes it to survive, such functions will supervene on the structure and its relation to the environment.

This propensity theory has been accused of smuggling in historical factors in its notion of "natural habitat". Bigelow and Pargetter relativize their view to an environment, and the question is how it is to be determined what the "natural habitat" of a creature is. Neander (1991b) points out that the most common current context for a trait is not necessarily natural. Without a historical basis, it is hard to distinguish between cases where a trait is survival-enhancing yet now fatal in an unnatural habitat, or naturally lethal in a natural habitat.[3]

III. B. 4.   Is the historical approach the orthodox view?

The distinction between Cummins' functional analysis, causal role, approach to functions and the historical, etiological, evolutionary approach seems to be the major distinction among modern naturalistic theories of teleological function. Yet a number of people have prematurely declared the historical, etiological evolutionary side to be the dominant, even "orthodox", view. Boorse (2002: 74) describes Buller's declaration of the situation thus:

> Buller sees among 'current theories' of biological function a 'core of agreement' that is 'as great a consensus as has been achieved in philosophy': that 'a trait or organ has a function in virtue of its role in a selection process – either its role in a selection process that a lineage bearing that trait or organ actually has undergone, or in virtue of a selection process it is currently undergoing or is set to undergo'.

Boorse points out that this claim is exaggerated in excluding the propensity theory of Bigelow and Pargetter (1987), which focuses on functions as "current" evolutionary role. The claim also overlooks Godfrey-Smith's grounding functions on factors in modern history which maintain (not select) a trait. Buller's alleged "core of agreement" also excludes all the non-historical causal role theories, such as Cummins's (1975, 2002), which are live options that do not accept the historical orientation. In the present issue, we see again the historical evolutionary orientation labeled by Richard Cameron as the "orthodox" view, denouncing its dominance as undefended dogma. This assessment not only overlooks one entire branch of the literature of teleology and functions, but it oversimplifies the variety of etiological theories. Etiology can base teleological functions on selection of a trait as it acts in producers (Millikan 1984), or consumers (Millikan 1986), or can focus on the design aspects of selection. So I see the debate among naturalists

as having three main groups: Present-looking, Backward-looking, and Forward-looking, each with numerous subgroups. No main camp can as yet be declared the winner, much less is it "orthodox" dogma.

Beyond that, we should note the diverse bunch of theories which have been lumped together as *the* historical evolutionary *theory* of teleology. Evolution and natural selection are at work in many of them (but not all), in that only those causes which historically have served a biologically adaptive function are included in the account of the 'function' of an organ or mechanism in an organism. Exactly *which* function it is said to serve varies among the evolutionary theories. The function can be to indicate the presence of certain objects or events (Dretske 1986), or to help in the reproduction of the representational mechanism (Millikan 1984, 1986, 1989a), or to satisfy the desires of the organism (Papineau 1987). So to say there is a single historical, evolutionary, etiological theory that is the imposing orthodox view is to ignore the diversity of ways evolution and natural selection can be invoked in an account of teleology.

Moreover, as we have seen with the non-naturalistic and quasi-naturalistic theories, there are many who want out of the etiology project, who would take functions as primitive, emergent properties (Cameron 2004) or teleology as based on value as a basic property (Bedau 1990, 1991, 1992a, 1993). Yet it is still instructive that Buller and Cameron are calling the historical evolutionary naturalistic theories of teleology the standard view, even dogmatic orthodox view. It is a striking sign of the dramatic developments in taking teleology seriously. This is a far cry from the low philosophical regard teleology and functions had before the 1970s. It is this trend that makes the topic of ever growing concern and importance – teleology has come a long way toward rehabilitation.

Notes

1   For more discussion on this, see Fodor (1980).
2   Of course, this brings up all sorts of disputes about whether analysis destroys the subject of the analysis, or whether (as Quine (1960) says) "explication is elimination". I will also avoid detouring us into these problems here.

3 Godfrey-Smith (1994) also makes this point against propensity theories.

References

Allen, Collin and Mark Bekoff, 1995. "Function, Natural Design, and Animal Behavior: Philosophical and Ethological Considerations". *Perspectives in Ethology* 11: 1–47.

Ariew, André, 2002. "Platonic and Aristotelian Roots" in *Functions: New Essays in the Philosophy of Psychology and Biology*. Ed. by André Ariew, Robert Cummins, and Mark Perlman. New York: Oxford University Press, 2002.

Beckner, Morton, 1959. *The Biological Way of Thought*, New York: Columbia University Press.

Bedau, Mark A., 1990. "Against Mentalism in Teleology". *American Philosophical Quarterly* 27: 61–70.

Bedau, Mark A., 1991. "Can Biological Teleology be Naturalized?" *The Journal of Philosophy* 88: 647–655.

Bedau, Mark A., 1992a. "Goal Directed Systems and the Good". *The Monist* 75: 34–49.

Bedau, Mark A., 1992b. "Where's the Good in Teleology?" *Philosophy and Phenomenological Research* 52: 781–806. (Reprinted in Colin Allen, Marc Bekoff, George Lauder, eds., *Nature's Purposes: Analyses Of Function And Design In Biology*, MIT Press, 1997.)

Bedau, Mark A., 1993. "Naturalism and Teleology". In S. Wagner and R. Warner, eds., *Naturalism: A Critical Appraisal*, Notre Dame, IN: University of Notre Dame Press, pp. 23–51.

Bedau, Mark A., 1997. "Weak Emergence". In J. Tomberlin, ed., *Philosophical Perspectives: Mind, causation, and World*, Vol. 11, pp. 375–399, New York: Blackwell.

Bigelow, John and Robert Pargetter, 1987. "Functions". *Journal of Philosophy* 84: 181–96.

Boorse, Christopher, 1976. "Wright on Functions". *Philosophical Review* 85: 70–86.

Boorse, Christopher, 1977. "Health as a Theoretical Concept". *Philosophy of Science* 44: 542–73.

Boorse, Christopher, 2002. "A Rebuttal on Functions" in *Functions: New Essays in the Philosophy of Psychology and Biology*. Ed. by André Ariew, Robert Cummins, and Mark Perlman. New York: Oxford University Press, 2002.

Cameron, Richard, 2004. "How to be a Realist about *sui generis* Teleology – Yet Feel at Home in the 21st century". *The Monist*. 87: 1, pp. 72–95.

Carnap, Rudolf, 1950. *Logical Foundations of Probability*, 2nd ed., Chicago: University of Chicago Press.

Churchland, Patricia Smith, 1986. *Neurophilosophy: Toward a Unified Understanding of the Mind-Brain*. MIT Press, Cambridge, Mass.

Churchland, Paul M., 1979. *Scientific Realism and the Plasticity of the Mind*, Cambridge University Press, Cambridge.

Churchland, Paul M., 1981. "Eliminative materialism and the Propositional Attitudes". *Journal of Philosophy* 78: 2, pp. 67–90.

Churchland, Paul M., 1984. *Matter and Consciousness*, MIT Press, Cambridge, Mass.

Churchland, Paul M., 1989. *A Neurocomputational Perspective: The Nature of Mind and the Structure of Science*, MIT Press, Cambridge, Mass.

Cummins, Robert, 1975. "Functional Analysis". *The Journal of Philosophy* 72: 741–65.

Cummins, Robert, 2002. "Neo-Teleology" in *Functions: New Essays in the Philosophy of Psychology and Biology*. Ed. by André Ariew, Robert Cummins, and Mark Perlman. New York: Oxford University Press, 2002.

Dennett, Daniel C., 1981. 'Three Kinds of Intentional Psychology', in *Reduction, Time, and Reality*, edited by R. Healey, Cambridge University Press, Cambridge, reprinted in Dennett 1989, pp. 43–68.

Dennett, Daniel C., 1983. 'Intentional Systems in Cognitive Ethology: The "Panglossian Paradigm" Defended'. *Behavioral and Brain Sciences* 6, pp. 343–90, reprinted in Dennett 1987, 237–268.

Dennett, Daniel C., 1987. *The Intentional Stance*, MIT Press, Cambridge, Mass.

Dennett, Daniel C., 1995. *Darwin's Dangerous Idea: Evolution And The Meanings Of Life*, New York: Simon & Schuster.

Dipert, R. R., 1993. *Artifacts, Art Works, and Agency*, Philadelphia: Temple University Press.

Dretske, F. I., 1986. 'Misrepresentation' in R. Bogdan Ed. *Belief*. Oxford University Press, Oxford.

Enc, Berent, 2002. "Indeterminacy of Function Attributions", in *Functions: New Essays in the Philosophy of Psychology and Biology*. Ed. by André Ariew, Robert Cummins, and Mark Perlman. New York: Oxford University Press, 2002.

Fodor, Jerry, 1980. 'Methodological Solipsism Considered As a Research Strategy in Cognitive Psychology'. *Behavioral and Brain Sciences* 3: 1, reprinted in Fodor, *RePresentations*, MIT Press, Cambridge, Mass. 1981.

Fodor, Jerry, 1987. *Psychosemantics*, Bradford/MIT Press, Cambridge, Mass.

Fodor, Jerry, 1990. "A Theory of Content I" in *A Theory of Content and Other Essays*. Bradford/MIT Press, Cambridge, Mass.

Godfrey-Smith, Peter, 1993. "Functions: Consensus Without Unity" *Pacific Philosophical Quarterly* 74: 196–208.

Godfrey-Smith, Peter, 1994. "A Modern History Theory of Functions" (originally *Nôus*. 28: 344–362), in *Nature's Purposes: Analyses of Function and Design in Biology* Ed. by Colin Allen, Marc Bekoff, and George Lauder. Cambridge, MA: MIT Press. 1998.

Goldman, Alvin, 1976. "Discrimination and Perceptual Knowledge". *Journal of Philosophy*, 73: 771–791.

Goldman, Alvin, 1986. *Epistemology and Cognition*, Harvard University Press, Cambridge, Mass.

Griffiths, P. E., 1993. "Functional Analysis and Proper Functions". *The British Journal for the Philosophy of Science* 44: 409–422.

Kitcher, Philip, 1993. "Function and Design" *Midwest Studies in Philosophy XVIII*, Minneapolis: University of Minnesota Press, pp. 379–397.

Kripke, Saul, 1982. *Wittgenstein on Rules and Private Languages*, Harvard University Press, Cambridge, Mass.

Mayr, Ernst, 1974/1988. "The Multiple Meanings of Teleological" reprinted with a new postscript in E. Mayr 1988 *Towards a New Philosophy of Biology*, Cambridge, Mass: MIT Press.

Millikan, Ruth Garrett, 1984. *Language, Thought, and Other Biological Categories: New Foundations for Realism*, Bradford/MIT Press, Cambridge, Mass.

Millikan, Ruth Garrett, 1986. "Thoughts Without Laws: Cognitive Science Without Content". *Philosophical Review* 95: 47–80.

Millikan, Ruth Garrett, 1989a. "In Defense of Proper Functions". *Philosophy of Science* 56: 288–302. (also in Millikan 1993)

Millikan, Ruth Garrett, 1989b. "Biosemantics". *The Journal of Philosophy* 86: 281–297. (also in Millikan 1993)

Millikan, Ruth Garrett, 1989c. "An Ambiguity in the Notion of 'Function'". *Biology and Philosophy* 4: 2, pp. 172–176.

Millikan, Ruth Garrett, 1990. "Truth Rules, Hoverflies, and the Kripke-Wittgenstein Paradox", in *White Queen Psychology and Other Essays for Alice*. MIT Press, Cambridge, Mass. 1993.

Millikan, Ruth Garrett, 1993. *White Queen Psychology and Other Essays for Alice*, MIT Press, Cambridge, Mass.

Nagel, Ernst, 1961. *The Structure of Science: Problems in the Logic of Scientific Explanation*, New York: Harcourt, Brace & World.

Nagel, Ernst, 1977. "Teleology Revisited", in *Teleology Revisited and Other Essays in the Philosophy and History of Science*, Ed. by E. Nagel, New York: Columbia University Press, 1979.

Neander, Karen, 1991a. "The Teleological Notion of 'Function'". *Australasian Journal of Philosophy* 69: 454–468.

Neander, Karen, 1991b. "Functions as Selected Effects: The Conceptual Analyst's Defense". *Philosophy of Science* 58: 168–1–84.

Neander, Karen, 2002. "Types of Traits: The Importance of Functional Homologues" in *Functions: New Essays in the Philosophy of Psychology and Biology*. Ed. by André Ariew, Robert Cummins, and Mark Perlman. New York: Oxford University Press, 2002.

Nissen, Lowell, 1997. *Teleological Language in the Life Sciences*, Lanham, Md.: Rowman & Littlefield.

Papineau, David, 1987. *Reality and Representation*, Blackwell, Oxford.

Perlman, Mark, 1997. "The Trouble With Two-Factor Conceptual Role Theories". *Minds and Machines* 7: 495–513.

Perlman, Mark, 2000. *Conceptual Flux: Mental Representation, Misrepresentation, and Concept Change*, Kluwer Academic Publishers.

Perlman, Mark, 2002. "Pagan Teleology" in *Functions: New Essays in the Philosophy of Psychology and Biology*. Ed. by André Ariew, Robert Cummins, and Mark Perlman. New York: Oxford University Press.

Plantinga, Alvin, 1993a. *Warrant: The Current Debate*, New York: Oxford University Press.

Plantinga, Alvin, 1993b. *Warrant and Proper Function*, New York: Oxford University Press.

Plantinga, Alvin, 2000. *Warranted Christian Belief*, New York: Oxford University Press.

Pollock, John, 1989. *How to Build a Person: A Prolegomenon*, Cambridge, MA: MIT Press.

Pollock, John, 1995. *Cognitive Carpentry: A Blueprint for How to Build a Person*, Cambridge, MA: MIT Press.

Pollock, John, 1986. *Contemporary theories of knowledge*, Lanham, MD: Rowman and Littlefield.

Preston, Beth, 1998. "A Pluralist Theory of Functions". *Journal of Philosophy* XCV: 5, 215–254.

Quine, W. V. O., 1951. "Two Dogmas of Empiricism". reprinted in Quine (1953).

Quine, W. V. O., 1953. *From a Logical Point of View*, Harvard University Press, Cambridge, Mass.

Quine, W. V. O., 1960. *Word and Object*, MIT Press, Cambridge, Mass.

Quine, W. V. O., 1969. *Ontological Relativity and other essays*. Columbia University Press, New York.

Rosenbleuth, A., Wiener, N., and Bigelow, J., 1943. "Behavior, Purpose and Teleology". *Philosophy of Science* 10: 18–24.

Rowlands, Mark, 1997. "Teleological Semantics". *Mind* 106: 279–303.

Ruse, Michael, 1971. "Functional Statements in Biology". *Philosophy of Science* 38: 87–95.

Schiffer, Michael, 1992. *Technological Perspectives on Behavioral Change*, Tucson: University of Arizona Press.

Searle, J., 1990. "Collective Intentions and Actions". In Philip R. Cohen, Jerry Morgan, and Martha E. Pollack, eds., *Intentions in Communication*. Cambridge, MA: MIT Press.

Vermaas, Pieter E. and Wybo Houkes, 2003. "Ascribing Functions to Technical Artifacts: A Challenge to Etiological Accounts of Functions". *British Journal of the Philosophy of Science* 54: 2, pp. 261–289.

Walsh, Denis. and André Ariew, 1996. "A Taxonomy of Functions". *Canadian Journal of Philosophy* 26: 4.

Wimsatt, W.C. 1972. "Teleology and the logical structure of function statements". *Studies in the History and Philosophy of Science* 3: 1–80.

Wittgenstein, Ludwig, 1953. *Philosophical Investigations*. trans. G.E.M. Anscombe, 3rd edition. Macmillan, New York.

Woodfield, Andrew, 1976. *Teleology*, Cambridge: Cambridge University Press.

Woolfolk, Robert, 1999. "Malfunction and Mental Illness". *Monist* 82: 4, pp. 658–670.

Wright, L. 1973. "Functions". *Philosopical Review* 82: 139–168.

Wright, L. 1976. *Teleological Explanations*, Berkeley: University of California Press.

12

Neo-Teleology

Robert Cummins

1. Two Species of Functional Explanation

There are two subpopulations of functional explanation roaming the earth: teleological explanation, and functional analysis. The two are in competition. In this chapter, I hope to help select the latter, and nudge the former to a well-deserved extinction.

1.1. Teleology

Teleology is the idea that some things can and should be explained by appeal to their purpose or goal or function. It is, for example, the idea that one can explain why rocks fall and fire rises by appeal to the fact that the goal of matter is to go to its natural place, and that this is down for rocks and up for fire. It is also the idea that one can explain why (though not how) an acorn grows into an oak (rather than a beech or a clam) by appealing to the fact that the goal or function of a growing acorn is to become an oak tree. More plausibly, teleological explanation seeks to account for the existence or presence of a biological trait, or structure or behavior *by appeal to its function*.

It is said that animals that have hearts have them because of what hearts are for.[1] Hearts are for circulating the blood; they are not for generating a pulse. Therefore, circulating the blood is their function, and they are 'there' – animals have them – because they perform this function.

1.2. Functional Analysis

Teleological explanations and functional analyses have different explananda. The explanandum of a teleological explanation is the existence or presence of the object of the functional attribution: the eye has a lens because the lens has the function of focusing the image on the retina. Functional analysis instead seeks to explain the capacities of the system containing the object of functional attribution. Attribution of the function of focusing light is supposed to help us understand how the eye, and, ultimately, the visual system, works. In the context of functional analysis, a what-is-it-for question is construed as a question about the contribution 'it' makes to the capacities of some containing system.

While teleology seeks to answer a why-is-it-there question by answering a prior what-is-it-for

Robert Cummins, "Neo-Teleology," in André Ariew, Robert Cummins, and Mark Perlman (eds.) *Functions: New Essays in the Philosophy of Psychology and Biology*. Oxford: OUP (2002), pp. 157–172.

I received many hours of feedback and help with this chapter from the members of my weekly lab group: Pierre Poirier, Jim Blackman, David Byrd, and Martin Roth. I want also to thank Denise Cummins, André Ariew, and Mark Perlman for their extensive comments, as well as an anonymous reviewer for the Press.

question, functional analysis does not address a why-is-it-there question at all, but a how-does-it-work question. These last are answered by specifying the structure (design) of the system. Rube Goldberg devices are natural candidates for this sort of explanation. In my horse pasture, I have a device that opens, at a pre-set time, a gate dividing the pasture in two. Here is how it works. There is a wind-up alarm clock. When the alarm on the wind-up alarm clock goes off, a string wound on the key-stem unwinds, releasing a ratchet on a pully. A weight on one end of a rope over the pully falls, jerking open the gate latch attached to the other end of the rope. This is, if you like, a rather abstract mechanical description. It is also a functional analysis of the capacity to open the gate as this is realized in my Rube Goldberg device. The components are identified functionally, and their interactions are described in a way that, necessarily, abstracts away from the medium-dependent details.[2] When we understand how the thing works in the way provided by a functional analysis, we understand how others might be built – how other instantiations of the same design could do the same job, and, perhaps, do it better. This is possible, because the system and its components are specified functionally, and hence in a way that allows for multiple instantiations. By substituting functional equivalents at various points in the design, taking care to accommodate the need for adequate interfaces with other components, we can make incremental changes in the system while preserving its overall viability. This is precisely how we must understand a system to see how it could be incrementally improved, and hence how it could evolve.

Of the two forms of functional explanation, I suspect teleology is much the oldest. Teleology is a natural framework for thinking about tools, cooking and storage utensils, and shelters. These ideas extend quite naturally to the body: eyes are tools or instruments for seeing, ears for hearing, hands for grasping, teeth and jaws for chewing. Functional analysis, on the other hand, got a grip on the mind, I suspect, only with the invention of relatively complex artifacts. Carts and harnesses lend themselves to functional analysis. Machines such as catapults and water clocks are unthinkable without it. This kind of thinking extends naturally to social structures such as bureaucracies, and to complex anatomical systems: the digestive system, the circulatory system, the nervous system.

There is more to proto-teleology than attributing functions to tools and sense organs, however. What I am calling teleology is the idea that an appeal to something's function can explain 'why it is there': why there are hammers and hands. If having a function is to explain why a thing, or type of thing, exists, then there must be some background story about a mechanism or process that produces the items in question, and produces them because of their functions. It is this requirement for what I will call a *grounding process* that has proved to be the Achilles heel of teleology.

Different kinds of phenomena subject to teleological explanation have required different grounding processes. Teleological mechanics appealed to the selective attractiveness of natural places. The intentions, plans, and actions of designers, creators, and manufacturers have been rung in to support teleological explanations of quite literally everything, and remain popular as underpinnings of teleological explanations of artifacts. In the hands of Aristotle, and Hans Dreisch (1867–1941), teleological developmental biology appealed to the regulating capacites of entelochies, a sort of inner goal-directed agent. Finally, natural selection has become a popular grounding process for the teleological explanation of biological traits, and sometimes for traits of artifacts as well.

Teleological explanation of motion failed because the grounding processes were transparently insensitive to function. Even if one could make sense of natural places, any force or mechanical constraint that would get something to its natural place would get it there whether or not it was the function of the thing to go there. To take an example from Ptolemaic astronomy, if a star has its apparent motion because it is attached to a rigid moving sphere, centered at the earth, it will trace a circular orbit around the center of the sphere regardless of what its function happens to be. The same point holds of Newtonian gravitational explanations of planetary orbits. Teleological appeal to functions in mechanics therefore appears idle and misleading. Indeed, it no longer seems plausible to suppose that celestial bodies have mechanical functions at all. Like all teleology, teleological mechanics requires a grounding process. But the only grounding processes likely to satisfy render the appeal to functions utterly superfluous.[3]

Teleological explanation of growth and development fared even worse. The need for a grounding process spawned vitalism and the doctrine of entelochies. Entelochies could not be found. Moreover, appeals to them are regressive, since their own guiding and regulating behavior was itself teleologically explained, but without the hope of a corresponding grounding process. The whole misconceived enterprise rapidly became extinct when advances in cellular and molecular biology generated more adaptive theories competing for the same niche. Like post-teleological mechanics, those theories also appeal to factors that are transparently insensitive to the functions that were crucial to the teleological stories. So transparent is this, in fact, that it now seems silly to think it is the function of an acorn to develop into an oak rather than a birch, and that this explains why planting acorns never yields birch trees.

2.   Neo-Teleology

Nobody much likes teleological mechanics or teleological developmental biology any more. It has been eliminated root and branch from mechanics and the other non-life sciences, recalled only by vestigial forms such as least energy principles that are without exception explained away as mere *façons de parler*. And it is similarly absent from developmental biology. But teleology survives in evolutionary biology, or anyway in the philosophy of it, as the idea that one can explain why an organism has a biological trait or structure by appeal to the function of that trait or structure. According to neo-teleologists, as I shall call them, we *have* hearts because of what hearts are *for*. Hearts are *for* blood circulation, not the production of a pulse. Hence, hearts are there – animals have them – because their function is to circulate the blood.

It is important to read the neo-teleologist claim transparently. Neo-teleologists hold that mammals have hearts because of something special that hearts do (or did). The heart, of course, does (did) lots of things. Among the

things the heart does (did) is the thing we single out as its function, and it is *that* effect of heart presence – the one that counts as its function – that accounts for the presence of hearts. Neo-teleologists do not hold, as classical teleologists *did*, that circulation had its effect because it is or was the heart's function. They hold, rather, that the effect of heart presence that accounts for heart presence – call it e – accounts for heart presence because e has the property of being a circulating of the blood, not because e has the property of being the heart's function. Thus, neo-teleology is a mere shadow of the original (mistaken) idea, having only very limited aspirations. But, in spite of being widely influential, it is still mistaken in very much the same way that classical teleology was mistaken: the only plausible grounding processes render appeal to functions superfluous and misleading.

The idea behind neo-teleology is that evolutionary biology can provide the relevant grounding process and hence get you an answer to a why-is-it-there question from an answer to a what-is-it-for question. No doubt there is a sense of 'Why is that thing there?' that is just a way of asking what it is for. I point to a little rubber hemisphere on the carbureter of your lawnmower and ask you, 'Why is that thing there?' You reply by telling me its function – 'It is for priming the engine' – and this is an appropriate and satisfactory answer. But this only means that something that looks like teleology but is not can be had cheap.[4] What I am calling neo-teleology is more than this. It is the substantive thesis that, in some important sorts of cases at least, a thing's function – the effect we identify as its function – is a clue to its existence. If it is not to degenerate into the trivial thesis that 'why is it there?' can sometimes just mean 'what is it for?', neo-teleology must be the idea that, for example, there are eyes because they enable vision, wings because they enable flight, and opposable thumbs because they enable grasping.

3.   Neo-Teleology and Natural Selection

Neo-teleology as just construed has no lack of able defenders (Millikan 1984; Neander 1991; Griffiths 1993; Kitcher 1993; Godfrey-Smith

1994; Allen and Beckoff 1995).[5] Generally, these authors are associated with selectionist etiological accounts of functions. Notice, however, that a defense of a selectionist etiological account of functions is, in effect, a defense of neo-teleology, since selectionist etiological accounts of functions equate functional attributions with what I am calling neo-teleological explanations: to say the function of the heart is to circulate the blood is, on these accounts, to offer a neo-teleological explanation of the presence of hearts. I prefer to attack the position by attacking the explanations in question rather than a thesis about what functions are, since evolutionary theory bears directly on the viability of the explanations, and only indirectly on the thesis that functional attributions are equivalent to such explanations.

Contemporary defenses of neo-teleology all share a basic selectionist strategy. The underlying idea is that traits are selected for because of the effects that count as their functions, hence exist in organisms because they have (or had) the functions they do (or did). Neo-teleology is thus packaged as what appears to be an uncontroversial part of the theory of natural (or artificial) selection. No natural places, entelechies, designer's intentions or other skyhooks (Dennett 1995) appear to taint neo-teleology; it is selectionist through and through. Since selectionist explanations are clearly legitimate scientific explanations, how could anyone object to neo-teleology? Surely we have here a subspecies of teleology that has found a legitimate grounding process.

And yet I am unpersuaded. Biological traits once explained by a teleology grounded in appeals to the intentions, plans, and actions of a creator have, in discerning minds, given way to appeals to evolution generally, and to natural selection in particular.[6] Neo-teleologists want to read this as the discovery of a legitimate grounding process for a teleological explanation of these traits. I am inclined to read the same intellectual development as analogous to what happened in mechanics and developmental biology: not a vindication but a replacement. The grounding processes of evolution, rightly understood, do not ground neo-teleology, because they are insensitive to function. Functions, I believe, have a legitimate place to play in science generally, and in biology in particular. But neo-teleology has the role of

functions in selectionist explanations, and hence in biology, quite wrong: Biological traits, mechanisms, organs, etc., are not there because of their functions. They are there because of their developmental histories. Functions, I believe, enter into science legitimately as elements of functional analyses. Functional analysis is a powerful explanatory strategy that is widespread in all of the sciences. I have defended this view elsewhere (Cummins 1975, 1977, 1983), and will not comment on it further here. My focus in this chapter is rather to expose what I think are the vices of neo-teleology. An understanding of functional analysis will be relevant only because, as mentioned above, it appears to be precisely the framework we need to understand how complex systems could evolve.

4.   Against Neo-Teleology

The basic idea of my argument is quickly conveyed. Traits are acquired in a variety of ways. Some are learned. Some, like sunburn, limb loss, and the effects of disease, are the direct result of environmental influences. None of these is of interest here because they are not heritable, hence are not subject to (non-cultural[7]) selection. The traits that *are* subject to selection *develop*. For convenience, in the rest of this chapter, 'traits' will be restricted to heritable traits the expression of which is the result of development and hence highly canalized.

Development is determined by a complex interaction between genes and environment. It is utterly insensitive to the function of the trait developed. Selection, on the other hand, is sensitive to the effects that are functions, but is, in the sense relevant to neo-teleology, utterly incapable of producing traits. It can preserve them only by preserving the mechanisms that produce them. Nor can selection, in the sense relevant to neo-teleology, produce the mechanisms that underwrite a trait's development; it can preserve only whatever mechanisms it finds already there.

I say selection cannot produce traits *in the sense relevant to neo-teleology*, for there is a sense in which selection can assemble complex traits or structures. This is what gives selection its awesome explanatory power. But the creative power of selection is not the kind of process to which neo-teleologists appeal. I will return to this in a later section.

If the processes that produce traits are insensitive to their functions, how can functions account for why a trait is 'there' – that is, expressed in some specified population? The contemporary neo-teleologist answer is to concede that the processes that produce traits are insensitive to their functions, since, of course, traits do not have functions until *after* they are produced. But they argue that the processes that proliferate and preserve traits in a population are not insensitive to their functions. Certain traits spread through a population over time, and the mechanisms responsible are sensitive to function. Hence we can explain spread by appeal to function. Appeal to function thus gives us a handle on why a trait survived and proliferated, and hence a handle on why it is 'there'.

Imagine that crab grass invades a patch of Mendel's pea plants. The short ones will soon have trouble getting enough sunlight. The tall ones will do better. (They will all have trouble competing for root space underground.) The tall ones will reproduce more than the short ones, and will soon be far more common than the short ones, though they may be less common (per square foot) then either the tall or short ones previous to the crab grass invasion, and may eventually be crowded out altogether. In the meantime, tallness will, as we say, have spread through the population, and will be maintained.

This sort of story is supposed to explain why the pea plants in Mendel's crab-grass-infested garden are tall.[8] And it does. But how does neo-teleology get into the picture? Well, the idea is that the function of tallness in plants, or at least in these pea plants, is to achieve access to sunlight. Since gaining access to sunlight is what explains, via selection, why Mendel's pea plants are tall, we have explained why Mendel's pea plants are tall by appeal to the function of tallness in those pea plants.

5. Functions and Spread

The fundamental problem with neo-teleology is that traits do not spread because of (the effects that count as) their functions.

We can distinguish strong and weak variations of neo-teleology. The strong variation holds that any biological trait that has a function was selected for because it performed that function. The weak variation holds only that some traits were selected because of their functions.

A trait can be selected because of its function only if having that function counts as an adaptive variation in the population. For wings to be selected for because they enable flight, there must be a subpopulation in which wings enable flight, while wings in the rest of the population do not. For hearts to be selected for because they circulate the blood, there must be a subpopulation in which hearts circulate the blood, while hearts in the rest of the population do not. While it is plausible to suppose that there was a first flight-enabling wing somewhere among the ancestors of today's sparrows, those ancestors were not sparrows, nor was the wing in question anything like a contemporary sparrow wing. Similarly, somewhere in our ancestral line is to be found the first appearance of centralized blood circulation. But those ancestors were not even vertebrates, and the structures in question were nothing like our hearts.[9] It follows from these considerations that sparrow wings and human hearts were not selected because of their functions. Selection requires variation, and there was no variation in function in the structures in question, only variation in how well their functions were performed.[10]

Strong neo-teleology is refuted if there are legitimate targets of functional characterization that are not targets of selection. Strong neo-teleology must be rejected, since most, perhaps all, complex structures such as hearts, eyes, and wings patently have functions but were not selected because of (the effects that count as) their functions. And, since the selectionist etiological account of functions stands or falls with neo-teleology, it must be rejected as well, not because it is bad conceptual analysis (whatever that is), but because it equates functional attributions with bad evolutionary explanations.[11]

Weak neo-teleology survives this objection, but at a very considerable price. Weak neo-teleology comes out true only because of the rare though important cases in which the target of selection is also the bearer of a function that accounts for the selection of that trait. These will be cases in which genuine functional novelty is introduced; a trait present in a subpopulation that is not just better at performing some function that is also performed in competing subpopulations (though not as well), but a trait that performs a function that is not performed at all by any counterpart mechanism in competing subpopulations. This unquestionably happens, and the importance of such seeding events should not be underestimated. But complex structures such as sparrow wings and human hearts were not introduced in this way. They were selected because they were better[12] at performing some function that was also performed by the competition. It follows from the equivalence of neo-teleology and selectionist accounts of function that these accounts will limit function attribution to those traits for which neo-teleology comes out true – namely, traits in which selection was triggered by the fact that the trait in question had a function that was entirely novel in the relevant population.

This is not merely a defense of gradualism. You do not have to be the village gradualist to be skeptical of the idea that there was variability in the presence or absence of T whenever T is rightly said to have a function. The point is rather that whether or not something has a function, and what that function happens to be, is quite independent of whether it was selected and spread. When we look for a place for selection to act on wings, say, we need to be looking for variations in wing design. All of the variant wings will have the same function – to enable flight. Thus, one cannot look to differences in the function of the wings to predict or explain selection. One must look instead to how well the various wings are functioning, and this means looking at the functions, not of the wings, but of something else: feather design, bone structure, musculature, and so on. Moreover, this argument iterates. It is the better of two muscle attachment schemes that gets selected; both the better scheme and the inferior scheme have the same function. Functions just do not track the factors driving selection. No doubt there *are* cases in which one subpopulation acquires some structure or behavior that the rest of the population just does not have, a biological analogue of adding a governor to steam engines, or an escapement to clocks. But such cases must be quite rare.[13] If they exhaust the proper

domain of neo-teleology, then neo-teleology is insignificant at best. It comes out true as a kind of accident, a coincidence in the rare sort of case in which selective advantage happens to coincide with the introduction of something with a novel (in that context) function.

Selection can, to some degree at least, be explained by appeal to adaptiveness, although the connection between adaptiveness and selection is more indirect than is sometimes appreciated. What is uncontroversial is that a trait spreads because it is heritable and appears in a host that is more fit than the competition. Exactly the same thing can be said with equal truth about every trait of that host. Every trait of the winning host spreads, regardless of how adaptive it is – regardless, indeed, of whether it is adaptive (or has a function) at all. But this does not render adaptiveness irrelevant to selection, since the host in question was more fit than the competition because of some traits and in spite of others.[14] If H was a better design (in part) because of T, and all of H's traits spread because H was a better design, then T's positive contribution – its *adaptiveness*, in short – helps explain why it (and its neighbors) spread.

This suggests the possibility of saving neo-teleology by defining functions in terms of adaptiveness. This would turn neo-teleology into the idea that the proliferation and maintenence of some traits can be explained by appeal to the fact that they were adaptive. I certainly do not wish to take issue with that claim, though I think there are reasons for caution.[15] I do think, however, that there are good reasons to keep having a function and being adaptive distinct, and it is worth taking a brief detour to canvass these, for it will lead us back to the main point via another route.

Adaptiveness is a matter of degree; having a function is not. The more adaptive wing and its less adaptive competition both have the same function, but only the former is selected for. Functioning *better* is a matter of degree, and it is at least sometimes true that the more adaptive wing functions better. But this just makes it clear that functional analysis is prior to, and independent of, assessments of adaptiveness. When we have a system analyzed functionally, we are in a position to ask what sort of improvements could be made by substitution of functional equivalents.

The substitution of a functional equivalent that is (for example) more efficient, increases adaptiveness, but, by hypothesis, does not change anything's function.

The point here is not, as selectionist etiological accounts would have it, that only the selected wing has a function. After all, the worse wing was once the better one and was itself selected for. The point is rather that having a function is not what drives selection, but rather functioning better than the competition. What the function of a wing is should be distinguished from how well it performs it. The question of what function something has is evidently prior to the question of how well it is performed in a given organismic and environmental context, and hence prior to the question of how adaptive performing that function is for a given organism in a given environment. To repeat, the better and worse wings have the same function, but only the former spreads.

It might seem that there is a link of sorts between functions and adaptiveness, and hence between functions and selection. Knowing that the function of hearts is to circulate blood might be thought to constrain what sorts of variation in heart design would be adaptive, and hence what sorts of variations might be targets of selection. Indeed, I have been saying that it is the heart design that enables better circulation that gets selected. This suggests that when we identify the function of a trait, we have identified the dimension of performance that is relevant to assessing the adaptiveness of that trait. Circulation, not pulse production, is the function of the heart, and so it is variations in heart design that improve circulation, not variations that improve pulse production, that matter to adaptiveness.

Attractive as this line is, I do not think it will stand scrutiny. Better wing designs need not improve flight, but simply make it more efficient, or make development less error prone, or make the structure less fragile. Hence, selected changes in wing design that accumulate to yield the current design we seek to explain need not be related to the wing's function. Indeed, they may even compromise flight in the interest of other factors. Hence, if we are trying to understand why a given trait or structure is the way we find it, we cannot simply focus on variations that affect how well that trait or structure performs its function. We need, instead, to look at the complex economy of the whole unit of selection. This is precisely what a functional analysis of the whole unit facilitates, and is neglected when we focus on the function or functions of the trait in question.

6. Paley Questions

Even if we could make sense of the idea that things like wings and eyes – salient targets of functional attribution – spread through previously wingless and eyeless populations, the serious why-is-it-there question about such things as wings and eyes would remain untouched. How did there come to be such things in the first place? To harken back to Paley's famous example (1802), when we discover a watch in the wilderness, we are likely to infer a designer, not because we wonder why watches became so popular, but because we cannot otherwise understand how such a thing could come to exist at all.[16] And this is precisely the difficulty with eyes and wings. I propose to call this sort of why-is-it-there question a Paley question.

It is pretty generally conceded, I think, that Paley questions cannot be given neo-teleological answers (Godfrey-Smith 1994). Selection presupposes something to be selected. You cannot select for creatures with eyes unless eyes already exist. So it looks like selection cannot even address Paley questions. But, of course, this is much too quick.

Selection *can* address Paley questions, but only indirectly. The selection of eyes, or sighted organisms, is the wrong place to look. Selection builds a complex structure like a human eye or a sparrow wing by successive approximation (or what looks like it retrospectively) in relatively small steps beginning with an organism without an eye or wing and ending with what we observe today. Many of the fine details of such stories are unknown. Yet the in-principle possibility of the process is enough to provide the answer to Paley's original challenge: to explain how such things as the human eye came to exist in the first place without reference to the intentions, plans and actions of an intelligent creator and designer of eyes. Natural selection is clearly a central player in the sort of story that has successfully met this challenge. But it enters in by accounting for the spread of small modifications to precursor structures. To think of the modern human eye or sparrow wing as itself selected is, to repeat, to conjure up a scenario in which there is a population of sightless primates or wingless songbirds into which is born a sighted or winged variation whose progeny take over the land or air. No one, of course, really believes anything like this. Yet something very like this is implied by neo-teleology – by the idea that eyes are there because they enable sight and wings because they enable flight. The modern human eye or sparrow wing never spread through any population. Some small changes to earlier structures very like the modern human eye or sparrow wing may have spread. And small changes to those structures may have spread. And so on. In short, as we have already seen, targets of functional characterization and targets of selection just do not match.

To summarize: if we ask why some complex structure is 'there', in the sense in which this means how it came to exist, appeal to its function or functions, as teleology (neo and classical) requires, is only going to be misleading. Such stories either run into the fact, fatal to classical teleology, that the crucial details of evolutionary (or ontogenic) development predate anything with the function that is supposed to do the explaining, or they founder on the fact that competing traits in selection scenarios typically have the same function. Things do not evolve because of their functions any more than they develop because of their functions.

It *is* generally conceded that teleology does not address Paley questions. But we are now in a position to see that Paley questions are all the questions there are about the evolution of traits. The idea that, although eyes and wings did not come to exist because of their functions, they nevertheless *spread* because of their functions, leaves us with a distorted picture of the role of selection. It makes us think that selection can spread only what is already there. While this is true in a sense, it is seriously misleading when we focus on the kinds of traits that have salient functions. It makes us think that eyes – eyes like ours – came to be somehow (some massive mutation?), and then were selected for because they were so adaptive. When we explain how eyes like

ours came to be in the first place, we have said all there is to say about spread. When we have answered Paley's question, we have answered the evolutionary question. There is nothing left over for spread to do that it has not already done.

7. Conclusion

Let us consolidate our results. Neo-teleology, the idea that traits are there because of the effects that are their functions, is a non-starter when it comes to serious why-is-it-there questions: the questions I have called Paley questions. Appeals to function fail to address Paley questions, because nothing in the relevant lineage has the function in question until the trait in question is created. When it comes to Paley questions, neo-teleology has nothing to add to classical teleology. This is quite generally acknowledged. But neo-teleology fares no better as a story about why traits spread. Substantive neo-teleology misidentifies the targets of selection with the sort of complex generically defined traits – having eyes or wings – that have salient functional specifications.

Neo-teleology, I find, dies hard. Its rejection sounds to many like rejection of evolution by natural selection. But it is not. Darwin's brilliant achievement has no more need of neo-teleology than it has for its classical predecessor. What it needs is a conception of function that makes possession of a function logically independent of selection and adaptiveness. For it is only by articulating a reasonably illuminating functional analysis of a system that we can hope to understand *what* it is that evolution has created. If we want to understand *how* it was created as well, there is no avoiding the messy historical details by the cheap trick of assuming that all we have to do to understand trait proliferation and maintenance is to attribute a function. Neo-teleology thus amounts to a license to bypass the messy and difficult details, to jump over them in a way that makes it seem that the whole process was like the progress of a heat-sensing missile, arriving more or less inevitably at its goal regardless of the vicissitudes of wind and the meanderings of the target. The idea that evolution and development are goal oriented is precisely what makes classical teleology unacceptable. Neo-teleology creates the same impression while masquerading as good Darwinian science.

There is another nexus of reasons why neo-teleology hangs on, at least in Philosophy. Twentieth-century empiricist philosophers such as Hempel (1959) were worried about function talk in science because it smacked of (classical) teleology. They set out to determine whether functions have a legitimate role in science. For reasons I am not clear about, they took this to be an issue about functional explanation, and interpreted *that* as a question about whether things could be explained by appeal to their functions. Thus, one important strand in the debate over functions simply assumed that the legitimacy of functions and the legitimacy of neo-teleology were one and the same. In Cummins (1975) I argued that this was a mistake; that functional attribution and functional analysis could be, and often are, decoupled from explaining why things are there by appeal to their functions. Still, the idea that functional attributions are equivalent to neo-teleological explanations remains widespread.

However, even if you accept that functional explanation and functional description can be decoupled from teleological explanation (some do, some do not), it might seem that the original empiricist worry remains about functions. One might continue to think that they need, in current parlance, to be *naturalized*. But most, perhaps all, of the pressure to naturalize functions is really pressure to naturalize teleology. Once functions are separated from teleology, they do not look any more likely to offend empiricist scruples than any other dispositional properties. But this point is not widely appreciated, and therefore there still is, I think, a widespread feeling that functions need naturalizing, and that this amounts to naturalizing (neo-)teleology.

There is a different sort of philosophical problem that remains, however. It is pretty generally agreed that a thing's function (or functions) is some special class of its effects. The problem of analyzing functional attributions, then, seems to require some criterion for saying which effects count as functions. Why is blood circulation a function of the heart and not production of a pulse? Selectionist etiological accounts seem to many to provide an elegant solution to this problem: the functions of an X are those effects of an X that, historically, account for Xs having been selected.[17] I have, in effect, been arguing against the selectionist etiological account of functions in biology on the grounds that the targets of functional attribution are seldom the targets of selection. If I am right, then almost nothing has a function in the sense staked out by selectionist etiological accounts of what functions are. This, I think, is what Hempel (1959) did conclude. We are better off abandoning the selectionist etiological account of functions.

Notes

1 Paul Davies (2001) holds that something's function should not be identified with what it is for, since this builds an unacceptable sense of 'design' – one involving intentional considerations – into the concept of function. I have some sympathy with this, but am prepared, for the purposes of this chapter, to let "What is it for?" be a way of asking the same question as "What is its function?"

2 Most causal analysis is like this. When we describe causal interactions between functionally characterized components, the relevant causal generalizations pretty much come for free, since a functionally characterized component is a component identified by its relevant causal powers.

3 It might seem that there is mostly a difference in attitude between saying that masses follow geodesics unless disturbed and saying that their function is to follow their natural paths. The standard contemporary reply to this sort of worry is to say that functions are normative, and, since there is no question of non-geodesic motion being a malfunction, there is no place for functions in mechanics. But this misses the point. The point is rather that the grounding process winds up accounting for the motion by appeal to factors such as forces or mechanical constraints that could not be sensitive to function in any case.

4 It might also mean that teleology used to be uncontroversial, so that the two expressions seemed to mean the same thing.

5 See Buller (1999) for a collection of papers defending and elaborating some version or other of what I am calling neo-teleology. Notice that a defense of a selectionist account of functions is, in effect, a defense of neo-teleology, since selectionist accounts equate functional attribution with neo-teleological explanation.

6 Teleological appeal to designers, creators, and manufacturers to explain artifacts is still widespread.

The function of an escapement in a clock is said to explain its presence in a way that is grounded in the intentions of designers, and manufacturers. However, selectionist treatments are also popular: the escapement is said to be there because it solved a problem plaguing pre-escapement clocks, leading consumers to prefer escapement clocks. The resulting pressures of the marketplace then led to the (near) extinction of pre-escapement clocks. (These, in turn, are in the process of being replaced by electronic clocks that require no escapement, of course.)

7 I am going to ignore cultural evolution in this chapter. I think the points I make here against selectionist defenses of neo-teleology would apply to neo-teleological stories about cultural selection as well, but I have not investigated this issue.

8 It does not, in my view, explain why any particular plant is tall. See Sober (1984, 1995), and Pust (2001). Neander (1995a, b) argues for the opposing view. And it does not explain why all the pea plants in the garden are tall, since short ones will continue to occur, though they seldom reach maturity.

9 Even these scenarios are misleading in suggesting that flight or centralized circulators appeared suddenly on the scene. Circulation was probably centralized gradually, and early flight was no doubt a matter of short and ill-controled forays into the air.

10 Another way of putting this point is that complex structures such as human hearts and sparrow wings are not heritable traits. What is heritable, at most, are variations in these traits. This follows from the fact that heritability is a measure of how much of the variance is accounted for by genes. Hearts in humans and wings in sparrows are not heritable because there is no variance to account for.

11 One could deny the validity of neo-teleological explanation and still hold that functional attributions were disguised neo-teleological explanations. Presumably, someone holding this position would advocate abandoning functional attributions. Perhaps Hempel (1959) is an example. But contemporary defenders of selectionist etiological accounts of functions think of themselves as vindicating functional attribution by identifying if with what they take to be a form of viable evolutionary explanation.

12 Even this is too strong, and will be modified shortly.

13 Mutation, for example, is much more likely to change the size, density, shape, or attachment angle of a bone than to add a new bone. The altered bone will typically have the same function as its competitors.

14 This is Sober's distinction (1984a) between being selected and being selected for.

15 One needs to be careful with the idea that traits spread because they are adaptive. The underlying rationale is that adaptive traits are likely to give their hosts the kind of advantages that lead to greater reproductive success. Hence, over the long haul, the subpopulation that has the trait in question is likely to grow relative to the rest of the population. The resulting *spread* of the trait in question through the population is the essence of selection.

Two points need mentioning. First, for this story to be substantive, we require a conception of adaptiveness that makes it independent of fitness. Secondly, whether adaptive traits spread depends on the extent to which conditions approach what we might call 'full-shuffle' conditions – i.e. conditions under which there is a fixed pool of heritable traits that do not interact, and every combination of them gets tried out in the fullness of time in a fixed environment. (See Kaufman 1989, 1993, on the importance of trait interaction.) That these conditions are seldom if ever satisfied in complex organisms is evident. The wonder is that natural selection works at all, given the poor working conditions with which it is faced.

16 Of course, if watches are popular, you are more likely to find one. But Paley's beachcomber did not want an explanation of why a watch was found, but of why there were any watches to find.

17 This is sometimes confused with the idea that the functions of X are those effects of X that were adaptive – i.e. that contributed to the fitness of their hosts. This could be true, even though Xs were not selected because of those effects, or even though Xs were not selected at all. Selection presupposes variability; positive contributions to fitness do not.

References

Allen, Colin, and Beckoff, Marc (1995), 'Biological Function, Adaptation and Natural Design', *Philosophy of Science*, 62: 609–22.

Buller, David (1999) (ed.), *Function, Selection and Design: Philosophical Essays* (Albany, NY: SUNY Press).

Cummins, Robert (1975), 'Functional Analysis', *Journal of Philosophy*, 72/20: 741–65.

Cummins, Robert (1977), 'Programs in the explanation of behavior', *Philosophy of Science*, 44: 269–87.

Cummins, Robert (1983), *The Nature of Psychological Explanation* (Cambridge, Mass.: MIT Press).

Davies, Paul S. (2001), *Norms of Nature: Naturalism and the Nature of Functions* (Cambridge, Mass.: MIT Press).

Dennett, Daniel (1995), *Darwin's Dangerous Idea: Evolution and the Meanings of Life* (New York: Simon & Schuster).

Godfrey-Smith, Peter (1994), 'A Modern History Theory of Functions', *Noûs*, 28: 344–62.

Griffiths, Paul (1993), 'Functional Analysis and Proper Functions', *British Journal for the Philosophy of Science*, 44: 409–22.

Hempel, C. G. (1959), 'The Logic of Functional Analysis', repr. in C. G. Hempel, *Aspects of Scientific Explanation, and Other Essays in the Philosophy of Science* (New York: Macmillan, 1965), 297–330.

Kauffman, S. A. (1989), 'Origin of Order in Evolution: Self-Organization and Selection', in B. C. Goodwin and P. Saunders (eds.), *Theoretical Biology: Epigenetic and Evolutionary Order from Complex Systems* (Edinburgh: Edinburgh University Press).

Kauffman, S. A. (1993), *The Origins of Order: Self-Organization and Selection in Evolution* (New York: Oxford University Press).

Kitcher, Phillip (1993), 'Function and Design', *Midwest Studies in Philosophy*, 18: 379–97.

Millikan, Ruth Garrett (1984), *Language, Thought, and Other Biological Categories: New Foundations for Realism* (Cambridge, Mass.: MIT Press).

Neander, Karen (1991), 'The Teleological Notion of "Function"', *Australasian Journal of Philosophy*, 69: 454–68.

Neander, Karen (1995a), 'Pruning the Tree of Life', *British Journal for the Philosophy of Science*, 46: 59–80.

Neander, Karen (1995b), 'Explaining Complex Adaptations: A Reply to Sober's Reply to Neander', *British Journal for the Philosophy of Science*, 46: 583–87.

Paley, William (1802), *Natural Theology: or, Evidences of the Esistence and Attributes of the Diety, Collected from the Appearances of Nature* (London: Faulder).

Pust, Joel (2001), 'Natural Selection Explanation and Origin Essentialism', *Canadian Journal of Philosophy*, 31 (June), 201–20.

Sober, Elliot (1984a), 'Force and Disposition in Evolutionary Theory', in C. Hookway (ed.), *Minds, Machines and Evolution* (Cambridge: Cambridge University Press), 43–62.

Sober, Elliot (1984b), *The Nature of Selection: Evolutionary Theory in Philosophical Focus* (Cambridge, Mass.: MIT Press).

Sober, Elliot (1995), 'Natural Selection and Distributive Explanation: A Reply to Neander', *British Journal for the Philosophy of Science*, 46: 384–97.

13

A Modern History Theory
of Functions

Peter Godfrey-Smith

I. Introduction

Biological functions are dispositions or effects a trait has which explain the recent maintenance of the trait under natural selection. This is the "modern history" approach to functions. The approach is historical because to ascribe a function is to make a claim about the past, but the relevant past is the recent past; modern history rather than ancient.

The modern history view is not new. It is a point upon which much of the functions literature has been converging for the best part of two decades, and there are implicit or partial statements of the view to be found in many writers. This paper aims to make the position entirely explicit, to show how it emerges from the work of other authors, and to claim that it is the right approach to biological functions.

Adopting a modern history position does not solve all the philosophical problems about functions. It deals with a family of questions concerning time and explanation, but there are other difficulties which are quite distinct. The most important of these concern the extent to which functional characterization requires a commitment to some form of adaptationism (Gould and Lewontin 1978). These issues will not be addressed here. Further, as many writers note, "function" is a highly ambiguous term. It is used in a variety of scientific and philosophical theories, several domains of everyday discourse, and there is probably even a plurality of senses current within biology. This paper is concerned with one core biological sense of the term, which is associated with a particular kind of explanation. In this sense a function has some link to an explanation of why the functionally characterized thing exists, in the form it does.

Cummins (1975) argued that functions are properly associated with a different explanatory project, that of explaining how a component in a larger system contributes to the system exhibiting some more complex capacity. Following Millikan (1989b) I suggest that both kinds of functions should be recognized, each associated with a

Peter Godfrey-Smith, "A Modern History Theory of Functions," *Noûs* (1994) 28, 244–362. Reproduced with permission of Blackwell Publishing Ltd.

This work developed largely out of a series of discussions with Phillip Kitcher. Along with many of the ideas, the term "modern history theory" was his, though he should not be taken to endorse (or reject) the modern history view. I have also benefitted from discussions with Elisabeth Lloyd, Ruth Millikan, Sandra Mitchell and everyone at Kathleen Akins' Functions Reading Group. An anonymous referee for *Noûs* made a number of valuable criticisms of earlier drafts. I would also like to thank the University of Sydney for generous financial support during the period when most of this work was done.

different explanatory project. If it is claimed, for instance, that the function of the myelin sheaths round some brain cells is to make possible efficient long distance conduction of signals, it may not be obvious which explanatory project is involved – that of explaining why the sheath is there, or that of explaining how the brain manages to perform certain tasks. Often the same functions will be assigned by both approaches, but that does not mean the questions are the same.

The aim of this paper is to analyze an existing concept of function, which plays a certain theoretical role in biological science. So the aim is a certain sort of conceptual analysis, a conceptual analysis guided more by the demands imposed by the role the concept of function plays in science, the real weight it bears, than by informal intuitions about the term's application. Also, though I will defend the modern history view within the context of a particular theory of functions which draws on the work of Larry Wright and Ruth Millikan, the overall value of the modern history approach stands independently of many of the details of my theory.[1]

II.   The Wright Line

Our point of departure is a simple formula proposed by Larry Wright in 1973 and 1976: "The function of X is that particular consequence of its being where it is which explains why it is there" (1976 p. 78). That is:

The function of X is Z iff:
  (i)   Z is a consequence (result) of X's being there,
  (ii)   X is there because it does (results) in Z. (1976 p. 81)

Wright argued that his theory dealt with a broad range of cases, handling both the functions of artifacts and biological entities without significant modification. The function of spider webs is catching prey, because that's the thing they do that explains why they are there; the function of tyre tread is improving traction because that's also the thing it does that explains why its there; and the function of the newspaper under the door is to prevent a draft, for the same reason.

However, Wright's analysis covers more cases than these. Boorse 1976 notes that when a scientist sees a leak in a gas hose, but is rendered unconscious before it can be fixed, on Wright's schema the break has the function of releasing gas. The break is there because it releases gas, keeping the scientist immobilized, and the leaking gas is a consequence of the break in the hose. Similar cases take us even further from the plausible realm of purpose. One might see a small, smooth rock supporting a larger rock in a fast-flowing creek, and note that if it did not hold up that larger rock, it would be washed away, and no longer "be there." But it is not the function of the small rock to support the larger one. The problem here is with the broad range of "X" and "Z," with the need to restrict the kinds of things to which the schema can be applied. A restriction of this kind is a key component of Ruth Millikan's theory (1984, 1989a).

Before moving on however, it is important to recognize Wright's aims. Wright's strategy is to avoid convoluted analysis by trusting many details to pragmatic factors which will apply case by case. For Wright, function hinges directly on explanation, and explanation is pragmatically sensitive in a multitude of ways. There is a sense in which Wright's theory is not an "analysis" of function in the sense that earlier accounts are. Earlier writers were largely concerned with how it can ever be that something's existence can be inferred from its function, given that other things could often have done the same job (Hempel 1965). Without this inference, it was thought there could be no functional explanation. Wright simply insists that with a less demanding, more realistic picture of explanation, it becomes clear that people do explain the presence of things in terms of what they do, and a function is any effect that operates in such an explanation.

Wright also hopes, I suspect, that some natural slack in the notion of function will be mirrored and explained by corresponding slack in the notion of explanation, that the analysis will bend where the concept analyzed naturally bends. Wright's vague formulation of the relevant explanandum – "why its there" – is intended to wrap unsystematically around a variety of explanatory projects, in biology, engineering and everyday life. Nonetheless, counterexamples such as Boorse's do suggest that Wright has backed off too early, and a sensitivity to pragmatics should not prevent us from pushing an analysis as far as we profitably can.

Millikan's analysis, like Wright's, is historical. It locates functions in actual selective histories. The most important sophistication of the historical approach in Millikan 1984 is her detailed treatment of functional *categories*. The first concept she defines is that of a "reproductively established family." A reproductively established family is a group of things generated by a sort of copying. Family members can be copied one off the other, or be common copies off some template, or be generated in the performance of functions by members of another family. These different kinds of copying are all distinguished by Millikan, but the finer divisions are not important here. Call any entities which can be grouped as tokens of a type by these lines of descent by copying, members of a "family." Understand "copy" as a causal matter involving common properties and counter-factuals. The copy is like the copied in certain respects, though it is physically distinct, and if the copied had been different in certain ways, then, as a consequence of causal links from copied *to* copy, the copy would have been different in those ways too (1984 p. 20).[2] So two human hearts are members of the same family, as are two frill-necked lizard aggressive displays, two AIDS viruses, and two instances of the acronym "AIDS," assuming that acronym was hit upon only once. But two planets, and two time-slices of a rock or hose are not, as one was not copied off the other, nor are they produced off a common template, and so forth. Functions are only had by family members, and the performance of a function must involve the action of one of the properties copied, one of those properties defining the family.

This restriction deals with many of Boorse's counterexamples, such as the gas hose case. It also removes from the realm of function some cases Wright was concerned to capture, such as the newspaper under the door. However, our project here is to capture the biological usage. Preserving a continuity between biological cases and other domains can be sacrificed.[3]

The next step is to add to this an explanation-schema in the style of Wright. The explanandum is the existence of current members of the family. The explanans is a fact about prior members.

(F1) The function of m is to F iff:
  (i) m is a member of family T, and
  (ii)   among the properties copied between members of T is property or property cluster C, and
  (iii) one reason members of T such as m exist now is the fact that past members of T performed F, through having C.

Most simply, a family member's function is whatever prior members did that explains why current members exist (see also Brandon 1990 p. 188).

It is one of the strengths of the historical approach combined with an appeal to "families" that it can say without strain that some particular thing which is in principle unable to do F now, nonetheless has the function to do F. It has this function in virtue of its membership in a family which has that function. Whether this member can do F is irrelevant to its family membership, as long as it was produced by lines of copying that are generally normal enough. A genetic defect may produce a heart unable to ever pump blood, but if this token was produced in more-or-less the same way as others, it has the function characteristic of the family.

At this point we must confront an issue unrelated to history. It is striking that while analyses such as Wright's and Millikan's permit any activity or power explaining survival to qualify as a function, biologists apparently reserve "function" for activities or powers which are, in some intuitive sense, helpful and constructive. If being inconspicuous and avoiding attention by doing nothing is itself "doing something," then pieces of junk DNA, which sit idly on chromosomes and are never used to direct protein synthesis, have the function to do nothing. That is the thing past tokens of junk DNA types have done, which explains the survival of present tokens. If doing absolutely nothing is a behavior when an animal does it for concealment, why is it not something that junk DNA "does"? Perhaps the function of junk DNA is, alternatively, to be more expensive to get rid of than to retain. But biologists do not describe junk DNA like this; it is the paradigm of something with no function. Similarly, characters which hitch-hike genetically on useful traits or persist through developmental inevitability (like male nipples) might, in extended senses, be "doing" things which lead to their survival. So we might consider making some restriction on the selective processes relevant to functional status.

This will not be easy. A simple requirement that the trait do something positive, that the null power is not a power, will not suffice. Beside the cases where biological entities persist through doing nothing, there are positive and selectively salient powers which seem unlikely candidates for functions. As well as junk DNA, which does nothing, there is "selfish DNA" (Orgel and Crick 1980). Selfish DNA can move around within the genome, replicating itself as it goes, and proliferate in a population despite having deleterious effects on individuals carrying it.

Similarly, segregation distorter genes disrupt the special form of cell division (meiosis) which produces eggs and sperm (gametes). Meiosis usually results in a cell with two sets of chromosomes giving rise to four gametes with one set each, and on average a particular type of chromosome will be carried by half the gametes produced. Segregation distorters lever their way into more than their fair half share of gametes, by inducing sperm carrying the rival chromosome to self-destruct as they are formed (Crow 1979). Fruit flies, house mice, grasshoppers, mosquitoes and a variety of plants are known to have segregation distorters in their gene pools. Now, disrupting meiosis is something that segregation distorter genes do, that explains their survival (Lewontin 1962). Further, this explanation appeals to natural selection, at the gametic level; the problem can not be solved by disqualifying traits that survive for non-selective reasons. Disrupting meiosis is not generally claimed to be the genes' *function* though. Should we restrict the powers which can become functions, to exclude these subversive cases?[4]

There are two attitudes we might have to this issue. First, as a question of conceptual analysis, there is not much doubt that biologists typically restrict the powers that can qualify as functions. Many might say we should then change the selective theory of functions to include this factor. An obvious move is to bring in some reference to the goals of some larger system. Disrupting meiosis makes no contribution to the goals of individuals bearing segregation distorter genes, so this is not a function.

An appeal to goals is certainly a step backwards however. So we might consider a more aggressive attitude to the problem. It may be that many biologists reserve "function" for powers with some intuitively benign nature, and withhold it from more subversive activities, with there being no theoretically principled reason for this distinction. Some hold that biology since the 1960's has produced, for better or worse, an increasingly cynical view of the coalitions that make up organisms (Dawkins 1982, Buss 1987), families (Trivers 1974), and larger groups (Williams 1966, Hamilton 1971). The feeling that functions must involve harmonious interactions may, from this point of view, be a holdover from an earlier, more truly teleological view of nature. It might be claimed that the theoretically important category of properties, the category our concept of function should be tailored to, is simply the category of selectively salient powers and dispositions.[5] If so, we should remain with the simpler analysis that allows any survival-enhancing power, however subversive, to qualify as a function.

Although some may favor this more heartless approach I will adopt a third, intermediate position. Consider first another counterintuitive consequence of an unembellished selective account: whole organisms, like people, have functions. Past tokens of people did things – survived and reproduced – that explain why current tokens are here. Hence, we have the function to survive and reproduce. This usage seems odd – note that these are not functions people might have with respect to some social group, they are functions people just have, individually. One way to exclude both people as bearers of functions and also exclude disruption of meiosis as a function of segregation distorters is to stipulate that (i) the functionally characterized structure must reside within a larger biologically real system, and (ii) the explanation of the selection of the functionally characterized structure must go via a positive contribution to the fitness of the larger system. My account here resembles that of Brandon, who requires that a functional trait increase the "relative adaptedness of [its] possessors" (1990 p. 188). Brandon requires not just selective salience, but selective salience which goes via the fitness of a larger system "possessing" the trait.

The catalog of "real systems" is taken from biology, and clarifying the catalog is part of the units of selection problem.[6] Individuals, kin groups and perhaps populations and species might be examples of these systems. Thus hearts reside within people, and survive by aiding people's fitness. But people, considered individually, reside within no such systems. There may, however, be groups within which people do things which contribute to the selection of the group, and then people would have functions.

Similarly, segregation distorter genes do not have the function of disrupting meiosis, because their proliferation under selection does not occur through a positive contribution to the fitness of individuals bearing these genes. Indeed, many segregation distorters, when present in two copies, greatly impede the fitness of their carriers. On the other hand, as some readers may have felt earlier, there could well be functional characterization of *parts* of segregation distorter genes or gene combinations. Some part of the gene or combination might have its current presence explained by the fact that it has been selected for carrying out some part of the segregation distortion project. Crow (1979) distinguishes two genes which cooperate to produce segregation distortion in fruit flies. The "*S*" gene produces sabotage in sperm, and the "*R*" gene stops the chromosome that the *S* and *R* are on from sabotaging itself. So a chromosome with *S* but no *R* sabotages itself, and a chromosome with *R* but no *S* does not distort, but is immune to distortion by its rival. Here the segregation distorting chromosome is the larger system, and the selective explanation of *S* goes via the explanation of the success of the whole chromosome. *S* has the function of sperm sabotage, and it has this function with reference to the segregation distortion gene complex. The selection of *R* is only partly an explanation in terms of the selection of the distorter chromosome, as *R* is useful without *S*, once the population contains some chromosomes with *S*. So *R* has the function of preventing sabotage, and it has this function with reference to two larger units, the segregation distorter complex and the individual.

It is important that not all failures on the part of evolution to produce intuitively well-engineered animals disqualify selective episodes from bestowing functions. A question sometimes arises concerning the status of traits which are explained in terms of some forms of sexual selection. If it is true that sexual selection can operate through females favoring characteristics

in males which have no other benefit or use to the male (Fisher 1930, Lande 1981), then the explanation of a bird's long tail is not an explanation in terms of anything intuitively useful the tail does. The explanation is simply that females prefer long tails (Andersson 1982). Once a female preference gets established, for any reason, it can be sustained and made stronger through the association of the gene for the preference in females (unexpressed in males) and the gene for the preferred trait (unexpressed in females). The preference leads to the selection of long tails, and the selection of long tails leads to the strengthening of the associated preference. The long tail could be a hindrance elsewhere in life. Consequently, some biologists hesitate to describe the tail as an adaptation, and functional in the ordinary sense: "Runaway sexual selection is a fascinating example of how selection may proceed without adaptation" (Futayama 1986 p. 278). On the present account however the tail has the function to attract females. It has been selected because of that power, and this explanation goes via the augmentation of the individual's fitness.[7]

Here is an amended definition:

(F2) The function of *m* is to *F* iff:
  (i) *m* is a member of family *T*,
  (ii) members of family *T* are components of biologically real systems of type *S*,
  (iii) among the properties copied between members of *T* is property or property cluster *C*,
  (iv) one reason members of *T* such as *m* exist now is the fact that past members of *T* were successful under selection, through positively contributing to the fitness of systems of type *S*, and
  (v) members of *T* were selected because they did *F*, through having *C*.

III. Looking Forward

Although philosophers have discussed a variety of intuitive problems with the view that functions derive from a selective history (Boorse 1976), the most damaging charge against this view derives from the biological literature, from the wide acceptance of the distinctions made in "Tinbergen's Four Questions."

It is common in ethology and behavioral ecology to distinguish four questions "why?" we can ask about a behavior. Someone who asks why frill-necked lizards extend the skin around their necks so spectacularly might want an answer:

1. In terms of the physiological *mechanisms* and the physical stimuli that lead to the behavior.
2. In terms of the current *functions* of the behavior.
3. In terms of the evolutionary *history* of the behavior.
4. In terms of the *development* of the behavior in the life of the individual lizard.

This four-way distinction is usually attributed to Tinbergen 1963. Tinbergen in turn credits Julian Huxley with distinguishing questions 1–3, and adds question 4. Tinbergen, it must be admitted, uses the term "survival value" rather than "function" in the official formulation of question 2. But generally he uses these two expressions interchangeably (1963 p. 417, 420).

Tinbergen's distinctions are often endorsed in the opening pages of books about animal behavior (Krebs and Davies 1987 p. 5, Halliday and Slater 1983 p.vii, and see also Horan 1989). This is clearly an embarrassment for any historical theory of function which seeks to capture biological usage: on the historical view there should be three questions, not four, as the functional question *is* a question about evolutionary history, as long as the rest of (F2) above is satisfied. Related distinctions with this separation between function and history are found elsewhere in evolutionary writings as well. Mayr 1961 distinguishes "functional" from "evolutionary" biology, and Futuyma's widely used textbook echoes Mayr in dividing the study of biology into functional and historical "modes" (1986 p. 286).[8]

There are various ways to respond to this problem. Many ahistorical usages of "function" are probably best understood as referring to Cummins' functions. However, it is common for writers to both regard functions as ahistorical *and* regard them as intrinsically tied to natural

selection, sometimes via the expression "survival value." This supports the proposal of a number of writers that functions involve not actual selective histories, but probable futures of selective success, or atemporal dispositions to succeed. Tinbergen may have accepted such a view: "the student of survival value, so-to-speak, looks 'forward in time'" (1963 p. 418). Tinbergen (p. 428) also casts the question about a structure's function as a question about how deviations from the actual structure *would* lower the fitness of the bearer. John Staddon concurs (1987 p. 195). One way to develop this approach is with an appeal to propensities.

Bigelow and Pargetter (1987) develop a theory of functions modelled explicitly on the widely accepted propensity view of fitness (Mills and Beatty 1979). The propensity view of fitness claims that the fitness of an individual is not the actual fact of its reproductive success, but its propensity to have a certain degree of reproductive success. Similarly, Bigelow and Pargetter claim, functions should be understood as dispositions or propensities to succeed under natural selection. "Something has a (biological) function just when it confers a survival-enhancing propensity on a creature that possesses it" (1987 p. 192).

The propensity view is not satisfactory, though its failure performs the valuable service of narrowing the discussion down, along with Tinbergen's Four Questions, to a point where the modern history view will become compelling. I will discuss first some internal difficulties with the propensity view and then argue that the whole forward-looking approach is on the wrong track.[9]

The central internal problem is that as one tries to fill in some more details, the theory tends to go in one or other of three undesirable directions. It can become enmeshed in strong counterfactual commitments. Alternately, it draws on the historical facts it sought to avoid. Or thirdly it makes the wrong kinds of demands on the future. Putting it briefly: propensities to be selected and survive bestow functions, but, the questions swarm: survive where? be selected over *what*? Bigelow and Pargetter address the first question, admitting that their account "must be relativized to an environment" (p. 192). The context assumed is the creature's "natural habitat." "Natural habitat," it appears, is understood historically by Bigelow and Pargetter. The statistically most common context for a trait now might be odd and unnatural (Neander 1991b).

More worrying is the question of the competitors that have a propensity to be ousted from the population by the trait we are interested in. Bigelow and Pargetter make no mention of the fact that claims about propensities to do well under natural selection are surely always comparative claims. A trait does not have a propensity to be selected and survive simpliciter, but always a propensity to be selected over some range of alternatives. Evolution is driven by differences in *relative* fitness. Bigelow and Pargetter cannot claim that current useful traits would triumph over any possible alternatives. Which are the relevant ones? Those alternatives genetically attainable (given mutation rates, population structure, other constraints . . .) now? Those that could enter the fray during the next thousand years? Those that could enter the fray if the ozone layer goes and mutation rates are elevated? If Bigelow and Pargetter think there is a range of alternatives, and circumstances of selection, appropriate to the trait in question independently of history, they are making strong modal commitments. These might be avoided with an appeal to what is most likely to happen in the actual future, but then problems are created by (what appear to be) irrelevant contingent features of this future. If a trait is adaptive, but doomed because of linkage to something bad, then it is not likely to survive. But this should not make a trait itself non-functional.

So, though the propensity theory is tailored to avoid dragging up the past, the propensities involved must either make tacit reference to millennia gone by, inappropriate predictions about the future, or questionable modal commitments about relevant ranges of alternatives and circumstances of selection. These internal problems are important, because it is easy to think that propensity views are somehow more economical than analyses appealing to the past. Still, the propensity view has recommendations. It does seem to be a way to accommodate the intuition that functions derive from selection with the observation that many biologists keep functional and historical questions separated. In addition, I am often told that no matter how questionable philosophers may find the modal commitments

outlined above, many biologists constantly talk as if these facts are quite unproblematic and accessible. It is difficult to work out the right attitude to such a datum. Further, one principled way to deal with these internal problems is to fashion a mixed theory, using the basic propensity format with an appeal to history to answer the objections raised above. (This mixing was suggested to me by Elisabeth Lloyd).

The mixed theory claims that functions derive from propensities to be selected, but all the factors that Bigelow and Pargetter left vague are understood historically. The relevant ecological conditions are the actual ones that obtained during the development of the trait. The range of alternatives the trait has a propensity to be selected over are the ones it actually triumphed over, and continues to be selected over. The propensity that bestows functions is strictly atemporal; a trait is held to have a certain advantage under certain conditions over certain rivals. But these conditions and rivals are determined by the actual world. So it does seem likely that the propensity approach can be developed in a coherent way, at the price of narrowing the gap between it and the historical view. This is the general form of the contemporary functions debate: each theory is made more plausible by setting it on a course of convergence with its rivals.

There is, however, a more important problem with propensity theories, and other forward-looking views. These theories inevitably distort our understanding of functional explanation. In the first section I claimed that the sense of function under discussion is a sense linked in some way to explanations of why the functionally characterized entity exists, or exists in the form it does. The most straightforward way to envisage this link, which I have been assuming, is to say that functions are used in explanations of why the functionally characterized thing exists now. If this is granted, and the explanation is understood causally, then there is a simple argument against propensity views. The only events that can explain why a trait is around now are events in me past. Forward-looking accounts claim that functions are not bestowed by facts about the past, but rather by how things are in the present. But then appealing to a function cannot itself explain the fact that the trait exists now. If the environment is uniform, then present propensities to do well under selection may be a good guide to actual prior episodes of selection. But this epistemological point does not alter the fact that it is not the present propensities, but the prior episodes, that are causally responsible for how things are now (see also Millikan 1989b, Neander 1991a).[10]

I do not claim that Bigelow and Pargetter have missed this straightforward point. On their view, there is a problem with the background assumptions I have made about the explanatory role of functions, and which the argument above assumes. Bigelow and Pargetter claim that if the fact that some effect is a function itself depends on the fact that this effect explains the survival of the trait in question, if the assignment of a function is always retrospective in this way, "then it is no longer possible to explain why a char- acter has persisted by saying that the character has persisted because it serves a given function" (1987 p. 190). This vacuity problem can be solved, according to Bigelow and Pargetter, if functions are understood as propensities. These propensities can be used to explain the existence of a trait in the present if we claim, in addition, that the propensities in question did exist in the past, and were causally active in the past. This postulation of the past action of the propensities is an extra claim; it is not guaranteed by the mere fact that the effects in question are functions.

Bigelow and Pargetter's claims about explan- atory vacuity and the historical view have been criticized effectively by Sandra Mitchell (1993) She points out that if we say "Trait X persisted because it had a consequence responsible for its selection and consequent evolution," this is only vacuous if we read "persisted" as meaning "evolved by natural selection." That is, it is only vacuous if we assume that the *only* mechanism which could explain some trait being around today is natural selection, though in fact there are alternative evolutionary forces which could play this explanatory role (1993 pp. 253–54). This is correct, and it shows that Bigelow and Pargetter's argument about the vacuity of the historical view assumes an implausible adaptationism. There is also another objection to Bigelow and Pargetter's claim, which is compatible with even the strongest adaptationism. On the historical view and with the assumption of adaptationism, it will be truly vacuous to say that X persisted because it

serves *some* function, because we are assuming that this is the only possible type of explanation. But even against this background it will of course not be vacuous to say that X persisted because it provided effective camouflage, or because it attracted mates, or because it conserved heat. Neither is it vacuous to say that the trait persisted because some specific effect was its *function*. If the historical theorist says "X persisted because its function was to conserve heat," this is to be translated into something which is ungainly, and contains a redundancy – "X persisted because its actually-selected effect was that of conserving heat." But this is not vacuous; it does contain a real explanation, though to express it this way mentions the explanatoriness of the effect twice. So this is not the most natural mode of expres- sion for the historical view; on that view the sentence "The function of X is to conserve heat" is itself explanatory, and if someone is asked "Why is X there?" they can reply by simply citing the function. This is not possible at all on the propensity view. On the propensity view, a functional explanation must give a function and also make an additional claim that the function was causally active in the past.

So despite what Bigelow and Pargetter claim, as long as "a given function" is understood to refer to some specific task or benefit, it is not trivial to say that "the character has persisted because it serves a given function," even assuming adapta- tionism. This, along with Mitchell's argument, shows that there is no vacuity problem with the background assumptions about explanation that proponents of the historical view make. It is possible to retain the explanatory force of func- tion ascriptions, along with the philosophically attractive view, argued by Wright, that actual explanatory salience is exactly what *distinguishes* functions from mere effects.

A "forward-looking" approach to functions has also been endorsed by Barbara Horan (1989), but the claims she makes about explanation are more problematic than those of Bigelow and Pargetter. Horan says "questions about the func- tion of a given pattern of social behavior are a way of asking how that behavior enhances the fitness of an individual who engages in it" (1989 p. 135). Nevertheless, she claims soon after that the presence of a trait like a social behavior can be explained by an attribution of a function to that behavior. The model of explanation she applies, citing G. A. Cohen, is called a "consequence explanation." Consequence explanations use laws of the form: "If (if *C* then *E*), then *C*." In the present context: "if a behavior pattern would increase individual fitness, individuals will come to display that behavior" (1989 p. 136).

This is trying to have it both ways. It is true that useful things a behavior does now can lead to its prevalence in the future. So forward-looking functions may predict and explain the future prevalence of a trait. But if the explanandum is how things are now, nothing present or future can be the explanans. Only the past will do. Of course, traits that are useful now were often useful then, so we can often infer that a propensity existing now was also causally active then. But if so, it is explanatory with respect to the present *because* it was causally active then. To claim that present usefulness in itself explains the morphologies and behaviors organisms presently display, and to build this into an account of functions, is to distort the explanatory structure of evolutionary theory.

IV. The Modern History Theory

It might appear that we are painting ourselves into an analytical corner. Historical analyses are unacceptable because they fail to respect an apparently important distinction in biology between functional and evolutionary explanation. Forward-looking analyses are unacceptable because they distort our understanding of functions' explanatory role. In fact there are several options available at this point. Bechtel (1989) suggests that we retain a forward-looking account of functions while giving up our prior conception of functional explanation. We might, alternately, claim that functional explanation just is evolutionary explanation, and banish other notions of function (except for Cummins') as creatures of teleological darkness. A third option is to analyze functional explanation as a particular *kind* of evolutionary explanation. One alternative here is to regard a functional explanation as a selective explanation which satisfies (F2) above, hence a subset of evolutionary explanation. The option I prefer, however, is to construe functional explanation more narrowly still.

This brings us, at last, to the modern history view: functions are dispositions and powers which explain the recent maintenance of a trait in a selective context. Several people have already said, in effect, that this is the answer, but these people either make the suggestion in passing (Kitcher 1990), or more often, they only say it some of the time. Horan says "to explain the maintenance of a trait in a species, one gives a functional explanation" (1989 p. 135), but insists on an atemporal construal of this explanation. And consider this remark of Millikan's, in response to Horan:

> If natural selection accounts for a trait, that is something that happened in the past, but the past may have been, as it were, "only yesterday." Indeed, *usually* the relevant past is only yesterday: the *main business* of natural selection is steady maintenance of useful traits against new intruders in the gene pool. But only yesterday is not outside of time. (1989b p. 173)

We need not endorse the claim about the "main business" of natural selection; whether or not maintaining traits is the main business of selection, it is one important kind of selection. It might be important enough to make this a constitutive part of the concept of function. Millikan does not take this step; her historical account does not *build into* functions the historically recent nature of the relevant selective episodes. Indeed, in her 1984 treatment she explicitly allows powers which were important in ancient history, but not in modern history, to be functions (1984 p. 32). In the 1989b treatment her emphasis is different, and she claims the relevant past is "*usually*" only yesterday. But perhaps, as far as functions go, it *must* be only yesterday.

The modern history view does not respect the letter of Tinbergen's Four Questions, but it is faithful to their spirit. Tinbergen makes the modern/ancient history distinction himself (1963 pp. 428–29), but he regards *both* these explanations as "evolutionary" rather than functional. This puts two distinct questions under one head, however, as well as leaving the explanatory significance of functions in the dark. From the present viewpoint, the "evolutionary" question is the question about the forces which originally built the structure or trait in question. This may or may not be a selective explanation, and this explanation might be different from the explanation of why the trait has recently been maintained in the population.

Some might wonder how recent the selective episodes relevant to functional status have to be. The answer is not in terms of a fixed time – a week, or a thousand years. Relevance fades. Episodes of selection become increasingly irrelevant to an assignment of functions at some time, the further away we get. The modern history view does, we must recognize, involve substantial biological commitments. Perhaps traits are, as a matter of biological fact, retained largely through various kinds of inertia. Perhaps there is not constant phenotypic variation in many characters, or new variants are eliminated primarily for non-selective reasons. That is, perhaps many traits around now are not around because of things they have been doing. Then many modern-historical function statements will be false. If functions are to be understood as explanatory, in Wright's sense, there is no avoiding risks of this sort.

One way to support the modern history view of function is to demonstrate that the category of explanation it distinguishes is a theoretically principled one. This can be done by focusing on traits for which the modern historical explanation and the ancient historical explanation diverge, so the selective forces salient in the origin of the trait are different from those salient in the recent maintenance of the trait. Here is where a distinctively functional style of characterization – in the modern history sense – can be seen to be useful.

The importance of the distinction between modern and ancient evolutionary explanations is discussed, in support of an analysis of function quite opposed to mine, in Gould and Vrba 1982. The central concern of Gould and Vrba is a distinction between adaptations and "exaptations" (their coinage). They understand adaptations as characters shaped by natural selection for the role they perform now. Exaptations are characters built originally by selection for one job, or characters with no direct selective explanation at all, which have since been coopted for a new use. This analysis has consequences for their concept of function; only adaptations have functions, and exaptations have "effects." Gould and Vrba do not discuss the recent past, as distinct from

the present, so I am uncertain how they would classify modern-historical functions. Generally they seem to understand effects-of-exaptations as propensities (1982 p. 6). Their effects-of-exaptations correspond to the functions of Bigelow and Pargetter, and Horan. It should be clear why I think their way of dividing the cases is inadequate: modern history and ancient history can *both* furnish genuine explanations, which we should distinguish, for why something exists now, while present propensities cannot themselves furnish such explanations.

Gould and Vrba's central point is the importance of cases where a trait's original and current uses diverge. But these are also cases where the selective forces that built a character and those maintaining it in the recent past diverge, so they also illustrate the origin/function distinction as I understand it. Gould and Vrba make two claims about such cases. Firstly, there are many of them, and secondly, the cases are theoretically significant. The co-opting of existing traits for new uses is important in the development of complex and novel adaptive characters.

Feathers, it has been argued, did not originate as adaptations for flight. The earliest known bird *Archaeopteryx* did not have the skeleton for anything beyond very rudimentary flight, but was well-covered with feathers. It has been claimed that feathers originated as insulation, and only later were coopted for flight (Gould and Vrba 1982 p. 7 cite Ostrom 1979). Thus the question about the evolutionary origin of feathers is answered in terms of selection for effective insulation, but if we ask today about the function of feathers, in a sub-tropical bird for instance, the answer appeals to the reason feathers have recently been maintained – their facilitating flight.

A similar story can be told about the development of bone. Bone is essential as a support for land-dwelling vertebrates, but it developed in sea animals well before it could be put to its modern use. Gould and Vrba discuss the hypothesis that bone was developed as store of phosphates needed for metabolic activity (Halstead 1969). In this case, the original use continues, and bone functions in modern vertebrates as storage for mineral ions, including phosphate ions, as well as support.

Gould and Vrba's examples can be augmented easily. The electric eel's ability to kill prey and defend itself with electric shocks is a development of the weaker electric abilities of other fish, which generate electric fields as part of a perceptual system, used in orientation and communication (Futayama 1986 pp. 423–24). Shepherd (1988 p. 67) discusses a suggestion made by J. B. S. Haldane about the origin of neurotransmitters, the chemicals whose function now is passing signals between neurons in the brain. Haldane suggested that these chemicals may have developed originally as chemical messengers between individuals. There are a number of neurotransmitters which can induce effects on other organisms.

A final illustration of the importance of the distinction between originating and maintaining selection is found in some of the literature applying game theory to animal behavior (Maynard Smith 1982).[11] An ESS, or evolutionarily stable strategy, is a strategy which, once prevalent in a population, cannot be invaded by rival strategies. However, an ESS need not be a strategy that can evolve from scratch in any situation. Often a critical mass of like-minded individuals is needed before a strategy becomes stable. Thus to explain a behavior by showing it to be an ESS is not necessarily to explain how that behavior originally became established. Rather, it is to point to the selective pressures responsible for the recent maintenance of the strategy in the population.[12]

The point is not just the apparent commonality of a divergence between modern and ancient history, but the fact that this distinction has sufficient theoretical importance to justify its place in an analysis of functions.

One final problem must be discussed, which can be introduced with a feature of Wright's analysis. It is initially perplexing that Wright uses the present tense in the expression: "X is there because it does (results in) Z" (1976 p. 81). If his account is historical ("etiological"), why does he not make it explicit that the performances of Z that explain the presence of X's are in the past?

In general, when we explain something by appeal to a causal principle, the tense of the operative verb is determined by whether or not the principle still holds at the time the explanation is given. ... We might say, for example, "The Titanic sank because when you tear a hole that size in the bow of a ship it sinks," using the verb "to sink"

in the present tense even though the sinking in question took place in the past. ... If we were to throw the statement into the past tense it would imply that nowadays one could get away with tearing a hole that size in the bow of a ship without it sinking. (1976 pp. 89–90)

Wright requires that the effects appealed to in a functional explanation still exist at the time of the functional ascription, and these effects must still have the same causal efficacy that they have had in the past. If this means that the structure in question must now have a propensity to continue to be selected for the same reason that it was selected in the past, Wright's account converges with that of Gould and Vrba, who demand that functions presently "promote fitness" (1982 p. 6).

Should the modem history view include these extra requirements? In my view, there may be good reason to require that the trait still be able to do now what it was selected for doing, but we should not require that the trait also have the same propensity to succeed under selection that it has had in the past. This problem is less pressing for the modern history view than for other historical views. If a trait has very recently been selected for doing F, it will tend to still be able to do F now. As it is *possible* for it (the type) to be unable to do F now, no matter how recently it has been selected for doing F, it is probably reasonable to add an extra clause requiring the continuation of the disposition into the strict present.[13] Whichever way one goes here, it is an advantage of the modern history view that these uncooperative cases should be made very rare.

Here is my final attempt at a definition of function.

(F3)   The function of m is to F iff:
    (i) m is a member of family T,
    (ii)   members of family T are components of biologically real systems of type S,
    (iii)   among the properties copied between members of T is property or property cluster C, which can do F,
    (iv)   one reason members of T such as m exist now is the fact that past members of T were successful under selection in the recent past,

through positively contributing to the fitness of systems of type S, and

(v) members of T were selected because they did F, through having C.

Much of this definition is proposed tentatively. The most important part is the appeal to modern history, which can also be incorporated in other theories of functions. The central recommendation of the modern history view is the fact that it accounts for the explanatory force of function ascriptions, but does this while making sense of the biological distinction between "functional" and "historical" explanation. It is a theory which steers a principled middle course.

Notes

1 Neander (1991a), Mitchell (1989, 1993) and Brandon (1990) have defended theories of functions running along similar lines. Sober's (1984) analysis of adaptation is also a relative.

2 Those familiar with some units of selection debates in philosophy of biology will note that family members need not be replicators: see Dawkins 1982, Hull 1981.

My definition of copying is not supposed to be airtight, and may be too inclusive. Kim Sterelny suggested that it lets in molecular structures in a crystal lattice, for instance, though it is not so certain that this case should be kept out. See Millikan 1984 for more details.

3 Millikan presents her 1984 account as a stipulative definition, not an analysis of an existing concept, so this is not a problem for her. It is also important that Millikan's restrictions do not prevent the analysis being applied to artifacts generated by copying in the right ways.

4 The treatment in Millikan 1984 fudges here. Millikan's official definition of function begins with a stipulated function F, and explains why something has this function F. Can any activity or power qualify as function F, as long as it promotes survival? If not, Millikan owes us an account of what sorts of properties can be functions. If on the other hand she allows any power to be a function, then why does she take the indirect route, of starting with a function to be fulfilled and then explaining why one structure, rather than a rival, has this function as its own?

5 Most philosophical commentators on an earlier draft of this material inclined towards the heartless line on this question.

6 In the terms of the units of selection debate, the larger system needs to be a real *interactor* (Hull 1981, Lloyd 1988, Brandon 1990).

7 Wright 1976 discusses the possibility of an appeal to the broader system (p. 106). He dismisses it firmly (though this fails to prevent other writers from attributing such an appeal to him: Nagel 1977 p. 283, Hampe and Morgan 1988 p. 123, Wright however does not discuss examples like those causing trouble in the present discussion.

8 A puzzling case is George Williams (1966) Williams is usually regarded as an advocate of a Wright-style account of functions (Boorse 1976 p. 85, Wright 1976 pp. 92–93), as suggested by this well-known passage: "One should never imply that an effect is a function unless he can show that it is produced by design [natural selection] and not by happenstance" (1966 p. 261). But when Williams lays down principles for the general study of adaptation, he seems to imply that the basic fact of something's having a function is not a historical fact. It appears that the "prime" question asked about a character in such a study – "What is its function?" – is answered in terms of contributions to goals (1966 p. 258, citing Pittendrigh 1958). The *second* question asked is the historical one about selection (p. 259, see also p. 264). Williams does goes on to say that an activity is not a function unless it was produced by design rather than chance (p. 261, quoted above). So the ahistorical nature of the "prime" question might be merely epistemological.

9 The version of the propensity view I am discussing is based on the survival propensities of *character types*. The propensity is possessed by human hearts as a type, not by individual hearts, and not by individual people. Bigelow and Pargetter are not consistent here. Sometimes they talk about the survival of the individual bearing the functionally characterized trait (1987 p. 192). But later, when speaking more strictly, they focus explicitly on the character type (p. 195, see also p. 194). On my reading, their talk of the "survival" of individuals is really talk of individuals' inclusive fitness (in the biological cases at least). Sandra Mitchell pointed out to me that if their propensities are read as belonging to individual trait-bearers, their theory is more like a classical goal theory. Admittedly, they do regard their account as a "cousin" of goal theories (p. 182). Neither interpretation squares with everything they say, but this exegetical question is less important

than the theoretical issue of the viability of a propensity-based selective account.

10 Focusing on causal explanation in this way also makes it clear why the selective advantage relevant to functional status cannot be understood with reference to a range of counterfactual alternative traits, as opposed to actual ones, as some propensity views might maintain. Only competition with actual, past rivals is causally relevant in explaining why a trait exists today.

11 I am indebted to Philip Kitcher for this point.

12 The distinction between the original establishment and the maintenance of a strategy is stressed, for instance, in Axelrod and Hamilton's well-known discussion of the properties of tit-for-tat in the iterated prisoner's dilemma (1981).

13 This suggestion is made cautiously – perhaps all these additional requirements are ill-advised (Neander 1991b p. 183).

References

Andersson, M. (1982) Female Choice Selects for Extreme Tail Length in a Widowbird. *Nature* 299: 818–20.

Axelrod, R. and W. Hamilton (1981) The Evolution of Cooperation. *Science* 211: 1390–1396.

Bechtel, W. (1989) Functional Analyses and their Justification. *Biology and Philosophy* 4: 159–162.

Bigelow, J. and R. Pargetter (1987) Functions. *Journal of Philosophy* 84: 181–197.

Boorse, C. (1976) Wright on Functions. *Philosophical Review* 85: 70–86.

Brandon, R. (1990) *Adaptation and Environment*. Princeton: Princeton University Press.

Brandon, R. and R. Burian, eds. (1984) *Genes, Organisms, Populations: Controversies over the Units of Selection*. Cambridge, MA: MIT Press.

Buss, L. (1987) *The Evolution of Individuality*. Princeton: Princeton University Press.

Crow, J. (1979) Genes that Violate Mendel's Rules. *Scientific American* 240: 134–146.

Cummins, R. (1975) Functional Analysis. *Journal of Philosophy* 72: 741–765.

Dawkins, R. (1982) *The Extended Phenotype*. Oxford: Oxford University Press.

Fisher, R. A. (1930) *The Genetical Theory of Natural Selection*. Oxford: Clarendon.

Futayama, D. (1986) *Evolutionary Biology*. 2nd edition. Sunderland: Sinauer.

Gould, S. J. and R. C. Lewontin (1978) The Spandrells of San Marco and the Panglossian Paradigm: A Critique of the Adaptationist Program. *Proceedings of the Royal Society, London* 205: 581–598.

Gould, S. J. and E. Vrba (1982) Exaptation – a Missing Term in the Science of Form. *Paleobiology* 8: 4–15.

Halliday, T. R. and P. J. B. Slater, eds. (1983) *Animal Behavior, Vol.2: Communication*. New York: Freeman.

Halstead, L. B. (1969) *The Pattern of Vertebrate Evolution*. Edinburgh: Oliver and Boyd.

Hamilton, W. D. (1971) Geometry for the Selfish Herd. *Journal of Theoretical Biology* 31: 295–311.

Hampe, M. and S. R. Morgan (1988) Two Consequences of Richard Dawkins' View of Genes and Organisms. *Studies in the History and Philosophy of Science* 19: 119–138.

Hempel, C. G. (1965) The Logic of Functional Analysis. In *Aspects of Scientific Explanation*. New York: Free Press.

Horan, B. (1989) Functional Explanations in Sociobiology. *Biology and Philosophy* 4: 131–158.

Hull, D. (1981) Units of Evolution: a Metaphysical Essay. Reprinted in Brandon and Burian 1984.

Kitcher, P. S. (1990) Developmental Decomposition and the Future of Human Behavioral Ecology. *Philosophy of Science* 57: 96–117.

Krebs J. and N. Davies (1987) *An Introduction to Behavioural Ecology*, 2nd edition. Oxford: Blackwell.

Lande, R. (1981) Models of Speciation by Sexual Selection on Polygenic Traits. *Proceedings of the National Academy of the Sciences* 78: 3721–3725.

Lewontin, R. C. (1962) Interdeme Selection Controlling a Polymorphism in the House Mouse. *American Naturalist* 96: 65–78.

Lloyd, E. A. (1988) *The Structure and Confirmation of Evolutionary Theory*. New York: Greenwood Press.

Maynard Smith, J . (1982) *Evolution and the Theory of Games*. Cambridge: Cambridge University Press.

Mayr, E. (1961) Cause and Effect in Biology. *Science* 134: 1501–1506.

Millikan, R. G. (1984) *Language, Thought, and Other Biological Categories*. Cambridge, MA.: MIT Press.

Millikan, R. G. (1989a) In Defence of Proper Functions. *Philosophy of Science* 56: 288–302.

Millikan, R. G. (1989b) An Ambiguity in the Notion "Function." *Biology and Philosophy* 4: 172–176.

Mills, S. and J. Beatty (1979) The Propensity Interpretation of Fitness. *Philosophy of Science* 46: 263–286.

Mitchell, S. (1989) The Causal Background of Functional Explanation. *International Studies in the Philosophy of Science* 3: 213–229.

Mitchell, S. (1993) Dispositions or Etiologies? A Comment on Bigelow and Pargetter. *Journal of Philosophy* 90: 249–259.

Nagel, E. (1977) Teleology Revisited: The Dewey Lectures 1977. (1) Goal-directed Processes in Biology. (2) Functional Explanations in Biology. *Journal of Philosophy* 74: 261–301.

Neander, K. (1991a) The Teleological Notion of "Function." *Australasian Journal of Philosophy* 69: 454–468.

Neander, K. (1991b) Functions as Selected Effects: The Conceptual Analyst's Defence. *Philosophy of Science* 58: 168–184.

Orgel, L. E. and F. H. C. Crick (1980) Selfish DNA; the Ultimate Parasite. *Nature* 284: 604–606.

Ostrom, J. H. (1979) Bird flight: How Did it Begin? *American Scientist* 67: 46–56.

Pittendrigh, C. S. (1958) Adaptation, Natural Selection, and Behavior. In A. Roe and G. G. Simpson, eds., *Behavior and Evolution*. New Haven: Yale University Press.

Shepherd, G. M. (1988) *Neurobiology*, 2nd edition. Oxford: Oxford University Press.

Sober, E. (1984) *The Nature of Selection*. Cambridge, MA: MIT Press.

Staddon, J. E. R. (1987) Optimality Theory and Behavior. In J. Dupre, ed., *The Latest on the Best: Essays on Evolution and Optimality*. Cambridge, MA: MIT Press.

Tinbergen, N. (1963) On the Aims and Methods of Ethology. *Zeitschrift für Tierpsychologie* 20: 410–33.

Trivers, R. (1974) Parent-offspring Conflict. *American Zoologist* 14: 249–264.

Williams, G. C. (1966) *Adaptation and Natural Selection*. Princeton: Princeton University Press.

Wright, L. (1973) Functions. *Philosophical Review* 82: 139–168.

Wright, L. (1976) *Teleological Explanations*. Berkeley: University of California Press.

PART VI

EVOLUTIONARY DEVELOPMENTAL BIOLOGY

Introduction

Evolutionary biology and developmental biology are two sub-disciplines that came to be separated in the nineteenth century, especially as Darwinism was pulled in the direction of genetics. Both made impressive progress over the next 100 years, without however having much if any impact on one another.

Several biologists, including Steven J. Gould, long complained that Darwinian theory was impoverished by its isolation from developmental biology and that at least some of its adaptational excesses could be curbed once we recognize more clearly how development constraints restrict the appearance of adaptations. However, it was only with the advent of the molecular revolution that developmental biologists began to see exactly how important developmental mechanisms are for evolution, and to exploit the identity of germ line genes and somatic genes to reveal the way evolutionary change is precisely channeled through the genome.

We provide the details of these discoveries as reported by one of the founders of the intersecting sub-discipline, *evo-devo* (short for evolution and development), Sean Carroll (chapter 14). This paper is followed by a more recent article by Casper Breuker, Vincent Debat, and Christian Klingenberg (chapter 15) that further explains evo-devo and shows its compatibility with more traditional evolutionary views.

Evo-devo's basic claim is broadly threefold: First, gene mapping and manipulation shows that all animals are built following information from essentially the same structural genes. Second, differences in organism structure and behavior are largely the result of when and where these genes are switched on, as opposed to being the result of the presence or absence of different genes. Third, evolutionary differences are the result of changes in the switching on and off of these structural genes, and this process is determined by other so called *regulatory genes*, which do differ across species and whose differences in expression control the expression of the structural genes most organisms share in common.

If true, this is a fascinating discovery – although, in the *Origin*, Darwin (1859/1999) already recognized that "characters derived from the embryo should be of equal importance with those derived from the adult, for a natural classification of course includes all ages" (p. 342). The philosophically interesting questions focus on how these two bodies of theory and data are to be combined and what the meaning of developmental molecular biology is for issues like reductionism and genic selectionism as well as biological systematics.

Further Reading

de Beer, G. (1954). *Embryos and ancestors*. Oxford: Oxford University Press.

Carroll, S. (2005). *Endless forms most beautiful: The new science of evo devo and the making of the animal kingdom*. New York: W. W. Norton.

Carroll, S. (2006). *The making of the fittest: DNA and the ultimate forensic record of evolution*. New York: W. W. Norton.

Carroll, S., Grenier, J., & Weatherbee, S. (2004). *From DNA to diversity: Molecular genetics and the evolution of animal design*. New York: W. W. Norton.

Carroll, S., Prud'homme, B., & Gompel, N. (2008). Regulating evolution. *Scientific American*, 298, 60–67.

Darwin, C. (1859/1999). *The origin of species by natural selection: Or, the preservation of favoured races in the struggle for life*. New York: Bantam Books.

Duncan, D., Burgess, E., & Duncan, I. (1998). Control of distal antennal identity and tarsal development in *Drosophila* by *Spineless-aristapedia*: A homolog of the mammalian dioxin receptor. *Genes & Development*, 12, 1290–1303.

Gilbert, S. (2003). The morphogenesis of evolutionary developmental biology. *International Journal of Developmental Biology*, 47, 467–477.

Goodman, C., & Coughlin, B. (eds.) (2000). Special feature: The evolution of evo-devo biology. *Proceedings of the National Academy of Sciences*, 97, 4424–4456.

Gould, S. (1977). *Ontogeny and phylogeny*. Cambridge, MA: Harvard University Press.

Hall, B. (2000). Evo-devo or devo-evo – does it matter? *Evolution & Development*, 2, 177–178.

Kirschner, M., & Gerhart, J. (2005). *The plausibility of life: Resolving Darwin's dilemma*. New Haven, CT: Yale University Press.

Laubichler, M., & Maienschein, J. (eds.) (2007). *From embryology to evo-devo: A history of developmental evolution*. Cambridge, MA: MIT Press.

Minelli, A. (2003). *The development of animal form: Ontogeny, morphology, and evolution*. Cambridge, MA: Cambridge University Press.

Müller, G., & Newman, S. (eds.) (2003). *Origination of organismal form: Beyond the gene in developmental and evolutionary biology*. Cambridge, MA: MIT Press.

Müller, G., & Newman, S. (eds.) (2005). Special issue: Evolutionary innovation and morphological novelty. *Journal of Experimental Zoology Part B: Molecular and Developmental Evolution*, 304B, 485–631.

Oyama, S., Griffiths, P., & Gray, R. (eds.) (2001). *Cycles of contingency: Developmental systems and evolution*. Cambridge, MA: MIT Press.

Palmer, R. (2004). Symmetry breaking and the evolution of development. *Science*, 306, 828–833.

Prud'homme, B., Gompel, N., & Carroll, S. (2007). Emerging principles of regulatory evolution. *Proceedings of the National Academy of Sciences USA*, 104, 8605–8612.

Quiring, R., Walldorf, U., Kloter, U., & Gehring, W. (1994). Homology of the eyeless gene of *Drosophila* to the small eye gene in mice and *Aniridia* in humans. *Science*, 265, 785–789.

Raff, R. (1996). *The shape of life: Genes, development, and the evolution of animal form*. Chicago: University of Chicago Press.

Robert, J. (2004). *Embryology, epigenesis, and evolution: Taking development seriously*. Cambridge: Cambridge University Press.

Robert, J. (2008). Evo-devo. In M. Ruse (ed.), *The Oxford handbook of philosophy of biology* (pp. 291–309). Oxford: Oxford University Press.

Takeuchi, S., Suzuki, H., Yabuuchi, M., & Takahashi, S. (1996). A possible involvement of melanocortin 1-receptor in regulating feather color pigmentation in the chicken. *Biochimica et Biophysica Acta*, 1308, 164–168.

Valverde, P., Healy, E., Jackson, I., Rees, J., & Thody, A. (1995). Variants of the melanocyte-stimulating hormone receptor gene are associated with red hair and fair skin in humans. *Nature Genetics*, 11, 328–330.

West-Eberhard, M. (2003). *Developmental plasticity and evolution*. Oxford: Oxford University Press.

14

Endless Forms: The Evolution of Gene Regulation and Morphological Diversity

Sean B. Carroll

... we are always slow in admitting great changes of which we do not see the steps ... The mind cannot possibly grasp the full meaning of the term of even a million years; it cannot add up and perceive the full effects of many slight variations, accumulated during an almost infinite number of generations.

C. Darwin, *The Origin of Species* (1859)

Species diverge from common ancestors through changes in their DNA. One of the ultimate questions in biology, then, is which changes in DNA are responsible for the evolution of morphological diversity? The answers have eluded biologists for the half-century since the Modern Synthesis and the discovery of the structure of DNA. The reasons for this are many-fold. Foremost among them is that the genes that affect morphology had to be identified first.

The genetic basis of morphological diversity is now being attacked at two ends of the evolutionary spectrum – the large-scale differences in body patterns at higher taxonomical levels, and the smaller-scale differences in morphology within or between closely related species. Here, I will review the growing body of evidence that points to a central role for differences in developmental gene regulation in both intraspecific variation and the diversification of body plans and body parts. I will discuss why changes in

the *cis*-regulatory systems of genes more often underlie the evolution of morphological diversity than do changes in gene number or protein function. And, finally, I will address, from a developmental genetic perspective, the long-standing question of the sufficiency of evolutionary mechanisms observed at or below the species level ("microevolution") to account for the larger-scale patterns of morphological evolution ("macroevolution").

Animal Body Patterns Have Evolved around an Ancient Genetic Toolkit

One of the most surprising biological discoveries of the past two decades is that most animals, no matter how different in appearance, share several families of genes that regulate major aspects of body pattern. The discovery of this common genetic "toolkit" for animal development, containing many families of transcription factors and most signaling pathways, has provided the means to study the genetic basis of animal diversity by enabling comparisons of how the number, regulation, or function of genes within the toolkit has changed in the course of animal evolution.

The similarities in gene content among long-diverged phyla did not meet initial expectations that the expansion of gene families would track the evolution of later, more complex forms. For example, all four arthropod classes and the onychophora, a closely related phylum with a simpler body organization, share nearly identical sets of *Hox* genes, despite their great morphological diversity and the long span of time (>540 million years) since their divergence from a common ancestor. Similarly, most protostomes and deuterostomes, with the exception of the vertebrates (which possess four or more clusters), possess roughly equivalent clusters of *Hox* genes that must date back to at least their last common Precambrian bilaterian ancestor (de Rosa *et al.*, 1999). Since cnidarians (jellyfish, sea anemones) and sponges possess fewer *Hox* and other developmental genes, we can infer that the bilaterian genetic "toolkit" was assembled and expanded early in animal evolution, before the bilaterian radiation (Knoll and Carroll, 1999), and then significantly expanded again at the base of the vertebrates. The diversity of protostomes and lower deuterstomes, and of vertebrates (after genome expansions) has largely evolved then around ancient and fairly equivalent sets of regulatory genes.

A considerable body of evidence suggests that evolutionary changes in developmental gene regulation have shaped large-scale changes in animal body plans and body parts. In particular, many comparative analyses of *Hox* gene expression in arthropods, annelids, and vertebrates have revealed a consistent correlation between major differences in axial morphology and differences in the spatial regulation of *Hox* genes (Carroll *et al.*, 2000). For example, *Hox* genes are expressed at different relative positions along the rostrocaudal axis in the mouse, chick, and python (Belting *et al.*, 1998; Cohn and Tickle, 1999). Such shifts in *Hox* expression domains during evolution could arise from changes in the expression of *trans*-acting regulators and/or within the *cis*-regulatory

regions of *Hox* genes. Direct evidence that a *Hox cis*-regulatory element has functionally diverged during the course of bird and mammal evolution has been found to underlie a relative shift in the expression of the mouse and chick *Hoxc8* genes (Belting *et al.*, 1998).

The divergence of *cis*-regulatory element function is also implicated in the morphological diversification of structures such as insect hind-wings and vertebrate limbs. Both the regulation of *Hox* genes within insect (Stern, 1998) and vertebrate (Sordino *et al.*, 1995) appendage fields and the sets of target genes regulated by *Hox* or other selector genes that control appendage identity have diverged between lineages (Weatherbee and Carroll, 1999). The evolutionary divergence of both *Hox* and downstream target gene regulation appears to arise through selected changes in one or a subset of the many individual elements that independently regulate gene expression in different parts of animals.

Intraspecific Variation and the Response to Selection

Most comparisons of developmental gene expression and regulation have focused on large differences between groups at higher taxonomic levels. However, since morphological variation is the fuel for evolution, it is important to understand the genetic architecture and the molecular nature of morphological variation *within* species. The most successful approaches to understanding intraspecific variation exploit genetic methodologies for assessing the number and identity of genes involved in differences between character states. By utilizing detailed knowledge of the genetics of model organisms such as fruit flies and maize plants, specific differences have been localized to developmental loci involved in intra- and interspecific variation and the response to natural selection or artificial selection (e.g., domestication).

One of the better-analyzed examples of intraspecific variation is the bristle pattern of adult fruit flies. The number and pattern of sensory bristles on most fly body parts vary between individuals and populations. Quantitative genetic analyses of variation in bristle number have shown that many loci (quantitative trait loci, or

QTL) contribute, but a few are responsible for the bulk of the variation. QTL of "large effect" that account for 5%–10% or more of the variance in bristle number have been identified including the *achaete-scute, scabrous,* and *Delta* genes – all loci known to affect the development of bristle patterns. Importantly, the sites within these loci that are associated with differences in bristle patterns all map outside of coding regions (Mackay, 1996; Long *et al.,* 1998). They appear to exert their effects on gene expression, apparently by subtly altering the function of *cis*-regulatory elements in controlling the level, pattern, or timing of gene expression.

Artificial selection offers a different experimental approach to the genetics of morphological variation. Long applied to the domestication of plants and animals, it is possible in the laboratory to derive populations with much greater divergence in traits than natural populations through repeated selection over several generations for individuals with character states at either end of a continuum. Genetic analyses of selected populations can estimate the number of genes involved in trait divergence and identify candidate loci. The crucial point that has emerged from selection experiments on fly bristle number is that just as for natural variation, many known loci contribute to differences in such modest traits (Nuzhdin *et al.,* 1999). Importantly, the effects of genetic variation are not strictly additive. Combinatorial interactions, so fundamental to developmental processes, play powerful roles in morphological variation and evolution.

The most detailed picture of a developmental gene's response to artificial selection has come from the domestication of maize from the wild Mexican grass teosinte 5,000 to 10,000 years ago (Wang *et al.,* 1999). These plants differ dramatically from each other in morphology. Genetic analyses have identified the *teosinte-branched* (*tb1*) gene as the major locus that controls the difference in branch length and morphology. Analysis of various maize and teosinte *tb1* gene sequences revealed that no differences have been fixed over the entire amino acid sequence of the protein. Rather, analysis of polymorphisms and the levels of *tb1* mRNA indicate that, during domestication of maize from teosinte, selection acted primarily upon regulatory elements of the *tb1* gene.

The Evolution of Regulatory DNA and Morphological Diversity

Comparisons of developmental gene regulation between morphologically divergent animals, analyses of intraspecific variation, and the response of organisms and genes to selection all support the claim that regulatory DNA is the predominant source of the genetic diversity that underlies morphological variation and evolution. While there may be many factors contributing to the importance of *cis*-regulatory DNA in evolution, I emphasize three here. First and foremost is the modular organization of *cis*-regulatory systems (Arnone and Davidson, 1997). Individual elements can act, and therefore evolve, independently of others. The typical organization of the *cis*-regulatory regions of developmental regulatory genes, composed of many independent elements, is tacit evidence for the expansion and diversification of *cis*-regulatory systems in evolution.

Second, there is a greater degree of freedom in *cis*-regulatory sequences (as opposed to coding sequences) that imparts a tolerance of regulatory DNA to all varieties of mutational change. Regulatory elements need not maintain any reading frame, they can function at widely varying distances and in either orientation to the transcription units they control. This evolvability of regulatory DNA sequence means that it is a rich source of genetic and, potentially, phenotypic variation.

Finally, the combinatorial nature of transcriptional regulation, controlled by the diverse repertoire of transcription factors in animals, has important evolutionary ramifications. Most spatially regulated elements are controlled by a handful of transcription factors whose DNA binding specificities are sufficiently relaxed such that the affinity and number of sites for each factor can evolve at a significant rate, even in functionally conserved elements. Ludwig *et al.* (1998) have shown, for example, that essential sites for four regulators within the *even-skipped* stripe 2 element have diverged between closely related *Drosophila* species. The function of the element is conserved by compensatory changes in other sites. This dynamic picture of evolutionary change within *cis*-regulatory elements suggests that, as binding sites evolve in existing elements, variations in the level, pattern, or timing of gene expression may arise which are the raw material for morphological evolution.

Seeing the Steps: From Bristles to Body Plans

One of the longest running debates in evolutionary biology concerns the sufficiency of processes observed within populations and species for explaining macro-evolution. Explanations of large-scale evolutionary patterns, such as those evident in the fossil record, seek to encompass processes such as speciation, selection, drift, competition, environmental change, extinction, and more. While all of these forces have shaped history, they represent dimensions beyond the fundamental genetic and developmental questions considered here. From the perspective of developmental genetics, the global micro/macro evolutionary debate can be reduced to the question of whether the same genetic mechanisms underlying intraspecific variation and interspecific differences are sufficient to account for the large-scale changes in evolution. Several arguments can be made in support of the explanatory sufficiency of regulatory evolution and against the necessity for or the probability of dramatic large-scale "macromutations" playing a significant role in morphological evolution.

First, there is apparently abundant variation regulatory regions of developmental genes in natural populations. Some of this is expressed, as described for variation in bristle number. But, in addition, several recent studies have shown that there is also "cryptic" or latent variation that can be expressed under artificial selection or when developmental mutations are introduced into populations (Gibson and Hogness, 1996; Polaczyk et al., 1998; Gibson et al., 1999).

Second, variation exists within and between natural populations for characters that appear to be considerably more significant to the evolution of body plans and body parts than bristle number – such as segment number in centipedes (Arthur, 1999), body patterns in stickleback fish (Anh and Gibson, 1999), and limb morphology in salamanders (Shubin et al., 1995). In the latter example, detailed analysis of just a single population of newts revealed a surprising spectrum

of variation in limb skeleton morphologies in about 30% of individuals. Most interestingly, some variations were similar to standard limb skeletal morphologies found in ancestral and other species. The occurrence of both potentially novel and atavistic (reversions to ancestral states) forms in a single population suggests that complex morphologies can evolve through ordinary, readily available genetic variation in the myriad interactions that shape development.

Third, while the existence of homeotic mutations has been cited most often in support of the plausibility of macromutations, it is clear that such dramatic phenotypes are associated with disruptions of gene structure and reductions in fitness that make their fixation unlikely at best. There are more attractive alternative scenarios for homeotic gene involvement in morphological evolution that do not invoke macromutations. Incremental changes in the function of individual cis-regulatory elements can account for the shifts in Hox expression domains that have occurred during arthropod (Akam, 1998) and vertebrate evolution (Belting et al., 1998).

One such picture of how small differences in Hox expression have evolved has emerged from analysis of rather modest differences in leg morphology between certain Drosophila species. Stern (1998) has shown that differences in leg hair patterns between D. melanogaster and D. simulans, species that diverged about 2-3 million years ago, are due to differences in the regulation of the Ultrabithorax gene in just part of the second leg. One attractive implication of this work is that these subtle interspecific differences in Hox regulation and morphology potentially represent intermediates in a continuum from intraspecific variation to larger-scale morphological evolution.

It is not realistic to expect to reconstruct the entire sequence of genetic and developmental steps involved in the diversification of any group. But the successes of comparative approaches that correlate major genetic regulatory differences with body plan diversity and of genetic analyses that identify regulatory genes involved in inter- and intraspecific differences enable us to perceive the general mechanisms at work. In the near term, detailed analyses of cis-regulatory element variation and evolution will be of central importance in expanding our understanding of how gene expression and morphology evolve.

Selected Reading

Akam, M. (1998). *Int. J. Dev. Biol. 42*, 445–451.

Anh, D.-G., and Gibson, G. (1999). *Evol. Dev. 1*, 100–112.

Arnone, M. I., and Davidson, E. H. (1997). *Development 124*, 1851–1864.

Arthur, W. (1999). *Evol. Dev. 1*, 62–69.

Belting, H.-G., Shasshikant, C. S., and Ruddle, F. H. (1998). *Proc. Natl. Acad. Sci. USA 95*, 2355–2360.

Carroll, S. B., Grenier, J. K., and Weatherbee, S. D. (2000). In DNA to Diversity: Molecular Genetics and the Evolution of Animal Design (Malden, MA: Blackwell Scientific), in press.

Cohn, M. J., and Tickle, C. (1999). *Nature 399*, 474–479.

de Rosa, R., Grenier, J. K., Andreeva, T., Cook, C., Adoutte, A., Akam, M., Carroll, S. B., and Balavoine, G. (1999). *Nature 399*, 772–776.

Gibson, G., and Hogness, D. S. (1996). *Science 271*, 200–203.

Gibson, G., Wemple, M., and van Helden, S. (1999). *Genetics 151*, 1081–1091.

Knoll, A. H., and Carroll, S. B. (1999). *Science 284*, 2129–2137.

Long, A. D., Lyman, R. F., Langley, C. H., and Mackay, T. F. C. (1998). *Genetics 149*, 999–1017.

Mackay, T. F. C. (1996). *BioEssays 18*, 113–121.

Nuzhdin, S. V., Dilda, C. L., and Mackay, T. F. C. (1999). *Genetics 153*, 1317–1331.

Polaczyk, P. J., Gasperini, R., and Gibson, G. (1998). *Dev. Genes Evol. 207*, 462–470.

Shubin, N., Wake, D. B., and Crawford, A. J. (1995). *Evolution 49*, 874–884.

Sordino, P., van der Hoeven, F., and Duboule, D. (1995). *Nature 375*, 678–681.

Stern, D. L. (1998). *Nature 396*, 463–466.

Wang, R.-L., Stec, A., Hey, J., Lukens, L., and Doebley, J. (1999). *Nature 398*, 236–239.

Weatherbee, S. D., and Carroll, S. B. (1999). *Cell 97*, 283–286.

15

Functional Evo-devo

Casper J. Breuker, Vincent Debat and Christian Peter Klingenberg

Functional factors such as optimal design and adaptive value have been the central concern of evolutionary biology since the advent of the New Synthesis. By contrast, evolutionary developmental biology (evo-devo) has concentrated primarily on structural factors such as the ways in which body parts can be built. These different emphases have stood in the way of an integrated understanding of the role of development in evolution. Here, we try to bridge this gap by outlining the relevance of functional factors in evo-devo. We use modularity and the view of development as a flexible evolutionary system to outline a unified perspective that includes both structural and functional aspects.

Development as a factor in evolution

Whereas development has long been recognized as being important in evolution, its role as an evolutionary factor has only begun to be investigated relatively recently with the study of heterochrony [1] and developmental quantitative genetics [2]. The rise of evolutionary developmental biology (evo-devo) as a biological discipline has brought about several changes in perspective [3,4]. In addition to a new focus on the developmental mechanisms that generate new variation, the discovery of the widespread evolutionary conservation of genes with prominent roles in development (e.g. *Hox* genes [5]) has revived an interest in comparative studies at a large phylogenetic scale.

This shift of interest and emphasis has drawn attention away from the traditional focus of evolutionary studies, namely the adaptive value and functional significance of phenotypic traits. Here, we attempt to integrate functional considerations with the central concepts emerging from evo-devo. We hope that this will contribute to a more unified understanding of the role of development in adaptive evolution.

Structural and functional factors in evolution

The debate about the relative importance of intrinsic structural factors and external adaptation in biological evolution has a long history [6,7]. By the mid-20th century, the neo-darwinian

Casper Breuker, Vincent Debat, and Christian Klingenberg, "Functional Evo-Devo." Reprinted from *Trends in Ecology and Evolution*, Vol. 21, pp. 21, 488–492, Copyright 2006, with permission from Elsevier.

The authors thank M. Gibbs, P. Goodwyn, N. Navarro, J. Patterson, and three anonymous referees for discussion and comments on the manuscript. Financial support was provided by the European Commission, Marie Curie Intra-European Fellowships (MEIF–CT–2003–502052 and MEIF–CT–2003–502168).

New Synthesis had established adaptation as the central theme of evolutionary biology, such that the primary research emphasis was on the external factors that shaped organisms through natural selection. The discovery of ample genetic variation in natural populations suggested that the raw material for natural selection is plentiful. It was therefore expected that selection would produce optimal solutions in an engineering sense [8], where each organ is optimised for performing certain functions that confer maximum fitness jointly to the organism. The evolution of a trait could therefore be explained by its function.

Neo-darwinian theory has emphasized function at the expense of structural and historical concerns. When cladistics, the study of relationships among organisms through the branching of evolutionary lineages, became the dominant direction of systematics in the 1980s, historical considerations entered mainstream evolutionary biology under the headings of phylogenetics and the comparative method [9].

The discovery of the pervasive conservation of Hox genes [5] and their expression patterns across animal phyla was surprising because it was at odds with the expectation that genetic and developmental systems would evolve just as much as the morphological traits they generate [7]. This discovery of conserved developmental genes, along with similar findings for other families of genes involved in key developmental processes, provided an important impetus for the emergence of evo-devo as a discipline. Evo-devo also awakened a renewed interest in phylotypic stages [4], developmental stages shared by the species across entire phyla in spite of vast differences in the development and morphology before and after that stage, and coined the new concept of the zootype [10], a hypothetical ground plan for all bilaterian animals. These ideas were tied explicitly to the concept of the archetype, the idea of a common body plan that underlies the variation in a group (such as the vertebrates) that had been rejected vehemently by the main exponents of the New Synthesis [7]. Altogether, these discoveries have attracted new attention to structural factors.

Evo-devo has also revived structuralist arguments that emphasized the importance of generic physical factors [11,12], such as the forces driving morphogenetic movements, in the

development and evolution of organismal forms. The combination of such factors with findings from developmental genetics has made it possible to formulate general models of pattern formation [13]. Models of this kind have been applied to the variation and morphological innovation in the patterns of mammalian tooth cusps [14] and have subsequently been confirmed experimentally [15].

Given its primary focus on large-scale phylogenetic comparison and developmental mechanisms generating variation, evo-devo has emphasized a structural and partly historical perspective on evolution, but has not concerned itself with functional aspects. However, if the goal is to gain an integrated view of the role of development in evolution, the link to function is essential (Box 15.1).

Evo-devo researchers have begun to study the developmental basis of evolutionary changes with immediate adaptive significance such as the reduction of pelvic structures [16] and bony armour [17] in sticklebacks, the differences in beak shape among species of Darwin's finches [18], and divergence in jaw shape of cichlid fishes [19]. These are all examples of adaptive evolutionary change driven by natural selection, and therefore relate directly to the functions of the respective traits. A key challenge will be to make this relation more explicit.

Box 15.1 singles out biomechanics and functional morphology as disciplines where this kind of relation has been investigated already. These areas have a well established emphasis on performance as a measurable intermediate between morphology (or other phenotypic traits) and fitness [20]. Performance is an attribute of the organism and results from the function of a trait in a specific task. Performance can be measured in the laboratory or in the field, and established procedures exist for relating it to fitness [21]. The daunting challenge of relating development to function can therefore be rephrased as the more tractable task of relating development to performance. This approach has only been adopted recently in a study linking the biomechanics of jaw movement in cichlid fishes explicitly to the quantitative and developmental genetics of mandibular shape [19].

There are alternatives to this way to bring functional considerations into evo-devo. Here, we outline work on developmental modularity and

Box 15.1.   Explanations in evolutionary biology

What counts as an explanation in evolutionary biology depends on the specific context. Explanations offered by studies of adaptation are different from those derived from phylogenetic analyses. Gould [6] has developed a graphical framework that is useful for thinking about evolutionary causation and constraints. He distinguished three primary kinds of causation in evolution – functional, historical and structural – and arranged them in a triangular diagram (Figure 15.1). Different types of study put the emphasis more or less on one of the corners of the diagram or between them.

The main emphasis in neo-Darwinian evolutionary biology is on functional aspects (Figure 15.1a). The goal is to understand how traits evolved by natural selection and how they contribute to fitness. Historical and structural factors act as constraints by setting boundary conditions in these explanations. In areas such as life-history studies, these factors have a relatively minor role by comparison, with the main aim being to document the adaptedness of different life-history strategies. By contrast, structural and historical factors have a prominent role in biomechanics or in studies using the phylogenetic comparative method for setting the context for functional explanations.

In evo-devo, the primary emphasis is on structural explanations (Figure 15.1b). The main goal of evo-devo is to understand how developmental mechanisms influence evolution and how these mechanisms themselves have evolved. Structural considerations about embryos and developmental processes have a central role in this endeavour. Studies such as the reconstruction of the zootype [10] clearly have a strong historical component, whereas comparisons of gene expression in more or less closely related species involve functional and historical components to some degree.

Unifying evo-devo and functional studies puts new emphasis on the lower side of the triangle (Figure 15.1). A comparison of Figures 15.1a and b shows that functional evo-devo is placed closely to biomechanics and related disciplines such as functional morphology. These specialties all combine structural and functional considerations, and the link between them therefore provides a promising new perspective to bring functional aspects into evo-devo. Biomechanics and functional morphology have clear criteria for establishing the functional performance of morphological traits. The challenge will be to apply those criteria to a context that explicitly considers the developmental origin of the traits.

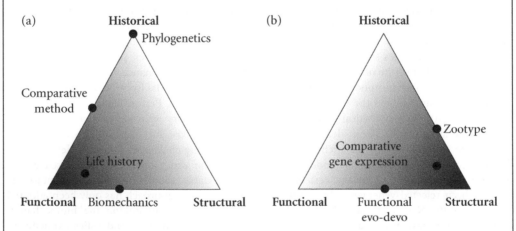

Figure 15.1   Differences in emphasis of explanatory factors (bold print) in different fields of evolutionary biology (dots). (a) Some of the traditional disciplines in evolutionary biology. (b) Some areas that have emerged as parts of evo-devo. The shading indicates the overall emphasis in neo-darwinian evolutionary biology (a) and in evo-devo (b).

flexibility that also address function, albeit in a less direct manner.

Modularity

Biological systems consist of parts that are recognizable because they are integrated internally and are relatively distinct from other such parts [22,23]. In general, the concept of modularity refers to this properly of integration within, and relative autonomy among, the parts or modules. Modularity is studied most often in a structural context, where it refers to the spatial arrangement of physical parts at different organizational levels from molecules to entire organisms. However, modularity also exists in contexts that are based on different kinds of interactions, such as pathways in metabolic networks, gene regulatory interactions, developmental and functional interactions among traits, and even behavioural interactions among individuals [24].

Developmental modules are integrated internally by developmental interactions between the components of the module, and the developmental processes within each module are relatively unaffected by the module's surroundings. In response to a developmental perturbation, component parts are therefore expected to covary only within a developmental module; the covariation among the resulting traits can therefore be used to infer developmental integration in various biological systems. This reasoning is used in studies inferring developmental modularity through the study of developmental mutants [25] or through the analysis of correlated asymmetry [26]. Because these developmental interactions also are involved in the expression of genetic variation, they are key determinants of patterns of pleiotropy [23].

A functional module is an integrated unit of traits serving a common function and is separable from other such units, which are associated with different functions. The interactions between traits that provide the coherence of modules are therefore of a functional nature, and usually are evolved by selection for optimal performance in that functional context. Functional interactions can be understood, for example, through biomechanical methods or by studying the arrangement of muscle insertions [19,27].

How functional and developmental modules relate to each other has been discussed extensively. One view is that developmental modules and the genetic architecture they determine are derived features that have been moulded by selection to match functional modularity [28,29]. Alternatively, developmental modules can be considered to be ancestral features that act potentially as developmental constraints influencing subsequent evolutionary changes [4,30]. These are not mutually exclusive alternatives, but the opposite ends of a spectrum of explanations. To distinguish between the possibilities, it is therefore important to compare developmental and functional modules (Box 15.2).

Modularity has often been used to illustrate the existence of developmental constraints, yet it can be argued just as easily that modularity in itself can be adaptive. The fact that developmental modules are independent developmentally and that functional modules are independent from surrounding traits in their fitness effects would enable rapid and specific adaptation to changing environmental conditions [28,31]. Modularity therefore contributes to the great flexibility of developmental processes and facilitates adaptive variation in developmental and functional units. The modular structure of butterfly wing patterns, for example, is often thought to have enabled not only the great diversity of colour patterns, but also the evolution of various adaptive features such as mimicry, mate selection, camouflage and seasonally polyphenic patterns [32].

Organisms as flexible functional systems

A common theme that has emerged from analyses in evo-devo and other areas of evolutionary biology is that organisms are flexible systems. If the surroundings of an organism change, its developmental systems provide the ability to adapt to achieve and maintain some function. This can be adaptation in either the physiological or evolutionary sense, and encompasses timescales from almost instantaneous physiological responses through reaction norms on an ecological timescale (the lifecycle of an individual) to adaptive responses of lineages over macroevolutionary timescales.

Box 15.2.   Functional versus developmental modularity

It is difficult to evaluate empirically the hypothesis that developmental modules evolve adaptively to match functional modules [28,29] (here called the 'matching hypothesis') against the alternative that developmental modules are evolutionarily conserved features acting primarily as constraints [4]. The test has to rely on evidence provided by the comparison of multiple case histories.

The matching hypothesis predicts that functional and developmental modules evolve to coincide. It will therefore be necessary to compile lists of cases with matches and mismatches between functional and developmental modules. Developmental modularity can be inferred from information about the developmental organization of or from the covariation of the parts [23,26]. Functional modularity has to be inferred from other sources of information such as biomechanical analyses or studies of selection.

In many cases, developmental and functional modules will match. For instance, the forewing of a butterfly is a single developmental module that serves as a unit for functions such as flight and signalling. This case is the one predicted by the matching hypothesis, but it is not informative by itself because it does not reveal the

sequence of evolutionary changes that led to this state. Many other cases will also be problematic because the function is unclear or the developmental modularity is ambiguous (e.g. in the example of the hemelytra of a true bug).

Informative cases involve a clear mismatch between functional and developmental modularity. Different parts of a single ancestral developmental module can have distinct functions. An example is the pitcher plant *Nepenthes*, where the tip of the leaf is transformed into the 'pitcher' for trapping insects. If developmental modularity (e.g. as revealed by morphometric covariation) also follows this derived two-part modular structure, then the matching hypothesis is supported.

Another informative case is where several distinct ancestral developmental modules perform a single function. An example is the blossom of *Dalechampia*, which is an inflorescence consisting of several separate flowers and bracts that function together in pollination. Integration among the parts of the blossoms is found, providing some support for the matching hypothesis [46]. Similarly, the scutellum and the basal part of the forewings of true bugs function together for protection.

In evo-devo, the aspect of this flexibility that has received the most attention is 'evolutionary tinkering' [33,34]. To produce a new trait, natural selection does not start from scratch, but from what is already available: existing organs, tissues and cells, as well as existing genes and gene networks. This raw material often leads to surprising and sometimes suboptimal solutions to engineering problems. The inverted structure of the vertebrate retina (Box 15.3) or the evolution of the mammalian middle-ear ossicles from the jaw bones [35] are two examples. Evo-devo has provided these case studies not only with developmental genetic details, but has also shown that a similar tinkering is occurring at the molecular level because genes can be co-opted in new contexts.

A gene (or a genetic network) can be co-opted independently in the formation of analogous

traits in phylogenetically distant taxa. Among the most famous examples are the roles of *engrailed* in segmentation [36] and *distal-less* in the patterning of appendages [37,38]. In these cases, selection appears to have co-opted an already existing genetic network to perform a similar function [39] in a novel context (i.e. developmental exaptation [40]). One of the consequences of this process is that non-homologous structures can share developmental features, thereby complicating the detection of homology [41].

Another aspect of developmental flexibility that is just starting to be investigated in evo-devo is the maintenance of functionality under varying environmental conditions on an ecological timescale. Phenotypic plasticity, reaction norms and genotype-by-environment interactions are the labels under which this phenomenon has been studied so far. The developmental component of

---

Box 15.3. Flexibility in eye development and evolution

The camera eyes of vertebrates and some cephalopod mollusks provide an example of the flexibility or development in arriving at a functional solution by transforming existing structures in different ways. Dibranchiate cephalopods, such as squid and octopus, have a camera-type eye that is similar to the vertebrate eye in its overall structure and function. But because the cephalopod eye develops differently from the vertebrate eye, it differs in one key aspect, namely the orientation of the photoreceptor cells [47,48]. The vertebrate eye develops from the neural plate as an evagination from the brain, whereas the cephalopod eye forms as an invagination of the ectoderm. This causes the photoreceptor cells in cephalopods to be directed towards the light source and those of vertebrate eyes to be facing in the opposite direction [47,48]. As a consequence of this arrangement in vertebrate eyes, the optic nerve passes through the retina and creates the 'blind spot', a suboptimal design that necessitated neurosensory and behavioral compensation [49]. This need for behavioural compensation of what appears to be a design flaw illustrates yet another form of flexibility associated with a functional structure:

co-evolution of the eye and the associated neurosensory system.

A key feature of a camera eye is its lens, which is an example of a remarkable convergence by evolutionary tinkering. The lenses of both cephalopods and vertebrates consist of cells filled with soluble crystallin proteins, and that are packed together to form a concentration gradient from the periphery to the centre of the lens that produces the refractive index gradient necessary for a lens to be functional [48]. Although crystallins were long thought to be unique to lens tissue and to have evolved for this special function, more recent research suggests that they are co-opted proteins that are not specialized structurally for this function and are used in various other functional contexts as well [48]. Cephalopod lens tissue consists nearly entirely of the enzyme glutathione S-transferase, whereas 11 different vertebrate lens proteins have been identified and found to correspond to molecular chaperone proteins and various enzymes [48]. It appears, therefore, that the proteins destined to be used as lens crystallins were co-opted opportunistically from the available range of existing proteins.

---

environmental reactions has begun to be fully integrated only recently [42], leading to the first cases of adaptive phenotypic plasticity that are well documented even at the molecular level. Examples of such integrations of developmental genetics and traditional evolutionary biology are the seasonal polyphenism in the butterfly *Bicyclus* [43] and the shade-avoidance syndrome in *Arabidopsis* [44].

Flexibility is a fundamental property of developmental and physiological systems that enables them to adapt to achieve new functions and to maintain them when the environment changes. Interestingly, these two aspects are not disconnected from each other; some authors have even argued that phenotypic plasticity, far from counteracting the effects of natural selection, provides selection with a wider spectrum of phenotypes to act upon, thereby facilitating adaptive evolution [45].

The way ahead for functional evo-devo

Combining structuralist and functionalist perspectives will facilitate a fuller understanding of evolutionary processes. To date, evo-devo has taken a mostly structuralist approach, whereas Neo-Darwinian evolutionary biology has taken the functionalist viewpoint. A full understanding of evolution requires the use of the entire conceptual space (see Box 15.1), and a fusion of functional aspects with evo-devo is therefore to be welcomed.

We have outlined two subjects, modularity and flexibility, in which the union of evo-devo with functional considerations is particularly straightforward. We do not imply that these two areas are the only ones in which a synthesis is possible, and we anticipate a wide range of research programmes exploring the interface between structural and functional aspects of evolution.

References

1 Gould, S. J. (1977) *Ontogeny and Phylogeny*, Harvard University Press.

2 Atchley, W. R. (1987) Developmental quantitative genetics and the evolution of ontogenies. *Evolution* 41, 316–330.

3 Hall, B. K. (1992) *Evolutionary Developmental Biology*, Chapman & Hall.

4 Raff, R. A. (1996) *The Shape of Life: Genes, Development and the Evolution of Animal Form*, University of Chicago Press.

5 de Rosa, R. *et al.* (1999) Hox genes in brachiopods and priapulids and protostome evolution. *Nature* 399, 772–776.

6 Gould, S. J. (2002) *The Structure of Evolutionary Theory*, Harvard University Press.

7 Amundson, R. (2005) *The Changing Role of the Embryo in Evolutionary Thought: Roots of Evo-Deve*, Cambridge University Press.

8 Parker, G. A. and Maynard Smith, J. (1990) Optimality theory in evolutionary biology. *Nature* 348, 27–33.

9 Felsenstein, J. (2004) *Inferring Phylogenies*, Sinauer Associates.

10 Slack, J. M. W. *et al.* (1993) The zootype and the phylotypic stage. *Nature* 361, 490–492.

11 Thompson, D. A. W. (1961) *On Growth and Form*, Cambridge University Press.

12 Goodwin, B. (1994) *How the Leopard Changed its Spots: the Evolution of Complexity*, Phoenix.

13 Salazar-Ciudad, I. *et al.* (2003) Mechanisms of pattern formation in development and evolution. *Development* 130, 2027–2037.

14 Jernvall, J. (2000) Linking development with generation of novelty in mammalian teeth. *Proc. Natl. Acad. Sci. U.S.A.* 97, 2641–2645.

15 Kassai, Y. *et al.* (2005) Regulation of mammalian tooth cusp patterning by ectodin. *Science* 309, 2067–2070.

16 Shapiro, M. D. *et al.* (2004) Genetic and developmental basis of evolutionary pelvic reduction in threespine sticklebacks. *Nature* 428, 717–723.

17 Colosimo, P. F. *et al.* (2005) Widespread parallel evolution in sticklebacks by repeated fixation of ectodysplasin alleles. *Science* 307, 1928–1933.

18 Abzhanov, A. *et al.* (2004) *Bmp4* and morphological variation of beaks in Darwin's finches. *Science* 305, 1462–1465.

19 Albertson, R. C. *et al.* (2005) Integration and evolution of the cichlid mandible: the molecular basis of alternate feeding strategies. *Proc. Natl. Acad. Sci. U.S.A.* 102, 16287–16292.

20 Kingsolver, J. G. and Huey, R. B. (2003) Introduction. The evolution of morphology, performance, and fitness. *Integr. Comp. Biol.* 43, 361–366.

21 Arnold, S. J. (2003) Performance surfaces and adaptive landscapes. *Integr. Comp. Biol.* 43, 367–375.

22 Wagner, G. P. (1996) Homologues, natural kinds and the evolution of modularity. *Am. Zool.* 36, 36–43.

23 Klingenberg, C. P. (2005) Developmental constraints, modules and evolvability. In *Variation: a Central Concept in Biology* (Hallgrímsson, B. and Hall, B. K., eds), pp. 219–247, Elsevier.

24 Schlosser, G. and Wagner, G. P. (2004) *Modularity in Development and Evolution*, The University of Chicago Press.

25 Monteiro, A. *et al.* (2003) Mutants highlight the modular control of butterfly eyespot patterns. *Evol. Dev.* 5, 180–187.

26 Klingenberg, C. P. (2003) Developmental instability as a research tool: using patterns of fluctuating asymmetry to infer the developmental origins of morphological integration. In *Developmental Insta-bility: Causes and Consequences* (Polak, M., ed.), pp. 427–442, Oxford University Press.

27 Badyaev, A. V. *et al.* (2005) Evolution of morphological integration: developmental accommodation of stress-induced variation. *Am. Nat.* 166, 382–395.

28 Wagner, G. P. and Altenberg, L. (1996) Complex adaptations and the evolution of evolvability. *Evolution* 50, 967–976.

29 Cheverud, J. M. (1984) Quantitative genetics and developmental constraints on evolution by selection. *J. Theor. Biol.* 110, 156–171.

30 Kirschner, M. and Gerhart, J. (1998) Evolvability. *Proc. Natl. Acad. Sci. U.S.A.* 95, 8420–8427.

31 Love, A. C. and Raff, R. A. (2003) Knowing your ancestors: themes in the history of evo-devo. *Evol. Dev.* 5, 327–330.

32 Nijhout, H. F. (2003) Development and evolution of adaptive polyphenisms. *Evol. Dev.* 5, 9–18.

33 Jacob, F. (1977) Evolution and tinkering. *Science* 196, 1161–1166.

34 Duboule, D. and Wilkins, A. S. (1998) The evolution of 'bricolage'. *Trends Genet.* 14, 54–59.

35 Martin, T. and Luo, Z. X. (2005) Homoplasy in the mammalian ear. *Science* 307, 861–862.

36 Minelli, A. and Fusco, G. (2004) Evo-devo perspectives on segmentation: model organisms, and beyond. *Trends Ecol. Evol.* 19, 423–429.

37 Angelini, D. R. and Kaufman, T. C. (2005) Insect appendages and comparative ontogenetics. *Dev. Biol.* 286, 57–77.

38 Lowe, C. J. and Wray, G. A. (1997) Radical alterations in the roles of homeobox genes during echinoderm evolution. *Nature* 389, 718–721.

39 True, J. R. and Carroll, S. B. (2002) Gene co-option in physiological and morphological evolution. *Annu. Rev. Cell Dev. Biol.* 18, 53–80.

40 Chipman, A. D. (2001) Developmental exaptation and evolutionary change. *Evol. Dev.* 3, 299–301.

41 Abouheif, E. *et al.* (1997) Homology and developmental genes. *Trends Genet.* 13, 432–433.

42 Gilbert, S. F. (2003) The morphogenesis of evolutionary developmental biology. *Int. J. Dev. Biol.* 47, 467–477.

43 Brakefield, P. M. *et al.* (1998) The regulation of phenotypic plasticity of eyespots in the butterfly *Bicyclus anynana*. *Am. Nat.* 152, 853–860.

44 Franklin, K. A. and Whitelam, G. C. (2005) Phytochromes and shade-avoidance responses in plants. *Ann. Bot. (London)* 96, 169–175.

45 West-Eberhard, M. J. (2005) Phenotypic accommodation: adaptive innovation due to developmental plasticity. *J. Exp. Zool. B Mol. Dev. Evol.* 304, 610–618.

46 Armbruster, W. S. *et al.* (2004) Floral integration, modularity, and accuracy. In *Phenotypic Integration: Studying the Ecology and Evolution of Complex Phenotypes* (Pigliucci, M. and Preston, K., eds), pp. 23–49, Oxford University Press.

47 Gehring, W. J. (2004) Historical perspective on the development and evolution of eyes and photoreceptors. *Int. J. Dev. Biol.* 48, 707–717.

48 Fernald, R. D. (2004) Evolving eyes. *Int. J. Dev. Biol.* 48, 701–705.

49 Burr, D. (2005) Vision: in the blink of an eye. *Curr. Biol.* 15, R554–R556.

50 Halder, G. *et al.* (1995) New perspectives on eye evolution. *Curr. Opin. Genet. Dev.* 5, 602–609.

PART VII

REDUCTIONISM AND THE BIOLOGICAL SCIENCES

Introduction

The history of physics and chemistry since Newton appears, to many, to be a history of older, narrower, less accurate theories being subsumed, deepened, corrected, and explained by newer, broader, and more accurate theories. This history is taken to vindicate 'reductionism' – that there is a fundamental theory in physics from which the rest of physical science, including chemistry, can be derived. The spectacular developments in biology in the second half of the twentieth century have held out the prospect that we can expect it, too, ultimately to be grounded in physical science.

In particular, the success of molecular biology in shedding light on many hitherto mysterious biological processes, especially in heredity and embryological development, disease, and illness, has led many to conclude that, ultimately, our understanding of macromolecular processes will enable us to explain all biological processes. This is a thesis that often goes by the name *reductionism*. The label is also employed by critics of the explanatory claims of molecular biologists to label the more radical thesis that molecular biology may someday replace or eliminate biological theories and explanations that focus on organisms and populations.

Reductionists look to the history of unification in physics and chemistry as a model for how biology should advance. In the physical sciences, the laws of earlier and narrower theories – the gas laws of Boyle and Charles in the seventeenth century, for example – have been explained by their derivation from the laws of later and more fundamental ones like the nineteenth century kinetic theory of gases. Similarly, reductionists believe that Watson and Crick's discoveries about the structure and function of DNA will explain the processes discovered by Mendel; in fact, they will deepen our understanding of them, while correcting mistakes in Mendel's theory and its early developments.

Kitcher's paper (chapter 16) is justly famous for showing that the comparison often made between reduction in physics and biology rests on false presuppositions about whether there are laws in biology, whether biological processes can be given molecular descriptions at all, and whether non-molecular explanations can be, or need to be, improved by deeper molecular explanations of the same process.

Much of Kitcher's argument turns on the fact that many biological processes can be and are realized by a large number of quite different molecular mechanisms: just as something's being a chair cannot be defined in terms of any particular composition and geometry – there are too many ways to make chairs out of too many different materials – biological processes and structures (Kitcher's example is meiosis) can be, and are, implemented in many different ways by many different molecules. This is *multiple realizability* and, because of it, biological terms cannot be linked to molecular terms in finitely long and biologically useful definitions. Kitcher argues that this multiple realizability is the source of profound obstacles to the program of macro-molecular reductionism in biology. He concludes that molecular biology's contributions to biology must be much more limited than reductionists hope it will be.

Elliot Sober is a philosopher who long agreed with Kitcher and employed quite similar arguments to support equally antireductionist conclusions. But his paper (chapter 17), written 15 years after Kitcher's, reflects the conclusion that anti-reductionists have drawn a stronger conclusion from multiple realizability about the limits to Reductionism than is warranted.

Further Reading

Arp, R. (2007). Homeostatic organization, emergence, and reduction in biological phenomena. *Philosophia Naturalis*, 44, 238–270.

Arp, R. (2008). Life and the Homeostatic Organization View of biological phenomena. *Cosmos and History: The Journal of Natural and Social Philosophy*, in press.

Bechtel, W. (2006). *Discovering cell mechanisms: The creation of modern cell biology*. Cambridge: Cambridge University Press.

Beurton, P., Falk, R., & Rheinberger, H. (eds.) (2000). *The concept of the gene in development and evolution: Historical and epistemological perspectives*. Cambridge: Cambridge University Press.

Craver, C., & Darden, L. (eds.) (2005). Studies in History and Philosophy of Biological and Bio-medical Sciences. *Mechanisms in Biology*, 31, 2.

Darden, L. (2006). *Reasoning in biological discoveries: Essays on mechanisms, interfield relations, and anomaly resolution*. Cambridge: Cambridge University Press.

Delehanty, M. (2005). Emergent properties and the context objection to reduction. *Biology and Philosophy*, 20, 715–734.

Griffiths, P. (2002). Molecular and developmental biology. In P. Machamer & M. Silberstein (eds.), *The Blackwell guide to the philosophy of science* (pp. 252–271). Malden, MA: Blackwell Publishers.

Hoyningen-Huene, P., & Wuketits, F. (eds.) (1989). *Reductionism and systems theory in the life sciences*. Dordrecht: Kluwer.

Keller, E. (2002). *The century of the gene*. Cambridge, MA: Harvard University Press.

Kimbrough, S. (1978). On the reduction of genetics to molecular biology. *Philosophy of Science*, 46, 389–406.

Machamer, P., Darden, L., & Craver, C. (2000). Thinking about mechanisms. *Philosophy of Science*, 67, 1–25.

Mayr, E. (1996). The autonomy of biology: The position of biology among the sciences. *The Quarterly Review of Biology*, 71, 97–106.

Moss, L. (2001). Deconstructing the gene and reconstructing molecular developmental systems. In S. Oyama, P. Griffiths, & R. Gray (eds.), *Cycles of contingency: Developmental systems and evolution* (pp. 85–97). Cambridge, MA: MIT Press.

Robinson, W. (2005). Zooming in on downward causation. *Biology and Philosophy*, 20, 117–136.

Rosenberg, A. (1997). Reductionism redux: Computing the embryo. *Biology and Philosophy*, 12, 445–470.

Rosenberg, A. (2001a). Reductionism in a historical science. *Philosophy of Science*, 68, 135–164.

Rosenberg, A. (2001b). On multiple realization and special sciences. *Journal of Philosophy*, 98, 365–373.

Rosenberg, A. (2006). *Darwinian reductionism, or how to stop worrying and love molecular biology*. Chicago: University of Chicago Press.

Rosenberg, A., & McShea, D. (2008). *Philosophy of biology: A contemporary introduction* (Chapter 4). London: Routledge.

Sarkar, S. (ed.) (1996). *The philosophy and history of molecular biology*. Dordrecht: Kluwer.

Sarkar, S. (1998). *Genetics and reductionism*. Cambridge: Cambridge University Press.

Sarkar, S. (2005). *Molecular models of life: Philosophical papers on molecular biology*. Cambridge, MA: MIT Press.

Schaffner, K. (1967). Approaches to reduction. *Philosophy of Science*, 34, 137–147.

Schaffner, K. (1969). The Watson–Crick model and reductionism. *British Journal for the Philosophy of Science*, 20, 325–348.

Schaffner, K. (2006). Reduction: The Cheshire cat problem and a return to roots. *Synthese*, 151, 377–402.

Silberstein, M. (2002). Reduction, emergence and explanation. In P. Machamer & M. Silberstein (eds.), *The Blackwell guide to the philosophy of science* (pp. 80–107). Malden, MA: Blackwell Publishers.

Silberstein, M., & McGreever, J. (1999). The search for ontological emergence. *Philosophical Quarterly*, 49, 182–200.

Waters, C. (1994). Genes made molecular. *Philosophy of Science*, 61, 163–185.

Waters, C. (2000). Molecules made biological. *Revue Internationale de Philosophie*, 4, 539–564.

Waters, C. (2004). What was classical genetics? *Studies in History and Philosophy of Science*, 35, 783–809.

Waters, C. (2008). Beyond theoretical reduction and layer-cake antireduction: How DNA retooled genetics and transformed biological practice. In M. Ruse (ed.), *The Oxford handbook of philosophy of biology* (pp. 238–262). Oxford: Oxford University Press.

Watson, J., & Crick, F. (1953a). A structure for deoxyribose nucleic acid. *Nature*, 171, 737–738.

Watson, J., & Crick, F. (1953b). Genetical implications of the structure of deoxyribonucleic acid. *Nature*, 171, 964–967.

16

1953 and All That:
A Tale of Two Sciences

Philip Kitcher

"Must we geneticists become bacteriologists, physiological chemists and physicists, simultaneously with being zoologists and botanists? Let us hope so."
H. J. Muller, 1922[1]

1. The Problem

Toward the end of their paper announcing the molecular structure of DNA, James Watson and Francis Crick remark, somewhat laconically, that their proposed structure might illuminate some central questions of genetics.[2] Thirty years have passed since Watson and Crick published their famous discovery. Molecular biology has indeed transformed our understanding of heredity. The recognition of the structure of DNA, the understanding of gene replication, transcription and translation, the cracking of the genetic code, the study of gene regulation, these and other breakthroughs have combined to answer many of the questions that baffled classical geneticists. Muller's hope – expressed in the early days of classical genetics – has been amply fulfilled.

Yet the success of molecular biology and the transformation of classical genetics into molecular genetics bequeath a philosophical problem. There are two recent theories which have addressed the phenomena of heredity. One, *classical genetics*, stemming from the studies of T. H. Morgan, his colleagues and students, is the successful outgrowth of the Mendelian theory of heredity rediscovered at the beginning of this century. The other, *molecular genetics*, descends from the work of Watson and Crick. What is the relationship between these two theories? How does the molecular theory illuminate the classical theory? How exactly has Muller's hope been fulfilled?

There used to be a popular philosophical answer to the problem posed in these three connected questions: classical genetics has been

Philip Kitcher, "1953 and All That: A Tale of Two Sciences," *The Philosophical Review*, Volume 93, pp. 335–373. Copyright 1984, The Sage School of Philosophy at Cornell University. All rights reserved. Used by permission of the publisher, Duke University Press.

Earlier versions of this paper were read at Johns Hopkins University and at the University of Minnesota, and I am very grateful to a number of people for comments and suggestions. In particular, I would like to thank Peter Achinstein, John Beatty, Barbara Horan, Patricia Kitcher, Richard Lewontin, Kenneth Schaffner, William Wimsatt, an anonymous reader and the editors of *The Philosophical Review*, all of whom have had an important influence on the final version. Needless to say, these people should not be held responsible for residual errors. I am also grateful to the American Council of Learned Societies and the Museum of Comparative Zoology at Harvard University for support and hospitality while I was engaged in research on the topics of this paper.

reduced to molecular genetics. Philosophers of biology inherited the notion of reduction from general discussions in philosophy of science, discussions which usually center on examples from physics. Unfortunately attempts to apply this notion in the case of genetics have been vulnerable to cogent criticism. Even after considerable tinkering with the concept of reduction, one cannot claim that classical genetics has been (or is being) reduced to molecular genetics.[3] However, the antireductionist point is typically negative.[4] It denies the adequacy of a particular solution to the problem of characterizing the relation between classical genetics and molecular genetics. It does not offer an alternative solution.

My aim in this paper is to offer a different perspective on intertheoretic relations. The plan is to invert the usual strategy. Instead of trying to force the case of genetics into a mold, which is alleged to capture important features of examples in physics, or resting content with denying that the material can be forced, I shall try to arrive at a view of the theories involved and the relations between them that will account for the almost universal idea that molecular biology has done something important for classical genetics. In so doing, I hope to shed some light on the general questions of the structure of scientific theories and the relations which may hold between successive theories. Since my positive account presupposes that something is wrong with the reductionist treatment of the case of genetics, I shall begin with a diagnosis of the foibles of reductionism.

2.   What's Wrong with Reductionism?

Ernest Nagel's classic treatment of reduction[5] can be simplified for our purposes. Scientific theories are regarded as sets of statements.[6] To reduce a theory T_2 to a theory T_1, is to deduce the statements of T_2 from the statements of T_1. If there are nonlogical expressions which appear in the statements of T_2, but do not appear in the statements of T_1, then we are allowed to supplement the statements of T_1 with some extra premises connecting the vocabulary of T_1 with the distinctive vocabulary of T_2 (so-called *bridge principles*). Intertheoretic reduction is taken to be important because the statements which are

deduced from the reducing theory are supposed to be explained by this deduction.

Yet, as everyone who has struggled with the paradigm cases from physics knows all too well, the reductions of Galileo's law to Newtonian mechanics and of the ideal gas laws to the kinetic theory do not exactly fit Nagel's model. Study of these examples suggests that, to reduce a theory T_2 to a theory T_1, it suffices to deduce the laws of T_2 from a suitably modified version of T_1, possibly augmented with appropriate extra premises.[7] Plainly, this sufficient condition is dangerously vague.[8] I shall tolerate its vagueness, proposing that we understand the issue of reduction in genetics by using the examples from physics as paradigms of what "suitable modifications" and "appropriate extra premises" are like. Reductionists claim that the relation between classical genetics and molecular biology is sufficiently similar to the intertheoretical relations exemplified in the examples from physics to count as the same type of thing: to wit, as intertheoretical reduction.

It may seem that the reductionist thesis has now become so amorphous that it will be immune to refutation. But this is incorrect. Even when we have amended the classical model of reduction so that it can accommodate the examples that originally motivated it, the reductionist claim about genetics requires us to accept three theses:

(R1)   Classical genetics contains general laws about the transmission of genes which can serve as the conclusions of reductive derivations.

(R2)   The distinctive vocabulary of classical genetics (predicates like '① is a gene', '① is dominant with respect to ②') can be linked to the vocabulary of molecular biology by bridge principles.

(R3)   A derivation of general principles about the transmission of genes from principles of molecular biology would explain why the laws of gene transmission hold (to the extent that they do).

I shall argue that each of the theses is false, offering this as my diagnosis of the ills of reductionism.

Before offering my criticisms, it may help to explain why reductionism presupposes (R1)–(R3).

If the relation between classical genetics and molecular biology is to be like that between the theory of ideal gases and the kinetic theory (say), then we are going to need to find general principles, identifiable as the central laws of classical genetics, which can serve as the conclusions of reductive derivations. (We need counterparts for the Boyle–Charles law.) These will be general principles about genes, and, because classical genetics seems to be a theory about the inheritance of characteristics, the only likely candidates are laws describing the transmission of genes between generations. [So reductionism leads to (R1).] If we are to derive such laws from molecular biology, then there must be bridge principles connecting the distinctive vocabulary figuring in the laws of gene transmission (presumably expressions like '① is a gene', and perhaps '① is dominant with respect to ②') with the vocabulary of molecular biology. [Hence (R2).] Finally, if the derivations are to achieve the goal of intertheoretical reduction then they must explain the laws of gene transmission. [(R3).]

Philosophers often identify theories as small sets of general laws. However, in the case of classical genetics, the identification is difficult and those who debate the reducibility of classical genetics to molecular biology often proceed differently. David Hull uses a characterization drawn from Dobzhansky: classical genetics is "concerned with gene differences; the operation employed to discover a gene is hybridization: parents differing in some trait are crossed and the distribution of the trait in hybrid progeny is observed."[9] This is not unusual in discussions of reduction in genetics. It is much easier to identify classical genetics by referring to the subject matter and to the methods of investigation, than it is to provide a few sentences that encapsulate the content of the theory.

Why is this? Because when we read the major papers of the great classical geneticists or when we read the textbooks in which their work is summarized, we find it hard to pick out *any* laws about genes. These documents are full of informative statements. Together, they tell us an enormous amount about the chromosomal arrangement of particular genes in particular organisms, about the effect on the phenotype of various mutations, about frequencies of recombination, and so forth.[10] In some cases, we might explain the absence of formulations of general laws about genes (and even of reference to such laws) by suggesting that these things are common knowledge. Yet that hardly accounts for the nature of the textbooks or of the papers that forged the tools of classical genetics.

If we look back to the pre-Morgan era, we do find two general statements about genes, namely Mendel's Laws (or "Rules"). Mendel's second law states that, in a diploid organism which produces haploid gametes, genes at different loci will be transmitted independently; so, for example, if A, a and B, b are pairs of alleles at different loci, and if an organism is heterozygous at both loci, then the probabilities that a gamete will receive any of the four possible genetic combinations, AB, Ab, aB, ab, are all equal.[11] Once it was recognized that genes are (mostly) chromosomal segments, (as biologists discovered soon after the rediscovery of Mendel's laws), we understand that the law will not hold in general: alleles which are on the same chromosome (or, more exactly, close together on the same chromosome) will tend to be transmitted together because (ignoring recombination)[12] one member of each homologous pair is distributed to a gamete.[13]

Now it might seem that this is not very important. We could surely find a correct substitute for Mendel's second law by restricting the law so that it only talks about genes on nonhomologous chromosomes. Unfortunately, this will not quite do. There can be interference with normal cytological processes so that segregation of nonhomologous chromosomes need not be independent.[14] However, my complaint about Mendel's second law is not that it is incorrect: many sciences use laws that are clearly recognized as approximations. Mendel's second law, amended or unamended, simply becomes irrelevant to subsequent research in classical genetics.

We envisaged amending Mendel's second law by using elementary principles of cytology, together with the identification of genes as chromosomal segments, to correct what was faulty in the unamended law. It is the fact that the application is so easy and that it can be carried out far more generally that makes the "law" it generates irrelevant. We can understand the transmission of genes by analyzing the cases that interest us from a cytological perspective – by proceeding from "first principles," as it were. Moreover, we can adopt this approach whether the organism is haploid, diploid or polyploid, whether it reproduces sexually or asexually, whether the genes with which we are concerned are or are not on homologous chromosomes, whether or not there is distortion of independent chromosomal segregation at meiosis. Cytology not only teaches us that the second law is false; it also tells us how to tackle the problem at which the second law was directed (the problem of determining frequencies for pairs of genes in gametes). The amended second law is a restricted statement of results obtainable using a general technique. What figures largely in genetics after Morgan is the technique, and this is hardly surprising when we realize that one of the major research problems of classical genetics has been the problem of discovering the distribution of genes *on the same chromosome*, a problem which is beyond the scope of the amended law.

Let us now turn from (R1) to (R2), assuming, contrary to what has just been argued, that we can identify the content of classical genetics with general principles about gene transmission. (Let us even suppose, for the sake of concreteness, that the principles in question are Mendel's laws – amended in whatever way the reductionist prefers.) To derive these principles from molecular biology, we need a bridge principle. I shall consider first statements of the form

(*)   (x) (x is a gene ↔ Mx)

where 'Mx' is an open sentence (possibly complex) in the language of molecular biology. Molecular biologists do not offer any appropriate statement. Nor do they seem interested in providing one. I claim that no appropriate bridge principle can be found.

Most genes are segments of DNA. (There are some organisms – viruses – whose genetic material is RNA; I shall henceforth ignore them.) Thanks to Watson and Crick, we know the molecular structure of DNA. Hence the problem of providing a statement of the above form becomes that of saying, in molecular terms, which segments of DNA count as genes.

Genes come in different sizes, and, for any given size, we can find segments of DNA of that size that are not genes. Therefore genes cannot be identified as segments of DNA containing a

particular number of nucleotide pairs. Nor will it do to give a molecular characterization of those codons (triplets of nucleotides) that initiate and terminate transcription, and take a gene to be a segment of DNA between successive initiating and terminating codons. In the first place, mutation might produce a *single* allele containing within it codons for stopping and restarting transcription.[15] Secondly, and much more importantly, the criterion is not general since not every gene is transcribed on mRNA.

The latter point is worth developing. Molecular geneticists recognize regulatory genes as well as structural genes. To cite a classic example, the operator region in the *lac* operon of *E. coli* serves as a site for the attachment of protein molecules, thereby inhibiting transcription of mRNA and regulating enzyme production.[16] Moreover, it is becoming increasingly obvious that genes are not always transcribed, but play a variety of roles in the economy of the cell.[17]

At this point, the reductionist may try to produce a bridge principle by brute force. Trivially, there are only a finite number of terrestrial organisms (past, present and future) and only a finite number of genes. Each gene is a segment of DNA with a particular structure and it would be possible, in principle, to provide a detailed molecular description of that structure. We can now give a molecular specification of the gene by enumerating the genes and disjoining the molecular descriptions.[18] The point made above, that the segments which we count as genes do not share any structural property can now be put more precisely: any instantiation of (*) which replaces 'M' by a structural predicate from the language of molecular biology will insert a predicate that is essentially disjunctive.

Why does this matter? Let us imagine a reductionist using the enumerative strategy to deduce a general principle about gene transmission. After great labor, it is revealed that all actual genes satisfy the principle. I claim that more than this is needed to reduce a *law* about gene transmission. We envisage laws as sustaining counterfactuals, as applying to examples that might have been but which did not actually arise. To reduce the law it is necessary to show how possible but nonactual genes would have satisfied it. Nor can we achieve the reductionist's goal by adding further disjuncts to the envisaged bridge principle. For although there are only finitely many *actual* genes, there are indefinitely many genes which *might* have arisen.

At this point, the reductionist may protest that the deck has been stacked. There is no need to produce a bridge principle of the form (*). Recall that we are trying to derive a general law about the transmission of genes, whose paradigm is Mendel's second law. Now the gross logical form of Mendel's second law is:

(1)   (x) (y) ((Gx & Gy) → Axy).

We might hope to obtain this from statements of the forms

(2)   (x) (Gx → Mx)

(3)   (x) (y) ((Mx & My) → Axy)

where 'Mx' is an open sentence in the language of molecular biology. Now there will certainly be true statements of the form (2): for example, we can take 'Mx' as 'x is composed of DNA v.x is composed of RNA'. The question is whether we can combine some such statement with other appropriate premises – for example, some instance of (3) – so as to derive, and thereby explain (1). No geneticist or molecular biologist has advanced any suitable premises, and with good reason. We discover true statements of the form (2) by hunting for weak necessary conditions on genes, conditions which have to be met by genes but which are met by hordes of other biological entities as well. We can only hope to obtain *weak* necessary conditions because of the phenomenon that occupied us previously: from the molecular standpoint, genes are not distinguished by any common structure. Trouble will now arise when we try to show that the weak necessary condition is jointly sufficient for the satisfaction of the property (independent assortment at meiosis) that we ascribe to genes. The difficulty is illustrated by the example given above. If we take 'Mx' to be 'x is composed of DNA v.x is composed of RNA' then the challenge will be to find a general law governing the distribution of all segments of DNA and RNA!

I conclude that (R2) is false. Reductionists cannot find the bridge principles they need, and the tactic of abandoning the form (*) for something weaker is of no avail. I shall now consider (R3). Let us concede both of the points that I have denied, allowing that there are general laws about the transmission of genes and that bridge principles are forthcoming. I claim that exhibiting derivations of the transmission laws from principles of molecular biology and bridge principles would not explain the laws, and, therefore, would not fulfill the major goal of reduction.

As an illustration, I shall use the envisaged amended version of Mendel's second law. Why do genes on nonhomologous chromosomes assort independently? Cytology provides the answer. At meiosis, chromosomes line up with their homologues. It is then possible for homologous chromosomes to exchange some genetic material, producing pairs of recombinant chromosomes. In the meiotic division, one member of each recombinant pair goes to each gamete, and the assignment of one member of one pair to a gamete is probabilistically independent of the assignment of a member of another pair to that gamete. Genes which occur close on the same chromosome are likely to be transmitted together (recombination is not likely to occur between them), but genes on nonhomologous chromosomes will assort independently.

This account is a perfectly satisfactory explanation of why our envisaged law is true to the extent that it is. (We recognize how the law could fail if there were some unusual mechanism linking particular nonhomologous chromosomes.) To emphasize the adequacy of the explanation is not to deny that it could be extended in certain ways. For example, we might want to know more about the mechanics of the process by which the chromosomes are passed on to the gametes. In fact, cytology provides such information. However, appeal to molecular biology would not deepen our understanding of the transmission law. Imagine a successful derivation of the law from principles of chemistry and a bridge principle of the form (*). In charting the details of the molecular rearrangements the derivation would only blur the outline of a simple cytological story, adding a welter of irrelevant detail. Genes on nonhomologous chromosomes assort independently because nonhomologous chromosomes are transmitted independently at meiosis, and, so long as we recognize this, we do not need to know what the chromosomes are made of.

In explaining a scientific law, L, one often provides a deduction of L from other principles. Sometimes it is possible to explain some of the principles used in the deduction by deducing them, in turn, from further laws. Recognizing the possibility of a sequence of deductions tempts us to suppose that we could produce a better explanation of L by combining them, producing a more elaborate derivation in the language of our ultimate premises. But this is incorrect. What is relevant for the purposes of giving one explanation may be quite different from what is relevant for the purposes of explaining a law used in giving that original explanation. This general point is illustrated by the case at hand. We begin by asking why genes on nonhomologous chromosomes assort independently. The simple cytological story rehearsed above answers the question. That story generates *further* questions. For example, we might inquire why nonhomologous chromosomes are distributed independently at meiosis. To answer this question we would describe the formation of the spindle and the migration of chromosomes to the poles of the spindle just before meiotic division.[19] Once again, the narrative would generate yet further questions. Why do the chromosomes "condense" at prophase? How is the spindle formed? Perhaps in answering these questions we would begin to introduce the chemical details of the process. Yet simply plugging a molecular account into the narratives offered at the previous stages would *decrease* the explanatory power of those narratives. What is relevant to answering our original question is the fact that nonhomologous chromosomes assort independently. What is relevant to the issue of why nonhomologous chromosomes assort independently is the fact that the chromosomes are not selectively oriented toward the poles of the spindle. (We need to eliminate the doubt that, for example, the paternal and maternal chromosomes become separated and aligned toward opposite poles of the spindle.) In neither case are the molecular details relevant. Indeed, adding those details would only disguise the relevant factor.

There is a natural reductionist response. The considerations of the last paragraphs presuppose far too subjective a view of scientific explanation.

After all, even if *we* become lost in the molecular details, beings who are cognitively more powerful than we could surely recognize the explanatory force of the envisaged molecular derivation. However, this response misses a crucial point. The molecular derivation forfeits something important.

Recall the original cytological explanation. It accounted for the transmission of genes by identifying meiosis as a process of a particular kind: a process in which paired entities (in this case, homologous chromosomes) are separated by a force so that one member of each pair is assigned to a descendant entity (in this case, a gamete). Let us call processes of this kind *PS-processes*. I claim first that explaining the transmission law requires identifying PS-processes as forming a natural kind to which processes of meiosis belong, and second that PS-processes cannot be identified as a kind from the molecular point of view.

If we adopt the familiar covering law account of explanation, then we shall view the cytological narrative as invoking a law to the effect that processes of meiosis are PS-processes and as applying elementary principles of probability to compute the distribution of genes to gametes from the laws that govern PS-processes. If the illumination provided by the narrative is to be preserved in a molecular derivation, then we shall have to be able to express the relevant laws as laws in the language of molecular biology, and this will require that we be able to characterize PS-processes as a natural kind from the molecular point of view. The same conclusion, to wit that the explanatory power of the cytological account can be preserved only if we can identify PS-processes as a natural kind in molecular terms, can be reached in analogous ways if we adopt quite different approaches to scientific explanation – for example, if we conceive of explanation as specifying causally relevant properties or as fitting phenomena into a unified account of nature.

However, PS-processes are heterogeneous from the molecular point of view. There are no constraints on the molecular structures of the entities which are paired or on the ways in which the fundamental forces combine to pair them and to separate them. The bonds can be forged and broken in innumerable ways: all that matters is that there be bonds that initially pair the entities in question and that are subsequently (somehow) broken. In some cases, bonds may be formed directly between constituent molecules of the entities in question; in others, hordes of accessory molecules may be involved. In some cases, the separation may occur because of the action of electromagnetic forces or even of nuclear forces; but it is easy to think of examples in which the separation is effected by the action of gravity. I claim, therefore, that PS-processes are realized in a motley of molecular ways. (I should note explicitly that this conclusion is independent of the issue of whether the reductionist can find bridge principles for the concepts of classical genetics.)

We thus obtain a reply to the reductionist charge that we reject the explanatory power of the molecular derivation simply because we anticipate that our brains will prove too feeble to cope with its complexities.[20] The molecular account objectively fails to explain because it cannot bring out that feature of the situation which is highlighted in the cytological story. It cannot show us that genes are transmitted in the ways that we find them to be because meiosis is a PS-process and because any PS-process would give rise to analogous distributions. Thus (R3) – like (R1) and (R2) – is false.

3. The Root of the Trouble

Where did we go wrong? Here is a natural suggestion. The most fundamental failure of reductionism is the falsity of (R1). Lacking an account of theories which could readily be applied to the cases of classical genetics and molecular genetics, the attempt to chart the relations between these theories was doomed from the start. If we are to do better, we must begin by asking a preliminary question: what is the structure of classical genetics?

I shall follow this natural suggestion, endeavoring to present a picture of the structure of classical genetics which can be used to understand the intertheoretic relations between classical and molecular genetics.[21] As we have seen, the main difficulty in trying to axiomatize classical genetics is to decide what body of statements one is attempting to axiomatize. The history of genetics makes it clear that Morgan, Muller,

Sturtevant, Beadle, McClintock, and others have made important contributions to genetic theory. But the statements occurring in the writings of these workers seem to be far too specific to serve as parts of a general theory. They concern the genes of particular kinds of organisms – primarily paradigm organisms, like fruit flies, bread molds, and maize. The idea that classical genetics is simply a heterogeneous set of statements about dominance, recessiveness, position effect, non-disjunction, and so forth, in *Drosophila*, *Zea mays*, *E. coli*, *Neurospora*, etc. flies in the face of our intuitions. The statements advanced by the great classical geneticists seem more like *illustrations* of the theory than *components* of it. (To know classical genetics it is not necessary to know the genetics of any particular organism, not even *Drosophila melanogaster*.) But the only alternative seems to be to suppose that there are general laws in genetics, never enunciated by geneticists but reconstructible by philosophers. At the very least, this supposition should induce the worry that the founders of the field, and those who write the textbooks of today, do a singularly bad job.

Our predicament provokes two main questions. First, if we focus on a particular time in the history of classical genetics, it appears that there will be a set of statements about inheritance in particular organisms, which constitutes the corpus which geneticists of that time accept: what is the relationship between this corpus and the version of classical genetic theory in force at the time? (In posing this question, I assume, contrary to fact, that the community of geneticists was always distinguished by unusual harmony of opinion; it is not hard to relax this simplifying assumption.) Second, we think of genetic theory as something that persisted through various versions: what is the relation among the versions of classical genetic theory accepted at different times (the versions of 1910, 1930, and 1950, for example) which makes us want to count them as versions of the same theory?

We can answer these questions by amending a prevalent conception of the way in which we should characterize the state of a science at a time. The corpus of statements about the inheritance of characteristics accepted at a given time is only one component of a much more complicated entity that I shall call the *practice* of classical genetics at that time. There is a common language used to talk about hereditary phenomena, a set of accepted statements in that language (the corpus of beliefs about inheritance mentioned above), a set of questions taken to be the appropriate questions to ask about hereditary phenomena, and a set of patterns of reasoning which are instantiated in answering some of the accepted questions; (also: sets of experimental procedures and methodological rules, both designed for use in evaluating proposed answers; these may be ignored for present purposes). The practice of classical genetics at a time is completely specified by identifying each of the components just listed.[22]

A pattern of reasoning is a sequence of *schematic sentences*, that is sentences in which certain items of nonlogical vocabulary have been replaced by dummy letters, together with a set of *filling instructions* which specify how substitutions are to be made in the schemata to produce reasoning which instantiates the pattern.[23] This notion of pattern is intended to explicate the idea of the common structure that underlies a group of problem-solutions.

The foregoing definitions enable us to answer the two main questions I posed above. Beliefs about the particular genetic features of particular organisms illustrate or exemplify the version of genetic theory in force at the time in the sense that these beliefs figure in particular problem-solutions generated by the current practice. Certain patterns of reasoning are applied to give the answers to accepted questions, and, in making the application, one puts forward claims about inheritance in particular organisms. Classical genetics persists as a single theory with different versions at different times in the sense that different practices are linked by a chain of practices along which there are relatively small modifications in language, in accepted questions, and in the patterns for answering questions. In addition to this condition of historical connection, versions of classical genetic theory are bound by a common structure: each version uses certain expressions to characterize hereditary phenomena, accepts as important questions of a particular form, and offers a general style of reasoning for answering those questions. Specifically, throughout the career of classical genetics, the theory is directed toward answering questions about the distribution of characteristics in successive

generations of a genealogy, and it proposes to answer those questions by using the probabilities of chromosome distribution to compute the probabilities of descendant genotypes.

The approach to classical genetics embodied in these answers is supported by reflection on what beginning students learn. Neophytes are not taught (and never have been taught) a few fundamental theoretical laws from which genetic "theorems" are to be deduced. They are introduced to some technical terminology, which is used to advance a large amount of information about special organisms. Certain questions about heredity in these organisms are posed and answered. Those who understand the theory are those who know what questions are to be asked about hitherto unstudied examples, who know how to apply the technical language to the organisms involved in these examples, and who can apply the patterns of reasoning which are to be instantiated in constructing answers. More simply, successful students grasp general patterns of reasoning which they can use to resolve new cases.

I shall now add some detail to my sketch of the structure of classical genetics, and thereby prepare the way for an investigation of the relations between classical genetics and molecular genetics. The initial family of problems in classical genetics, the family from which the field began, is the family of *pedigree problems*. Such problems arise when we confront several generations of organisms, related by specified connections of descent, with a given distribution of one or more characteristics. The question that arises may be to understand the given distribution of phenotypes, or to predict the distribution of phenotypes in the next generation, or to specify the probability that a particular phenotype will result from a particular mating. In general, classical genetic theory answers such questions by making hypotheses about the relevant genes, their phenotypic effects and their distribution among the individuals in the pedigree. Each version of classical genetic theory contains one or more problem-solving patterns exemplifying this general idea, but the detailed character of the pattern is refined in later versions, so that previously recalcitrant cases of the problem can be accommodated.

Each case of a pedigree problem can be characterized by a set of *data*, a set of *constraints*, and a question. In any example, the data are statements describing the distribution of phenotypes among the organisms in a particular pedigree, or a diagram conveying the same information. The level of detail in the data may vary widely: at one extreme we may be given a full description of the interrelationships among all individuals and the sexes of all those involved; or the data may only provide the numbers of individuals with specific phenotypes in each generation; or, with minimal detail, we may simply be told that from crosses among individuals with specified phenotypes a certain range of phenotypes is found.

The constraints on the problem consist of general cytological information and descriptions of the chromosomal constitution of members of the species. The former will include the thesis that genes are (almost always)[24] chromosomal segments and the principles that govern meiosis. The latter may contain a variety of statements. It may be pertinent to know how the species under study reproduces, how sexual dimorphism is reflected at the chromosomal level, the chromosome number typical of the species, what loci are linked, what the recombination frequencies are, and so forth. As in the case of the data, the level of detail (and thus of stringency) in the constraints can vary widely.

Lastly, each problem contains a question that refers to the organisms described in the data. The question may take several forms: "What is the expected distribution of phenotypes from a cross between *a* and *b*?" (where *a*, *b* are specified individuals belonging to the pedigree described by the data), "What is the probability that a cross between *a* and *b* will produce an individual having *P*?" (where *a*, *b* are specified individuals of the pedigree described by the data and *P* is a phenotypic property manifested in this pedigree), "Why do we find the distribution of phenotypes described in the data?" and others.

Pedigree problems are solved by advancing pieces of reasoning that instantiate a small number of related patterns. In all cases the reasoning begins from a *genetic hypothesis*. The function of a genetic hypothesis is to specify the alleles that are relevant, their phenotypic expression, and their transmission through the pedigree. From that part of the genetic hypothesis that specifies the genotypes of the parents in any mating that occurs in the pedigree, together with

the constraints on the problem, one computes the expected distribution of genotypes among the offspring. Finally, for any mating occurring in the pedigree, one shows that the expected distribution of genotypes among the offspring is consistent with the assignment of genotypes given by the genetic hypothesis.

The form of the reasoning can easily be recognized in examples – examples that are familiar to anyone who has ever looked at a textbook or a research report in genetics.[25] What interests me is the style of reasoning itself. The reasoning begins with a genetic hypothesis that offers four kinds of information: (a) Specification of the number of relevant loci and the number of alleles at each locus; (b) Specification of the relationships between genotypes and phenotypes; (c) Specification of the relations between genes and chromosomes, of facts about the transmission of chromosomes to gametes (for example, resolution of the question whether there is disruption of normal segregation) and about the details of zygote formation; (d) Assignment of genotypes to individuals in the pedigree. After showing that the genetic hypothesis is consistent with the data and constraints of the problem, the principles of cytology and the laws of probability are used to compute expected distributions of genotypes from crosses. The expected distributions are then compared with those assigned in part (d) of the genetic hypothesis.[26]

Throughout the career of classical genetics, pedigree problems are addressed and solved by carrying out reasoning of the general type just indicated. Each version of classical genetic theory contains a pattern for solving pedigree problems with a method for computing expected genotypes which is adjusted to reflect the particular form of the genetic hypotheses that it sanctions. Thus one way to focus the differences among successive versions of classical genetic theory is to compare their conceptions of the possibilities for genetic hypotheses. As genetic theory develops, there is a changing set of conditions on admissible genetic hypotheses. Prior to the discovery of polygeny and pleiotropy (for example), part (a) of any adequate genetic hypothesis was viewed as governed by the requirement that there would be a one-one correspondence between loci and phenotypic traits.[27] After the discovery of incomplete dominance and epistasis, it was recognized that part (b) of an adequate hypothesis might take a form that had not previously been allowed: one is not compelled to assign to the heterozygote a phenotype assigned to one of the homozygotes, and one is also permitted to relativize the phenotypic effect of a gene to its genetic environment.[28] Similarly, the appreciation of phenomena of linkage, recombination, nondisjunction, segregation distortion, meiotic drive, unequal crossing over, and crossover suppression, modify conditions previously imposed on part (c) of any genetic hypothesis. In general, we can take each version of classical genetic theory to be associated with a set of conditions (usually not formulated explicitly) which govern admissible genetic hypotheses. While a general form of reasoning persists through the development of classical genetics, the patterns of reasoning used to resolve cases of the pedigree problem are constantly fine-tuned as geneticists modify their views about what forms of genetic hypothesis are allowable.

So far I have concentrated exclusively on classical genetic theory as a family of related patterns of reasoning for solving the pedigree problem. It is natural to ask if versions of the theory contain patterns of reasoning for addressing other questions. I believe that they do. The heart of the theory is the theory of *gene transmission*, the family of reasoning patterns directed at the pedigree problem. Out of this theory grow other subtheories. The theory of *gene mapping* offers a pattern of reasoning which addresses questions about the relative positions of loci on chromosomes. It is a direct result of Sturtevant's insight that one can systematically investigate the set of pedigree problems associated with a particular species. In turn, the theory of gene mapping raises the question of how to identify mutations, issues which are to be tackled by the *theory of mutation*. Thus we can think of classical genetics as having a central theory, the theory of gene transmission, which develops in the ways I have described above, surrounded by a number of satellite theories that are directed at questions arising from the pursuit of the central theory. Some of these satellite theories (for example, the theory of gene mapping) develop in the same continuous fashion. Others, like the theory of mutation, are subject to rather dramatic shifts in approach.

4. Molecular Genetics and Classical Genetics

Armed with some understanding of the structure and evolution of classical genetics, we can finally return to the question with which we began. What is the relation between classical genetics and molecular genetics? When we look at textbook presentations and the pioneering research articles that they cite, it is not hard to discern major ways in which molecular biology has advanced our understanding of hereditary phenomena. We can readily identify particular molecular explanations which illuminate issues that were treated incompletely, if at all, from the classical perspective. What proves puzzling is the connection of these explanations to the theory of classical genetics. I hope that the account of the last section will enable us to make the connection.

I shall consider three of the most celebrated achievements of molecular genetics. Consider first the question of *replication*. Classical geneticists believed that genes can replicate themselves. Even before the experimental demonstration that all genes are transmitted to all the somatic cells of a developing embryo, geneticists agreed that normal processes of mitosis and meiosis must involve gene replication. Muller's suggestion that the central problem of genetics is to understand how mutant alleles, incapable of performing wild-type functions in producing the phenotype, are nonetheless able to replicate themselves, embodies this consensus. Yet classical genetics had no account of gene replication. A molecular account was an almost immediate dividend of the Watson–Crick model of DNA.

Watson and Crick suggested that the two strands of the double helix unwind and each strand serves as the template for the formation of a complementary strand. Because of the specificity of the pairing of nucleotides, reconstruction of DNA can be unambiguously directed by a single strand. This suggestion has been confirmed and articulated by subsequent research in molecular biology.[29] The details are more intricate than Watson and Crick may originally have believed, but the outline of their story stands.

A second major illumination produced by molecular genetics concerns the characterization of mutation. When we understand the gene as a segment of DNA we recognize the ways in which mutant alleles can be produced. "Copying errors" during replication can cause nucleotides to be added, deleted or substituted. These changes will often lead to alleles that code for different proteins, and which are readily recognizable as mutants through their production of deviant phenotypes. However, molecular biology makes it clear that there can be *hidden* mutations, mutations that arise through nucleotide substitutions that do not change the protein produced by a structural gene (the genetic code is redundant) or through substitutions that alter the form of the protein in trivial ways. The molecular perspective provides us with a general answer to the question, "What is a mutation?" namely that a mutation is the modification of a gene through insertion, deletion or substitution of nucleotides. This general answer yields a basic method for tackling (in principle) questions of form, "Is *a* a mutant allele?" namely a demonstration that *a* arose through nucleotide changes from alleles that persist in the present population. The method is frequently used in studies of the genetics of bacteria and bacteriophage, and can sometimes be employed even in inquiries about more complicated organisms. So, for example, there is good biochemical evidence for believing that some alleles which produce resistance to pesticides in various species of insects arose through nucleotide changes in the alleles naturally predominating in the population.[30]

I have indicated two general ways in which molecular biology answers questions that were not adequately resolved by classical genetics. Equally obvious are a large number of more specific achievements. Identification of the molecular structures of particular genes in particular organisms has enabled us to understand why those genes combine to produce the phenotypes they do. One of the most celebrated cases is that of the normal allele for the synthesis of human hemoglobin and the mutant allele that is responsible for sickle-cell anemia.[31] The hemoglobin molecule – whose structure is known in detail – is built up from four amino-acid chains (two "α-chains" and two "β-chains"). The mutant allele results from substitution of a single nucleotide with the result that one amino acid is different (the sixth amino acid in the β-chains). This slight modification causes a change in the interactions of hemoglobin molecules: deoxygenated mutant

hemoglobin molecules combine to form long fibres. Cells containing the abnormal molecule become deformed after they have given up their oxygen, and because they become rigid, they can become stuck in narrow capillaries, if they give up their oxygen too soon. Individuals who are homozygous for the mutant gene are vulnerable to experience blockages of blood flow. However, in heterozygous individuals, there is enough normal hemoglobin in blood cells to delay the time of formation of the distorting fibres, so that the individual is physiologically normal.

This example is typical of a broad range of cases, among which are some of the most outstanding achievements of molecular genetics. In all of the cases, we replace a simple assertion about the existence of certain alleles which give rise to various phenotypes with a molecular characterization of those alleles from which we can derive descriptions of the phenotypes previously attributed.

I claim that the successes of molecular genetics which I have just briefly described – and which are among the accomplishments most emphasized in the biological literature – can be understood from the perspective on theories that I have developed above. The three examples reflect three different relations among successive theories, all of which are different from the classical notion of reduction (and the usual modifications of it). Let us consider them in turn.

The claim that genes can replicate does not have the status of a central law of classical genetic theory.[32] It is not something that figures prominently in the explanations provided by the theory (as, for example, the Boyle–Charles law is a prominent premise in some of the explanations yielded by phenomenological thermodynamics). Rather, it is a claim that classical geneticists took for granted, a claim presupposed by explanations, rather than an explicit part of them. Prior to the development of molecular genetics that claim had come to seem increasingly problematic. If genes can replicate, how do they manage to do it? Molecular genetics answered the worrying question. It provided a theoretical demonstration of the possibility of an antecedently problematic presupposition of classical genetics.

We can say that a theory presupposes a statement *p* if there is some problem-solving pattern of the theory, such that every instantiation of the pattern contains statements that jointly imply the truth of p. Suppose that, at a given stage in the development of a theory, scientists recognize an argument from otherwise acceptable premises which concludes that it is impossible that p. Then the presupposition p is problematic for those scientists. What they would like would be an argument showing that it is possible that p and explaining what is wrong with the line of reasoning which appears to threaten the possibility of p. If a new theory generates an argument of this sort, then we can say that the new theory gives a theoretical demonstration of the possibility of an antecedently problematic presupposition of the old theory.

A less abstract account will help us to see what is going on in the case of gene replication. Very frequently, scientists take for granted in their explanations some general property of entities that they invoke. Their assumption can come to seem problematic if the entities in question are supposed to belong to a kind, and there arises a legitimate doubt about whether members of the kind can have the property attributed. A milder version of the problem arises if, in all cases in which the question of whether things of the general kind have the property can be settled by appealing to background theory, it turns out that the answer is negative. Under these circumstances, the scientists are committed to regarding their favored entities as unlike those things of the kind which are amenable to theoretical study with respect to the property under discussion. The situation is worse if background theory provides an argument for thinking that *no* things of the kind can have the property.

Consider now the case of gene replication. For any problem-solution offered by any version of the theory of gene transmission (the central subtheory of classical genetic theory), that problem-solution will contain sentences implying that the alleles which it discusses are able to replicate. Classical genetics presupposes that a large number of identifiable genes can replicate. This presupposition was always weakly problematic because genes were taken to be complicated molecules and, in all cases in which appeal to biochemistry could be made to settle the issue of whether a molecular structure was capable of replication, the issue was decided in the negative. Muller exacerbated the problem by suggesting that mutant alleles are damaged molecules (after

all, many of them were produced through x-ray bombardment, an extreme form of molecular torture). So there appeared to be a strong argument against the possibility that any mutant allele can replicate. After the work of Watson, Crick, Kornberg, and others, there was a theoretical demonstration of the allegedly problematic possibility. One can show that genes can replicate by showing that any segment of DNA (or RNA) can replicate. (DNA and RNA are the genetic materials. Establishing the power of the genetic material to replicate bypasses the problem of deciding which segments are genes. Thus the difficulties posed by the falsity of [R2] are avoided.) The Watson–Crick model provides a characterization of the (principal) genetic material, and when this description is inserted into standard patterns of chemical reasoning one can generate an argument whose conclusion asserts that, under specified conditions, DNA replicates. Moreover, given the molecular characterization of DNA and of mutation, it is possible to see that although mutant alleles are "damaged" molecules, the kind of damage (insertion, deletion or substitution of nucleotides) does not affect the ability of the resultant molecule to replicate.

Because theoretical demonstrations of the possibility of antecedently problematic presuppositions involve derivation of conclusions of one theory from the premises supplied by a background theory, it is easy to assimilate them to the classical notion of reduction. However, on the account I have offered, there are two important differences. First, there is no commitment to the thesis that genetic theory can be formulated as (the deductive closure of) a conjunction of laws. Second, it is not assumed that all general statements about genes are equally in need of molecular derivation. Instead, one particular thesis, a thesis that underlies all the explanations provided by classical genetic theory, is seen as especially problematic, and the molecular derivation is viewed as addressing a specific problem that classical geneticists had already perceived. Where the reductionist identifies a general benefit in deriving all the axioms of the reduced theory, I focus on a particular derivation of a claim that has no title as an axiom of classical genetics, a derivation which responds to a particular explanatory difficulty of which classical geneticists were acutely aware. The reductionist's global relation between theories does not obtain between classical and molecular genetics, but something akin to it does hold between special fragments of these theories.[33]

The second principal achievement of molecular genetics, the account of mutation, involves a conceptual refinement of prior theory. Later theories can be said to provide conceptual refinements of earlier theories when the later theory yields a specification of entities that belong to the extensions of predicates in the language of the earlier theory, with the result that the ways in which the referents of these predicates are fixed are altered in accordance with the new specifications. Conceptual refinement may occur in a number of ways. A new theory may supply a descriptive characterization of the extension of a predicate for which no descriptive characterization was previously available; or it may offer a new description which makes it reasonable to amend characterizations that had previously been accepted.[34] In the case at hand, the referent of many tokens of 'mutant allele' was initially fixed through the description "chromosomal segment producing a heritable deviant phenotype." After Bridges's discovery of unequal crossing-over at the *Bar* locus in *Drosophila*, it was evident to classical geneticists that this descriptive specification covered cases in which the internal structure of a gene was altered and cases in which neighboring genes were transposed. Thus it was necessary to retreat to the less applicable description "chromosomal segment producing a heritable deviant phenotype as the result of an internal change within an allele." Molecular genetics offers a precise account of the internal changes, with the result that the description can be made more informative: mutant alleles are segments of DNA that result from prior alleles through deletion, insertion, or substitution of nucleotides. This re-fixing of the referent of 'mutant allele' makes it possible in principle to distinguish cases of mutation from cases of recombination, and thus to resolve those controversies that frequently arose from the use of 'mutant allele' in the later days of classical genetics.[35]

Finally, let us consider the use of molecular genetics to illuminate the action of particular genes. Here we again seem to find a relationship that initially appears close to the reductionist's ideal. Statements that are invoked as premises

in particular problem-solutions – statements that ascribe particular phenotypes to particular genotypes – are derived from molecular characterizations of the alleles involved. On the account of classical genetics offered in Section 3, each version of classical genetic theory includes in its schema for genetic hypotheses a clause which relates genotypes to phenotypes (clause [b] in the description of a genetic hypothesis on p. 356 above). Generalizing from the hemoglobin example, we might hope to discover a pattern of reasoning within molecular genetics that would generate as its conclusion the schema for assigning phenotypes to genotypes.

It is not hard to characterize the relation just envisioned. Let us say that a theory T′ provides an *explanatory extension* of a theory T just in case there is some problem-solving pattern of T one of whose schematic premises can be generated as the conclusion of a problem-solving pattern of T′. When a new theory provides an explanatory extension of an old theory, then particular premises occurring in explanatory derivations given by the old theory can themselves be explained by using arguments furnished by the new theory. However, it does not follow that the explanations provided by the old theory can be improved by replacing the premises in question with the pertinent derivations. What is relevant for the purposes of explaining some statement S may not be relevant for the purposes of explaining a statement S′ which figures in an explanatory derivation of S.

Even though reductionism fails, it may appear that we can capture part of the spirit of reductionism by deploying the notion of explanatory extension. The thesis that molecular genetics provides an explanatory extension of classical genetics embodies the idea of a global relationship between the two theories, while avoiding two of the three troubles that were found to beset reductionism. That thesis does not simply assert that some specific presupposition of classical genetics (for example, the claim that genes are able to replicate) can be derived as the conclusion of a molecular argument, but offers a general connection between premises of explanatory derivations in classical genetics and explanatory arguments from molecular genetics. It is formulated so as to accommodate the failure of (R1) and to honor the picture of classical genetics developed in Section 3. Moreover, the failure of (R2) does not affect it. If we take the hemoglobin example as a paradigm, we can justifiably contend that the explanatory extension does not require any general characterization of genes in molecular terms. All that is needed is the possibility of deriving phenotypic descriptions from molecular characterizations of the structures of *particular* genes. Thus, having surmounted two hurdles, our modified reductionist thesis is apparently within sight of success.

Nevertheless, even born-again reductionism is doomed to fall short of salvation. Although it is true that molecular genetics belongs to a cluster of theories which, taken together, provide an explanatory extension of classical genetics, molecular genetics, on its own, cannot deliver the goods. There are some cases in which the ancillary theories do not contribute to the explanation of a classical claim about gene action. In such cases, the classical claim can be derived and explained by instantiating a pattern drawn from molecular genetics. The example of human hemoglobin provides one such case. But this example is atypical.

Consider the way in which the hemoglobin example works. Specification of the molecular structures of the normal and mutant alleles, together with a description of the genetic code, enables us to derive the composition of normal and mutant hemoglobin. Application of chemistry then yields descriptions of the interactions of the proteins. With the aid of some facts about human blood cells, one can then deduce that the sickling effect will occur in abnormal cells, and, given some facts about human physiology, it is possible to derive the descriptions of the phenotypes. There is a clear analogy here with some cases from physics. The assumptions about blood cells and physiological needs seem to play the same role as the boundary conditions about shapes, relative positions and velocities of planets that occur in Newtonian derivations of Kepler's laws. In the Newtonian explanation we can see the application of a general pattern of reasoning – the derivation of explicit equations of motion from specifications of the forces acting – which yields the general result that a body under the influence of a centrally directed inverse square force will travel in a conic section; the general result is then applied to the motions of the planets by

incorporating pieces of astronomical information. Similarly, the derivation of the classical claims about the action of the normal and mutant hemoglobin genes can be seen as a purely chemical derivation of the generation of certain molecular structures and of the interactions among them. The chemical conclusions are then applied to the biological system under consideration by introducing three "boundary conditions": first, the claim that the altered molecular structures only affect development to the extent of substituting a different molecule in the erythrocytes (the blood cells that transport hemoglobin); second, a description of the chemical conditions in the capillaries; and third, a description of the effects upon the organism of capillary blockage.

The example is able to lend comfort to reductionism precisely because of an atypical feature. In effect, one concentrates on the *differences* among the phenotypes, takes for granted the fact that in all cases development will proceed normally to the extent of manufacturing erythrocytes – which are, to all intents and purposes, simply sacks for containing hemoglobin molecules – and compares the difference in chemical effect of the cases in which the erythrocytes contain different molecules. *The details of the process of development can be ignored.* However, it is rare for the effect of a mutation to be so simple. Most structural genes code for molecules whose presence or absence make subtle differences. Thus, typically, a mutation will affect the distribution of chemicals in the cells of a developing embryo. A likely result is a change in the timing of intracellular reactions, a change that may, in turn, alter the shape of the cell. Because of the change of shape, the geometry of the embryonic cells may be modified. Cells that usually come into contact may fail to touch. Because of this, some cells may not receive the molecules necessary to switch on certain batteries of genes. Hence the chemical composition of these cells will be altered. And so it goes.[36]

Quite evidently, in examples like this, (which include most of the cases in which molecular considerations can be introduced into embryology) the reasoning that leads us to a description of the phenotype associated with a genotype will be much more complicated than that found in the hemoglobin case. It will not simply consist in a chemical derivation adapted with the help of a few boundary conditions furnished by biology. Instead, we shall encounter a sequence of subarguments: molecular descriptions lead to specifications of cellular properties, from these specifications we draw conclusions about cellular interactions, and from these conclusions we arrive at further molecular descriptions. There is clearly a pattern of reasoning here which involves molecular biology and which extends the explanations furnished by classical genetics by showing how phenotypes depend upon genotypes – but I think it would be folly to suggest that the extension is provided by molecular genetics alone.

In Section 2, we discovered that the traditional answer to the philosophical question of understanding the relation that holds between molecular genetics and classical genetics, the reductionist's answer, will not do. Section 3 attempted to build on the diagnosis of the ills of reductionism, offering an account of the structure and evolution of classical genetics that would improve on the picture offered by those who favor traditional approaches to the nature of scientific theories. In the present section, I have tried to use the framework of Section 3 to understand the relations between molecular genetics and classical genetics. Molecular genetics has done something important for classical genetics, and its achievements can be recognized by seeing them as instances of the intertheoretic relations that I have characterized. Thus I claim that the problem from which we began is solved.

So what? Do we have here simply a study of a particular case – a case which has, to be sure, proved puzzling for the usual accounts of scientific theories and scientific change? I hope not. Although the traditional approaches may have proved helpful in understanding some of the well-worn examples that have been the stock-in-trade of twentieth century philosophy of science, I believe that the notion of scientific practice sketched in Section 3 and the inter-theoretic relations briefly characterized here will both prove helpful in analyzing the structure of science and the growth of scientific knowledge *even in those areas of science where traditional views have seemed most successful.*[37] Hence the tale of two sciences which I have been telling is not merely intended as a piece of local history that fills a small but troublesome gap in the orthodox chronicles. I hope that it introduces concepts of general significance in the project of understanding the growth of science.

5. Anti-Reductionism and the Organization of Nature

One loose thread remains. The history of biology is marked by continuing opposition between reductionists and anti-reductionists. Reductionism thrives on exploiting the charge that it provides the only alternative to the mushy incomprehensibility of vitalism. Anti-reductionists reply that their opponents have ignored the organismic complexity of nature. Given the picture painted above, where does this traditional dispute now stand?

I suggest that the account of genetics which I have offered will enable reductionists to provide a more exact account of what they claim, and will thereby enable anti-reductionists to be more specific about what they are denying. Reductionists and anti-reductionists agree in a certain minimal physicalism. To my knowledge, there are no major figures in contemporary biology who dispute the claim that each biological event, state or process is a complex physical event, state, or process. The most intricate part of ontogeny or phylogeny involves countless changes of physical state. What anti-reductionists emphasize is the organization of nature and the "interactions among phenomena at different levels." The appeal to organization takes two different forms. When the subject of controversy is the proper form of evolutionary theory, then anti-reductionists contend that it is impossible to regard all selection as operating at the level of the gene.[38] What concerns me here is not this area of conflict between reductionists and their adversaries, but the attempt to block claims for the hegemony of molecular studies in understanding the physiology, genetics, and development of organisms.[39]

A sophisticated reductionist ought to allow that, in the current practice of biology, nature is divided into levels which form the proper provinces of areas of biological study: molecular biology, cytology, histology, physiology, and so forth. Each of these sciences can be thought of as using certain language to formulate the questions it deems important and as supplying patterns of reasoning for resolving those questions. Reductionists can now set forth one of two main claims. The stronger thesis is that the explanations provided by any biological theories can be reformulated in the language of molecular biology and be recast so as to instantiate the patterns of reasoning supplied by molecular biology. The weaker thesis is that molecular biology provides explanatory extension of the other biological sciences.

Strong reductionism falls victim to the considerations that were advanced against (R3). The distribution of genes to gametes is to be explained, not by rehearsing the gory details of the reshuffling of the molecules, but through the observation that chromosomes are aligned in pairs just prior to the meiotic division, and that one chromosome from each matched pair is transmitted to each gamete. We may formulate this point in the biologists' preferred idiom by saying that the assortment of alleles is to be understood at the cytological level. What is meant by this description is that there is a pattern of reasoning which is applied to derive the description of the assortment of alleles and which involves predicates that characterize cells and their large-scale internal structures. That pattern of reasoning is to be objectively preferred to the molecular pattern which would be instantiated by the derivation that charts that complicated rearrangements of individual molecules because it can be applied across a range of cases which would look heterogeneous from a molecular perspective. Intuitively, the cytological pattern makes connections which are lost at the molecular level, and it is thus to be preferred.

So far, anti-reductionism emerges as the thesis that there are *autonomous levels of biological explanation*. Anti-reductionism construes the current division of biology not simply as a temporary feature of our science stemming from our cognitive imperfections but as the reflection of levels of organization in nature. Explanatory patterns that deploy the concepts of cytology will endure in our science because we would foreswear significant unification (or fail to employ the relevant laws, or fail to identify the causally relevant properties) by attempting to derive the conclusions to which they are applied using the vocabulary and reasoning patterns of molecular biology. But the autonomy thesis is only the

beginning of anti-reductionism. A stronger doctrine can be generated by opposing the weaker version of sophisticated reductionism.

In Section 4, I raised the possibility that molecular genetics may be viewed as providing an explanatory extension of classical genetics through deriving the schematic sentence that assigns phenotypes to genotypes from a molecular pattern of reasoning. This apparent possibility fails in an instructive way. Anti-reductionists are not only able to contend that there are autonomous levels of biological explanation. They can also resist the weaker reductionist view that explanation always flows from the molecular level up. Even if reductionists retreat to the modest claim that, while there are autonomous levels of explanation, descriptions of cells and their constituents are always explained in terms of descriptions about genes, descriptions of tissue geometry are always explained in terms of descriptions of cells, and so forth, anti-reductionists can resist the picture of a unidirectional flow of explanation. Understanding the phenotypic manifestation of a gene, they will maintain, requires constant shifting back and forth across levels. Because developmental processes are complex and because changes in the timing of embryological events may produce a cascade of effects at several different levels, one sometimes uses descriptions at higher levels to explain what goes on at a more fundamental level.

For example, to understand the phenotype associated with a mutant limb-bud allele, one may begin by tracing the tissue geometry to an underlying molecular structure. The molecular constitution of the mutant allele gives rise to a nonfunctional protein, causing some abnormality in the internal structures of cells. The abnormality is reflected in peculiarities of cell shape, which, in turn, affects the spatial relations among the cells of the embryo. So far we have the unidirectional flow of explanation which the reductionist envisages. However, the subsequent course of the explanation is different. Because of the abnormal tissue geometry, cells that are normally in contact fail to touch; because they do not touch, certain important molecules, which activate some batteries of genes, do not reach crucial cells; because the genes in question are not "switched on" a needed morphogen is not produced; the result is an abnormal morphology in the limb.

Reductionists may point out, quite correctly, that there is some very complex molecular description of the entire situation. The tissue geometry is, after all, a configuration of molecules. But this point is no more relevant than the comparable claim about the process of meiotic division in which alleles are distributed to gametes. Certain genes are not expressed because of the geometrical structure of the cells in the tissue: *the pertinent cells are too far apart.* However this is realized at the molecular level, our explanation must bring out the salient fact that it is the presence of a gap between cells that are normally adjacent that explains the nonexpression of the genes. As in the example of allele transmission at meiosis, we lose sight of the important connections by attempting to treat the situation from a molecular point of view. As before, the point can be sharpened by considering situations in which radically different molecular configurations realize the crucial feature of the tissue geometry: situations in which heterogeneous molecular structures realize the breakdown of communication between the cells.

Hence, embryology provides support for the stronger anti-reductionist claim. Not only is there a case for the thesis of autonomous levels of explanation, but we find examples in which claims at a more fundamental level (specifically, claims about gene expression) are to be explained in terms of claims at a less fundamental level (specifically, descriptions of the relative positions of pertinent cells). Two anti-reductionist biologists put the point succinctly:

> . . . a developmental program is not to be viewed as a linearly organized causal chain from genome to phenotype. Rather, morphology emerges as a consequence of an increasingly complex dialogue between cell populations, characterized by their geometric continuities, and the cells' genomes, characterized by their states of gene activity.[40]

A corollary is that the explanations provided by the "less fundamental" biological sciences are not extended by molecular biology alone.

It would be premature to claim that I have shown how to reformulate the anti-reductionist appeals to the organization of nature in a completely precise way. My conclusion is that, to the extent that we can make sense of the present explanatory structure within biology – that division of the field into subfields corresponding to levels of organization in nature – we can also understand the anti-reductionist doctrine. In its minimal form, it is the claim that the commitment to several explanatory levels does not simply reflect our cognitive limitations; in its stronger form, it is the thesis that some explanations oppose the direction of preferred reductionistic explanation. Reductionists should not dismiss these doctrines as incomprehensible mush unless they are prepared to reject as unintelligible the biological strategy of dividing the field (a strategy which seems to me well understood, even if unanalyzed).

The examples I have given seem to support both anti-reductionist doctrines. To clinch the case, further analysis is needed. The notion of explanatory levels obviously cries out for explication, and it would be illuminating to replace the informal argument that the unification of our beliefs is best achieved by preserving multiple explanatory levels with an argument based on a more exact criterion for unification. Nevertheless, I hope that I have said enough to make plausible the view that, despite the immense value of the molecular biology that Watson and Crick launched in 1953, molecular studies cannot cannibalize the rest of biology. Even if geneticists must become "physiological chemists" they should not give up being embryologists, physiologists, and cytologists.

Notes

1 "Variation due to change in the individual gene," reprinted in J. A. Peters ed., *Classic Papers in Genetics* (Englewood Cliffs, N.J.: Prentice-Hall, 1959), pp. 104–116. Citation from p. 115.

2 "Molecular Structure of Nucleic Acids," *Nature* 171 (1953), pp. 737–738; reprinted in Peters, op. cit., pp. 241–243. Watson and Crick amplified their suggestion in "Genetic Implications of the Structure of Deoxyribonucleic Acid" *Nature* 171 (1953), pp. 934–937.

3 The most sophisticated attempts to work out a defensible version of reductionism occur in articles by Kenneth Schaffner. See, in particular, "Approaches to Reduction," *Philosophy of Science* 34 (1967), pp. 137–147; "The Watson–Crick Model and Reductionism," *British Journal for the*

Philosophy of Science 20 (1969), pp. 325–348; "The Peripherality of Reductionism in the Development of Molecular Biology," *Journal of the History of Biology* 7 (1974), pp. 111–139; and "Reductionism in Biology: Prospects and Problems," R. S. Cohen *et al.* eds., *PSA 1974*, (Boston: D. Reidel, 1976), pp. 613–632. See also Michael Ruse, "Reduction, Replacement, and Molecular Biology," *Dialectica* 25 (1971), pp. 38–72; and William K. Goosens, "Reduction by Molecular Genetics," *Philosophy of Science* 45 (1978), pp. 78–95. A variety of anti-reductionist points are made in David Hull, "Reduction in Genetics – Biology or Philosophy?" *Philosophy of Science* 39 (1972), pp. 491–499; and Chapter 1 of *Philosophy of Biological Science* (Englewood Cliffs, N.J.: Prentice-Hall, 1974); in Steven Orla Kimbrough, "On the Reduction of Genetics to Molecular Biology," *Philosophy of Science* 46 (1979), pp. 389–406 and in Ernst Mayr, *The Growth of Biological Thought* (Cambridge, Mass.: Harvard University Press, 1982), pp. 59–63.

4   Typically, though not invariably. In a suggestive essay, "Reductive Explanation: A Functional Account," (R. S. Cohen *et al.* op. cit. pp. 671–710), William Wimsatt offers a number of interesting ideas about intertheoretic relations and the case of genetics. Also provocative are Nancy Maull's "Unifying Science Without Reduction," *Studies in the History and Philosophy of Science* 8 (1977), pp. 143–171; and Lindley Darden and Nancy Maull, "Interfield Theories," *Philosophy of Science* 44 (1977), pp. 43–64. My chief complaint about the works I have cited is that unexplained technical notions – "mechanisms," "levels," "domain," "field," "theory" – are invoked (sometimes in apparently inconsistent ways), so that no precise answer to the philosophical problem posed in the text is ever given. Nevertheless, I hope that the discussion of the later sections of this paper will help to articulate more fully some of the genuine insights of these authors, especially those contained in Wimsatt's rich essay.

5   E. Nagel, *The Structure of Science* (New York: Harcourt Brace, 1961), Chapter 11. A simplified presentation can be found in Chapter 8 of C. G. Hempel, *Philosophy of Natural Science* (Englewood Cliffs, N.J.: Prentice-Hall, 1966).

6   Quite evidently, this is a weak version of what was once the "received view" of scientific theories, articulated in the works of Nagel and Hempel cited in the previous note. A sustained presentation and critique of the view is given in the Introduction to F. Suppe ed., *The Structure of Scientific Theories* (Urbana: University of Illinois Press, 1973). The fact that the standard model of reduction presupposes the thesis that theories are reasonably regarded as sets of statements has been noted by Clark Glymour, "On Some Patterns of Reduction," *Philosophy of Science* 36 (1969), pp. 340–353, 342; and by Jerry Fodor (*The Language of Thought*, New York: Crowell, 1975, p. 11, footnote 10). Glymour endorses the thesis; Fodor is skeptical about it.

7   Philosophers often suggest that, in reduction, one derives *corrected* laws of the reduced theory from an *unmodified* reducing theory. But this is not the way things go in the paradigm cases: one doesn't correct Galileo's law by using Newtonian mechanics; instead, one neglects "insignificant terms" in the Newtonian equation of motion for a body falling under the influence of gravity; similarly, in deriving the Boyle–Charles law from kinetic theory (or statistical mechanics), it is standard to make idealizing assumptions about molecules, and so obtain the exact version of the Boyle–Charles law; subsequently, corrected versions are generated by "subtracting" the idealizing procedures. Although he usually views reduction as deriving a corrected version of the reduced theory, Schaffner notes that reduction might sometimes proceed by modifying the reducing theory ("Approaches to Reduction," p. 138; "The Watson–Crick Model and Reductionism" p. 322). In fact, the point was already made by Nagel, op. cit.

8   In part, because modification might produce an inconsistent theory that would permit the derivation of anything. In part, because of the traditionally vexing problem of the proper form for bridge principles in heterogeneous reductions in physics. The former problem is discussed in Glymour, op. cit., p. 352 and in Dudley Shapere, "Notes towards a Post-Positivistic Interpretation of Science" (P. Achinstein and S. Barker, eds. *The Legacy of Logical Positivism*, Baltimore: Johns Hopkins University Press, 1971). For discussion of the latter issue, see Larry Sklar, "Types of Inter-Theoretic Reduction," *British Journal for the Philosophy of Science* 18 (1967), pp. 109–124; Robert Causey, "Attribute Identities in Microreductions," *Journal of Philosophy* 69 (1972), pp. 407–422; and Berent Enc, "Identity Statements and Micro-reductions," *Journal of Philosophy* 73 (1976), pp. 285–306. The concerns that I shall raise are orthogonal to these familiar areas of dispute.

9   Hull, *Philosophy of Biological Science*, p. 23, adapted from Theodosius Dobzhansky, *Genetics of the Evolutionary Process* (New York: Columbia University Press, 1970), p. 167. Similarly molecular genetics is said to have the task of "discovering

how molecularly characterized genes produce proteins which in turn combine to form gross phenotypic traits" (Hull *ibid.*; see also James D. Watson, *Molecular Biology of the Gene*, Menlo Park, Ca., W. A. Benjamin, 1976, p. 54).

10 The phenotype/genotype distinction was introduced to differentiate the observable characteristics of an organism from the underlying genetic factors. In subsequent discussions the notion of phenotype has been extended to include properties which are not readily observable (for example, the capacity of an organism to metabolize a particular amino acid). The expansion of the concept of phenotype is discussed in my paper "Genes," *British Journal for the Philosophy of Science* 33 (1982), pp. 337–359.

11 "A *locus* is the place on a chromosome occupied by a gene. Different genes which can occur at the same locus are said to be *alleles*. In diploid organisms, chromosomes line up in pairs just before the meiotic division that gives rise to gametes. The matched pairs are pairs of *homologous chromosomes*. If different alleles occur at corresponding loci on a pair of homologous chromosomes, the organism is said to be *heterozygous* at these loci.

12 *Recombination* is the process (which occurs before meiotic division) in which a chromosome exchanges material with the chromosome homologous with it. Alleles which occur on one chromosome may thus be transferred to the other chromosome, so that new genetic combinations can arise.

13 Other central Mendelian claims also turn out to be false. The Mendelian principle that if an organism is heterozygous at a locus then the probabilities of either allele being transmitted to a gamete are equal falls afoul of cases of meiotic drive. (A notorious example is the *t*-allele in the house mouse, which is transmitted to 95% of the sperm of males who are heterozygous for it and the wild-type allele; see R. C. Lewontin and L. C. Dunn, "The Evolutionary Dynamics of a Polymorphism in the House Mouse," *Genetics* 45 (1960), pp. 705–722. Even the idea that genes are transmitted across the generations, unaffected by their presence in intermediate organisms, must be given up once we recognize that intra-allelic recombination can occur.

14 To the best of my knowledge, the mechanisms of this interference are not well understood. For a brief discussion, see J. Sybenga, *General Cytogenetics* (North-Holland, 1972), pp. 313–314. In this paper, I shall use "segregation distortion" to refer to cases in which there is a propensity for nonhomologous chromosomes to assort together.

"Meiotic drive" will refer to examples in which one member of a pair of homologous chromosomes has a greater probability of being transmitted to a gamete. The literature in genetics exhibits some variation in the use of these terms. Let me note explicitly that, on these construals, both segregation distortion and meiotic drive will be different from *nondisjunction*, the process in which a chromosome together with the whole (or a part) of the homologous chromosome is transmitted to a gamete.

15 This point raises some interesting issues. It is common practice in genetics to count a segment of DNA as a single gene if it was produced by mutation from a gene. Thus many mutant alleles are viewed as DNA segments in which modification of the sequence of bases has halted transcription too soon, with the result that the gene product is truncated and nonfunctional. My envisaged case simply assumes that a second mutation occurs further down the segment so that transcription starts and stops in two places, generating two useless gene products. The historical connection with the original allele serves to identify the segment as one gene.

Conversely, where there is no historical connection to any organism, one may have qualms about counting a DNA segment as a gene. Suppose that, in some region of space, a quirk of nature brings together the constituent atoms for the white eye mutant in *Drosophila melanogaster*, and that the atoms become arranged in the right way. Do we have here a *Drosophila* gene? If the right answer is "No" then it would seem that a molecular structure only counts as a gene given an appropriate history. I hasten to add that "appropriate histories" need not simply involve the usual biological ways in which organisms transmit, replicate and modify genes: one can reasonably hope to synthesize genes in the laboratory. The case seems analogous to questions that arise about personal identity. If a person's psychological features are replicated by a process that sets up the "right sort of causal connection" between person and product, then we are tempted to count the product as the surviving person. Similarly, if a molecular structure is generated in a way that sets up "the right sort of causal connection" between the structure and some prior gene then it counts as a gene. In both cases, causal connections of "the right sort" may be set up in everyday biological ways and by means of deliberate attempts to replicate a prior structure.

16 So called *structural genes* direct the formation of proteins by coding for RNA molecules. They are

"transcribed" to produce *messenger* RNA (mRNA) which serves as a more immediate "blueprint" for the construction of the protein. Transcription is started and stopped through the action of regulatory genes. In the simplest regulatory system (that of the *lac* operon) an area adjacent to the structural gene serves as a "dumping ground" for a molecule. When concentration of the protein product becomes too high, the molecule attaches to this site and transcription halts; when more protein is required, the cell produces a molecule that removes the inhibiting molecule from the neighborhood of the structural gene, and transcription begins again. (For much more detail, see Watson, op. cit., Chapter 14, and M. W. Strickberger, *Genetics* (New York: Macmillan, 1976), Chapter 29.)

17   The situation is complicated by the existence of "introns" – segments within genes whose products under transcription are later excised – and by the enormous amount of repetitive DNA that most organisms seem to contain. Moreover, the regulatory systems in eukaryotes appear to be much more complicated than the prokaryote systems (of which the *lac* operon is *one* paradigm). For a review of the situation, as of a few years ago, see Eric H. Davidson, *Gene Expression in Early Development* (New York: Academic Press, 1976).

18   The account will be even more complicated if we honor the suggestion of footnote 15, and suppose that, for a molecular structure to count as a gene, it must be produced in the right way.

19   Early in the process preceding meiotic division the chromosomes become more compact. As meiosis proceeds, the nucleus comes to contain a system of threads that resembles a spindle. Homologous chromosomes line up together near the center of the spindle, and they are oriented so that one member of each pair is slightly closer to one pole of the spindle, while the other is slightly closer to the opposite pole.

20   The point I have been making is related to an observation of Hilary Putnam's. Discussing a similar example, Putnam writes: "The same explanation will go in any world (whatever the microstructure) in which those *higher level structural features* are present"; he goes on to claim that "explanation is superior not just subjectively but methodologically, . . . if it brings out relevant laws." (Putnam, "Philosophy and our Mental Life," in *Mind, Language, and Reality*, (Cambridge, Cambridge University Press, 1975), pp. 291–303, p. 296). The point is articulated by Alan Garfinkel (*Forms of Explanation*, New Haven: Yale University Press, 1981), and William Wimsatt has also raised

analogous considerations about explanation in genetics.

It is tempting to think that the independence of the "higher level structural features" in Putnam's example and in my own can be easily established: one need only note that there are worlds in which the same feature is present without any molecular realization. So, in the case discussed in the text, PS-processes might go on in worlds where all objects were perfect continua. But although this shows that PS-processes form a kind which could be realized without molecular reshufflings, we know that all *actual* PS-processes do involve such reshufflings. The reductionist can plausibly argue that *if* the set of PS-processes with molecular realizations is itself a natural kind, then the explanatory power of the cytological account can be preserved by identifying meiosis as a process of this narrower kind. Thus the crucial issue is not whether PS-processes form a kind with non-molecular realizations, but whether those PS-processes which have molecular realizations form a kind that can be characterized from the molecular point of view. Hence, the easy strategy of responding to the reductionist must give way to the approach adopted in the text. (I am grateful to the editors of *The Philosophical Review* for helping me to see this point.)

21   It would be impossible in the scope of this paper to do justice to the various conceptions of scientific theory that have emerged from the demise of the "received view." Detailed comparison of the perspective I favor with more traditional approaches (both those that remain faithful to core ideas of the "received view" and those that adopt the "semantic view" of theories) must await another occasion.

22   My notion of a practice owes much to some neglected ideas of Sylvain Bromberger and Thomas Kuhn. See, in particular, Bromberger, "A Theory about the Theory of Theory and about the Theory of Theories," (W. L. Reese ed., *Philosophy of Science, The Delaware Seminar*, New York, 1963); and "Questions," (*Journal of Philosophy* 63 (1966), pp. 597–606); and Kuhn, *The Structure of Scientific Revolutions* (Chicago: University of Chicago Press, 1962) Chapters II–V. The relation between the notion of a practice and Kuhn's conception of a paradigm is discussed in Chapter 7 of my book *The Nature of Mathematical Knowledge* (New York: Oxford University Press, 1983).

23   More exactly, a general argument pattern is a triple consisting of a sequence of schematic sentences (a *schematic argument*), a set of filling

instructions (directions as to how dummy letters are to be replaced), and a set of sentences describing the inferential characteristics of the schematic argument (a *classification* for the schematic argument). A sequence of sentences instantiates the general argument pattern just in case it meets the following conditions: (i) the sequence has the same number of members as the schematic argument of the general argument pattern; (ii) each sentence in the sequence is obtained from the corresponding schematic sentence in accordance with the appropriate filling instructions; (iii) it is possible to construct a chain of reasoning which assigns to each sentence the status accorded to the corresponding schematic sentence by the classification. For some efforts at explanation and motivation, see my "Explanatory Unification," *Philosophy of Science* 48 (1981), pp. 507–531.

24 Sometimes particles in the cytoplasm account for hereditary trails. See Strickberger, op. cit., pp. 257–265.

25 For examples, see Strickberger op. cit. Chapters 6–12, 14–17, especially Chapter 11; Peters, op. cit.; and H. L. K. Whitehouse, *Towards an Understanding of the Mechanism of Heredity* (London: Arnold, 1965).

26 The comparison will make use of standard statistical techniques, such as the chi-square test.

27 *Polygeny* occurs when many genes affect one characteristic; *pleiotropy* occurs when one gene affects more than one characteristic.

28 *Incomplete dominance* occurs when the phenotype of the heterozygote is intermediate between that of the homozygotes; *epistasis* occurs when the effect of a particular combination of alleles at one locus depends on what alleles are present at another locus.

29 See Watson, op. cit., Chapter 9; and Arthur Kornberg *DNA Synthesis* (San Fransisco: W. H. Freeman, 1974).

30 See. G. P. Georghiou, "The Evolution of Resistance to Pesticides," *Annual Review of Ecology and Systematics* 3 (1972), pp. 133–168.

31 See Watson, op. cit., pp. 189–193; and T. H. Maugh II, "A New Understanding of Sickle Cell Emerges," *Science* 211 (1981), pp. 265–267.

32 However, one might claim that "Genes can replicate" is a law of genetics, in that it is general, lawlike, and true. This does not vitiate my claim that the structure of classical genetics is not to be sought by looking for a set of general laws, for the law in question is so weak that there is little prospect of finding supplementary principles which can be conjoined with it to yield a representation of genetic theory. I suggest that "Genes can replicate" is analogous to the thermodynamic "law," "Gases can expand," or to the Newtonian "law," "Forces can be combined." If the only laws that we could find in thermodynamics and mechanics were weak statements of this kind we would hardly be tempted to conceive of these sciences as sets of laws. I think that the same point goes for genetics.

33 A similar point is made by Kenneth Schaffner in a book on theory structure in the biomedical sciences (K. Schaffner, ed., *Logic of Discovery and Diagnosis in Medicine*, Berkeley: University of California Press, 1985). Schaffner's terminology is different from my own, and he continues to be interested in the prospects of global reduction, but there is considerable convergence between the conclusions that he reaches and those that I argue for in the present section.

34 There are numerous examples of such modifications from the history of chemistry. I try to do justice to this type of case in "Theories, Theorists, and Theoretical Change," *The Philosophical Review* 87 (1978), pp. 519–547 and in "Genes."

35 Molecular biology also provided significant refinement of the terms 'gene' and 'allele'. See "Genes."

36 For examples, see N. K. Wessels *Tissue Interactions and Development* (Menlo Park, Ca.: W. A. Benjamin, 1977), especially Chapters 6, 7, 13–15; and Donald Ede, *An Introduction to Developmental Biology* (London: Blackie, 1978) especially Chapter 13.

37 I attempt to show how the same perspective can be fruitfully applied to other examples in "Explanatory Unification," Sections 3 and 4; *Abusing Science* (Cambridge: MIT Press, 1982) Chapter 2; and "Darwin's Achievement," N. Rescher, ed. in a volume of the *Pittsburgh Studies in the Philosophy of Science Reasons and Rationality in Natural Science*, 1985, pp. 127–189.

38 The extreme version of reductionism is defended by Richard Dawkins in *The Selfish Gene* (New York: Oxford University Press, 1976) and *The Extended Phenotype* (San Francisco: W. H. Freeman, 1982). For an excellent critique, see Elliott Sober and Richard C. Lewontin, "Artifact, Cause, and Genic Selection," *Philosophy of Science* 49 (1982), pp. 157–180. More ambitious forms of anti-reductionism with respect to evolutionary theory are advanced in S. J. Gould, "Is a new and general theory of evolution emerging?" *Paleobiology*, 6 (1980), pp. 119–130; N. Eldredge and J. Cracraft, *Phylogenetic Patterns and the Evolutionary Process* (New York: Columbia University Press, 1980); and Steven M. Stanley,

Macroevolution (San Francisco: W. H. Freeman, 1979). A classic early source of some (but not all) later anti-reductionist themes is Ernst Mayr's *Animal Species and Evolution* (Cambridge, Harvard University Press, 1963) especially Chapter 10.

39 Gould's *Ontogeny and Phylogeny* (Harvard, 1977) provides historical illumination of both areas of debate about reductionism. Contemporary anti-

reductionist arguments about embryology are expressed by Wessels (op. cit.) and Ede (op. cit.). See also G. Oster and P. Alberch, "Evolution and Bifurcation of Developmental Programs," *Evolution* 36 (1982), pp. 444–459.

40 Oster and Alberch, op. cit., p. 454. The diagram on p. 452 provides an equally straightforward account of their anti-reductionist position.

17

The Multiple Realizability Argument Against Reductionism

Elliott Sober

Reductionism is often understood to include two theses: (1) every singular occurrence that the special sciences can explain also can be explained by physics; (2) every law in a higher-level science can be explained by physics. These claims are widely supposed to have been refuted by the multiple realizability argument, formulated by Putnam (1967, 1975) and Fodor (1968, 1975). The present paper criticizes the argument and identifies a reductionistic thesis that follows from one of the argument's premises.

1. Introduction

If there is now a received view among philosophers of mind and philosophers of biology about reductionism, it is that reductionism is mistaken. And if there is now a received view as to why reductionism is wrong, it is the multiple realizability argument.[1] This argument takes as its target the following two claims, which form at least part of what reductionism asserts:

1. Every singular occurrence that a higher-level science can explain also can be explained by a lower-level science.
2. Every law in a higher-level science can be explained by laws in a lower-level science.

The "can" in these claims is supposed to mean "can in principle," not "can in practice." Science is not now complete; there is a lot that the physics of the present fails to tell us about societies, minds, and living things. However, a completed physics would not thus be limited, or so reductionism asserts (Oppenheim and Putnam 1958).

The distinction between higher and lower of course requires clarification, but it is meant to evoke a familiar hierarchical picture; it runs (top to bottom) as follows – the social sciences, individual psychology, biology, chemistry, and physics. Every society is composed of individuals who have minds; every individual with a mind is alive;[2] every individual who is alive is an individual in which chemical processes occur; and every system in which chemical processes occur is one

Elliot Sober, "The Multiple Realizability Argument against Reductionism," *Philosophy of Science* (Philosophy of Science Association, 1999), 66, 542–564. Reprinted by permission of the publisher, the University of Chicago Press.

My thanks to Martin Barrett, John Beatty, Tom Bontly, Ellery Kells, Berent Enç, Branden Fitelson, Jerry Fodor, Martha Gibson, Daniel Hausman, Dale Jamieson, Andrew Levine, Brian McIaughlin, Terry Penner, Larry Shapiro, Chris Stephens, Richard Teng, Ken Waters, Ann Wolfe, and an anonymous referee for *Philosophy of Science* for comments on earlier drafts.

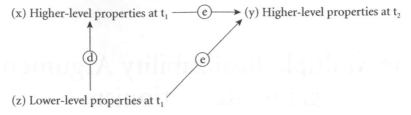

Figure 17.1   Relations of synchronic determination *(d)* and diachronic explanation (e) that may connect higher- and lower-level properties.

in which physical processes occur. The domains of higher-level sciences are sub-sets of the domains of lower-level sciences. Since physics has the most inclusive domain, immaterial souls do not exist and neither do immaterial vital fluids. In addition, since the domains are (properly) nested, there will be phenomena that lower-level sciences can explain, but that higher-level sciences cannot. Propositions (1) and (2), coupled with the claim of nested domains, generate an asymmetry between higher-level and lower-level sciences.

Reductionism goes beyond what these two propositions express. Events have multiple causes. This means that two causal explanations of the same event may cite different causes. A car skids off the highway because it is raining, and also because the tires are bald (Hanson 1958). Proposition (1) says only that if there is a psychological explanation of a given event, then there is also a physical explanation of that event. It does not say how those two explanations are related, but reductionism does. Societies are said to have their social properties *solely in virtue of* the psychological properties possessed by individuals; individuals have psychological properties *solely in virtue of* their having various biological

properties; organisms have biological properties *solely in virtue of* the chemical processes that occur within them; and systems undergo chemical processes *solely in virtue of* the physical processes that occur therein. Reductionism is not just a claim about the explanatory capabilities of higher- and lower-level sciences; it is, in addition, a claim to the effect that the higher-level properties of a system are determined by its lower-level properties.[3]

These two parts of reductionism are illustrated in Figure 17.1. The circled e represents the relation of diachronic explanation; the circled d represents the relation of synchronic determination. Reductionism says that if (x) explains (y), then (z) explains (y); it also asserts that (z) determines (x). The multiple realizability argument against reductionism does not deny that higher-level properties are determined by lower-level properties. Rather, it aims to refute propositions (1) and (2) – (z) does not explain (y), or so this argument contends.

2.   Multiple Realizability

Figure 17.2 is redrawn from the first chapter, entitled "Special Sciences," of Fodor's 1975

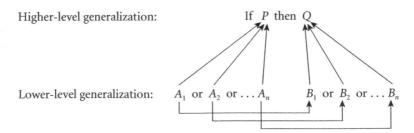

Figure 17.2   The lower-level properties A_i and B_i provide multiple realizations of the higher-level properties P and Q respectively. One higher-level law and n lower-level laws are depicted, following Fodor 1975.

book, *The Language of Thought*. It describes a law in a higher-level science and how it might be related to a set of laws in some lower-level science. The higher-level law is couched in its own proprietary vocabulary; P and Q are higher-level properties and the higher-level law says that everything that has P also has Q. The lower-level science provides n laws, each of them connecting an A predicate to a B predicate; the lower-level laws say that everything that has A_i also has B_i (for each $i = 1, 2, \ldots, n$).

The higher-level property P is said to be multiply realizable; $A_1, A_2, \ldots, A_n$ are the different (mutually exclusive and collectively exhaustive) realizations that P might have. Similarly, Q has $B_1, B_2, \ldots, B_n$ as its alternative realizations. What does multiple realizability mean? First, it entails the relation of simultaneous determination; necessarily, if something has A_i at time t, then it has P at t, and if it has B_i at time t, then it has Q at t. But there is something more, and it is this second ingredient that is supposed to ensure that the multiple realizability relation is antisymmetric. An individual that has P has that property *solely in virtue* of the fact that it has whichever A_i it possesses. Because the higher-level properties are multiply realizable, the mapping from lower to higher is many-to-one. You cannot tell which of the A_i properties is exhibited by a system just from knowing that it has property P, and you cannot tell which of the B_i properties the system has just from knowing that it has Q.[4]

Two examples will make the intended meaning of multiple realizability sufficiently clear. Suppose that different types of physical system can have minds; minds can be built out of neurons, but perhaps they also can be built out of silicon chips. An individual mind – you, for example – will have its psychological properties in virtue of the physical properties that the system possesses. But if you and someone else have some psychological properties in common, there is no guarantee that the two of you also will share physical properties; you and this other person may deploy different physical realizations of the same psychological properties. The same point can be made with respect to biological properties – you have various biological properties, and each of these is present in virtue of your possessing this or that set of physical properties. However, you

and some other organism may share a given biological property even though you are physically quite different; this will be true if you and this other organism deploy different physical realizations of the same biological properties.

Since the multiple realizability relation obtains between simultaneously instantiated properties, the relation is not causal (assuming as I will that cause must precede effect). However, the diachronic laws I want to consider *are* causal – they say that a system's having one property at one time causes it to exhibit another property sometime later. The reason I will focus on causal diachronic laws is not that I think that all diachronic laws are causal, but that these provide the clearest cases of scientific explanations.[5] Thus, returning to propositions (1) and (2), we can ask the following two questions about the multiple realizability relations depicted in the second figure:

(1′) If an individual's having property P explains its having property Q, is it also true that its having property A_i explains its having property Q?

(2′) Do lower-level laws of the form "if A_i then B_i" explain the higher-level law "if P then Q"?

Let us assume that the properties described in higher-level sciences are multiply realized by properties discussed in a lower-level science. What consequences follow from this concerning reductionism?

3. The Explanation of Singular Occurrences – Putnam's Peg

Suppose a wooden board has two holes in it. One is circular and has a 1-inch diameter; the other is square and is 1 inch on a side. A cubical peg that is $^{15}/_{16}$ of an inch on each side will fit through the square hole, but not the circular one. What is the explanation? Putnam (1975) says that the explanation is provided by the *macro*-properties just cited of the peg and the holes. He denies that the *micro*-properties of molecules or atoms or particles in the peg and the piece of wood explain this fact. The micro-description is long and complicated and it brings in a welter of irrelevant detail. To explain why the peg goes through one hole but not the other, it does not matter what micro-properties the molecules have, as long as the peg and board have the macro-properties I mentioned. The macro-properties are explanatory; the micro-properties that realize those macro-properties are not. Hence, reductionism is false.

This is a delightfully simple example and argument, but it is possible to have one's intuitions run in the opposite direction. Perhaps the micro-details do not interest *Putnam*, but they may interest *others*, and for perfectly legitimate reasons. Explanations come with different levels of detail. When someone tells you more than you want to hear, this does not mean that what is said fails to be an explanation. There is a difference between explaining too much and not explaining at all.

Compare the micro-story that Putnam derides with a quite different story. Suppose someone suggested that the reason the peg goes through one hole but not the other is that the peg is *green*. Here it is obvious that a mistake has been made. If we demand that explanations be *causal* explanations, it will be quite clear why the color of the peg is not explanatory. It is causally irrelevant. This is an objective feature of the system under consideration and has nothing to do with our desire for brevity or detail.

It is possible to be misled by a superficial similarity that links the micro-story about the particles in the peg and board and the pseudo-explanation that cites the peg's color. Both of the following counterfactuals are true:

> If the particles in the peg and board had been different, the peg still would have passed through one hole but not the other, as long as the macro-dimensions were as described.
>
> If the peg had not been green, it still would have passed through one hole but not the other, as long as the macro-dimensions were as described.

If we say that causes are *necessary* for their effects (as does Lewis 1973a), we might be tempted to use these counterfactuals to conclude that the system's micro-features and the peg's color are both causally irrelevant, and hence should not be cited in a causal explanation. This proposal

should be understood to mean that the effect would not have happened if the cause had not, *in the specific circumstances that actually obtained*; striking a match is not always necessary to get the match to light, but it may be necessary in various specific circumstances.

There are general questions that may be raised about the adequacy of this account of causation.[6] However, even if we waive these questions, it is important to examine more closely how the counterfactual test connects with Putnam's argument. Let us suppose that the micro-properties of the peg and board's molecules are not necessary for the peg to go through one hole but not the other, if we hold fixed the macro-dimensions. But are the macro-dimensions necessary, if we hold fixed the micro-properties? That is, are we prepared to affirm the following counterfactual?

> If the macro-dimensions of the peg and board had been different, while the micro-properties were as described, the peg would not have passed through the one hole but not the other.

This counterfactual has a nomologically impossible antecedent. Many of us simply draw a blank when asked to assign a truth value to such assertions. The semantics of Stalnaker (1968) and Lewis (1973b) does not; it says that the counterfactual is vacuously true. However, before we interpret this as vindicating Putnam's argument, we also should note that the same semantic theory says that the following counterfactual is true as well:

> If the macro-dimensions of the peg and board had been different, while the micro-properties were as described, the peg still would have passed through the one hole but not the other.

It is hard to see how such counterfactuals can vindicate the judgment that the macro-properties are causally efficacious while their micro-realizations are not.[7]

I very much doubt that the concept of explanatory relevance means what Putnam requires it to mean in this argument. When scientists discover why smoking causes cancer, they are finding out which ingredients in cigarette smoke are carcinogenic. If smoking causes cancer, this is presumably because the micro-configuration of cigarette smoke is doing the work. If there turn out to be several carcinogenic ingredients and different cigarettes contain different ones, this does not make the molecular inquiry explanatorily irrelevant to the question of why people get cancer. The fact that P is multiply realizable does not mean that P's realizations fail to explain the singular occurrences that P explains. A smoker may not want to hear the gory details, but that does not mean that the details are not explanatory.[8]

Putnam says he does not care whether we call the micro-story about the peg and the board a non-explanation, or simply describe it as a "terrible" explanation (Putnam 1975, 296). He thinks that the "goodness" of an explanation "is not a subjective matter." According to the objective concept of good explanation that Putnam has in mind, "an explanation is superior if it is more general" and he quotes with approval a remark by Alan Garfinkel – that "a good explanation is invariant under small perturbations of the assumptions" (301). What makes a more general (more invariant) explanation *objectively* better than one that is less? Putnam's answer is that "one of the things we do in science is to look for laws. Explanation is superior not just subjectively, but *methodologically*, in terms of facilitating the aims of scientific inquiry, if it brings out relevant laws" (301). My reply is that the goal of finding "relevant" laws cuts both ways. Macro-generalizations may be laws, but there also may be laws that relate micro-realizations to each other, and laws that relate micro- to macro- as well. Although "if P then Q" is more general than "if A_i then B_i,"[9] the virtue of the micro-generalization is that it provides more details. Science aims for depth as well as breadth. Some good explanations are fox-like; others are hedgehogian (Berlin 1953). There is no objective rule concerning which is better.

The claim that the preference for breadth over depth is a matter of taste is consistent with the idea that the difference between a genuine explanation and a nonexplanation is perfectly objective. In fact, it also is consistent with Hempel's (1965) view that the concept of scientific explanation should be explicated in terms of the notion of an ideally complete explanation, and that this is an objective notion. Perhaps an ideally complete scientific explanation of a singular occurrence in which an individual (or set of individuals) exhibits a multiply realizable property (or relation) would include the macro-story, the micro-story, and an account of how these are connected, If this is right, then reductionists and antireductionists alike are mistaken if they think that only part of this multilevel account deserves mention. But whatever the merits are of the idea of an ideally complete scientific explanation, we need to recognize that science in its currently incomplete state still is able to offer up "explanations." Perhaps these should be termed "explanation sketches," since they fall short of the Hempelian ideal. In any case, it remains true that science provides a plurality of such accounts of a given event. They vary in how detailed they are and in the level of organization described.[10]

Returning to Putnam's example, let us imagine that we face *two* peg-plus-board systems of the type that he describes. If we opt for the macro-explanation of why, in each case, the peg goes through one hole but not the other, we will have provided a *unified explanation*. We will have explained similar effects by describing similar causes. However, if we choose a micro-explanation, it is almost inevitable that we will describe the two systems as being physically different, and thus our explanation will be *dis*unified. We will have explained the similar effects by tracing them back to different types of cause. Putnam uses the terms "general" and "invariant" to extol the advantages of macro-explanation, but he might just as well have used the term "unified" instead. In claiming that it is a matter of taste whether we prefer the macro- or the micro-explanation, I am claiming that there is no objective reason to prefer the unified over the disunified explanation. Science has room for both lumpers and splitters. Some people may not be interested in hearing that the two systems are in fact different; the fact that they have the same macro-properties may be all they wish to learn. But this does not show that discerning differences is less explanatory. Indeed, many scientists would find it more illuminating to be shown how the same effect is reached by different causal pathways.

In saying that the preference for unified explanation is merely a matter of taste, I seem to be contradicting a fundamental fact about scientific inference – that it counts in favor of the plausibility of a theory that the theory unifies disparate phenomena. Actually, no such consequence follows from what I am saying. Here, it is essential to distinguish the *context of justification* from the *context of explanation*.[11] When two theories are evaluated in the light of the evidence available, the fact that one is unified and the other is disunified is epistemologically relevant. In a wide range of circumstances, the unified theory can be expected to be more predictively accurate than the theory that is disunified, when they fit the data about equally well (Forster and Sober 1994). Whether a theory is unified is relevant to deciding whether we should accept it. However, the problem addressed by the multiple realizability argument is not about acceptance. We are supposed to assume that the macro-story and the micro-story are both *true*. Given this, we now are asked to decide which provides the better explanation of why the systems behave similarly. Unification is relevant to acceptance, but unification is not objectively relevant to deciding which accepted statements we should use in formulating explanations. The latter is simply a matter of taste – do we want more details or fewer? The context of justification and the context of explanation are different.

4. The Explanation of Laws – Fodor's Horror of Disjunctions

Whereas Putnam discusses the explanation of singular occurrences, Fodor uses the idea of multiple realizability to argue that laws in a higher-level science are not explained by laws in a lower-level science. This shift introduces some new considerations. Although many, if not all, explanations of singular occurrences are causal, the most familiar cases of explaining laws do not involve tracing effects back to their causes. Laws are usually explained by deriving them from "deeper" laws and initial condition statements; the explained laws and the explaining laws are true at the same time, so it is hard to think of the one as causing the other.

To understand Fodor's antireductionist position, let us consider the following derivation of a higher-level law:

If A_i then B_i (for each $i = 1, 2, \ldots, n$).
If A_1 or A_2 or $\ldots$ or A_n, then B_1 or B_2 or $\ldots$ or B_n.
P iff A_1 or A_2 or $\ldots$ or A_n.
Q iff B_1 or B_2 or $\ldots$ or B_n.

If P then Q.

The first premise describes a set of lower-level laws; the second premise follows from the first. The third and fourth premises state bridge principles that connect a property discussed in a higher-level science with its multiple, lower-level, realizations. By assumption, the premises are true and the conclusion follows from the premises. Why, then, is this derivation not an explanation of the higher-level law?

Fodor's answer is not that the premises involve concepts that come from the higher-level science. Given that the higher-level science and the lower-level science use different vocabularies, any derivation of the one from the other must include bridge principles that bring those different vocabularies into contact (Nagel 1961). Rather, Fodor's reason is that laws cannot be disjunctive. Although he grants that each statement of the form "if A_i then B_i" is a law, he denies that the second premise expresses a law. For the same reason, the third and fourth premises also fail to express laws. To reduce a law, one must explain why the proposition is not just true, but is a law; this is supposed to mean that one must derive it solely from lawful propositions. This is why Fodor thinks that multiple realizability defeats reductionism.

Even if laws cannot be disjunctive, why does the above derivation fail to explain why "if P then Q" is a law? After all, the conclusion will be nomologically necessary if the premises are, and Fodor does not deny that the premises are necessary. Are we really prepared to say that the truth and lawfulness of the higher-level generalization is *inexplicable*, just because the above derivation is peppered with the word "or"? I confess that I feel my sense of incomprehension and mystery palpably subside when I contemplate this derivation. Where am I going wrong?

It also is not clear that laws must be nondisjunctive, nor is it clear what this requirement really amounts to. Take a law that specifies a quantitative threshold for some effect – for example, the law that water at a certain pressure will boil if the ambient temperature exceeds 100°C. This law seems to be disjunctive – it says that water will boil at 101°C, at 102°C, and so on. Of course, we have a handy shorthand for summarizing these disjuncts; we just say that any temperature "above 100°C" will produce boiling water. But if this strategy suffices to render the law about water nondisjunctive, why can't we introduce the letter α to represent the disjunction "A_1 or A_2 or ... or A_n" and β to represent the disjunction "B_1 or B_2 or ... or B_n"? It may be replied that the different disjuncts in the law about water all bring about boiling by the same type of physical process, whereas the different physical realizations A_i that the higher-level property P might have are heterogeneous in the way they bring about the B_i's that are realizations of Q.[12] The point is correct, but it remains unclear why this shows that laws cannot be disjunctive.

Disjunctiveness makes sense when it is understood as a *syntactic* feature of sentences. However, what does it mean for a proposition to be disjunctive, given that the same proposition can be expressed by different sentences? The problem may be illustrated by way of a familiar example. Suppose that the sentence "every emerald is green" and the sentence "every emerald is grue and the time is before the year 2000, or every emerald is bleen and the time is after the year 2000" are equivalent by virtue of the definitions of the terms "grue" and "bleen" (Goodman 1965). If laws are language-independent propositions of a certain type, and if logically equivalent sentences pick out the same proposition, then both sentences express laws, or neither does. Nothing changes if green is a natural kind whereas grue and bleen are not.

Although Fodor (1975) does not mention grue and bleen, it is fairly clear that his thinking about natural kinds – and his horror of disjunctions – both trace back to that issue.[13] Goodman (1965) held that law-like generalizations are confirmed by their positive instances, whereas accidental generalizations are not. The statement "all emeralds are green" is supposed to be lawlike, and hence instance confirmable, in virtue of the fact that "emerald" and "green" name natural kinds (or are "projectible"); "all emeralds are grue," on the other hand, is supposed to be non-lawlike, and so not confirmable by its instances, because it uses the weird predicate "grue." However, sub-

sequent work on the confirmation relation has thrown considerable doubt on the idea that all and only the lawlike statements are instance confirmable (see, e.g., Sober 1988).

If P and (A_1 or A_2 or ... or A_n) are known to be nomologically equivalent, then any probabilistic model of confirmation that takes that knowledge into account will treat them as *confirmationally* equivalent. For example, if a body of evidence confirms the hypothesis that a given individual has P, then that evidence also confirms the hypothesis that the individual has (A_1 or A_2 or ... or A_n). This is a feature, for example, of Bayesian theories of confirmation (on which, e.g., see Howson and Urbach 1989 and Earman 1992). Disjunctiveness has no special meaning within that framework.

Fodor (1975, 21) concedes that the claim that laws must be nondisjunctive is "not strictly mandatory," but then points out that "one denies it at a price." The price is that one loses the connection between a sentence's expressing a law and the sentence's containing kind predicates. "One thus inherits the need for an alternative construal of the notion of a kind"; I am with Fodor when he says that he does not "know what that alternative would be like" (22). Fodor is right here, but his argument is prudential, not evidential. Like Pascal, Fodor is pointing out the disutility of denying a certain proposition, but this is not to show that the proposition is true.

The multiple realizability argument against the reducibility of laws is sometimes formulated by saying that the disjunctions that enumerate the possible realizations of P and Q are "open-ended." This would defeat the derivation described above – the third and fourth premises would be false – but it is important to see that the rules of the game now have changed. The mere fact that P and Q are multiply realizable would no longer be doing the work. And if the point about "openendedness" is merely epistemological (we now do not *know* all of the physical realizations that P and Q have), it is irrelevant to the claim that higher-level sciences are reducible *in principle*.[14]

5. Probabilistic Explanations

The multiple realizability argument is usually developed by considering deterministic laws.

However, laws in many sciences are probabilistic. How would the argument be affected by assuming that P and Q are probabilistically related, and that the A_i and the B_i are too?

Suppose that A_1 and A_2 are the only two possible realizations that P can have, and that B_1 and B_2 are the only two realizations that Q can have (the points I'll make also hold for $n > 2$). Suppose further that the probabilistic law connecting P to Q has the form

$$\Pr(Q \mid P) = p.$$

Then it follows that

$$p = \Pr(Q \mid P) = \Pr(Q \mid A_1)\Pr(A_1 \mid P) + \Pr(Q \mid A_2)\Pr(A_2 \mid P).$$

If we substitute $p_1 = \Pr(Q \mid A_1)$ and $p_2 = \Pr(Q \mid A_2)$ into this expression, we obtain

$$p = (p_1)\Pr(A_1 \mid P) + (p_2)\Pr(A_2 \mid P).$$

The probability (p) described in the higher-level law is a *weighted average* of the two probabilities p_1 and p_2; the weighting is determined simply by how often systems with P happen to deploy one micro-realization rather than the other.

It is not inevitable that $p = p_1 = p_2$. For example, suppose that smoking (P) makes lung cancer (Q) highly probable and that cigarette smoke always contains one of two carcinogenic ingredients (A_1 or A_2), which are found only in cigarette smoke. It can easily turn out that one of these ingredients is more carcinogenic than the other.[15] This means that there can be an important difference between higher-level and lower-level explanations of the same event – they may differ in terms of the probabilities that *explanans* confers on *explanandum*. To see why, let us add one more detail to the example. Suppose that lung cancer can be realized by one of two types of tumor (B_1 or B_2) growing in the lungs. Given this, consider an individual who has lung cancer. How are we to explain why this person has that disease? One possible reply is to say that the person smoked cigarettes. A second possibility is to say that the cancer occurred because the person inhaled ingredient A_1. Putnam's multiple realizability argument entails that the second

suggestion is either no explanation at all, or is a "terrible" explanation. I suggest, however, that it should be clear to the unjaundiced eye that the second explanation may have its virtues. Perhaps A_1 confers on lung cancer a different probability from the one entailed by A_2 ($p_1 \neq p_2$), and so the first account entails a different probability of cancer than the second ($p \neq p_1$). Furthermore, perhaps A_1 and A_2 confer different probabilities on the two tumors B_1 and B_2 and these tumors respond differently to different treatments. The additional details provided by the micro-explanation are not stupid and irrelevant. They make a difference – to the probability of the *explanandum*, and to much else.[16] Perhaps it is a good thing for cancer research that the multiple realizability argument has not won the hearts of oncologists.

6.   Inference to the Best Explanation

I suspect that the multiple realizability argument has exerted so much influence because of a widespread misunderstanding concerning how *inference to the best explanation* works. The rough idea behind this mode of inference is that one should accept or reject hypotheses by deciding whether they are needed to explain observed phenomena. This inferential procedure seems to bear on the issue of reductionism as follows; We *now* need statements formulated in higher-level sciences because present day physics is not able to tell us how to understand societies, minds, and living things. However, if reductionism is correct, then these higher-level statements will not be needed once we have an ideally complete physics, and so they *then* should be rejected. But surely an ideally complete physics would not make it reasonable to reject all statements in higher-level sciences. This means that those statements must be needed to explain something that statements in an ideal physics could not explain. The multiple realizability argument presents itself as a diagnosis of why this is so.

This line of argument rests on a misunderstanding of inference to the best explanation. If you think that A_1 is one of the micro-realizations that P has, then you should not view "P causes

Q" and "A_1 causes Q" as competing hypotheses (Sober 1999). The evidence you have may justify accepting both. Inference to the best explanation is a procedure that belongs to the context of justification. Once you have used that technique to accept a variety of different hypotheses, it is perfectly possible that your set of beliefs will furnish several explanations of a given phenomenon, each perfectly compatible with the others. Some of those explanations will provide more details while others will provide fewer. Some may cite proximal causes while others will cite causes that are more distal. The mistake comes when one applies the principle of inference to the best explanation a *second* time – to the set of hypotheses one *already* believes, and rejects hypotheses that one does not "need" for purposes of explanation. Inference to the best explanation is a rule for deciding what to believe; it is not a principle for retaining or eliminating beliefs that one already has perfectly good evidence for accepting. If hypotheses in higher-level sciences can be accepted on the basis of evidence, they will not be cast into the outer darkness simply because physics expands.

It is worth bearing in mind that the phrase "inference to 'the' best explanation" can be misleading. The hypothesis singled out in such inferences is not the best of all explanations (past, present, and future) that could be proposed; it is merely the best of the competing hypotheses under evaluation. Hypothesis testing is essentially a contrastive activity; a given hypothesis is tested by testing it *against* one or more alternatives (Sober 1994). When psychological hypotheses compete against each other, inference to the best explanation will select the best of the competitors; of necessity, the winner in this competition will be a psychological hypothesis, because all the competitors are. Likewise, when physicalistic explanations of a behavior compete against each other, the resulting selection will, of course, be a physicalistic explanation. It is perfectly consistent with these procedures that a given phenomenon should have a psychological *and* a physicalistic explanation. Both reductionists and antireductionists go wrong if they think that the methods of science force one to choose among hypotheses that, in fact, are not in competition at all.[17]

7. Two Other Criticisms of the Multiple Realizability Argument

The multiple realizability argument, when it focuses on the explanation of singular occurrences, has three premises:

> Higher-level sciences describe properties that are multiply realizable and that provide good explanations.

> If a property described in a higher-level science is multiply realizable at a lower level, then the lower-level science will not be able to explain, or will explain only feebly, the phenomena that the higher-level science explains well.

> If higher-level sciences provide good explanations of phenomena that lower-level sciences cannot explain, or explain only feebly, then reductionism is false.

> Reductionism is false.

I have criticized the second premise, but the first and third have not escaped critical scrutiny (see, e.g., Lewis 1969, Churchland 1982, Enç 1983, and Kim 1989; Bickle 1998 provides a useful discussion). I will consider these other objections separately.

Philosophers with eliminitivist leanings have criticized the first premise. They have suggested that if "pain," for example, is multiply realizable, then it probably does not have much explanatory power. Explanations that cite the presence of "pain" will be decidedly inferior to those that cite more narrow-gauged properties, such as "human pain," or "pain with thus-and-such a neural realization." Philosophers who advance this criticism evidently value explanations for being deep, but not for being general. I disagree with this one-dimensional view, just as I disagree with the multiple realizability argument's single-minded valuation of generality at the expense of depth. Higher-level explanations often provide fewer explanatory details, but this does not show that they are inferior *tout court*.

It might interest philosophers of mind who have these worries about multiply realized psychological properties to consider the multiply realized properties discussed in evolutionary biology. In cognitive science, it is difficult to point to many present-day models that are well-confirmed and that are articulated by describing multiply realizable properties; this is mostly a hoped-for result of scientific advance. However, in evolutionary biology, such models are extremely common. Models of the evolution of altruism (Sober and Wilson 1998), for example, use the concept of fitness and it is quite clear that fitness is multiply realizable. These models have a useful generality that descriptions of the different physical bases of altruism and selfishness would not possess.

The third premise in the multiple realizability argument also has come in for criticism. Perhaps *pain* is multiply realizable, but *human pain* may not be. And if *human pain* is multiply realizable, then some even more circumscribed type of pain will not be. What gets reduced is not pain in general, but specific physical types of pain (Nagel 1965). The multiple realizability argument is said to err when it assumes that reductionism requires *global* reduction; *local* reduction is all that reductionism demands. To this objection, a defender of the multiple realizability argument might reply that there are many questions about reduction, not just one. If human pain gets reduced to a neurophysiological state, but pain in general does not, then reductionism is a correct claim about the former, but not about the latter. If psychology provides explanations in which pain – and not just *human* pain – is an *explanans*, then reductionism fails as a claim about *all* of psychology.

Scientists mean a thousand different things by the term "reductionism." Philosophers have usually been unwilling to tolerate this semantic pluralism, and have tried to say what reductionism "really" is. This quest for univocity can be harmless as long as philosophers remember that what they call the "real" problem is to some degree stipulative. However, philosophers go too far when they insist that reductionism requires local reductions but not global reductions. There are many reductionisms – focusing on one should not lead us to deny that others need to be addressed.

8. A Different Argument Against a Different Reductionism

Although the multiple realizability argument against reductionism began with the arguments by Putnam and Fodor that I have reviewed, more recent appeals to multiple realizability sometimes take a rather different form. The claim is advanced that higher-level sciences "capture patterns" that would be invisible from the point of view of lower-level science. Here the virtue attributed to the higher-level predicate "P" is not that it *explains* something that the lower-level predicate "A_1" cannot explain, but that the former *describes* something that the latter does not. The predicate "P" describes what the various realizations of the property P have in common. The disjunctive lower-level predicate "A_1 or A_2 or ... or A_n," does not do this in any meaningful sense. If I ask you what pineapples and prime numbers have in common and you reply that they both fall under the disjunctive predicate "pineapple or prime number," your remark is simply a joke. As a result, "if P then Q" is said to describe a regularity that "if (A_1 or A_2 or ... or A_n) then (B_1 or B_2 or ... or B_n)" fails to capture.

Whether or not this claim about the descriptive powers of higher- and lower-level sciences is right, it involves a drastic change in subject. Putnam and Fodor were discussing what higher- and lower-level sciences are able to *explain*. The present argument concerns whether a lower-level science is able to *describe* what higher-level sciences *describe*. I suspect that this newer formulation of the multiple realizability argument has seemed to be an elaboration, rather than a replacement, of the old arguments in part because "capturing a pattern" (or a generalization) has seemed to be more or less equivalent with "explaining a pattern" (or a generalization). However, there is a world of difference between describing a fact and explaining the fact so described. This new argument does not touch the reductionist claim that physics can explain everything that higher-level sciences can explain.

9. Concluding Comments

Higher-level sciences often provide more general explanations than the ones provided by lower-level sciences of the same phenomena. This is the kernel of truth in the multiple realizability argument – higher-level sciences "abstract away" from the physical details that make for differences among the micro-realizations that a given higher-level property possesses. However, this does not make higher-level explanations "better" in any absolute sense. Generality is one virtue that an explanation can have, but a distinct – and competing – virtue is depth, and it is on this dimension that lower-level explanations often score better than higher-level explanations. The reductionist claim that lower-level explanations are *always* better and the antireductionist claim that they are *always* worse are both mistaken.

Instead of claiming that lower-level explanations are always better than higher-level explanations of the same phenomenon, reductionistic might want to demure on this question of better and worse, and try to build on the bare proposition that physics in principle can explain any singular occurrence that a higher-level science is able to explain. The level of detail in such physical explanations may be more than many would want to hear, but a genuine explanation is provided nonetheless, and it has a property that the multiple realizability argument has overlooked. For reductionistic, the interesting feature of physical explanations of social, psychological, and biological phenomena is that they use the same basic theoretical machinery that is used to explain phenomena that are nonsocial, nonpsychological, and nonbiological. This is why reductionism is a thesis about the *unity* of science. The special sciences unify by abstracting away from physical details; reductionism asserts that physics unifies because everything can be explained, and explained *completely*, by adverting to physical details. It is ironic that "unification" is now a buzz word for antireductionists, when not so long ago it was the *cri de coeur* of their opponents.

To say that physics is capable in principle of providing a complete explanation does not mean that physical explanations will mention everything that might strike one as illuminating. As noted above, the explanations formulated by higher-level sciences can be illuminating, and physics will not mention *them*. Illumination is to some degree in the eye of the beholder; however, the sense in which physics can provide complete explanations is supposed to be perfectly objective. If we focus on *causal* explanation, then an objective notion of explanatory completeness is provided by the concept of *causal completeness*:

$$\text{Pr(higher-level properties at } t_2 \mid \text{physical properties at } t_1 \text{ \& higher-level properties at } t_1)$$
$$= \text{Pr(higher-level properties at } t_2 \mid \text{physical properties at } t_1).$$

To say that physics is causally complete means that (a complete description of) the physical facts at t_1 *determines* the probabilities that obtain at t_1 of later events; adding information about the higher-level properties instantiated at t_1 makes no difference.[18] In contrast, multiple realizability all but guarantees that higher-level sciences are causally incomplete:

$$\text{Pr(higher-level properties at } t_2 \mid \text{physical properties at } t_1 \text{ \& higher-level properties at } t_1)$$
$$\neq \text{Pr(higher-level properties at } t_2 \mid \text{higher-level properties at } t_1).$$

If A_1 and A_2 are the two possible realizations of P, then one should not expect that $\text{Pr}(Q \mid P \& A_1) = \text{Pr}(Q \mid P \& A_2) = \text{Pr}(Q \mid P)$ (Sober 1999).

Is physics causally complete in the sense defined? It happens that causal completeness follows from the thesis of simultaneous determination described earlier (Sober 1999). This fact does not settle whether physics *is* causally complete, but merely pushes the question back one step. Why think that the physical facts that obtain at a given time determine all the nonphysical facts that obtain at that time? This is a question I will not try to answer here. However, it is worth recalling that defenders of the multiple realizability argument usually assume that the lower-level physical properties present at a time determine the higher-level properties that are present at that same time. This commits them to the thesis of the causal completeness of physics. If singular occurrences can be explained by citing their causes, then the causal completeness of physics insures that physics has a variety of explanatory completeness that other sciences do not possess. This is reductionism of a sort, though not the sort that the multiple realizability argument aims to refute.

Notes

1   Putnam (1967, 1975) and Fodor (1968, 1975) formulated this argument with an eye to demonstrating the irreducibility of psychology to

physics. It has been criticized by Lewis (1969), Churchland (1982), Enç (1983), and Kim (1989), but on grounds distinct from the ones to be developed here. Their criticisms will be discussed briefly towards the end of this paper.

The multiple realizability argument was first explored in philosophy of biology by Rosenberg (1978, 1985), who gave it an unexpected twist; he argued that multiple realizability entails a kind of reductionism (both about the property of fitness and also about the relation of classical Mendelian genetics to molecular biology). In contrast, Sober (1984) and Kitcher (1984) basically followed the Putnam/Fodor line. The former work argues that the multiple realizability of fitness entails the irreducibility of theoretical generalizations about fitness; the latter argues for the irreducibility of classical Mendelian genetics to molecular biology. Waters (1990) challenges the specifics of Kitcher's argument; much of what he says is consonant with the more general criticisms of the multiple realizability argument to be developed here. Sober (1993) defends reductionism as a claim about singular occurrences, but denies that it is correct as a claim about higher-level laws.

2   If some computers (now or in the future) have minds, then the reducibility of psychology to biology may need to be revised (if the relevant computers are not "alive"); the obvious substitute is to have reductionism assert that psychology reduces to a physical science. Similarly, if some societies are made of mindless individuals (consider, for example, the case of the social insects), then perhaps the reduction will have to "skip a level" in this instance also.

3   Reductionism should not be formulated so that it is committed to individualism of the sort discussed in philosophy of mind. For example, if wide theories of content are correct, then the beliefs that an individual has at a time depend not just on what is going on inside the skin of that individual at that time, but on what is going on in the individual's environment, then and earlier.

4   Although multiple realizability induces an asymmetry between P and each A_1 it does not entail that there is an asymmetry between P and the disjunctive property (A_1 or A_2 or . . . or A_n). Fodor would say that this disjunctive predicate fails to pick out a natural kind, a point that will be discussed later.

5   Here I waive the question of whether *all* explanations are causal explanations, on which see Sober 1983 and Lewis 1986.

6   I will mention two. The first concerns how this theory of causation analyzes putative cases of overdetermination by multiple actual causes.

Suppose Holmes and Watson each simultaneously shoot Moriarty through the heart. The theory entails that Holmes did not cause Moriarty's death, and Watson did not either. Rather, the cause is said to be disjunctive – Holmes shot him or Watson did. The second question comes from thinking about the possibility of indeterministic causation. Just as the totality of the antecedent causal facts need not suffice for the effect to occur, so the effect could have happened even if the causes had been different.

7   I am grateful to Brian Mclaughlin for drawing my attention to this line of argument.

8   It is worth considering a curious remark that Putnam makes in a footnote before he introduces the example of the peg and board. He says:

> Even if it were not physically possible to realize human psychology in a creature made of anything but the usual protoplasm, DNA, etc., it would still not be correct to say that psychological states are identical with their physical realizations. For, as will be argued below, such an identification has no *explanatory* value in *psychology*. (1975, 293)

He then adds the remark: "on this point, compare Fodor, 1968," presumably because Fodor thought that antireductionism depends on higher-level properties being *multiply* realizable.

If we take Putnam's remark seriously, we must conclude that he thinks that the virtue of higher-level explanations does not reside in their greater generality. If a higher-level predicate (P) has just one possible physical realization (A_1), then P and A_1 apply to exactly the same objects. Putnam presumably would say that citing A_1 in an explanation provides extraneous information, whereas citing P does not. It is unclear how this concept of explanatory relevance might be explicated. In any event, I have not taken this footnote into account in describing the "multiple realizability argument," since Putnam's point here seems to be that *multiple* realizability does not bear on the claims he is advancing about explanation. This is not how the Putnam/Fodor argument has been understood by most philosophers.

9   I grant this point for the sake of argument, but it bears looking at more closely. Intuitively, "if P then Q" is more general than "if A_1 then B_1," because the extension of P properly contains the extension of A_1. However, each of these conditionals is logically equivalent with its contrapositive, and it is equally true that the extension of not-B_1 properly contains the extension of not-Q. This point is not a mere logical trick, to be swept aside by saying that the "right" formulation of a law is one that uses predicates that name natural kinds. After all,

some laws (specifically, zero force laws) are typically stated as conditionals but their applications usually involve the predicates that occur in the contrapositive formulation. For example, the Hardy–Weinberg Law in population genetics describes how gamete frequencies will be related to genotype frequencies when no evolutionary forces are at work; its typical applications involve noting a departure from Hardy–Weinberg genotype frequencies, with the conclusion being drawn that some evolutionary forces are at work (Sober 1984). To say that the Hardy–Weinberg law has zero generality because every population is subject to evolutionary forces is to ignore the standard way in which the law is applied, and applied frequently, to nature.

10   Putnam's argument also has implications about the explanatory point of citing distal and proximate causes of a given effect. Imagine a causal chain from C_d to C_p to E. Suppose that C_d suffices for the occurrence of C_p, but is not necessary, and that the only connection of C_d to E is through C_p. Then Putnam's argument apparently entails that C_p explains E, and that C_d is either not an explanation of E, or is a terrible explanation of that event. But surely there can be an explanatory point to tracing an effect more deeply into the past. And surely it does not automatically increase explanatory power to describe more and more proximate causes of an effect.

11   The distinction between justification and explanation was clearly drawn by Hempel (1963), who points out that why-questions can be requests for evidence or requests for explanation. This distinction supplements the familiar logical empiricist distinction between the *context of discovery* and the *context of justification*.

12   Fodor (1998, 16) says that a disjunction may occur in a bridge law if and only if the disjunction is "independently certified," meaning that "it also occurs in laws at its own level." The disjunction in the law about boiling presumably passes this test.

13   See, for example, Davidson's (1966) discussion of "all emeroses are gred" and also Davidson 1970.

14   Moreover, the multiple realizability argument is not needed to show that the thesis of *reducibility in practice* is false; one can simply inspect present-day science to see this.

15   If laws must be time-translationally invariant, then it is doubtful that "$\Pr(Q \mid P) = p$" expresses a law, if P is multiply realizable (Sober 1999).

16   This argument would not be affected by demanding that a probabilistic explanation must cite the positive and negative causal factors that raise and lower the probability of the *explanandum* (see, e.g., Salmon 1984). Cigarette smoke may raise the

probability of lung cancer to a different extent than inhaling A_1 does, and so the two explanations will differ in important ways.

17   This point bears on an argument that Fodor (1998) presents to supplement his (1975) argument against reductionism. I am grateful to Fodor for helping me to understand this new argument. Fodor compares two hypotheses (which I state in the notation I have been using): (i) "if (A_1 or A_2 or ... or A_n), then Q" and (ii) "if (A_1 or A_2 or ... or A_n) then P (because the A_i's are possible realizations of P), and if P then Q." Fodor points out that the latter generalization is logically stronger (19); he then claims that it is sound inductive practice to "prefer the strongest claim compatible with the evidence, all else being equal" (20). Since we should accept the stronger claim instead of the weaker one, Fodor concludes that reductionism is false.

I have three objections to this argument. First, I do not think that the two generalizations are in competition with each other. If one thinks that the first conditional is true, and wants to know whether, in addition, it is true that the A_i's are realizations of P, then the proper competitor for this conjecture is that at least one of the A_i's is *not* a realization of P. Second, even if the two hypotheses were competitors, Fodor's Popperian maxim is subject to the well-known "tacking problem" – that irrelevant claims can be conjoined to a well-confirmed hypothesis to make it logically stronger. Fodor, of course, recognizes that $H \& I$ is not always preferable to H, ceteris paribus; however, he thinks that a suitably clarified version of the maxim he describes is plausible and that it will have the consequence he says it has for the example at hand. I have my doubts. It is illuminating, I think, to compare this inference problem to a structurally similar problem concerning intervening variables. If the A_i's are known to cause Q, should one postulate a variable (P) that the A_i's cause, and which causes Q? I do not think that valid inductive principles tell one to prefer the intervening variable model over one that is silent on the question of whether the intervening variable exists, when both models fit the data equally well (see Sober 1998 for further discussion). Third, even if the stronger hypothesis should be accepted in preference to the weaker one, I do not see that this refutes reductionism (though it does refute "eliminativist reductionism"). After all, the reductionist can still maintain that "if P then Q" is explained by theories at the lower level.

Notice that Fodor's argument does not depend on whether the A_i's listed are some or all of the possible realizations that P can have; it also does not matter whether the modality involved is metaphysical or nomological. Notice, finally, that this argument concerns inductive inference (the "context of justification," mentioned earlier), not explanation, which is why it differs from the argument of Fodor 1975.

18 Let M = all the higher-level properties a system has at time t_1. Let P = all the physical properties that the system has at t_1. And let B = some property that the system might have at the later time t_2. We want to show that

$$\Pr(M \mid P) = 1.0$$

entails

$$\Pr(B \mid P) = \Pr(B \mid P \, \& \, M).$$

First note that $\Pr(B \mid P)$ can be expanded as follows:

$$
\begin{aligned}
\Pr(B \mid P) &= \Pr(B \, \& \, P)/\Pr(P) \\
&= [\Pr(B \, \& \, P \, \& \, M) + \Pr(B \, \& \, P \, \& \\
&\quad \text{not-}M)]/\Pr(P) \\
&= [\Pr(B \mid P \, \& \, M)\Pr(P \, \& \, M) + \\
&\quad \Pr(B \, \& \, \text{not-}M \mid P)\Pr(P)]/\Pr(P) \\
&= \Pr(B \mid P \, \& \, M)\Pr(M \mid P) + \Pr(B \, \& \\
&\quad \text{not-}M \mid P)
\end{aligned}
$$

From this last equation, it is clear that if $\Pr(M \mid P) = 1.0$, then $\Pr(B \mid P) = \Pr(B \mid P \, \& \, M)$.

References

Berlin, Isaiah (1953), *The Hedgehog and the Fox*. New York: Simon and Shuster.

Bickle, John (1998), *Psychoneural Reduction: The New Wave*. Cambridge, MA: MIT Press.

Churchland, Paul (1982), "Is 'Thinker' a Natural Kind?", *Dialogue* 21: 223–238.

Davidson, Donald (1966), "Emeroses by Other Names", *Journal of Philosophy* 63: 778–780. Reprinted in *Essays on Actions and Events*. Oxford: Oxford University Press, 1980, 225–227.

Davidson, Donald (1970), "Mental Events", in L. Foster and J. Swanson (eds.), *Experience and Theory*. London: Duckworth. Reprinted in *Essays on Actions and Events*. Oxford: Oxford University Press, 1980, 207–225.

Earman, John (1992), *Bayes or Bust?: A Critical Examination of Bayesian Confirmation Theory*. Cambridge, MA: MIT Press.

Enç, Berent (1983), "In Defense of the Identity Theory", *Journal of Philosophy* 80: 279–298.

Fodor, Jerry (1968), *Psychological Explanation*. Cambridge, MA: MIT Press.

Fodor, Jerry (1975), *The Language of Thought*. New York: Thomas Crowell.

Fodor, Jerry (1998), "Special Sciences – Still Autonomous After All These Years", in *In Critical Condition: Polemical Essays on Cognitive Science and the Philosophy of Mind*. Cambridge, MA: MIT Press.

Forster, Malcolm and Elliott Sober (1994), "How to Tell When Simpler, More Unified, or Less *Ad Hoc* Theories Will Provide More Accurate Predictions", *British Journal for the Philosophy of Science* 45: 1–35.

Goodman, Nélson (1965), *Fact, Fiction, and Forecast*. Indianapolis: Bobbs-Merrill.

Hanson, N. Russell (1958), *Patterns of Discovery*. Cambridge: Cambridge University Press.

Hempel, Carl (1965), "Aspects of Scientific Explanation", in *Aspects of Scientific Explanation and Other Essays in the Philosophy of Science*. New York: Free Press.

Howson, Colin and Peter Urbach (1989), *Scientific Reasoning: The Bayesian Approach*. La Salle: Open Court.

Kim, Jaegwon (1989), "The Myth of Nonreductive Materialism", *Proceedings and Addresses of the American Philosophical Association* 63: 31–47. Reprinted in *Supervenience and Mind*. Cambridge: Cambridge University Press, 1993.

Kim, Sungsu (unpublished), "Physicalism, Supervenience, and Causation – a Probabilistic Approach".

Kitcher, Philip (1984), "1953 and All That: A Tale of Two Sciences", *Philosophical Review* 93: 335–373. Reprinted in E. Sober (ed.), *Conceptual Issues in Evolutionary Biology*. Cambridge, MA: MIT Press, 1994, 379–399.

Lewis, David (1969), "Review of *Art, Mind, and Religion*", *Journal of Philosophy* 66: 22–27. Reprinted in N. Block (ed.), *Readings in Philosophy of Psychology*, vol. 1. Cambridge, MA: Harvard University Press, 1983, 232–233.

Lewis, David (1973a), "Causation", *Journal of Philosophy* 70: 556–567. Reprinted with a "Postscript" in D. Lewis, *Philosophical Papers*, vol. 2. Oxford: Oxford University Press, 1986, 159–213.

Lewis, David (1973b), *Counterfactuals*. Cambridge, MA: Harvard University Press. Revised edition 1986.

Lewis, David (1986), "Causal Explanation", in D. Lewis, *Philosophical Papers*, vol. 2. Oxford: Oxford University Press, 214–240.

Nagel, Ernest (1961), *The Structure of Science*. New York: Harcourt Brace.

Nagel, Thomas (1965), "Physicalism", *Philosophical Review* 74: 339–356.

Oppenheim, Paul, and Hilary Putnam (1958), "Unity of Science as a Working Hypothesis", in H. Feigl, G. Maxwell, and M. Scriven (eds.), *Minnesota Studies in the Philosophy of Science*, Minneapolis: University of Minnesota Press, 3–36.

Putnam, Hilary (1967), "Psychological Predicates", in W. Capitan and D. Merrill (eds.), *Art, Mind, and Religion*. Pittsburgh: University of Pittsburgh Press, 37–48. Reprinted as "The Nature of Mental States" in *Mind, Language, and Reality*. Cambridge: Cambridge University Press, 1975, 429–440.

Putnam, Hilary (1975), "Philosophy and our Mental Life", in *Mind, Language, and Reality*. Cambridge: Cambridge University Press, 291–303.

Rosenberg, Alexander (1978), "The Supervenience of Biological Concepts", *Philosophy of Science* 45: 368–386.

Rosenberg, Alexander (1985), *The Structure of Biological Science*. Cambridge: Cambridge University Press.

Salmon, Wesley (1984), *Scientific Explanation and the Causal Structure of the World*. Princeton: Princeton University Press.

Sober, Elliott (1983), "Equilibrium Explanation", *Philosophical Studies* 43: 201–210.

Sober, Elliott (1984), *The Nature of Selection: Evolutionary Theory in Philosophical Focus*. Cambridge, MA: MIT Press. 2nd edition, University of Chicago Press, 1994.

Sober, Elliott (1988), "Confirmation and Lawlikeness", *Philosophical Review* 97: 93–98.

Sober, Elliott (1994), "Contrastive Empiricism", in *From a Biological Point of View*. Cambridge: Cambridge University Press, 114–135.

Sober, Elliott (1998), "Black Box Inference: When Should an Intervening Variable be Postulated?", *British Journal for the Philosophy of Science* 49: 469–498.

Sober, Elliott (1999), "Physicalism from a Probabilistic Point of View", *Philosophical Studies* 95: 135–174.

Sober, Elliott and David S. Wilson (1998), *Unto Others: The Evolution and Psychology of Unselfish Behavior*. Cambridge, MA: Harvard University Press.

Stalnaker, Robert (1968), "A Theory of Conditionals", in N. Rescher (ed.), *Studies in Logical Theory*. Oxford: Blackwell, 98–112.

Waters, Kenneth (1990), "Why the Antireductionist Consensus Won't Survive the Case of Classical Mendelian Genetics", *PSA 1990*. E. Lansing, MI: Philosophy of Science Association, 125–139. Reprinted in E. Sober (ed.), *Conceptual Issues in Evolutionary Biology*. Cambridge, MA: MIT Press, 1994, 402–417.

PART VIII

SPECIES AND CLASSIFICATION PROBLEMS

Introduction

From Aristotle to Darwin, it was held that species were fixed, unchanging, and had a set of essential properties that identify conditions for species membership. This species essentialism was reflected in the classification system invented a century before Darwin by Carl Linnaeus (1707–78), and still in use among biologists even after Darwin showed that species were not fixed nor had any essential properties. This, of course, produces a tension between biology's chief explanatory theory and its most fundamental taxonomy.

The first selection in this section (chapter 18) includes parts of a paper by Marc Ereshefsky which deals with both the nature of particular species and definition of the concept of species in general. In the paper, he addresses species essentialism and the case for taxonomic pluralism – the provision of many different taxonomies for different biological purposes, as well as alternatives to the Linnaean system of classification. As Ereshefsky notes, consistent with the work of others presented here in this anthology, "conceptual issues in biological taxonomy often straddle the boundary between biology and philosophy."

Without a doubt, the most influential contemporary account of what a species is in general is due to Ernst Mayr, who elaborated the so-called *biological species concept* (BSC). According to the BSC, a species is a population of organisms that can interbreed with one another but are reproductively isolated from – cannot breed with – other organisms. Despite its power, this definition of a species suffers from obvious problems. To begin with, it is inapplicable to asexual species. And there are many problems of borderline cases as one species anagenically evolves into another one. In the second selection in this section (chapter 19), Jerry Coyne and H. Allen Orr explore the many alternatives to the BSC which vie for attention in the biological sciences. The outcome of this debate must have considerable influence on how philosophy treats natural kinds throughout the sciences.

Further Reading

Baker, J. (2005). Adaptive speciation: The role of natural selection in mechanisms of geographic and non-geographic speciation. *Studies in History and Philosophy of Biological and Biomedical Sciences*, 36, 303–326.

Coyne, J., & Orr, H. (2004). *Speciation*. Sunderland, MA: Sinauer Associates, Inc.

Ereshefsky, M. (1992). Eliminative pluralism. *Philosophy of Science*, 59, 671–690.

Ereshefsky, M. (2001). Names, numbers and indentations: A guide to post-Linnaean taxonomy. *Studies in the History and Philosophy of Biology and Biomedical Sciences*, 32, 361–383.

Ereshefsky, M. (2007). *The poverty of the Linnaean hierarchy: A philosophical study of biological taxonomy*. Cambridge: Cambridge University Press.

Ereshefsky, M., & Matthen, M. (2005). Taxonomy, polymorphism and history: An introduction to population structure theory. *Philosophy of Science*, 72, 1–21.

Mallet, J. (2004). Poulton, Wallace and Jordan: How discoveries in Papilio butterflies initiated a new species concept 100 years ago. *Systematics and Biodiversity*, 1, 441–452.

Mallet, J. (2005). Speciation in the 21st Century. *Heredity*, 95, 105–109.

Rice, W., & Hostert, E. (1993). Laboratory experiments on speciation: What have we learned in forty years? *Evolution*, 47, 1637–1653.

Richards, R. (2008). Species and taxonomy. In M. Ruse (ed.), *The Oxford handbook of philosophy of biology* (pp. 161–188). Oxford: Oxford University Press.

Rosenberg, A., & McShea, D. (2008). *Philosophy of biology: A contemporary introduction* (Chapters 2 and 6). London: Routledge.

Schluter, D., & Nagel, L. (1995). Parallel speciation by natural selection. *American Naturalist*, 146, 292–301.

Sepkoski, D. (2008). Macroevolution. In M. Ruse (ed.), *The Oxford handbook of philosophy of biology* (pp. 211–237). Oxford: Oxford University Press.

Sterelny, K., & Griffiths, P. (1999). *Sex and death: An introduction to philosophy of biology*. Chicago: University of Chicago Press.

Weinberg, J., Starczak, V., & Jora, P. (1992). Evidence for rapid speciation following a founder event in the laboratory. *Evolution*, 46, 1214–1220.

Wilson, R. (ed.) (1999). *Species: New interdisciplinary essays*. Cambridge, MA: MIT Press.

18

Species, Taxonomy, and Systematics

Marc Ereshefsky

1. Introduction

Conceptual issues in biological taxonomy often straddle the boundary between biology and philosophy. Consider questions over the nature of species. Are species natural kinds or individuals? Does 'species' refer to a category in nature or is it merely a useful device for organizing our biological knowledge? These questions involve elements of metaphysics, epistemology, and philosophy of science. These questions also turn on empirical information and biological theory. When we ask about the nature of species are we asking a philosophical or biological question? The answer is 'both.'

Three conceptual issues in biological taxonomy and systematics will be discussed in this chapter. One is the ontological status of species. Many if not most philosophers believe that species are natural kinds – classes of organisms with theoretically significant similarities. Other philosophers, and many biologists, believe that species are individuals akin to particular entities. The outcome of this debate has implications beyond biological taxonomy, for example, it has implications for our conception of human nature. Is *Homo sapiens* a kind defined by similarities among its members, or is *Homo sapiens* an evolving lineage defined

by genealogy? If species are individuals, then no qualitative trait is necessary and sufficient for being a human.

Another conceptual issue is taxonomic pluralism. A common scientific and philosophical view is that, from a God's eye perspective, there is a single correct classification of the organic world. A number of authors respond that this view is wrong. They argue that there is no single correct classification of the organic world and that the organic world itself is pluralistic. This issue has implications outside of taxonomy. In philosophy of science it raises the question of whether philosophers should promote scientific unity as an aim of science. If taxonomic pluralism occurs in biology then biology, and science as a whole, lacks unity.

A third conceptual issue concerns the Linnaean hierarchy: should biologists keep using the Linnaean hierarchy or should they adopt a new system of classification? The Linnaean hierarchy was developed prior to the Darwinian revolution. Many biologists and some philosophers believe that the Linnaean hierarchy faces pressing problems. Some even suggest junking the Linnaean hierarchy. Again, important issues outside of classification hinge on this debate. Most biodiversity studies are given in terms of Linnaean ranks,

Marc Ereshefsky, "Species, Taxonomy, and Systematics," in Mohan Matthen and Christopher Stephens (eds.) *Handbook for the Philosophy of Science, Philosophy of Biology* (Elsevier, 2007), pp. 403–427.

for example, the rank of family. If those ranks do not correspond to categories in nature, as many critics of the Linnaean hierarchy argue, then most biodiversity studies are based on faulty measures.

1.1. Terms and distinctions

We have already used the terms *classification, taxonomy*, and *systematics*, but what do they mean? What distinctions do they highlight? We are familiar with classifications: organisms are sorted into species, species are sorted into genera, genera into families, and so on up the Linnaean hierarchy. A classification is a hypothesis concerning how organisms or taxa are related. Taxonomy is the discipline that tells us how to sort organisms into taxa and taxa into more inclusive taxa. Taxonomic theory offers principles for constructing classifications. Systematics does not tell us how to construct classifications but studies how organisms and biological taxa are related in world. Classification, taxonomy, and systematics line up in the following way. Systematics, the study of the relations among organisms and taxa, guides our choice of taxonomic theory; taxonomic theory guides biologists in constructing classifications of the organic world. The path from systematics to classification, however, is far from smooth. Biologists do not agree on systematic theories, so they posit different taxonomic theories (i.e., different principles for constructing classifications). Moreover, even when there is agreement on taxonomic theory, how to implement that theory can be controversial. For example, biologists tend to prefer more parsimonious classifications, but they disagree on how to measure parsimony.

Another distinction is between *species taxa* and the *species category*. Species taxa are groups of organisms; *Homo sapiens* and *Drosophila melanogaster* (a fruit fly) are examples of species taxa. The species category is a more inclusive entity: it is the class of all species taxa. When biologists and philosophers discuss the definition of 'species' they are discussing the definition of the species category. In Section 2, we focus on the ontological status of species *taxa*: is a species taxon a kind or an individual? In Section 3, we focus on the species *category*: is there a single, unified species category? In Section 4 we move up the Linnaean hierarchy and discuss the reality of the other Linnaean categories – genera, families, and so on. Does the Linnaean hierarchy reflect a hierarchy of natural categories, or are the Linnaean ranks merely useful instruments for organizing information about the organic world?

2. Species

2.1. Essentialism

Traditionally philosophers have treated biological species as natural kinds with essences. This approach to species is found in the work of Kripke (1972) and Putnam (1975), and has its roots in Aristotle and Locke. Biologists have been essentialists as well. John Ray, Maupertuis, Bonnet, Linnaeus, Buffon, and Lamark all talk of classifying species taxa according to their species-specific essences (Hull 1965; Sober 1980; Mayr 1982).[1] The view that species are kinds with essences is part of a general view – kind essentialism. A natural kind is a class of entities that share a kind-specific essence. Such essences capture the fundamental structure of the world; or using Plato's phrase, they "carve nature at its joints."

Kind essentialism has two crucial tenets. First, all and only the members of kind share a common essential property. If an entity has a certain essence, it is member of a particular kind. If, for example, a piece of metal has a certain atomic structure, it is a member of the kind gold. However, some properties occur in all and only the members of kind yet they are not that kind's essence. The Aristotelian example is humans and the property featherless biped. All and only humans are featherless bipeds, nevertheless that property is not the essence of humans. Here is where the second tenet of essentialism comes into play. The essence of a kind is a property that plays a fundamental role in explaining the occurrence of other properties typically found among the members of a kind. For example, the atomic structure of gold causes pieces of gold to dissolve in acid and conduct electricity. That atomic structure, not any property that occurs in all and only pieces of gold, is the essence of the kind gold.

Both philosophers and biologists have applied kind essentialism to species. There are few essentialist biologists today, but essentialism remains a common view among philosophers. A number of philosophers and biologists have challenged species essentialism – the view that species are natural kinds with essences (Mayr 1959, Hull 1965, Ghiselin 1974, Sober 1980, Dupré 1981). Here are two lines of argument against species essentialism.

Recall the first tenet of essentialism: a property must occur in all and only the members of a kind. Biologists have been hard-pressed to find a biological trait that occurs in all and only the members of a particular species. Evolutionary theory explains why. Suppose a genetically based trait were found in all the members of a species. Mutation, random drift, and recombination can cause that trait not to occur in a future member of that species. All it takes is the disappearance of a trait in a single member of a species to show that it is not essential. The universality of a trait among the members of a species is quite fragile.

Evolutionary forces also undermine the uniqueness of a trait among the members of a species. Different species frequently live in similar habitats that cause the parallel evolution of similar traits in different species. Birds and bats, for example, each have wings, but the evolutionary path for each type of wing is distinct. Organisms in different species also share common ancestors, so they draw on common stores of genetic material and developmental constraints. These common genetic and developmental resources cause the members of different species to share similar characteristics. Just think of all those organisms in the world with four limbs. Just as the forces of evolution work against the universality of a trait in a species they also work against the uniqueness of trait in a species.

The above considerations pose a strong challenge to the essentialist requirement that a trait must occur in all and only the members of a species. This does not mean that the occurrence of a biological trait in all and only the members of species is empirically impossible. Nonetheless, consider what conditions must be met for that requirement to hold. For essentialism to work, a trait must occur in all the members of species, for the entire life of that species – from its initial speciation event to its extinction. Moreover, for a trait to qualify as an essential property of a species, it must occur only in that species. That

is, the trait in question cannot occur in any member of any other species for the entire span of life on this planet, indeed, for the entire span of life in this universe, This is a tall order! The occurrence of a trait in all and only the members of species is an empirical possibility. But given current biological theory, that possibility is unlikely.

Sober (1980), following Mayr (1969), offers a different objection to species essentialism. Sober illustrates how essentialist explanations have been replaced by evolutionary ones in biology, thereby rendering essentialist explanations theoretically obsolete. Suppose we want to explain variation in a population, say variation in height. The essentialist explanation cites the essence of the organisms in a species, and then cites ontogenetic interference that prevents the manifestation of that essence in some or all members. Organisms of that species would have the same height if they were exposed to the same environment, but they are exposed to different environments, hence there is variation from a common type. Variation is explained by essence and interference. In contrast, the evolutionist explains variation in a population without positing essences. Evolutionists cite gene frequencies within a population and the evolutionary forces that affect what frequencies occur in the next generation. Variation in height is explained by citing the relevant gene frequencies in one generation plus the occurrence of such evolutionary forces as selection and drift. No essences are posited. In contemporary biology, the positing of species essences has become theoretically superfluous.

2.2. Species as individuals

If species taxa are not natural kinds with essences, what is their ontological status? The prevailing view is that species are not natural kinds but individuals. Ghiselin (1974) and Hull (1978), the most prominent advocates of this view, contrast natural kinds and individuals in terms of spatiotemporal restrictedness. Membership in a kind requires that the members of a kind share a set of qualitative properties. A drop of liquid is water so long as it has the molecular structure H_2O. It does not matter where that liquid is located, whether it is on Earth now or in a distant galaxy in a million years. So long as that drop has the molecular structure H_2O it is water. Individuals, unlike kinds, consist of parts that are spatiotemporally restricted. Consider a paradigmatic individual, a mammalian organism. The parts of that organism cannot be scattered around the universe at diverse times if they are parts of a single living organism. Various biological processes, such as respiration and digestion, require those parts to be causally and spatiotemporally connected. The parts of a mammal must exist in a particular space-time region. Generalizing from these examples, the parts of an individual are spatiotemporally restricted, whereas the members of a kind are spatiotemporally unrestricted.

According to Ghiselin and Hull, membership in a species is spatiotemporally restricted, thus species are individuals and not natural kinds. What is their argument that species are spatiotemporally restricted entities? The argument starts with the assumption that 'species' is a theoretical term in evolutionary theory, so the ontological status of species is determined by the role 'species' plays in evolutionary theory. According to Hull (1978), species are units of evolution in evolutionary biology, and as such they are individuals. A number of processes can cause a species to evolve, natural selection is just one such process. Minimally, selection can cause a species to evolve by changing its gene frequencies from one generation to the next. More significantly, selection will cause a species to evolve by causing a rare trait to become prominent after a number of generations. Suppose the second type of evolution occurs. For such evolution to occur, the selected trait must be passed down through the generations of a species. Traits are not inherited unless some causal connection exists between the members of a species. Sex and reproduction require that organisms, or their parts (gametes, DNA), come into contact. Thus, evolution by selection requires the generations of a species to be causally and spatiotemporally connected. The organisms of a species cannot be scattered throughout the universe, but must occupy a spatiotemporally restricted region. Given that species are units of evolution, they are individuals and not kinds.

The species are individuals thesis is the prevalent ontological view of species. Nevertheless, that view has been contested on a number of grounds. Ruse (1987) and Ereshefsky (1991) argue that many species are not individuals because they fail to have the requisite cohesion required of individuals. Many species consist of asexual organisms. Such organisms are connected through parent-offspring relations, but no exchange of genetic material exists among the members of a species in a single generation. Consequently, asexual species are merely historical entities – their members are connected to a common ancestor, but there is no ongoing causal connection among contemporaneous members. This is certainly right, but suggesting that it shows that species are not individuals misses the importance of the individuality thesis. The important distinction is between being a natural kind and being an individual. Even as mere historical entities species are spatiotemporally restricted entities and not kinds. The idea that species are defined by causal connections rather than qualitative similarity is the important claim of the species are individuals thesis.

Another criticism of the individuality thesis challenges the assertion that all species are spatiotemporally restricted entities. Kitcher (1984), for example, believes that some species taxa are spatiotemporally continuous entities, while other species taxa are spatiotemporally scattered groups whose members share structural similarities. Such species are defined by their members having theoretically important genetic, chromosomal, and developmental similarities. Thus, Kitcher believes that there are two types of species: "historical" species and "structural" species. Historical species are defined by genealogy; structural species are defined by significant similarity. A problem with Kitcher's account of species is that it is out of step with contemporary biological practice. Biologists do not posit structural species. A quick glance at a biology text reveals that historical approaches to species, not structural approaches, are the going concern in biology. The question of whether species are kinds or individuals is a question about how the theoretical term 'species' functions in biology. Kitcher's argument against all species being individuals posits an account of species that is outside of contemporary biology.

2.3. Homeostatic Property Cluster kinds

Another response to the species are individuals thesis is offered by proponents of an alternative approach to natural kinds. According to Boyd (1999a, 1999b), Griffiths (1999), Wilson (1999), and Millikan (1999), species are natural kinds on a proper conception of natural kinds. These authors adopt Boyd's Homeostatic Property Cluster (HPC) theory of natural kinds. HPC theory assumes that natural kinds are groups of entities that share stable similarities. HPC theory does not require that species are defined by traditional essential properties. The members of *Canis familaris*, for example, tend to share a number of common properties – having four legs, two eyes, and so on, but given the forces of evolution, no biological property is essential for membership in that species. For HPC theory, the similarities among the members of a kind must be stable enough to allow better than chance prediction about various properties of a kind. Given that we know that Sparky is a dog, we can predict with greater than chance probability that Sparky will have four legs.

HPC kinds are more than groups of entities that share stable clusters of similarities. HPC kinds also contain "homeostatic causal mechanisms" that are responsible for the similarities found among the members of a kind. The members of a biological species interbreed, share common developmental programs, and are exposed to common selection regimes. These "homeostatic mechanisms" cause the members of a species to have similar features. Dogs, for instance, tend to have four legs and two eyes because they share genetic material and are exposed to common environmental pressures. An HPC kind consists of entities that share similarities induced by that kind's homeostatic mechanisms. According to Boyd, species are HPC kinds and thus natural kinds because "species are defined . . . by . . . shared properties and by the mechanisms (including both 'external' mechanisms and genetic transmission) which sustain their homeostasis" (1999b, 81).

HPC theory provides a more promising account of species as natural kinds than essentialism. HPC kinds need not have a common essential property, so the criticisms of species essentialism are avoided. Furthermore, HPC allows that external relations play a significant role in inducing similarity among the members of a kind. Traditional essentialism assumes that the essence of a kind is an internal or intrinsic property of a kind's members, such as the atomic

structure of gold or the DNA of tigers. Such intrinsic essences are ultimately responsible for a kind's similarities. HPC theory is more inclusive because it recognizes that both the internal properties of organisms and the external relations of organisms are important causes of species-wide similarities. For instance, HPC theory but not essentialism cites interbreeding as a fundamental cause of similarity among the organisms of many species.

While HPC theory is better at capturing the features of species than essentialism, a question remains: Does HPC theory provide an adequate account of species as natural kinds? Some argue that it does not (Ereshefsky and Matthen 2005). Here are two problems with HPC theory. HPC theory's objective is to explain the existence of stable similarities within groups of entities. However, species are also characterized by persistent differences. While HPC theory gives an account of the similarities among the members of a species, it provides no account of the persistent differences among a species' members. As such, it provides an impoverished account of species. Polymorphism – stable variation within a species – is an important feature of nearly every species. Species polymorphisms is easy to find. Consider sexual dimorphism. Within any mammalian species there are pronounced differences between males and females. Or consider polymorphism in the life cycles of organisms. The lives of organisms consist of dramatically different life stages, such as the difference between the caterpillar and butterfly stages of a single organism. HPC theorists recognize the existence of polymorphism, but they do not recognize polymorphism as a central feature of species in need of explanation. HPC theorists only privilege and attempt to explain similarities. In addition to Boyd's 'homeostatic' mechanisms we need to recognize 'heterostatic' mechanisms that maintain species variation.

A second problem with an HPC account of species concerns the identity conditions of species. The members of a species vary in their traits. Moreover, they vary in their homeostatic mechanisms. Over time and across geographic regions, the members of a single taxon are often exposed to different homeostatic mechanisms (Ereshefsky and Matthen 2005). Given such variation, what causes organisms with different traits and exposed to different homeostatic mechanisms to be members of the same species? The common answer is genealogy: the members of a species taxon form a continuous genealogical entity on the tree of life. A species' homeostatic mechanisms are mechanisms of a one species because they affect organisms that form a unique lineage.

Boyd and promoters of HPC theory recognize that the importance of genealogy and see historical relations as one type of homeostatic mechanism. However, Boyd does not see genealogy as the defining aspect of species taxa, and this goes against a fundamental assumption of biological systematics: that species are first and foremost continuous genealogical entities. Consider Boyd's example of a species that arose through multiple hybridization events. Boyd (1999b, 80) characterizes this species as containing "distinct lineages" of organisms that have significant similarities. This species is not a single historical entity, but a collection of separate lineages with separate origins. Boyd is quite clear that similarity and not genealogical connectedness is the final arbitrator of species sameness. This assumption makes sense given that Boyd believes that species are kinds and kinds are ultimately similarity-based classes that play a role in induction. But this view of the identity conditions of species taxa conflicts with the standard view in biological systematics that species are continuous genealogical lineages. Boyd, like other participants in the debate over the ontological status of species, assumes that 'species' is a theoretical term in evolutionary biology, so the status of species is determined by their role in biological theory. Nevertheless, HPC theory's preference for similarity-based kinds conflicts with the biological assumption that species are genealogical entities. Consequently, HPC theory fails to capture the proper identity conditions for species taxa.

2.4. Population Structure Theory

An alternative approach to species is offered by Ereshefsky and Matthen's (2005) "Population Structure Theory" (PST). PST treats similarity as just one type of trait distribution in species. PST does not privilege similarity over polymorphism, so PST offers a more inclusive account of trait distributions in species than HPC theory.

In addition, PST highlights a common type of explanation in biology – one that cites the population and inter-population structures of species. Such *population structure explanations* explain trait distributions in species, whether those distributions involve similarity or dissimilarity.

Population structure explanations are pervasive in biology. Consider E. O. Wilson's (1968) explanation of different social castes in some insect species. The fitness of an insect colony is enhanced by its ability to respond to certain sorts of contingency. There are several types of such contingency, and specialization is required to deal with each of them, so there is selection for different castes. In addition, there is selection for an optimal mix of castes. The optimal mix of castes is calculated by figuring out how to keep the combined cost of various contingencies at a tolerable level. Wilson's theory explains the distribution of difference and why some castes occur in small numbers and others occur in large numbers. Wilson attributes the distribution of castes to group selection: different colonies with different phenotypic distributions compete against one another, and the ones with more optimal distributions are selected. Wilson's account explains population variation not uniformity. Furthermore, it explains such variation by citing the structures of populations – here the distributions of castes in a colony.

Wilson's explanation depends on group selection, but explanations citing population structure need not appeal to group selection. Consider a population structure explanation that does not involve group selection and is aimed at explaining similarity within a species. A prime example is sexual dimorphism. Male elk have a number of similarities, including having large, fuzzy antlers. What explains that similarity? One cause – the proximal cause – is the individual development of each male elk. A deeper explanation – the distal cause – turns on relationships between male and female elks. Male antlers are the result of sexual selection. Such selection requires the participation of both male and female elk. Looked at in this way, we see that the existence of similarities within lower level groups, here within the genders, depend on higher level groups (here species) and the diversity within them. That is, polymorphism at the higher level, and the population structure that binds polymorphism, is essential in explaining lower level similarities within the genders and other sub-groups of a species.

Population structure explanations are common, and arguably essential, for understanding diversity and similarity within species. Such explanations are also essential for understanding the identity conditions of species. As we have seen, species are first and foremost genealogical entities. Genealogy is an inter-population structure – species are lineages of populations. So according to biological systematics, species identity is defined in terms of population and inter-populational structures, not organismic similarity. PST theory, unlike HPC theory, properly captures the identity conditions of species. Stepping back, we see that PST has three virtues. First, it accounts for both similarity and polymorphism within species. Second, by citing population and inter-population structures, PST provides a fundamental explanatory schema for explaining trait distributions in species. Third, that explanatory schema provides the proper basis for understanding species identity.

3. Taxonomic Pluralism

So far, we have discussed the nature of species taxa. Let us now move up one level and ask about the nature of the species category. Typically, biologists and philosophers believe that there is one correct definition of that category; they believe that there is one correct 'species concept,' as biologists call it. However, the biological literature contains over a dozen species concepts (Claridge, Dawah, and Wilson 1997). These concepts are not fringe or crank concepts, but concepts proposed and investigated by prominent biologists. Biologists and philosophers have taken two stances to this plethora of species concepts. Monists believe that one concept is the correct species concept, or one concept is to be preferred over all others. Monists believe that biologists need to sift through the various proposed species concepts and determine which concept gives the proper definition of 'species.' Pluralists take a different stance. They believe that more than one species concept is worthy of acceptance. This section explores the arguments for and against species pluralism.

3.1. The case for pluralism

Let us start by introducing three prominent species concepts in biology. There are many more prominent species concepts, but introducing three is sufficient for providing the argument for pluralism. The most common species concept in the biological literature is Mayr's (1970) Biological Species Concept. The Biological Species Concept defines a species taxon as a group of organisms that can successfully interbreed and produce fertile offspring. According to that concept, a species' integrity is maintained by interbreeding within a species as well as by reproductive barriers between organisms in different species. The Ecological Species Concept defines a species taxon as a lineage of organisms maintained and segmented by ecological forces (Van Valen 1976). Stabilizing selection maintains a species' integrity, while disruptive selection can lead to new species. The Phylogenetic Species Concept (which has multiple versions) defines a species taxon as a basal monophyletic lineage (Mishler and Brandon 1987). A monophyletic lineage contains all and only the descendants of a common ancestor. Because monophyletic lineages occur up and down the Linnaean hierarchy, species are defined as *basal* monophyletic lineages – the smallest lineages represented in Linnaean classifications.

These species concepts, the biological, ecological, and phylogenetic, not only provide different definitions of 'species,' their use gives rise to different classifications of the organic world. This is confirmed by numerous empirical investigations. The most glaring discrepancy is between the Biological Species Concept (BSC) and the other two concepts. BSC requires that the organisms of a species exchange genetic information through interbreeding. That requires sexual reproduction. BSC does not require that every member of a species successfully interbreed, but it does require that a sufficient number of the organisms sexuality reproduce to maintain a species' integrity. The problem is that most of life on this planet does not reproduce sexually but asexually, through cloning or vegetative means. Asexual organisms do not form species according to the BSC. Nevertheless, asexual organisms do form species according to the Phylogenetic Species Concept (PSC) and Ecological Species Concept

(ESC). For the PSC, species are defined genealo-gically, independent of mode of reproduction. For the ESC, species are defined as lineages of organisms maintained by selection forces. PSC's and ESC's classifications of the organic world include asexuals, while BSC's classifications exclude asexuals. These species concepts carve up the world in different ways.

Other cases of species pluralism are more complicated. For example, the BSC and the PSC sort the very same organisms into different species. Consider the case of ancestral species. Many supporters of the BSC believe that a stand-ard form of speciation occurs when a popula-tion of a species becomes isolated from the main body of a species and undergoes a 'genetic revolution.' The parental species, or 'ancestral species,' remains intact. For proponents of the BSC, two species are present in such cases: the ances-tral species consisting of A and B, and the new species C. However, the PSC cannot allow the existence of two species in this case. Recall for the PSC, a species must contain all and only the descendents of a common ancestor. The ancestral species consisting of A and B violates that requirement on species taxa: some of its descendents belong to the new species C. Thus, according to the PSC, either there is one species present (the combination of A, B, and C), or there are three species (the ancestral species A, which went extinct, and two new species, B and C). Either way, the PSC and the BSC cross-classify the very same group of organisms. Take an organism, X, in B. According to the BSC, X belongs to a species consisting of A and B. According to the PSC, X either belongs to a species containing only B or a species containing A, B, and C. Each species concept places X into two different taxa.

The above examples are just the tip of the iceberg of examples where species concepts pro-vide different classifications of the same group of organisms. Generalizing from these examples, different species concepts give rise to different classifications of the organic world. Pluralists believe that examples like these show that we should take a pluralistic approach to biological classification: different species concepts provide different but equally legitimate classifications. Monists disagree. Before turning to monist responses, let us focus on the various brands of pluralism in the literature. This will help further articulate the pluralist's argument.

3.2.   Varieties of pluralism

The argument for pluralism suggested above is motivated by ontological and not epistemolo-gical considerations. Some authors (for example, Rosenberg 1994) suggest that we adopt plural-ism because of our epistemological limitations. The world is exceedingly complex and we have limited cognitive abilities, so we should accept a plurality of simplified classifications of the world. The taxonomic pluralism advocated by Kitcher (1984), Ereshefsky (1992), and Dupré (1993) is not epistemologically driven. It is motivated by the idea that evolutionary theory, a well substantiated theory, tells us that the organic world is multifaceted. Pluralism is an ontological implication of one of our best scientific theories.

Though Kitcher, Ereshefsky, and Dupré agree that species pluralism should be adopted for ontological reasons, they adopt different versions of pluralism. Ereshefsky's form of pluralism, as outlined in the previous section, asserts that the tree of life is divided into different types of lineages – interbreeding species, ecological species, and phylogenetic species. Furthermore, these lineages cross-classify organisms on the tree of life (as illus-trated by the examples in the previous section). Why are there different types of species lineages? They are the result of different evolutionary forces: interbreeding species are the result of interbreeding; ecological species are caused by natural selection; and, phylogenetic species are the result of genealogy. To highlight only one or two of these types of lineages is to give an impoverished account of evolution. Of course, this picture of evolution could be mistaken, but it is our current best picture of evolution.

Kitcher and Dupré adopt a different version of ontological pluralism than the pluralism advocated by Ereshefsky. Both Kitcher and Dupré recognize the three species concepts explored here – the biological, ecological, and phylogenetic species concepts. Kitcher and Dupré also accept species concepts based on shared similarities among the members of a species even when such species do not form continuous genealogical entities. For example, Kitcher posits 'structural' species concepts that allow species to be spatiotemporally disconnected entities (see Section 2.2). As should be familiar by now, 'species' is a theoretical term in biology, thus the ontological status of species taxa is determined by the role that term plays in biological theory. According to evolutionary biology, species taxa are first and foremost genealogical entities. A quick search of the biological literature shows that the species concepts suggested by biologists assume that species taxa form continuous genealogical entities. Given these considerations, the form of species pluralism advocated by Kitcher and Dupré is too liberal. It posits species concepts outside of contemporary biology.

Another type of species pluralism is suggested by Mishler and Brandon (1987). While Kitcher and Dupré's form of pluralism is too liberal, Mishler and Brandon's is too conservative. Mishler and Brandon advocate a version of the phylogenetic species concept. They require that all species form monophyletic units – lineages containing all and only the descendents of a common ancestor. Furthermore, Mishler and Brandon suggest that different evolutionary forces cause monophyletic lineages to be species taxa. Some phylogenetic species are maintained by interbreeding, others by ecological forces, and others by developmental constraints. The result is one classification of the tree of life, but different forces affect different branches on that tree. This form of pluralism is too conservative because it is wedded to a phylogenetic approach to species and rules out the possibility of interbreeding species that are not monophyletic. As we saw in the case of ancestral species, many non-monophyletic interbreeding species meet population genetic standards for being cohesive evolutionary units. Mishler and Brandon's pluralism leaves out a significant kind of basal lineage on the tree of life.

3.3. Monist responses

Needless to say, monists are not happy with pluralism. They offer many responses to species pluralism. One objection is that pluralism is an overly liberal approach to science (Sober 1984; Ghiselin 1987; Hull 1987, 1999). Hull and others ask how pluralists determine which species concepts, among the plurality of suggested concepts, should be accepted as legitimate. Should

any concept proposed by a biologist be accepted? If pluralism offers no criteria for discerning among species concepts, then pluralism, according to Hull (1987), boils down to the position of 'anything goes.'

Ereshefsky (1992) and Dupré (1993) respond to this objection by suggesting criteria for accepting a proposed species concept. Those criteria are such epistemic virtues as empirical testability, internal consistency, and intertheoretic consistency. These are standard epistemic virtues that scientists and philosophers use for judging theories. In judging a species concept, one might ask, for example, if the concept's theoretical assumptions are testable. BSC, for instance, assumes that interbreeding is an important factor in maintaining the existence of stable groups of organisms. Whether interbreeding causes such stability is empirically testable, and biologists do test that hypothesis in the field and in the lab. Pluralists do not subscribe to a position of 'anything goes.' Species pluralists believe that legitimate species concepts must meet specific epistemic standards.

A recent response to species pluralism is that all well-accepted species concepts are captured by a more inclusive species concept. De Queiroz (1999) and Mayden (2002) observe that all prominent concepts assume that species taxa are lineages. Consequently, de Queiroz and Mayden offer a Lineage Concept of Species that, according to Mayden (2002, 191), "serves as the logical and fundamental over-arching conceptualization of what scientists hope to discover in nature behaving as species. As such, this concept . . . can be argued to serve as the primary concept of diversity." De Queiroz and Mayden believe that the species concepts currently accepted describe different types of species lineages – for example, interbreeding lineages, ecological lineages, and phylogenetic lineages, and the Lineage Species Concept provides the proper account of all species lineages. De Queiroz and Mayden believe that their species concept is a monist answer to pluralism because it provides the one correct account of species.

A problem with de Queiroz and Mayden's Lineage Species Concept is that it fails to highlight a unique group of entities that should be called 'species.' Recall that Mayden believes that the Lineage Species Concept captures that "over-arching conceptualization of what scientists hope to discover in nature behaving as species" (ibid.). Such a conceptualization should capture what is common and *unique* to all species taxa. Interbreeding, ecological, and phylogenetic species are all lineages, so the lineage species concept captures an important similarity of species taxa. However, all genera, families, and other Linnaean taxa are also lineages. The lineage concept is too inclusive because it captures all Linnaean taxa. In an attempt to provide an over-arching conceptualization of species taxa, de Queiroz and Mayden have cast their net too widely.

Consider a final monist response to pluralism inspired by advances in molecular genetic sequencing. Alex Rosenberg (in conversation) suggests that perhaps the correct species concept could be based on genetic similarity. As more molecular studies are performed, we may discover the distinctive genome of each species. We could then use that information to construct a single classification of the organic world. The problem with this suggestion is that classifying by molecular data would not unify biological taxonomy, instead it would add further classifications. As Ferguson (2002) observes, classifications based on overall genetic similarity and classifications based on the ability to interbreed do not coincide. The result is two different classifications: one classifying organisms by interbreeding, another classifying organisms by overall genetic similarity. The same sort of disunity is found when classifications based on overall genetic similarity and ecological adaptiveness are compared. Wu and Ting (2004) cite cases where classifications based on genes for ecological adpativeness fail to coincide with classifications based on overall genetic similarity. Bringing molecular data to the table does not reduce the number of classifications but increases their number and brings further disunity to biological taxonomy.

4.   The Linnaean Hierarchy

The Linnaean hierarchy should be familiar to all readers of this volume. The hierarchy contains a series of ranks – species, genus, family, and so on – that serve as the primary categories for biological classification. The Linnaean ranks are also cited in biological theory. Prey-predator models

in ecology refer to species, and hypotheses concerning the tempo and mode of macroevolution often refer to families, classes and more inclusive ranks. There is more to the Linnaean hierarchy than just a series of ranks. The Linnaean hierarchy also includes rules of nomenclature that tell us how to name taxa. For example, the Linnaean hierarchy tells us to give all and only species taxa binomial names, such as the name of our species, *Homo sapiens*.

Given the pivotal role of the Linnaean hierarchy in biology, one might be surprised to learn that a number of biologists and philosophers question the usefulness of that hierarchy. Critics of the Linnaean hierarchy offer a long list of problems (Hull 1966, Hennig 1969, de Queiroz and Gauthier 1992, and Ereshefsky 2001). Some critics believe that the problems facing the Linnaean hierarchy are so severe that the Linnaean hierarchy should be replaced with an alternative system of classification (Hennig 1969, de Queiroz and Gauthier 1992, Ereshefsky 2001, and Cantino *et al.* 2003). The Linnaean hierarchy has its defenders as well. Defenders are well aware of the problems facing the Linnaean hierarchy, but they argue that alternative systems have their own problems (Mayr 1969, Wiley 1981, Forey 2002, Brummit 2002). This section introduces the problems facing the Linnaean hierarchy as well as alternatives to that system.

4.1.   Linnaean ranks

Linnaeus's original hierarchy consisted of five ranks: subspecies (variety), species, genus, order and class. Taxonomists in the early 20th Century found Linnaeus's 5 ranks insufficient for representing life's diversity, so they increased the number of ranks to 21. A persistent question concerning the Linnaean ranks is how to define them. We have seen that biologists and philosophers spend a considerable amount of time defining the species category. What of the other Linnaean ranks? How are they defined? Two schools of taxonomy dominated 20th Century biology: Evolutionary Taxonomy and Cladism. Biologists in each school proposed different definitions of the higher Linnaean ranks (those ranks above the rank of species). However, none of those definitions have withstood criticism.

Before getting to those definitions we need a quick introduction to Evolutionary Taxonomy and Cladism. Evolutionary Taxonomy was founded by Ernst Mayr (1969) and Gaylord Simpson (1961). Evolutionary taxonomists attempt to capture two types of phenomena in taxa. First, a taxon is a single genealogical entity. Second, the members of a taxon inhabit a common adaptive zone and share a common way of life. Birds, for example, form a taxon for evolutionary taxonomists because birds are a single genealogical entity, and birds, for the most part, share a common adaptive zone that causes them to have a relatively distinct way of life. The other major taxonomic school of the 20th Century, Cladism, was founded by Willi Hennig (1966). Cladists attempt to capture only one type of biological phenomena – genealogy. For cladists, a taxon contains all and only the descendents of a common ancestor. Such taxa are monophyletic. Cladists do not recognize birds as a taxon because doing so would make reptilia a non-monophyletic taxon. Crocodiles are more closely related to birds than to other reptiles, yet crocodiles are recognized as part of reptilia. If birds are recognized as a taxon separate from reptilia, then reptilia would not be monophyletic – some of the reptilia's descendents, birds, would not be included in reptilia. Only genealogy matters for cladists, and removing birds from reptilia has no genealogical justification.

Because evolutionary taxonomists and cladists disagree on what counts as a taxon, they disagree on how to define the higher Linnaean ranks. For the evolutionary taxonomist, the higher Linnaean ranks are defined in terms of phenotypic diversity and ecological breadth: the greater the phenotypic diversity and ecological breadth of a taxon, the more inclusive the taxon. All families, for example, contain a certain degree of phenotypic diversity; and, all families occupy an adaptive zone of a certain width. The adaptive zone of a family will be smaller than the adaptive zone of a tribe, which is why, according to evolutionary taxonomists, families are less inclusive taxa than tribes.

The concepts of 'phenotypic diversity' and 'adaptive zone' were widely used in the 20th Century. But now many taxonomists object to their use in determining a taxon's Linnaean rank (Eldredge and Cracraft 1980). They argue that the concepts of 'phenotypic diversity' and 'adaptive zone' are imprecise and used ambiguously. What counts as a distinct adaptive zone varies from phylum to phylum. When it comes to phenotypic diversity, Hennig (1966, 156) playfully asks, "whether the morphological divergence between an earthworm and a lion is more or less than between a snail and a chimpanzee?" Most taxonomists now believe that the concepts of 'phenotypic diversity' and 'adaptive zone' are too malleable to serve as measures of a taxon's rank.

Hennig (1965) offered his own definitions of the higher Linnaean ranks. He suggested that taxa of the same rank originate during the same time period. Just as geological strata are organized according to time of origin, biological taxa should be assigned Linnaean ranks according to their time of origin. All taxa assigned the rank of class, for example, are taxa that originated during the Late Cretaceous. Orders would be defined as those taxa that originated during a more recent time. Hennig's suggested definitions of the Linnaean ranks are more precise than the definitions offered by evolutionary taxonomists. Unfortunately, Hennig's definitions are problematic from a cladistic perspective. (Recall that Hennig is the founder of the taxonomic school cladism.) Taxa that originate during the same period often have different phylogenetic or branching structures. Some taxa originating during the Late Cretaceous are very successful and contain a number of subtaxa. Other taxa originating during the same time are monotypic and contain only a single basal taxon: they are phylogenetic twigs. From a cladistic perspective, Hennig's definitions place different types of taxa under a single rank. Consequently, cladists, including later Hennig, abandoned the idea of correlating the ranks of taxa with times of origin.

Neither evolutionary taxonomists nor cladists have established a universal criterion for defining the higher Linnaean ranks. Instead, biologists use a patchwork of criteria for defining the higher ranks. As a result, taxa of the same rank can vary dramatically. Families can vary in their age, their phylogenetic structure, their phenotypic diversity, and the breadth of their adaptive zone. Calling a taxon a 'family' merely means that within a particular classification that taxon is more inclusive than a genus and less inclusive than a class. There is no definition for 'family' that applies to all or even most classifications of families. 'Family,' like the other Linnaean ranks, refers to a heterogeneous set of taxa. Some authors have suggested that the Linnaean hierarchy of ranks is a fictitious grid we place on nature (de Queiroz and Gauthier 1992, Ereshefsky 1994).

The heterogeneity of the Linnaean ranks has practical implications. Consider the use of Linnaean ranks in biodiversity studies. Biologists tend to measure biodiversity in Linnaean terms; an area or higher taxon is surveyed for the number of species or families present. However, the Linnaean ranks mask important biological differences. Suppose we want to measure the biodiversity of two groups of organisms and we measure that diversity by numbers of families. One group consists of snail families, and the other contains mammalian families. Snail families and mammalian families are biologically different. Snail families have denser phylogenetic structures than mammalian families; and, mammalian families exhibit more ecological diversity than snail families. Snail and mammalian families are not comparable. Counting biodiversity by the numbers of families present in a group masks important biological differences. Biodiversity studies should instead use parameters that capture such biological phenomena as phylogenetic structure or ecological breadth. Then we would have proper measures of the biodiversity.

4.2. Linnaean names

As mentioned earlier, the Linnaean hierarchy is a general system of classification containing both hierarchical ranks and rules for naming taxa. The Linnaean ranks are not the only aspect of the Linnaean hierarchy that is problematic, so are the Linnaean rules of nomenclature. Critics believe that the Linnaean rules of nomenclature lead to a number of practical problems in taxonomy (de Queiroz and Gauthier 1992, Ereshefsky 2001). What follows is a quick introduction to those problems.

Linnaeus's best-known naming rule is his binomial rule. Each species' name has two parts: a generic name and a specific name. In *Homo sapiens,* *Homo* is the name of our genus and *sapiens* is the specific name of our species. Binomials clearly indicate the classification of a species: *Homo* is the name of our species' genus. *Sapiens,*

the specific name, distinguishes our species from other species in *Homo.* Linnaeus's motivation for assigning binomial names was his belief that a biologist should memorize the classification of all species in a kingdom. Linnaeus realized that there were too many species to do that; for example, he recognized approximately 10,000 plant species. He believed that the number of plant genera in the world was much smaller, approximately 300 genera. Furthermore, he did not think that number would greatly increase given his assumption of how much of the world had been explored. With not too much difficulty, a biologist could memorize all the names and taxonomic positions of genera in a kingdom. Once those names and positions were memorized, a biologist would know the classification of each species in a kingdom by reading the generic name in a species' binomial name. Linnaeus's binomials, thus, served as guides for memorizing the taxonomic positions of species in a kingdom.

Binomials may have served their intended purpose in Linnaeus's time when it was assumed that there was approximately 300 plant genera and 300 animal genera and those numbers would not increase greatly. The problem is that Linnaeus did not envision the extent of biodiversity in the world, and so he grossly underestimated the number of genera in the world. Conservative estimates put the numbers of plant genera in the tens of thousands. Estimates for animal genera are in the tens of thousands as well. Given these numbers, there are too many generic names in a kingdom to memorize. As Mayr (1969, 334) writes, "a generic name no longer tells much to a zoologist except in a few popular groups of animals." Binomials have lost their original motivation: they do not serve as guides for memorizing the classification of all the species in a kingdom. Nevertheless, biologists are still required to assign a species both a generic and a specific name. The binomial rule remains in place even though it fails to achieve its original aim.

One might wonder if the continued use of the binomial rule is problematic. Though the original motivation is gone, there seems no harm in incorporating a generic name in a species name. However, the use of binomials has its costs. The binomial rule can place a biologist in the awkward position of having to assign a species to a genus before having the proper empirical information for making that assignment. The binomial rule, in other words, can lead to hasty classification. Recall that the binomial rule requires that a species be assigned to a genus before it can be named. But in some situations, a biologist lacks the proper information for assigning a species to a genus. In such situations, if a biologist wants to name a newly discovered species, she must assign that species to a genus on the basis of insufficient information. This is no small problem. Cain (1959, 242) writes that "the necessity of putting species into a genus before it can be named at all is responsible for the fact that a great deal of uncertainty is wholly cloaked and concealed in modern classification." Thus, the use of the binomial rule is not so innocuous. It causes biologists to assign species to genera without having sufficient data.

There are further problems with the Linnaean rules of nomenclature, and these involve not just Linnaeus's binomial rule but also other Linnaean rules. Many of these rules of nomenclature were introduced by taxonomists in the 20th Century. For example, one rule is that the names of taxa have rank-specific endings that indicate the Linnaean rank of a taxon. For example, the suffix -idae in the names 'Hominidae' and 'Tipulidae' indicate that such taxa are families. The suffix -ini shows that Hominini is a tribe. The inclusion of rank-specific endings in the names of higher taxa causes further problems with the Linnaean hierarchy.

Consider the case of taxonomic revision. Taxonomic revision is the activity of revising a classification. Such revision is the norm not the exception in biological taxonomy. A taxon may be reassigned to another higher taxon, for example, a species may be reassigned to another genus. Or revision may involve giving a taxon a new rank, for example, a taxon thought to be a tribe is reassigned as a family. Taxonomic revision occurs for a couple of reasons. New empirical information may be gathered, such as new DNA evidence or the discovery of new fossils, that causes revision. Or revision may occur when taxonomic theory occurs and old classifications are updated to cohere with the new theory. Taxonomic revision causes instability in classification, and this is to be expected. Classifications are hypotheses and are open to revision in light of new evidence or theoretical considerations.

Unfortunately, the Linnaean rules of nomenclature are themselves a cause of instability because they require that the names of taxa be changed when revision occurs. Recall that in the Linnaean hierarchy, the names of taxa reflect a taxon's rank and taxonomic position. A change in a taxon's classification requires a change in a taxon's name. This may not sound like much of an inconvenience, but it is. A case of taxonomic revision can involve the renaming of hundreds of taxa (Ereshefsky 2001). The Linnaean rules themselves are a source of instability.

Another area where the Linnaean rules of nomenclature make classifying taxa harder than need be is that of taxonomic disagreement. In some cases biologists disagree over the classification of a taxon. The disagreement may be over the rank of a taxon, or the disagreement may be over the taxonomic placement of a taxon (does it belong to this or that more inclusive taxon). The Linnaean rules require that the rank and placement of a taxon be reflected in its name. If biologists disagree over a taxon's rank, for instance, the Linnaean rules require those biologists to assign different names to what they agree is the same taxon. Consider a disagreement between E. O. Wiley and G. G. Simpson concerning the placement of our genus, Homo. They agree that Homo belongs to a more inclusive taxon, call it 'X,' but they disagree on X's Linnaean rank. They agree on the members and properties of X, but because they belong to different taxonomic schools they assign X different ranks. Wiley thinks X is a tribe. Simpson thinks it is a family. Following the Linnaean rules, Wiley calls X Hominini, and Simpson calls X Hominidae. The Linnaean rules force Simpson and Wiley to give different names to what they agree is the same taxon. The Linnaean rules are, thus, a source of semantic confusion.

4.3. Alternatives to the Linnaean hierarchy

In light of the problems facing the Linnaean hierarchy, several alternative systems of classification have been suggested in the last 40 years. Hull (1966) and Hennig (1969) offered alternative systems, and more recently a system of phylogenetic nomenclature called the "Phylocode" has been suggested (de Queiroz and Gauthier 1992;

Cantino *et al.* 2003). All of these alternatives to the Linnaean hierarchy recommend eliminating Linnaean ranks from biological classification. They promote a 'rankless taxonomy.' They offer a couple of alternative methods for indicating the hierarchical relations among taxa. Here we will consider just one suggested method.

According to Hennig (1969), Linnaean ranks should be replaced with positional numbers. Consider a standard Linnaean classification.

Subclass Reptilomorpha
Infraclass Aves
Infraclass Mammalia
Division Monotremata
Division Theria
Cohort Metaheria
Cohort Eutheria.

A classification of the same taxa using Hennig's positional numbers is the following:

2.4. Reptilomorpha
2.4.1. Aves
2.4.2. Mammalia
2.4.2.1. Monotremata
2.4.2.2. Theria
2.4.2.2.1. Metaheria
2.4.2.2.2. Eutheria.

Positional numbers indicate a couple of things. First, positional numbers indicate hierarchical relations among taxa. Eurtheria is a part of Mammalia and this is shown by Eurtheria's positional number containing the positional number of the more inclusive Mammalia. Second, positional numbers indicate the degree of inclusiveness of a taxon: the fewer the digits in a taxon's positional number, the more inclusive the taxon.

Hennig's positional number system overcomes the problems associated with Linnaean ranks and names. Recall that a problem with the Linnaean ranks is the assumption that taxa of the same Linnaean rank are comparable across classifications (Section 4.1). Positional numbers carry no such assumption; positional numbers are merely notational devices to indicate the hierarchical relations of taxa within specific classifications. Positional numbers have no meaning outside of particular classifications.

Consequently, no suspect ontological categories are associated with positional numbers, as is the case with the Linnaean ranks.

Another feature of Hennig's positional number system is that it avoids the naming problems facing the Linnaean rules of nomenclature. Recall that under the Linnaean rules, a taxon's name is not merely a name but also a device for indicating a taxon's classification. Taxon names in Hennig's system do not play that dual role: taxon names are merely names; positional numbers perform the task of indicating a taxon's placement in a classification. In Hennig's system, the activity of naming taxa is divorced from the activity of classifying taxa. By keeping the activities of naming and classifying separate, Hennig's system overcomes the naming problems highlighted earlier.

Consider the affect taxonomic revision has on Linnaean names. In the Linnaean hierarchy, a name indicates a taxon's classification, so a change in classification requires a change in name. In the positional system, taxonomic revision requires a change in a taxon's positional number, but the name of the taxon remains the same. Unlike the Linnaean rules, names remain stable during taxonomic revision. The positional system also avoids the problem posed by taxonomic disagreement. In the Linnaean hierarchy, when two biologists disagree on the rank of a taxon, they must assign that taxon different names, each indicating the different ranks assigned to the taxon. With positional numbers, a taxon has a single name, and biologists display their disagreement over that taxon's placement by assigning that taxon different positional numbers. Hence, the use of positional number avoids the semantic confusion caused by the Linnaean rules. Finally, recall the problem of hasty classification caused by the Linnaean binomial rule. In the Linnaean hierarchy, a biologist must first determine the genus of a species before she can name it, even though she may lack adequate empirical evidence for assigning a species to a genus. In Hennig's system, a species can be named before knowing that species' classification; a name is assigned to the species, and its positional number is determined later.

The positional number system nicely overcomes the ranking and naming problems outlined above. However, some proponents of non-Linnaean systems think that further changes in nomen- clature are necessary. For example, a current controversy among proponents of non-Linnaean systems is what to do with the names of species. Some post-Linnaeans suggest converting species binomial to uninomials (Cantino *et al.* 1999). There are two motivations for doing so. One is the belief that because there is no species category in nature, those taxa we call 'species' should not be given a special notational device; all taxa should have uninomial names. Another motivation for eliminating binomials has to do with revision. Taxa with binomial names may be assigned to more inclusive positions in classifications. If that occurs, then taxa with binomial names occur at various hierarchical levels in a classification. Classifications would then have the confusing feature of binomials referring to non-basal taxa.

How, then, should binomials be converted to uninomials? According to one suggestion, a binomial should be converted to a uninomial by placing the specific and generic name together. The binomial *Canis Familaris* would become the uninomial 'Canisfamilaris.' According to another proposal, the generic name of a species should be dropped and a registration number should be added to the specific name. *Canis Familaris* would become 'Familaris5732,' for example. The registration number is added to avoid homonyms.

4.4. A Middle ground

Defenders of the Linnaean hierarchy worry that the proposed alternative systems are too radical (Forey 2002, Brummitt 2002). They worry that the replacing the Linnaean hierarchy would be overly disruptive to biological practice. For example, suppose that the Linnaean ranks were replaced by positional numbers. The Linnaean ranks are well entrenched both in and outside of biology. The ranks of species and genera occur in text books, field guides, environmental legislation, and elsewhere. Arguably, replacing the Linnaean ranks both in and outside of biology would be too disruptive. A similar case is made against changing the names of taxa, for example, replac- ing binomial names with uninomials. Switching binomials to uninomials would require rewriting countless classifications, textbooks, and field guides. Critics of non-Linnaean systems think

that changing the names of taxa would be too disruptive and impractical.

Still, the Linnaean hierarchy has its problems, as illustrated in Sections 4.1 and 4.2. Keeping the Linnaean hierarchy in place avoids disrupting biological practice, but it also keeps in place the problems caused by the Linnaean hierarchy. Given the pros and cons of replacing the Linnaean hierarchy, what should be done? There is a middle ground between replacing the Linnaean hierarchy and keeping the Linnaean hierarchy as it stands. We could keep the Linnaean ranks and names as they are, but rid them of their Linnaean meaning.

When it comes to the names of taxa, we could keep the current taxon names yet deny that such names having any classificatory meaning. A binomial name, for example, would just be a name, it would no longer indicate the rank of a taxon. Thus during taxonomic revision a taxon with a binomial name would keep its name even if that taxon were reclassified as a more inclusive taxon. Similarly, two biologists that disagreed on the rank of a taxon could continue to use its binomial name because that name has no classificatory meaning – it is merely a name. Keeping taxon names constant but eliminating any classificatory meaning associated with such names avoids the problems facing the Linnaean rules while at the same time keeping names stable.

A similar approach can be applied to the Linnaean ranks. We can keep the Linnaean ranks and restrict the meaning of those ranks to being indicators of the hierarchical relations among taxa within a *particular* classification. At the same time, we would rid the Linnaean ranks of their metaphysical connotations: the Linnaean ranks would not highlight any categories in nature; and, the assumption that all taxa of the same rank, for example, all genera, share an ontological similarity would be dropped. This approach avoids the problems associated with the Linnaean ranks. For example, false compar- isons among taxa of the same Linnaean rank in biodiversity studies would be eliminated. At the same time, the Linnaean ranks would still remain in place, so the disruption that would be caused by junking the Linnaean ranks is avoided. This approach to Linnaean nomencla- ture and ranks charts a middle ground between critics and defenders of the Linnaean hierarchy. It acknowledges and avoids the problems facing the Linnaean hierarchy while at the same time keeping biological taxonomy as stable as possible.[2]

Stepping back, in this chapter we have seen a variety of conceptual issues at the forefront of biological taxonomy and systematics. Species may not be qualitative kinds but historical entities defined by genealogy. The idea of single correct classification of the organic world may need to be replaced with a more pluralistic approach to classification. The Linnaean ranks that pervade biology may have no basis in nature but are merely instruments for organizing life's diversity. Other radical changes in how we classify and bring order to the organic world may be just around the corner. For example, life on this planet may not be a single genealogical tree, but a tangled bush (Doolittle 1999). If life is a tangled bush, then hierarchical classifications may be the wrong way to represent life's diversity. Biological systematics is rife with conceptual issues in need of philosophical analysis.

Notes

1 Windsor (2003) has challenged the claim that prominent pre-Darwinian biologists were essentialists. This raises an interesting issue concerning taxonomic thought. Linnaeus, for example, wrote that organisms should be classified according to their taxon-specific essences. However, he realized we do not have access to those properties, so he classified according to clusters of similarities (Ereshefsky 2001). Was Linnaeus an essentialist or not? Linnaeus advocated essentialism but could not implement it in practice. (Biologists often advocate taxonomic theory that they cannot yet use.) So perhaps the proper answer is, in some ways Linneaus was an essentialist, in other ways he was not.

2 Wiley (1981) also suggests a revised Linnaean hiearchy, what he calls the "annotated Linnaean hierarchy." Wiley's annotated Linnaean hierarchy is different than the revised Linnaean hierarchy offered here. To cite one difference, Wiley's revision does not divorce the activity of naming taxa from the activity of classifying taxa. Thus, Wiley's annotated Linnaean hierarchy does not address or solve any of the naming problems mentioned in Section 4.3.

References

Boyd, R. 1999a: Homeostasis, species, and higher taxa. In R. Wilson (ed.), *Species: New Interdisciplinary Essays*, 141–185. Cambridge: MIT Press.

Boyd, R. 1999b: Kinds, complexity and multiple realization: comments on Millikan's "Historical Kinds and the Special Sciences". *Philosophical Studies* 95, 67–98.

Brummitt, R. 2002: How to chop up a tree. *Taxon* 51, 31–41.

Cain, A. 1959: The Post-Linnaean Development of Taxonomy. *Proceedings of the Linnaean Society of London* 170: 234–244.

Cantino, P. D. *et al.* 1999: Species names in phylogenetic nomenclature. *Systematic Biology* 48: 790–807.

Cantino, P. D. *et al.* 2003: Phylocode: a phylogenetic code of biological nomenclature. http:/www.ohiou. edu/phylocode/.

Claridge, M. F., A. Dawah, M. R. Wilson, editors, 1997: *Species: The Units of Biodiversity*. London: Chapman and Hall.

de Queiroz, K. and Gauthier, J. 1992: Phylogenetic taxonomy. *Annual Review of Ecology and Systematics* 23, 480–499.

de Queiroz, K. 1999: The general lineage concept of species and the defining properties of the species category. In R. Wilson (ed.), *Species: New Interdisciplinary Essays*, 49–90. Cambridge: MIT Press.

Doolittle, W. F. 1999: Phylogenetic classification and the universal tree. *Science* 284 (June), 2124–2128.

Dupré, J. 1981: Natural kinds and biological taxa. *Philosophical Review* 90, 66–90.

Dupré, J. 1993: *The Disorder of Things: Metaphysical Foundations of the Disunity of Science*. Cambridge: Harvard University Press.

Eldredge, N. and Cracraft, J. 1980: *Phylogenetic Patterns and the Evolutionary Process*. New York: Columbia University Press.

Ereshefsky, M. 1991: "Species, Higher Taxa, and the Units of Evolution. *Philosophy of Science* 58: 84–101.

Ereshefsky, M. 1992: Eliminative Pluralism. *Philosophy of Science* 59, 671–690.

Ereshefsky, M. 1994: Some problems with the Linnaean hierarchy. *Philosophy of Science* 61, 186–205.

Ereshefsky, M. 2001: *The Poverty of the Linnaean Hierarchy: A Philosophical Study of Biological Taxonomy*. Cambridge: Cambridge University Press.

Ereshefsky M. and Matthen, M. 2005: Taxonomy, polymorphism and history: an introduction to population structure theory.

Ferguson, J. 2002: On the use of genetic divergence for identifying species. *Biological Journal of the Linnean Society* 75, 509–519.

Forey, P. 2002: Phylocode – pain, no gain. *Taxon* 51, 43–54.

Ghiselin, M. 1974: A radical solution to the species problem. *Systematic Zoology* 23, 536–544.

Griffiths, P. 1999: Squaring the circle: natural kinds with historical essences. In R. Wilson (ed.), *Species: New Interdisciplinary Essays*, 209–228. Cambridge, Massachusetts: MIT Press.

Hennig, W. 1965: Phylogenetic systematics. *Annual Review of Entomology* 10, 97–116.

Hennig, W. 1966: *Phylogenetic Systematics*. Urbana: University of Illinois Press.

Hennig, W. 1969: *Insect Phylogeny*. John Wiley Press, New York.

Hull, D. 1965: The effect of essentialism on taxonomy: two thousand years of stasis. *British Journal for the Philosophy of Science* 15, 314–326.

Hull, D. 1966: Phylogenetic numericlature. *Systematic Zoology* 15: 14–17.

Hull, D. 1978: A matter of individuality. *Philosophy of Science* 45, 335–360.

Hull, D. 1987: Genealogical Actors in Ecological Roles. *Biology and Philosophy* 2: 168–183.

Hull, D. 1999: On the Plurality of Species: Questioning the Party Line. In R. Wilson (ed.), *Species: New Interdisciplinary Essays*, 23–48. Cambridge, Massachusetts: MIT Press.

Kitcher, P. S. 1984: Species. *Philosophy of Science* 51, 308–333.

Kripke, S. 1982: Naming and Necessity. In D. Davidson and G. Harmon (eds.), *Semantics of Natural Language*, 253–355. Reidel: Dordrecht.

Mayden, R. 2002: On biological species, species concepts and individuation in the natural world. *Fish and Fisheries* 3, 171–196.

Mayr, E. 1959: Typological versus Population Thinking. In *Evolution and Anthropology: A Centennial Appraisal*. The Anthropological Society of Washington, Washington D.C.

Mayr, E. 1969: *Principles of Systematic Zoology*. Harvard University Press, Cambridge, Massachusetts.

Mayr, E. 1970: *Populations, Species, and Evolution*. Cambridge, Massachusetts: Harvard University Press.

Mayr, E. 1982: *The Growth of Biological Thought*. Cambridge, Massachusetts: Harvard University Press.

Millikan, Ruth 1999: Historical Kinds and the "Special Sciences". *Philosophical Studies* 95: 45–65.

Mishler, B. and Brandon, R. 1987: Individuality, Pluralism, and the Phylogenetic Species Concept. *Biology and Philosophy* 2: 397–414.

Putnam, H. 1975: *Mind Language, and Reality. Philosophical Papers*, Volume 2. Cambridge University Press, Cambridge.

Rosenberg, A. 1994: *Instrumental Biology or the Disunity of Science*, Chicago: Chicago University Press.

Ruse, M. 1987: Biological Species: Natural Kinds, Individuals, or What? *British Journal for the Philosophy of Science* 38: 225–242.

Simpson, G. 1961: *The Principles of Animal Taxonomy*. Columbia University Press, New York.

Sober, E. 1980: Evolution, population thinking and essentialism. *Philosophy of Science* 47, 350–383.

Sober, E. 1984: Sets, Species, and Natural Kinds. *Philosophy of Science* 51, 334–41.

Van Valen, L. 1976: Ecological Species, Multispecies, and Oaks. *Taxon* 25: 233–239.

Wiley, E. 1981: *Phylogenetics: The Theory and Practice of Phylogenetic Systematics*. Wiley and Sons, New York.

Wilson, Edward O. 1968: The Ergonomics of Caste in the Social Insects. *American Naturalist* 102: 41–66.

Wilson, R. 1999: Realism, essence, and kind: resuscitating species essentialism? In R. Wilson (ed.), *Species: New Interdisciplinary Essays*, 187–208. Cambridge, Massachusetts: MIT Press.

Winsor, M. P. 2003: The Non-Essentialist Methods in Pre-Darwinian Taxonomy, *Biology and Philosophy* 18: 387–400.

Wu, C. and Ting, C. 2004: Genes and speciation. *Nature Genetics* 5, 114–122.

19

Speciation: A Catalogue and Critique of Species Concepts

Jerry A. Coyne and H. Allen Orr

Here we describe and evaluate eight species concepts that are considered serious competitors to the biological species concept (BSC) (Table 19.1). We describe the reasons why each concept was proposed (i.e., the "species problem" it was designed to solve), explain why its proponents see it as superior to the BSC, and assess its advantages and disadvantages. We also show how each concept deals with issues that are problematic for the BSC: allopatric populations, gene exchange between taxa that remain distinct, and uniparental organisms. We note how closely each concept coincides with the BSC – that is, whether it identifies the same species in sympatry. Finally, we note what process constitutes "speciation" under each concept. Throughout the discussion, we adhere to our version of the BSC, which allows limited gene exchange, rather than to the strict version that demands complete reproductive isolation between taxa.

We will not deal with strictly typological species concepts – those that define species by specifying an arbitrary degree of morphological or genetic difference. Mayr (1942, 1963) has explained the problems with such concepts. We do, however, discuss two somewhat typological concepts: the "genotypic cluster" species concept and several versions of the phylogenetic species concept.

Genotypic Cluster Species Concept (GCSC)

A species is a [morphologically or genetically] distinguishable group of individuals that has few or no intermediates when in contact with other such clusters.

(Mallet 1995)

This concept was proposed in response to the observation that, while the BSC *defines* species by the presence or absence of interbreeding, it *recognizes* them as distinguishable clusters in sympatry. (These clusters can be seen in phenotypic data as a bimodal distribution of traits, and in genetic data as a deficit of heterozygotes or the presence of linkage disequilibrium among genes.) For advocates of both the BSC and the GCSC, the species problem is identical: understanding the origin of discrete entities in sympatry. Unlike the BSC, however, the GCSC defines species solely by the features used to recognize them. The GCSC does not specify how many traits and/or genes are required to diagnose sympatric clusters as species.

Advocates of the GCSC claim that it has several advantages over the BSC. First, the GCSC is supposedly independent of theories about speciation: it is presented as a way to recognize

Jerry A. Coyne and H. Allen Orr, Appendix from *Speciation* (Sinauer, 2004), pp. 27, 447–472. Reprinted by permission of Sinauer Associates, Inc.

Table 19.1 The biological species concept and some recently proposed alternatives

Basis of concept	Concept	Definition
1. Interbreeding	Biological Species Concept (BSC)	Species are groups of interbreeding natural populations that are reproductively isolated from other such groups (Mayr 1995).
2. Genetic or phenotypic cohesion	Genotypic Cluster Species Concept (GCSC)	A species is a (morphologically or genetically) distinguishable group of individuals that has few or no intermediates when in contact with other such clusters (Mallet 1995).
	Recognition Species Concept (RSC)	A species is that most inclusive population of individual biparental organisms which shares a common fertilization system (Paterson 1985).
	Cohesion Species Concept (CSC)	A species is the most inclusive population of individuals having the potential for phenotypic cohesion through intrinsic cohesion mechanisms (Templeton 1989).
3. Evolutionary cohesion	Ecological Species Concept (EcSC)	A species is a lineage (or a closely related set of lineages) which occupies an adaptive zone minimally different from that of any other lineage in its range and which evolves separately from all lineages outside its range (Van Valen 1976).
	Evolutionary Species Concept (EvSC)	A species is a single lineage of ancestral descendant populations or organisms which maintains its identity from other such lineages and which has its own evolutionary tendencies and historical fate (Wiley 1978, modified from Simpson, 1961).
4. Evolutionary history	Phylogenetic Species Concept 1 (PSC1)	A phylogenetic species is an irreducible (basal) cluster of organisms that is diagnosably distinct from other such clusters, and within which there is a paternal pattern of ancestry and descent (Cracraft 1989).
	Phylogenetic Species Concept 2 (PSC2)	A species is the smallest [exclusive] monophyletic group of common ancestry (de Queiroz and Donoghue 1988).
	Phylogenetic Species Concept 3 (PSC3) or Genealogical Species Concept (GSC)	A species is a basal, exclusive group of organisms all of whose genes coalesce more recently with each other than with those of any organisms outside the group, and that contains no exclusive group within it (Baum and Donoghue 1995; Shaw 1998).

species rather than understand how they evolved. Defining clusters on the basis of interbreeding is said to lead the BSC into circularity: "Since theories of speciation involve a reduction in ability or tendency to interbreed, species cannot themselves be defined by interbreeding without confusing cause and effect" (Mallet 1995, p. 295; all quotations and page numbers refer to this paper). Mallet feels that the GCSC allows one to consider other causes of clustering besides reproductive isolation: "Gene flow is not the only factor maintaining a cluster; stabilizing selection will also be involved, as well as the historical inertia of the set of populations belonging to the cluster. . . . Clusters can remain distinct under relatively high levels of gene flow provided that there is strong selection against intermediates" (p. 296).

Proponents of the GCSC view the BSC's emphasis on isolating barriers as not only intellectually vacuous, but misleading: "To include such a number of different effects under a single label must be one of the most extraordinary pieces of philosophical trickery ever foisted successfully on a community of intelligent human beings" (pp. 297–298). The BSC is also considered unscientific: "Mayr has repeatedly stressed that the biological concept cannot be refuted by practical difficulties in its application; this means it is untestable" (p. 296). Moreover, the notion of species as reproductively isolated entities is said to impede our understanding of speciation: "Because no gene flow between species is conceptually possible under interbreeding concepts, it is extremely hard to imagine how speciation, which must often involve a gradual cessation of gene flow, can occur" (p. 295). Mallet notes that allopatric speciation is one such mode of speciation, but also argues that the BSC is biased against other modes of cluster formation – such as sympatric and parapatric speciation – that involve gene flow between incipient species. Finally, Mallet echoes the criticism of Sokal and Crovello (1970) that it is impossible to apply the BSC in practice, as this requires making or observing the crosses needed to test the reproductive compatibility of every pair of individuals.

It is important to recognize that, despite its emphasis on species recognition rather than reproductive isolation, the GCSC and our version of the BSC identify nearly the same set of species in sympatry. The real disparity between these concepts is in the amount of genetic difference required for species status. In principle, the GCSC could diagnose sympatric clusters that differ in only one or two genes or traits while exchanging alleles freely throughout the rest of the genome. Such clusters could be maintained by habitat-related selection. For example, Wilson and Turelli (1986) show that density- and frequency-dependent selection acting on alleles at one or two loci can lead to the stable coexistence of distinct genotypes, even though heterozygotes have the lowest fitness and appear in less-than-expected frequencies. This will produce statistically distinguishable clusters that might be considered species under the GCSC. If one claims, however, that such polymorphisms do not diagnose species because they represent only intraspecific variation, then one is reverting to the BSC. The failure to specify how many loci (or what degree of heterozygote deficit and linkage disequilibrium) are required to diagnose species is a problem for the GCSC. Setting such a threshold would involve an arbitrary decision.

In contrast, the BSC diagnoses species only if there is evidence that gene flow between them is strongly limited. This involves either observing many genetic differences between sympatric taxa – a degree of difference too large to be explained by disruptive selection alone – or observing isolating barriers so strong that gene flow is almost zero. This can also involve an arbitrary decision if there is any introgression between sympatric groups, but clearly genetic differentiation must be higher for recognizing biological than for recognizing genotypic-cluster species.

Another problem for the GCSC involves the *level* of clustering. Because of the hierarchical nature of evolution, genotypic clusters occur at many levels. These clusters can involve intrapopulation polymorphisms, local host races, species, or higher-level groups such as genera. Trying to apply the GCSC to the *Rhagoletis pomonella* complex of tephritid flies, Berlocher (1999, p. 661) observed a "continuum of decreasing degree of cluster overlap as level of genetic divergence increases from host face to distinct species. . . . No species threshold is apparent." Since the GCSC sees no fundamental distinction between species and higher-level groups (p. 296: "Whether species do have a greater 'objective' reality than lower or higher taxa is either wrong or at least debatable; the idea that taxa are qualitatively different from other taxa is therefore best not included within their definition"), the definition of species as "genotypic clusters" must be recast as "clusters that do not include other sub-clusters." But this would lead one to diagnose as species polymorphic forms such as beak morphs in

Pyrenestes finches (Smith 1987) or Batesian mimicry phenotypes in the butterfly *Papilio memnon* (Clarke and Sheppard 1969). To get around this problem, one must then include a reproductive criterion: such polymorphisms do not diagnose species because their carriers readily interbreed. This, however, defeats the GCSC's goal of avoiding criteria based on reproductive compatibility.

As noted in Chapter 1, the presence of sympatric clusters involving several genes implies the existence of isolating barriers. Thus, disruptive selection that creates and maintains distinct clusters involves a form of reproductive isolation – extrinsic postzygotic isolation. This is recognized in statements such as, "The maintenance of sympatric species is not just due to reproductive traits, but also due to ordinary within-species, stabilizing ecological adaptations that select disruptively against intermediates or hybrids" (Mallet 1995, p. 296). There is no real difference between positive selection for alternative phenotypes and negative selection against intermediates between those phenotypes.

As for other processes that can create and maintain species, we do not understand how "historical inertia" can play such a role. The maintenance of distinct clusters in sexually reproducing taxa must always involve selection against intermediates (i.e., reproductive isolation). We also fail to see why the BSC is circular. If species are regarded simply as an advanced stage in the evolution of reproductive isolation, then no circularity ensues. We are baffled by the claim that lumping diverse phenomena under the category "isolating mechanisms" involves "philosophical trickery." As Harrison (1998, p. 24) argues: "[It] is the common effect of all these differences (limiting or preventing gene exchange) that provides the rationale for grouping them. I see no reason not to adopt a single term (e.g., 'barriers to gene exchange') to refer to the set of differences that have this very important effect."

Moreover, the fact that the BSC is not theory-free – that it immediately suggests a *process* of speciation – seems to us an advantage, not a problem. A theory-free definition of identical twins might be given as "two individuals, born of one mother at the same time, who are exceedingly similar morphologically." But this definition is surely less useful than one that incorporates process, such as "identical twins are the products of splitting of a single fertilized egg." The claim that the BSC is untestable holds, as Brookfield (2002) notes, for *all* species concepts: none can be falsified by experiment or observation. A species concept is a tool for research, not a hypothesis subject to refutation.

The claim that under the BSC "it is extremely hard to imagine how speciation, which must often involve a gradual cessation of gene flow, can occur," seems unfounded. Over the last 60 years, biologists have had no problem imagining how biological speciation can occur. There are well-established genetic and ecological models – both verbal and mathematical – for the origin of isolating barriers in sympatry, allopatry, and parapatry.

Finally, the view that the BSC is not useful because one cannot do breeding tests seems misguided. While breeding tests can be useful for identifying species (e.g., Dobzhansky and Epling 1944), biological species can also be identified by the concordance of many characters and genes that show the existence of isolating barriers, or by the consistent correlation between a group of traits on one hand and reproductive compatibility on the other. Once one has described such a group, even a single trait can then be used to diagnose species. Traditionally, this has been done with great success: genitalia are reliable indicators of biological species status in many organisms (Eberhard 1986), and chromosomal and molecular characters have served equally well.

How does the GCSC differ from the BSC? The most important aspect is how these concepts deal with sympatric taxa showing moderate to substantial gene exchange. Such taxa include sympatric host races of insects such as the apple and hawthorn races of *Rhagoletis pomonella* (Feder *et al.* 1998), which form GCSC species but would probably not be accorded species status by the BSC. Hybrid zones are another example. If one considers a wide area including the zone of hybridization, one can recognize two genetic clusters with various intermediate individuals (hybrids) falling between the clusters. The two clusters would be considered species under the GCSC but not the BSC unless there was little gene flow beyond the hybrid zone. However, introgression outside hybrid zones is often limited (Barton and Hewitt 1985), allowing one to diagnose two biological species that are identical to the two genotypic-cluster species.

Both the BSC and the GCSC have difficulty diagnosing *allopatric* taxa that are morphologically distinguishable. (Genotypic clusters are recognizable only in sympatry.) Here, however, the BSC has something of an edge: if allopatric populations form either inviable or sterile hybrids when artificially hybridized, one can say with assurance that they are biological species.

There are arguably two advantages of the GCSC over the BSC. First, the GCSC is less ambiguous at diagnosing species in problematic situations such as taxa that hybridize with limited gene flow. Such cases constitute a gray area for the BSC, but offer no problem to the GCSC if one can observe distinct clusters. But because we are more concerned with process than with diagnosis, we do not consider this a particularly meaningful advantage, especially because GCSC clusters may involve only one or two genetic differences. Second, the GCSC can also be applied to largely or completely asexual taxa, groups where the BSC is impotent. But strict use of the GCSC would diagnose each asexual clone as a different species. For such groups it seems preferable to adopt neither the BSC nor the GCSC, but an ecological species concept.

The GCSC is one of the few species concepts that come with an explicit definition of speciation: "Speciation is the formation of a genotypic cluster that can overlap without fusing with its sibling" (Mallet 1995, p. 298). This differs from our own notion of speciation only in that we define "clusters" as "groups between which reproductive barriers are very strong." However, Mallet adds (p. 298), "To understand speciation, we need to understand when disruptive selection can outweigh gene flow between populations." This applies only to parapatric and sympatric speciation, because the conflict between selection and gene flow does not exist during allopatric speciation.

The GCSC is the most serious competitor to the BSC because the two concepts share many features. But by concentrating on the identification rather than the origin of species, the GCSC does not yield a particularly fruitful program of research.

Recognition Species Concept (RSC)

Species are the most inclusive population of individual biparental organisms, which share a common fertilization system.

(Paterson 1985; see also Lambert and Paterson 1984, Lambert *et al.* 1987, and Masters *et al.* 1987)

This concept is also motivated by the problem of organic discontinuity. The RSC resembles the GCSC and the cohesion species concept (see below) in that species are defined by those factors that hold populations together rather than by those that isolate them.

Under the RSC, these cohesive factors constitute their shared "fertilization system": the set of biological features that "contribute to the ultimate function of bringing about fertilization while the organism occupies its normal habitat" (Paterson 1985, p. 24). Important aspects of the fertilization system are included in the Specific-Mate-Recognition-System (SMRS), which includes all features by which organisms "recognize" each other as mates. This recognition can be either active (as in courtship signals and responses), or passive (as in biochemical processes of gamete fusion). Paterson's working definition of a species is a "field for gene recombination" (1985, p. 21). The RSC thus explicitly excludes ecological and temporal isolating barriers, as well as all postzygotic barriers.

Paterson and co-authors claim that the RSC remedies many of the weaknesses they find in the BSC (which they call the "isolation concept") because of the BSC's presumed concentration on species distinctness rather than cohesion. Detailed analysis and criticisms of this theory have been published elsewhere by ourselves and others (Butlin 1987a; Raubenheimer and Crowe 1987; Coyne *et al.* 1988; Templeton 1989). We refer the reader to these papers and to the counter-arguments of Spencer *et al.* (1987) and Masters and Spencer (1989).

As noted by Coyne *et al.* (1988), the RSC can be considered a subset of the BSC that involves a limited set of isolating barriers (behavioral, pollinator, and gametic). The RSC excludes other barriers that can create and maintain discrete clusters in sympatry. We have shown that these excluded barriers have clearly played a major role in speciation. In polyploidy, for example, new taxa are created by a combination of intrinsic postzygotic and ecological isolation. We see no advantage, and considerable disadvantage, in concentrating on only the subset of isolating barriers that involve mating and fertilization. Situations that are problematic for the BSC are equally problematic for the RSC. Moreover, the RSC faces additional problems, such as how to deal with cases in which there is some hybridization but no introgression because hybrids are sterile or inviable.

Cohesion Species Concept (CSC)

A species is the most inclusive population of individuals having the potential for phenotypic cohesion through intrinsic cohesion mechanisms.

(Templeton 1989)

The CSC gives a mechanistic underpinning to the GCSC by attempting to include all the factors that *preserve* morphological and genetic clusters in sexual and asexual organisms. It thus takes as its species problem the existence of discrete clusters, but, like the recognition concept, the CSC emphasizes factors keeping members of a cluster together more than those keeping members of different clusters apart.

The CSC is sometimes considered superior to the BSC for two reasons. First, it sees reproductive isolation as a misleading way to think about speciation (Templeton 1989, p. 5; all quotations and page numbers refer to this paper):

> For example, under the classic allopatric model of speciation, speciation occurs when populations are totally separated from each other by geographic barriers. The intrinsic isolating mechanisms given in Table 1 are obviously irrelevant as isolating barriers during speciation because they cannot function as isolating mechanisms during allopatry. Hence the evolutionary forces responsible for this allopatric speciation process have nothing to do with 'isolation'.

This is said to cause confusion for adherents to the BSC (p. 6):

> This is not to say that [reproductive] isolation is not a product of the speciation process in some

cases, but the product (i.e., isolation) should not be confused with the process (i.e., speciation). The isolation concept has been detrimental to studies of speciation precisely because it has fostered that confusion. (Paterson 1985)

We do not understand the rationale for separating the process of speciation (the evolution of barriers to gene exchange) from its product (species themselves). Under the BSC, speciation cannot be equated with simple differentiation of populations, because without the evolution of barriers to gene exchange, distinct taxa cannot coexist in sympatry. It is not difficult for us to see species as simply an advanced stage in the evolution of such barriers. And we cannot point to a single case in which research on speciation has been hindered by confusion between process and product.

Second, the CSC is deemed superior to the BSC because it can diagnose species in two difficult cases: asexuality, and hybridization between sympatric groups that nevertheless maintain their distinctness as clusters.

The difference between the GCSC and the CSC is that the latter incorporates explanations for why individuals within clusters remain genetically and phenotypically *similar*. Templeton describes a number of factors, called "cohesion mechanisms," that enforce this similarity. These mechanisms fall into two classes. "Genetic exchangeability" mechanisms include all factors "that define the limits of the spread of new genetic variants through *gene flow*" (p. 13). These include not only the complete list of reproductive isolating barriers characterizing the BSC, but also mechanisms facilitating gene flow *within* clusters, such as a common fertilization system ("the organisms are capable of successfully exchanging gametes") and a common developmental system ("the products of fertilization are capable of giving rise to viable and fertile adults"). But all these mechanisms are simply a different way of describing isolating barriers. A species can be seen as a group of interbreeding populations (i.e., conspecific individuals share "fertilization and developmental systems"), while *different* species can be seen as groups of populations whose fertilization and developmental systems are sufficiently diverged to prevent gene exchange.

The novel aspect of the CSC is the emphasis on cohesion mechanisms that enforce demographic exchangeability. This includes all factors "that define the fundamental niche and the limits of spread of new genetic variants through genetic drift and natural selection" (p. 13). (The fundamental niche is considered "the intrinsic [i.e., genetic] tolerances of the individuals to various environmental factors that determine the range of environments in which the individuals are potentially capable of surviving and reproducing" [pp. 14–15].) Within sexually reproducing populations, of course, selection and genetic drift promote genetic homogeneity of a species. But understanding how these forces operate within a cluster does not explain how *distinct* sympatric clusters arise.

Demographic exchangeability becomes more important when dealing with asexual or uniparental populations, because this factor – and not reproductive isolation – may limit the spread of alleles by natural selection and genetic drift. . . . The origin of a new adaptive mutation in a population of bacteria produces a periodic selection event, during which the mutant clone replaces all other clones having similar ecological properties. Such replacement can also occur through the asexual equivalent of genetic drift: random differences in reproductive rates among demographically exchangeable clones. As Templeton notes (p. 15), "Every individual in a demographically exchangeable population is a potential common ancestor to the entire population at some point in the future."

One might therefore use the CSC to demarcate species or taxa in asexually reproducing groups. It has been so used by Cohan (2001), who connects this species concept to an explicit mechanism for speciation in bacteria. (. . . We do not object to applying the words "species" and "speciation" to asexual groups, so long as one recognizes that these words have a different meaning in sexual groups.) We are less convinced that the CSC works better than the BSC in sexually reproducing organisms. Nor do we feel that a species concept is better when it applies to both sexual and asexual groups rather than to sexual groups alone. If the processes of cluster formation differ between these two groups, as we suspect they do, then adopting a single species concept for both groups may impede rather than promote progress.

In many cases, especially those involving sympatric sexual taxa, both the BSC and the CSC identify the same clusters, for the CSC considers isolating barriers to be "cohesion mechanisms." In other situations, however, the CSC encounters the same difficulties as does the BSC. When forced crosses show that allopatric populations have complete intrinsic postmating isolation, they would presumably be regarded as good species by both the CSC and BSC. But if isolation is not complete, there is no way to diagnose these populations under either concept, for it is impossible to determine the "fundamental niche" of allopatric taxa. Thus, the CSC also faces problems with allopatric populations. Both concepts also have difficulties when dealing with groups, such as host races, that show some gene exchange but that nevertheless remain distinct. Such hybridizing entities show genetic exchangeability (an adaptive allele can spread between races), but not demographic exchangeability (one group cannot ecologically displace the other).

There are other situations in which the BSC can diagnose species but where the CSC fails because the criteria of genetic and demographic exchangeability conflict. Consider, for example, two sympatric, reproductively isolated species that compete for resources. A new mutation may arise in one species that allows it to outcompete the other, driving it to extinction. In such cases – and in any case in which an invader outcompetes a local species – the two groups are genetically nonexchangeable but demographically exchangeable. Under the BSC they are good species, but under the CSC their status is unclear.

The main problem with the CSC, however, is that it causes confusion, especially through its emphasis on "cohesion mechanisms." As Harrison (1998, pp. 24–25) notes:

> Many (perhaps most) biological properties of organisms that confer "cohesion" did not arise for that purpose. They are also effects not functions! Thus, life cycles that result in adults appearing at the same season, or habitat/resource associations which lead to aggregation of individuals in particular places, facilitate fertilization or lead to genetic and/or demographic cohesion. But in most cases, life cycles and habitat associations have not been molded by selection for the purpose of "cohesion."

It is hard to regard forms of natural selection that can create isolating barriers as "cohesion mechanisms." The fixation of adaptive alleles that cause reproductive isolation as a byproduct do not involve selection for cohesion. Rather, it is the reproductive cohesion of the group that allows such alleles to spread. In such cases the CSC reverses cause and effect. Moreover, not all aspects of sexual reproduction can be regarded as "cohesion mechanisms." Antagonistic sexual selection, produced by differing reproductive interests of males and females, may be important in speciation. But such selection is a manifestly *non-cohesive* evolutionary force.

Under the BSC, it is fairly clear when speciation has occurred – substantial barriers to gene flow exist. Under the CSC, however, speciation is seen as "the process by which new genetic systems of cohesion mechanisms evolve within a population," or as "the genetic assimilation of altered patterns of genetic and demographic exchangeability into intrinsic cohesion mechanisms" (p. 24). But how can one know whether these processes have caused speciation unless one observes isolating barriers between a population and its relatives?

Finally, the CSC does not seem to lead naturally to a research program that reveals the causes of clustering in sexually reproducing groups. The concept of demographic exchangeability, however, may give insight into the origin of clusters in asexual organisms.

Evolutionary Species Concept (EvSC)

A species is a single lineage of ancestral descendant populations or organisms, which maintains its identity from other such lineages and which has its own evolutionary tendencies and historical fate.

(Wiley 1978, modified from Simpson 1961)

The EvSC differs from the BSC and other concepts discussed above by endowing species with broader evolutionary significance. Under the EvSC, the species problem becomes the recognition of evolutionarily independent entities, and the species is the unit that evolves independently of other species. Wiley asserts that the EvSC is more universally applicable than the BSC because the EvSC deals with both sexual and asexual taxa. Moreover, the EvSC, unlike the BSC, is said to be "capable of dealing with species as spatial, temporal, genetic, epigenetic, ecological, physiological, phenetic, and behavioral entities" (Wiley 1978, p. 18; all quotations and page numbers refer to this paper).

The EvSC differs from the BSC by including no explicit mention of genetic interchange or reproductive isolation. Nevertheless, the arguments in favor of the EvSC show that in most cases it is equivalent to the BSC, at least for diagnosing species in sympatry.

Indeed, this equivalence is recognized by Wiley: "Separate evolutionary lineages (species) must be reproductively isolated from one another to the extent that this is required for maintaining their separate identities, tendencies, and historical fates" (p. 20). But the notion of "separate identities," which implies recognizable genotypic or phenotypic clusters, can conflict with the notion of separate "tendencies" and "historical fates." Two species that hybridize, for instance, may maintain separate identities, but even a small amount of hybridization can allow a generally advantageous mutation to spread from one group to the other, so that their evolutionary fates are connected. The grass *Agrostis tenuis*, for example, has developed local races that can survive high concentrations of lead on mine tailings, while adjacent populations lack the genes for tolerance. Tolerant and non-tolerant populations are often adjacent, and, being wind-pollinated, freely exchange most genes (McNeilly and Antonovics 1968). The populations have diverged in a few traits, but still hybridize pervasively. Are they different evolutionary species? The EvSC gives no clue. With gene flow, taxa may be evolutionarily independent at some loci and not others.

Allopatric populations that are genetically differentiated pose as many problems for the EvSC as for the BSC. While geographic isolation may seem to confer separate evolutionary fates, Wiley (1978, p. 23) notes that "we have no corroboration that this particular geographic event will lead to separate evolutionary paths and thus we have no reason to recognize two evolutionary species." Such recognition becomes possible only when "significant evolutionary divergence" occurs between allopatric populations (p. 23). Wiley, however, gives no idea of what constitutes significant divergence. If "significant" means "divergence that prevents the populations from exchanging genes were they to become sympatric," then the EvSC becomes the BSC.

Asexually reproducing taxa may be resolvable by the EvSC if significant divergence occurs between them. But the meaning of "significant" is again unclear. Are distinct clusters of many loci necessary to diagnose asexual species, or can they differ at only one or two loci?

The unique aspect of the EvSC is that it can deal with a single lineage evolving through time. According to Wiley, such a lineage is considered to be a single species so long as it does not branch, no matter how much evolutionary change it undergoes. This, of course, may result in some taxonomic confusion, as the same species name will often be used for very different organisms (consider the lineage leading to modern humans). But applying names to stages of a single evolving lineage is always an exercise in subjectivity.

The major problem with the EvSC, then, is that it cannot deal with gene flow between populations of sexually reproducing organisms. Unless one is precise about the meaning of "separate evolutionary tendencies and historical fates," decisions about species status become arbitrary. Clearly, greater evolutionary independence is conferred by stronger barriers to gene flow. In this sense, the EvSC approximates the BSC. Given the choice, we prefer the BSC because it is more useful: in sexually reproducing organisms, this concept explains the evolutionary independence of taxa (whose origin is a mystery under the EvSC) as a byproduct of isolating barriers.

Ecological Species Concept (EcSC)

A species is a lineage (or a closely related set of lineages), which occupies an adaptive zone minimally different from that of any other lineage in its range and which evolves separately from all lineages outside its range.

(Van Valen 1976)

Van Valen proposed the EcSC to remedy the problem of ecologically differentiated entities that still exchange genes. He specifically mentions hybridizing oaks as "cutting across the frame of reference of the now usual concept of species" (Van Valen 1976, p. 233; all quotations and page numbers refer to this paper). The EcSC resembles the EvSC except mat the independently-evolving lineages are also characterized as occupying "minimally different adaptive zones." ("Minimal" is used so that higher taxa are not considered ecological species.) The species problem again seems to be that of explaining discontinuities among sympatric groups.

Requiring that different species occupy different adaptive zones imposes a severe burden on the EcSC. According to Van Valen, adaptive zones are defined a priori, independent of the organisms that inhabit them: "An adaptive zone is some part of the resource space together with whatever predation and parasitism occurs on the group considered. It is a part of the environment, as distinct from the way of life of a taxon that may occupy it, and exists independently of any inhabitants it might have" (p. 234). Van Valen suggests that occupants of different adaptive zones can be recognized by observing "a difference in the ultimately regulating factor, or factors, of population density" (p. 234).

This definition conflicts with the view that niches cannot be defined independently of their occupants, since many organisms, such as beavers and moles, change the environment to suit their needs (Lewontin 1983). Moreover, some groups can coexist as distinct entities in sympatry without gene flow, even though their adaptive zones are identical or nearly so. This is true, for example, of some temporally isolated species, such as periodical cicadas or the even- and odd-year races of pink salmon.

More important, it is often hard to determine whether two sympatric relatives occupy different adaptive zones, much less "minimally different" ones. The haplochromine cichlids of Lake Victoria, for instance, are often considered almost ecologically identical. In such cases, Van Valen suggests using a surrogate criterion: the coexistence of species in sympatry *proves* that they occupy minimally different adaptive zones. But this notion makes the EcSC operationally identical to either the GCSC or BSC, depending on the amount of hybridization. It is questionable, however, whether sympatric coexistence always constitutes evidence for "minimally different adaptive zones." Ecologists have suggested several ways that ecologically identical species can coexist.

While we believe that differential resource use is widespread among closely related sympatric species, it may not be necessary.

Further, very different taxa may nonetheless occupy the same adaptive zone, as shown by competitive exclusion. Criticizing ecological species concepts, Wiley (1978, p. 24) notes,

> In the case where resources are limiting, one of the species could replace the other through interspecific competition from that portion of the range where they are sympatric, or entirely via extinction. Indeed, if interspecific competition causes at least some extinctions, it can work only where the niches of the competing species are similar enough for competition to occur or where one species' niche completely overlaps the other's . . . one might argue that a species forced to extinction through interspecific competition was not a species at all.

Such situations are common in nature, especially with introduced species. Should the Argentine ant, *Linepithema humile*, be considered conspecific with the unrelated *Pheidole megacephala* because the former outcompeted the latter in Bermuda (Crowell 1968)? It seems better to regard ecological difference as a criterion for species *persistence* than for species status.

Van Valen suggests that it is a matter of taste whether differentiated allopatric populations are considered different species, although it is, in principle, possible to determine whether such populations are regulated by different ecological factors. But he further argues that "reproductive isolation of allopatric populations is of minor evolutionary importance and hence needs little consideration" (p. 234). Thus, allopatric taxa that yield completely inviable or sterile hybrids in forced crosses might not be considered different ecological species. However, sympatric taxa that exchange genes but maintain phenotypic distinctness, such as oaks, *would* be considered separate species because their different traits imply different niches. Here the EcSC resembles the GSC if there is substantial introgression, but resembles the BSC if introgression is limited.

Finally, Van Valen suggests that the EcSC could be useful for distinguishing species in asexual groups (p. 235): "Species are maintained for the most part ecologically, not reproductively.

Completely asexual communities would perhaps be as diverse as sexual ones, with numerous subcontinuities and even discontinuities. This suggests but does not require that the main criterion of species be ecological." We agree that the EcSC (and the CSC) might be more useful than the BSC in dealing with asexual groups, although, as noted by Van Valen, the EcSC encounters difficulties in agamic complexes.

Phylogenetic Species Concepts (PSCs)

PSCs differ markedly from the BSC and the five concepts discussed above, which take as their species problem the origin of discrete groups in nature. In contrast, PSCs are concerned with identifying historically related groups, and their species problem is reconstructing the history of life. Systematists are thus the main proponents of PSCs and the most severe critics of the BSC.

Systematists can be quite caustic when comparing the PSC to the BSC (e.g., Nelson 1989). This acrimony does not derive from their view that reproductive isolation is unimportant – for most of them admit that it is – but from the belief that it is largely irrelevant to reconstructing history. As Baum (1992, p. 1) notes, "The potential for gene exchange is only loosely coupled to historical relatedness – the central consideration of systematics." Wheeler and Nixon (1990, p. 79) state this position forcefully:

> The militant view that systematists need to embrace is that the responsibility for species concepts lies *solely* with systematists. If we continue to bow to the study of process over pattern, then our endeavors to elucidate pattern become irrelevant.

While most advocates of the BSC recognize that reproductive isolation may sometimes be inconsistent with evolutionary history (see below), they consider historical relationships as largely irrelevant to understanding the discreteness of nature.

Most modern systematists infer phylogenetic relationships using quantitative methods (Felsenstein 2004). One widely used method, cladistics, involves using shared derived characters, or *synapomorphies*. These characters can be either organismal traits or genes. When two or more species share a synapomorphy relative to an outgroup – a taxon known from independent evidence to be a more distant relative – these species are placed together in a monophyletic group (i.e., a group descended from one ancestral species). Thus, more closely related groups are those sharing more recently evolved synapomorphies.

It is reasonable to suppose that most sympatric species diagnosed by the BSC will be similarly diagnosed as monophyletic groups whose members have synapomorphies. But evolutionary history and reproductive compatibility need not coincide. Perhaps the most common cause of such discordance involves peripatric speciation: colonists originating in only one population of a species invade a new area, and their descendants evolve into a new species. Figure 19.1 (A) gives an example of a phylogeny resulting from this scenario. Here, a common ancestor, species A, give rise to three taxa (B_1, B_2, and C). Taxa B_2 and C are the most closely related because they share a derived character or characters, X, that evolved in their own common ancestor and not in the direct ancestor of population B_1. Taxon C, however, may have invaded a new habitat and evolved traits causing reproductive isolation (RI) from the two taxa B_1 and B_2, which themselves can interbreed. The BSC would recognize two species, C and [B_1 + B_2], with B_1 and B_2 considered conspecific populations. Most phylogenetic species concepts, however, would recognize a *different* pair of species, B_1 and [B_2 + C]. Using cladistics, one would not unite populations B_1 and B_2 based on their reproductive compatibility, for this compatibility is not a shared derived character but a *primitive* character retained from the common ancestor A (a "symplesiomorphy"). In technical terms, biological species [B_1 + B_2] is *paraphyletic* with respect to taxon C. That is, within the group [B_1 + B_2], members of B_2 are more closely related to members of C than to members of B_1.

In this simple case, taxa B_1 and B_2, although *capable* of interbreeding, do not. But this situation is unrealistic. After all, populations B_1 and B_2 are members of the same biological species and will exchange genes, erasing their distinctness. When this occurs, *some* genes will show the phylogeny depicted in Figure 19.1 (A), while other

genes will show B_1 and B_2 to be sister taxa, with C an outgroup. If speciation has occurred recently, different loci or traits will not yield congruent phylogenies, and the true history of populations cannot be reconstructed. The discrepancy between the histories of populations and the histories of genes within those populations is the biggest problem afflicting phylogenetic species concepts.

Reproductive isolation and evolutionary history can also conflict when two or more reproductively compatible populations arise independently from different evolutionary lineages, a situation known as "parallel speciation." Limnetic morphs of the threespine stickleback are often cited as an example. Figure 19.1 (C) shows a phylogeny

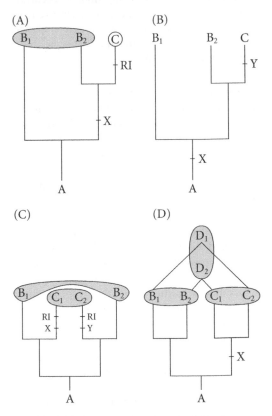

Figure 19.1 Phylogenies showing disparities between evolutionary history and reproductive isolation (see text for discussion). (A) Speciation in a peripheral isolate. (B) Species diagnosed by a trait difference using Phylogenetic Species Concept 1 (PSC 1). (C) Parallel speciation yielding a polyphyletic biological species. (D) Allopolyploidy yielding a polyphyletic biological species.

resulting from parallel speciation. Here, populations B_1 and B_2 are reproductively compatible, but each has a closer relative, C_1 and C_2, respectively. The latter two populations have evolved different derived traits (X and Y) but are reproductively compatible with each other and reproductively incompatible with B_1 and B_2. The BSC would diagnose two species, $[B_1 + B_2]$ and $[C_1 + C_2]$. However, $[C_1 + C_2]$ would not be recognized as a single species by most PSCs because it is *polyphyletic* (i.e., a taxon in which different populations or individuals have different common ancestors that reside outside the group). $[B_1 + B_2]$ also fails to constitute a phylogenetic species because this entity is paraphyletic. Despite the reproductive relationships among these groups, most PSCs would recognize only a single species comprising the group $[B_1 + B_2 + C_1 + C_2]$.

Polyphyly might seem to be rare given the implausibility that two independently evolved species would nevertheless be reproductively compatible with each other. But it may be common in one situation: polyploidy. Figure 19.1 (D) shows a phylogeny in which hybridization occurs between two biological species, B and C, eventually producing the allopolyploid species D. Two independent hybridizations between different individuals or populations of B (B_1 and B_2) and C (C_1 and C_2) can produce two allotetraploid populations (D_1 and D_2) that are reproductively compatible with each other but incompatible with species B and C. The reproductive compatibility between individuals of species D conceals the fact that this polyploid species includes two groups with independent historical origins.

But recognizing the polyphyly of populations may be as difficult as recognizing the paraphyly of populations. We are not able to directly witness the history of populations, and so must infer it from gene-based phylogenies. In both cases discussed above (Figures 19.1 C, D), interbreeding between populations of a biological species can quickly destroy our ability to reconstruct the history of populations, and thus our ability to show that this history is inconsistent with reproductive relationships.

Finally, some systematists dismiss the BSC because they view reproductive isolation as an *apomorphy* – a trait unique to one species – rather than as a synapomorphy that allows

cladistic analysis. But reproductive isolation differs from traditional traits used by cladists, for it is not diagnosable in individuals of one taxon. Rather, reproductive isolation is an *interaction*, or joint property of two taxa. Such interactions cannot be incorporated into cladistic studies, although the traits underlying them can.

There are three main versions of the PSC:

1. *PSC1* A phylogenetic species is an irreducible (basal) cluster of organisms that is diagnosably distinct from other such clusters, and within which there is a parental pattern of ancestry and descent (Cracraft 1989; see also Wheeler and Nixon 1990).

PSC1 is essentially a typological species concept that diagnoses species based on fixed differences in traits. (The term "irreducible" means that a species does not contain other diagnostic groups within it, so that large groups sharing diagnostic traits [e.g., mammals] are not deemed a single species. "Parental pattern of ancestry and descent" is included so that species status is not determined by sex differences or segregating polymorphic traits.)

Although advocates of PSC1 are not explicit on the point, in principle *any* trait can serve to diagnose a new species, even one as trivial as a small difference in color or a single nucleotide difference in DNA sequence. Applying PSC1 would thus tremendously increase the number of named species. *Homo sapiens*, for example, might be divided into several species based on diagnostic differences in morphology, molecules, or a combination of these features. (For diagnostic purposes, combinations of characters can be considered as single "traits.") Applying PSC1 to the birds of paradise, Cracraft (1992) increased the number of named species from about 40 to 90, often diagnosing as a new species an allopatric population having a slight difference in plumage color.

Like all phylogenetic species concepts, PSC1 cannot help us understand why organisms occur in discrete units, whether those units are defined phylogenetically or morphologically. However, its main difficulty is that its use may distort evolutionary history, the very problem it was meant to solve. Such distortion can occur because, under PSC1, species diagnosis is based not on shared derived traits, but on simple diagnostic traits. Figure 19.1 (B) gives an example of such distortion in three taxa whose true evolutionary history is shown by the phylogeny. The common ancestor of three taxa, B_1, B_2 and C, evolves a trait X. This state is retained in the descendant species B_1 and B_2. In species C, however, the trait has changed to state Y. Using this trait, the PSC1 would diagnose two species: $[B_1 + B_2]$ and C. This distorts the evolutionary history of the group because B_2 and C are more closely related to each other than either is to B_1. Similarly, using novel traits may diagnose a polyploid taxon as a phylogenetic species, even if it had a polyphyletic origin. There is no reason to expect that diagnostic traits will always mirror evolutionary history. Other criticisms of PSC1 are raised by Avise and Ball (1990), Baum (1992), and Baum and Donoghue (1995).

How well does PSC1 handle situations that are problematic for the BSC? In many cases, these two concepts pick out identical species in sympatry, especially when several traits are used. (For an example of this concordance, see Dettman et al.'s [2003] comparison of the BSC with PSC2.) Coordinated sets of diagnostic traits cannot be maintained in sympatry without some form of reproductive isolation. Confronting allopatric populations, PSC1 considers them different species if they differ in any trait. Likewise, PSC1 diagnoses each recognizable clone in an asexual group as a different species. Finally, under PSC1, speciation consists of the fixation of a diagnostic character in a lineage, making the process identical to divergent evolution. This type of speciation will occur faster than biological speciation: fixation of one new allele is undoubtedly faster than the evolution of reproductive isolation, which usually requires changes at several loci.

2. *PSC2* A species is the smallest (exclusive) monophyletic group of common ancestry (de Queiroz and Donoghue 1988; see also Rosen 1979, Mishler and Brandon 1987, and Baum and Donoghue 1995).

PSC2 goes back to Ronald Fisher, who suggested that all members of a sexually reproducing species should share "the effective identity of . . . remote ancestry" (1930, p. 124). This concept, updated in light of cladistics by de Queiroz and Donoghue (1988) differs from PSC1 by basing species recognition not on diagnostic characters, but on synapomorphies – shared derived characters that define monophyletic groups.

According to PSC2, a taxon is a species if cladistic analysis shows that it is monophyletic, exclusive (i.e., a group whose members are more closely related to each other than to those of any other group), and includes no other exclusive monophyletic groups within it.

When characterized properly, the units diagnosed by PSC2 will usually be congruent with evolutionary history. The main problem with this concept is operational: how can one *determine* whether a group is monophyletic and exclusive?

The problem arises from population genetics. One wants to know whether populations are exclusive groups sharing a common ancestry, but such a diagnosis can be made only using genes or genetically based traits. Increasingly, systematists rely on gene sequences to reconstruct this ancestry. DNA-based traits have two advantages over traditionally used morphological traits. First, genetic markers are more likely to be selectively neutral and thus to change in a more time-dependent fashion, making them useful for historical reconstruction. Second, gene sequences offer a nearly infinite number of characters, with each nucleotide potentially yielding information about ancestry.

However, the wealth of genetic data also creates a serious problem for the PSC2, because the ancestry of populations must be inferred from the ancestry of genes, and, as has been emphasized many times, *gene trees need not correspond to species trees*. That is, the historical branching pattern of taxa themselves need not coincide with the historical branching pattern of their genes (Avise and Ball 1990; Hey 1994; Avise and Wollenberg 1997).

This problem is demonstrated in Figure 19.2 (A), which shows three taxa of haploid organisms, A, B, and C, derived from a common ancestor. The phylogeny of these taxa is represented by the "fat branches" of the tree. The problem is to use the phylogenies of genes to reconstruct the phylogeny of the fat branches – the "true" population history. To illustrate the problems, we consider a single gene, designating each allele present in an individual with a dot. (The different gene copies do not necessarily differ in sequence, but we assume in this diagram that they can be individually identified.)

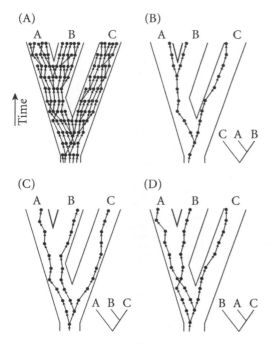

Figure 19.2   Gene sorting occurring in a phylogeny whose true population history is shown in (A). Depending on which gene copies are picked for analysis, one can derive all three possible phylogenies, shown by the small diagrams to the lower right of diagrams (B), (C), and (D). (After Hey 1994.)

At the two successive "fat branching" events, gene flow between lineages is instantaneously prevented by geographic isolation or the rapid evolution of reproductive isolation. At each such split, the two descendant branches contain many different copies of each gene. Through the process of "gene sorting" via drift and selection over generations, these gene copies create their own genealogies – "thin branches" – with some copies leaving no descendants, and others copies leaving varying numbers of descendants. ("Splitting" of a gene phylogeny in Figure 19.2 (A), reflects passage of a gene copy to more than one descendant, not new mutations or recombination events, which we ignore.) Gene copies present in the common ancestor can persist in descendants, often for long periods after the populations branch. Eventually, selection and drift will cause all gene copies within a lineage to descend from a single ancestral copy occurring within that lineage (that is, a *coalescence* occurs). When this

happens, the gene has become monophyletic within the fat branch.

Until coalescence takes place, however, there can be substantial disparity between the true genealogy of the populations (i.e., A and B are sister groups with respect to the outgroup C), and the genealogy inferred from genes. Figure 19.2 (B), (C), and (D) show that, using a single gene, one can obtain all three possible phylogenies between populations, only one of which gives the true population history. Although the populations have become evolutionarily independent taxa at the moment of isolation, in the sense that each now contains a nonoverlapping set of ancestors and descendants, one cannot genetically *demonstrate* that they are monophyletic until considerable time has passed. As noted by Avise and Ball (1990), after two populations become isolated, their genes will go through successive stages of polyphyly and paraphyly before finally becoming *reciprocally monophyletic* – the stage when all gene copies in each population are more closely related to each other than to copies in the other population.

This problem cannot be remedied by using larger samples of alleles or genes, because until reciprocal monophyly is attained, one can obtain conflicting phylogenies using different genes. This can be seen in Figure 19.3 for two genes, *zeste* and *YP2*, sampled in four species of *Drosopltila* (Hey and Kliman 1993). This group had a common ancestor that existed about 2.5 million years ago. *D. melanogaster* (mel) is an outgroup to the three species *D. simulans* (sim), *D. sechellia* (sec), and *D. mauritiana* (mall). All four species are distinguishable morphologically and show either substantial or complete reproductive isolation. Like *D. melanogaster*, *D. simulans* is cosmopolitan, while *D. sechellia* and *D. mauritiana* are endemic to the islands of the Seychelles and Mauritius, respectively. The endemics presumably arose after colonization of the islands by their common ancestor with *D. simulans*.

While the phylogeny of *D. melanogaster*, *D. sechellia*, and *D. mauritiana* is resolved under PSC2 using both genes, some sequences from *D. simulans* are more closely related to sequences found in *D. mauritiana* or *D. sechellia* than to other *D. simulans* sequences. That is, *D. simulans* is a paraphyletic species. Some systematists suggest that in such cases the paraphyletic taxon should

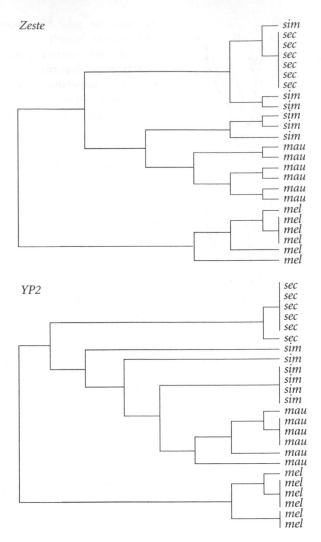

Figure 19.3 Phylogenies based on sequences of copies of two genes (*zeste* and *Yolk protein 2 [YP2]*) in four related species of *Drosophila* (*sim* = *D. simulans*, *sec* = *D. sechellia*, *mau* = *D. mauritiana*, and *mel* = *D. melanogaster*). Both sequences show that D. *simulans* is "paraphyletic," but the paraphyly is almost certainly of these genes and not of the populations themselves. (After Hey and Kliman 1993.)

be called a "metaspecies," so that individuals of *D. simulans* would not be recognized as belonging to *any* species (de Queiroz and Donoghue 1988; Baum and Shaw 1995). A study of 12 additional loci in this group gave similar results, failing to resolve the branching order of the two island colonizations (Kliman *et al.* 2000).

The most likely explanation for the discordant genealogies among genes is that *D. simulans* is still polymorphic for ancestral alleles that have become monophyletic in its island relatives. If geographic and genetic isolation persists, these species will eventually become monophyletic at all loci, but, as we show in the next section, this may take a very long time. Moreover, if there is any hybridization between the taxa (and there is some evidence for this in the *D. simulans* group), or if balancing selection maintains identical polymorphisms in different species, some genes will *never* become monophyletic within a lineage, and PSC2 status will never be attained.

The PSC2 is thus problematic because it ignores the distinction between monophyly of species and monophyly of genes. The latter is required to diagnose the former, but because of the long period required for different genes to show concordant phylogenies, the PSC2 will fail to diagnose (and resolve the history of) many species that are recognized using other concepts. This is why systematists introduced a third version of the PSC, the *genealogical species concept*.

3. *PSC3*   (Also called the "genealogical species concept" or GSC.) A species is a basal, exclusive group of organisms all of whose genes coalesce more recently with each other than with those of any organisms outside the group, and that contains no exclusive group within it (Baum and Donoghue 1995; Shaw 1998).

The GSC was proposed as a way to diagnose the phylogenetic status of populations using genes; it is, in fact, an operational definition of PSC2. Avise and Ball (1990) were the first to consider genetically based monophyly as a way to reconstruct the evolutionary history of taxa, but did not deem it a good way to diagnose species.

Like PSC2, the GSC recognizes species as exclusive groups whose members are more closely related to each other than to individuals of other species. The GSC also recognizes species as basal groups that contain no exclusive subgroups within them. The difference between the two concepts is that the GSC explicitly defines the monophyly of taxa as the monophyly of the genes carried by its members. Although diagnosing a group as a genealogical species (GS) should in principle involve many loci – after all, it is individual organisms and not genes that are members of a species – in practice monophyly is determined using a limited sample of genes.

The main task facing the GSC is specifying *how many loci* must be monophyletic to diagnose a group as a genealogical species. The original formulation by Baum and Donoghue (1995) requires *all* loci to be monophyletic. But this demand is too extreme, as balancing selection that preserves two or more alleles in an ancestral species can keep identical sets of alleles polymorphic in descendants. This is the situation at the MHC locus in humans versus chimps, and in rats versus mice, with both pairs of species having a divergence between 5 and 10 million years (Figueroa *et al.* 1988; Ayala and Escalante 1996). One also sees ancient polymorphisms of self-incompatibility alleles among species in the genus *Brassica* (Uyenoyama 1995).

Advocates of the GSC recognize the problem with demanding complete monophyly of all genes, but have avoided the question of what proportion of surveyed loci must be monophyletic to allow GS status. Shaw (2001), however, suggests that GS status might be recognized if most loci were monophyletic. Setting this "> 50%" threshold makes judgments about GS status somewhat arbitrary, but no more arbitrary than diagnosing biological species when reproductive isolation is incomplete.

Using coalescent theory, Hudson and Coyne (2002) studied the time to attain GS status when an ancestral species divides into two descendants and the only evolutionary forces operating are mutation and genetic drift. For a single descendant, attaining *complete* monophyly requires a long time, especially if many loci are sampled. To attain a 95% probability of observing monophyly at every sampled gene, where N is the effective size of the population, one requires 1.8 N generations to reach GS status for a single mitochondrial or chloroplast gene, 7.3 N generations for a single nuclear gene, and 26.3 N generations for 11,500 nuclear loci (roughly the number of genealogically independent units within the *Drosophila melanogaster* genome). Attaining *reciprocal* monophyly for both descendant populations takes roughly 10%–30% longer. Directional selection, which speeds the fixation of alleles, will shorten these times; but balancing selection, which retards fixation, will lengthen them. Surprisingly, the number of alleles sampled per gene has little effect on the time required to attain GS status.

Presumably, one uses a sample of loci to infer the GS status of the entire genome. Such a goal requires one to use a large sample of loci and to avoid diagnosing GSs based on single mitochondrial or chloroplast loci. Organelle genes have only one-fourth as many copies as any autosomal locus, and so will become monophyletic well before the rest of the genome. All genes in an organelle are also completely linked, so no additional information about ancestry is gained by using more than one such gene. Nevertheless, genealogical species have been diagnosed on the basis of a single mtDNA haplotype or allozyme locus (e.g., Young and Crother 2001; Leaché and Reeder 2002).

When one relaxes the criteria for GS status, so that only 50% or 95% of sampled loci need be monophyletic, the time to genealogical speciation is reduced. In a sample of 25 nuclear genes, for example, one observes complete reciprocal monophyly with 95% probability after 15.2 N generations. This drops to 4.7 N generations if only 50% of the loci need be monophyletic, and to 11.3 N generations if 95% of the loci need be monophyletic. In the limit, with an infinite sample of loci, one observes complete reciprocal monophyly with 100% probability after 3.8 N generations using the 50% criterion mid 8.7 N generations using the 95% criterion.

The conclusion is that one should not adopt a GS criterion requiring *complete* reciprocal monophyly for a large number of loci. Under this extreme view, humans and chimps would be considered one species, and rats and mice another. Moreover, attaining genome-wide monophyly takes so long that, before it occurs, taxa will be recognized as species using nearly every other species concept; indeed, many *additional* branching events might have occurred.

What is the relationship between biological and genealogical speciation? While both processes are accelerated by divergent natural selection and geographic isolation, there is no necessary correspondence between the times when species status is attained under the BSC and the GSC. However, biological speciation is almost certain to precede genealogical speciation if GS status requires complete reciprocal monophyly at many loci. *Drosophila simulans*, for example, is not a genealogical species with respect to *D. mauritiana* under even the "50% monophyly" criterion, and these taxa (although allopatric) have diverged in several morphological traits and show substantial reproductive isolation.

While a "relaxed" version of the GSC seems the most reasonable of all phylogenetic species concepts, we favor the BSC over the GSC for several reasons. First, applying the GSC will often involve designating taxa as metaspecies: large groups of individuals, such as *D. simulans*, will be not be recognized as belonging to any species. Unlike many doubtful cases in the BSC, the term "metaspecies" describes an *ontological* situation (organisms that are not members of any species) rather than an epistemological one "(groups that cannot be assigned to recognized species due to

a lack of evidence)" (Baum and Shaw 1995, p. 297; our italics). At the moment when an isolated population becomes monophyletic, every individual in every other population instantly loses its status as belonging to any species. It seems odd that, without any change in its own genetic composition, a group can lose species status based on what happens in a remote population. It should be added, however, that systematists disagree on whether the term "metaspecies" should be used, or which entities should be so characterized.

Second, little of biological import occurs at the completion of genealogical speciation. What significance, for example, can one impute to the moment at which the proportion of loci showing exclusivity rises from 50% to 50.1% – the completion of one type of genealogical speciation? In contrast, the completion of biological speciation – the moment when gene flow between sister taxa is no longer possible – corresponds to a biologically meaningful event. It is the moment when taxa become evolutionarily independent (Coyne 1994a). The termination of gene flow also allows genealogies to coalesce without pollution by genes from other taxa. Thus, these reproductive barriers, along with geographical barriers, provide the isolation that *permits* the monophyly required for genealogical speciation. In this sense, reproductive isolation is more fundamental than genetic coalescence.

Finally, genealogical speciation will often be transitory, for the coalescence of genes does not guarantee that geographically isolated populations will remain distinct when they become sympatric. One can envision many small, isolated populations quickly attaining genealogical species status, But range shifts or the disappearance of geographic barriers will quickly eliminate these genealogical species; they will hybridize with other populations and their exclusivity will vanish. In contrast, some forms of reproductive isolation are permanent. It is the permanence of reproductive isolation that guarantees the independence of genealogies among taxa.

The BSC and evolutionary history

Applying the BSC is an exercise not in reconstructing the history of taxa, but in identifying reproductively isolated groups. But implicit in this exercise is the idea that populations of a single biological species are more closely related to each other than to populations of a different species. To justify systematists' assertions that the BSC frequently distorts evolutionary history, one should be able to show many cases in which that history conflicts with reproductive compatibility. The most commonly cited examples are paraphyletic species and polyphyletic species. We have already noted some of the difficulties with using "thin branch" phylogenies of genes to determine whether population phylogenies – "fat branch" phylogenies – ate paraphyletic or polyphyletic.

Paraphyly

In principle, species with a paraphyletic *origin* should be common. There must be many cases (e.g., peripatric speciation) in which a new biological species, B, originates from only one population of ancestral species A. If species B evolves reproductive isolation from all populations of A, which themselves remain reproductively compatible, then species A is *historically* paraphyletic. That is, if we were present at the moment when the population destined to become species B was geographically isolated, we would see that it derived from only one population of species A.

But we were not present at this moment, and so cannot directly witness the history of taxa. We must rely on the thin branches – the phylogenies of genes – to reconstruct the pattern of fat branches. But gene-based phylogenies can yield false diagnoses of paraphyly for several reasons.

One reason, mentioned above, is that populations of the ancestral species do not remain genetically isolated. Their interbreeding will quickly erase the genetic differences between populations that can be used to diagnose paraphyly. *Recognizable* paraphyly is therefore likely to be a transitory phenomenon. What we expect is that some loci will show paraphyly of alleles in species A relative to those in species B, but that this pattern will not be consistent across all genes. In other words, different traits or loci may yield different patterns of relatedness, with some showing paraphyly and others not. Moreover, as seen in Figure 19.2, one expects this discordance even if species A is *not* paraphyletic, for discordance is the expected result when ancestral polymorphisms are sorted into descendant species. Such paraphyly is an ineluctable part of

the speciation process and therefore cannot conflict with the BSC.

How many genes, then, must show concordant phylogenies before we are confident that the taxa themselves are paraphyletic? This crucial issue has been almost completely ignored by systematists. In fact, we do not know of any case in which species paraphyly is demonstrated by concordant genealogies of many (or even several) genes. Claims for species paraphyly are almost always based on one or a few loci, usually on the mitochondria (e.g., Melnick *et al.* 1993; Patton and Smith 1994; Omland 1997; Omland *et al.* 2000). Such paraphyly tells us little about the evolutionary history of populations because organelle genes may not accurately mirror the rest of the genome.

Indeed, many studies show that mitochondrial and chloroplast DNA introgress between taxa much more readily than does nuclear DNA (e.g., Ferris *et al.* 1983; Smith 1992; Bernatchez *et al.* 1995; Howard *et al.* 1997; Taylor and McPhail 2000; Martinsen *et al.* 2001; Shaw 2002). The reasons for this are unclear, but may be due to the nature of mitochondrial genes. Most of these genes are constitutively expressed and perform internal metabolic "housekeeping" or protein-synthetic functions, such as producing tRNA or respiratory enzymes. Such functions may be largely divorced from external selective pressures, making mtDNA less responsive than nuclear genes to local environmental differences. This may also be true for cpDNA, which contains genes for photosynthesis, tRNA, and rRNA. Thus, organelle genes, unlike nuclear genes, may function fairly well in the genetic background of a related species. In addition, the spread of adaptive mutations in organelle DNA is not impeded by their linkage to nuclear genes that are divergently adapted between taxa or cause intrinsic postzygotic isolation in hybrids.

The consequence of introgression and linkage is that organelle DNA may appear paraphyletic even when the species themselves are not. Phylogenies based solely on organelle DNA can also distort history in other ways. For example, Shaw (1996) showed that a mtDNA-based phylogeny of Hawaiian crickets (*Laupala*) was discordant with traditional phylogenies based on morphology and biogeography. The mtDNA phylogenies showed that the most closely related species were sympatric, implying sympatric speciation. However, later phylogenies based on nuclear DNA were concordant with the traditional ones, supporting allopatric speciation following colonization of new islands (Shaw 2002). The most likely reason for this discordance is the introgression of mitochondria between sympatric taxa, which can hybridize. (One would predict that in species having heterogametic females, such as birds and Lepidoptera, mtDNA would introgress less readily. Because Haldane's rule holds in these groups, F_1 hybrid females, which pass on mtDNA, are often sterile.)

Under the "fat branch" approach, the evolutionary history of populations is usually represented by trees with bifurcating branches. However, the techniques used to reconstruct this history involve genes whose diverse genealogies can yield a complicated set of reticulations instead of a definitive phylogeny. Because there is no unitary *genetic* history at the population level, it is almost impossible to recognize true paraphyly among closely related taxa using genetically based phylogenies.

Polyphyly

A polyphyletic species includes individuals or populations having independent evolutionary origins from common ancestors residing outside that species. As noted above, polyphyletic species include independently formed polyploid individuals that interbreed with one another, as well as cases of parallel speciation.

There is strong genetic evidence for a polyphyletic origin of some auto- and allopolyploid plant species, and of at least one species of terrestrial snail (Ueshima and Asami 2003). These cases indeed show genuine discordance between the evolutionary history of populations and their reproductive relationships. And because hybridization can form new polyploids repeatedly, this discordance may persist for long periods. In some cases, independently derived polyploids co-occur in nature and interbreed, showing that they are indeed members of the same biological species. But phylogenetic species concepts are unable to deal with such polyphyletic species, as they combine the genes of two ancestral species.

Parallel speciation has a similar effect, except that the independent origins of a single species

involve convergent evolution rather than repeated hybridization. There are two possible cases of nonhybrid diploid species having multiple evolutionary origins: limnetic morphs of the three-spine stickleback, *Gasterosteus aculeatus* (Rundle et al. 2000), and host races of the stick insect *Timema cristinae* (Nosil *et al.* 2002). However, in both cases there is ongoing gene flow between sympatric taxa. This can yield inaccurate phylogenies, making speciation events appear independent when they are not.

We conclude that while the BSC may occasionally identify species that are not monophyletic, it is not clear that phylogenetic species concepts are better at dealing with this problem.

References

Avise, J. C. and R. M. Ball. 1990. Principles of genealogical concordance in species concepts and biological taxonomy. Pp. 45–67 *in* D. J. Futuyma and J. Antonovics (eds.) *Oxford Surveys in Evolutionary Biology*. Oxford University Press, Oxford.

Avise, J. C. and K. Wollenberg. 1997. Phylogenetics and the origin of species. *Proc. Natl. Acad. Sci. USA* 94: 7748–7755.

Ayala, F. J. and A. A. Escalante. 1996. The evolution of human populations – a molecular perspective. *Mol. Phylogenet. Evol.* 5: 188–201.

Barton, N. H. and G. M. Hewitt. 1985. Analysis of hybrid zones. *Annu. Rev. Ecol. Syst.* 16: 113–148.

Baum, D. 1992. Phylogenetic species concepts. *Trends Ecol. Evol.* 7: 1–2.

Baum, D. A. and M. J. Donoghue. 1995. Choosing among alternative phylogenetic species concepts. *Syst. Bot.* 20: 560–573.

Baum, D. A. and K. L. Shaw. 1995. Genealogical perspectives on the species problem. Pp. 289–303 *in* P. C. Hoch and A. C. Stephenson (eds.) *Experimental and Molecular Approaches to Plant Biosystematics*. Missouri Botanical Garden, St. Louis, MO.

Berlocher, S. H. 1999. Host race or species? Allozyme characterization of the 'flowering dogwood fly,' a member of the *Rhagoletis pomonella* complex. *Heredity* 83: 652–662.

Bernatchez, L., L. Glemete, C. C. Wilson, and R. G. Danzmann. 1995. Introgression and fixation of Arctic char (*Salvenlinus alpinus*) mitochondrial genome in an allopatric population of brook trout (*Salvelinus fontinalis*). *Can. J. Fish. Aquat. Sci.* 52: 179–185.

Brookfield, J. 2002. Review of *Genes, Categories, and Species* by Jody Hey. *Genet. Res.* 79: 107–108.

Butlin, R. K. 1987a. Speciation by reinforcement. *Trends Ecol. Evol.* 2: 8–13.

Butlin, R. K. 1987b. Species, speciation, and reinforcement. *Am. Nat.* 130: 461–464.

Clarke, C. A. and P. M. Sheppard. 1969. Further studies on the genetics of the mimetic butterfly *Papilio memnon*. *Philos. Trans. R. Soc. Lond. B* 263: 35–70.

Cohan, F. M. 2001. Bacterial species and speciation. *Syst. Biol.* 50: 513–524.

Coyne, J. A. 1994a. Ernst Mayr on the origin of species. *Evolution* 48: 19–30.

Coyne, J. A., H. A. Orr, and D. J. Futuyma. 1988. Do we need a new species concept? *Syst. Zool.* 37: 190–200.

Cracraft, J. 1989. Speciation and its ontology: The empirical consequences of alternative species concepts for understanding patterns and processes of differentiation. Pp. 28–59 *in* D. Otte and J. A. Endler (eds.) *Speciation and Its Consequences*. Sinauer Associates, Sunderland, MA.

Crowell, K. 1968. Rates of competitive exclusion by the Argentine ant in Bermuda. *Ecology* 49: 551–555.

de Queiroz, K. and M. J. Donoghue. 1988. Phylogenetic systematics and the species problem. *Cladistics* 4: 317–338.

Dettman, J. R., D. J. Jacobson, E. Turner, A. Pringle, and J. W. Taylor. 2003. Reproductive isolation and phylogenetic divergence in Neurospora: Comparing methods of species recognition in a model eukaryote. *Evolution* 57: 2721–2741.

Dobzhansky, T. and C. Epling. 1944. Contributions to the genetics, taxonomy, and ecology of *Drosophila pseudoobscura* and its relatives. Washington, DC: Pub. 54: 1–46, Carnegie Institute Washington.

Eberhard, W. G. 1986. *Sexual Selection and Animal Genitalia*. Harvard University Press. Cambridge, MA.

Feder, J. L., S. H. Berlocher, and S. B. Opp. 1998. Sympatric host race formation and speciation in *Rhagoletis* (Diptera: Tephritidae): A tale of two species for Charles Darwin. Pp. 408–441 *in* S. Mopper and S. Strauss (eds.) *Genetic Structure and Local Adaptation in Natural Insect Populations: Effects of Ecology, Life History, and Behavior*. Chapman and Hall, New York.

Felsenstein, J. 2004. *Inferring Phylogenies*. Sinauer Associates, Sunderland, MA.

Ferris, S. D., R. D. Sage, C.-M. Huang, J. T. Nielsen, U. Ritte, and A. C. Wilson. 1983. Flow of mitochondrial DNA across a species boundary. *Proc. Natl. Acad. Sci. USA* 80: 2290–2294.

Figueroa, F., E. Günter, and J. Klein. 1988. MHC polymorphism pre-dating speciation. *Nature* 335: 265–267.

Fisher, R. A. 1930. *The Genetical Theory of Natural Selection*. Clarendon Press, Oxford.

Harrison, R. G. 1998. Linking evolutionary pattern and process: The relevance of species concepts for the study of speciation. Pp. 19–31 *in* D. J. Howard and S. H. Berlocher (eds.) *Endless Forms: Species and Speciation*. Oxford University Press, New York.

Hey, J. H. 1994. Bridging phylogenetics and population genetics with gene tree models. Pp. 435–449 *in* B. Schierwater, B. Streit, G. P. Wagner, and R. DeSalle (eds.) *Molecular Ecology and Evolution: Approaches and Applications*. Birkhäuser-Verlag, Basel.

Hey, J. and R. M. Kliman. 1993. Population genetics and phylogenetics of DNA sequence variation at multiple loci within the *Drosophila melanogaster* species complex. *Mol. Biol. Evol.* 10: 804–822.

Howard, D. J., R. W. Preszler, J. Williams, S. Fenchel, and W. J. Boecklen. 1997. How discrete are oak species? Insights from a hybrid zone between *Quercus grisea* and *Quercus gambelii*. *Evolution* 51: 747–755.

Hudson, R. R. and J. A. Coyne. 2002. Mathematical consequences of the genealogical species concept. *Evolution* 56: 1557–1565.

Kliman, R. M., P. Andolfatto, J. A. Coyne, F. Depaulis, M. Kreitman, A. J. Berry, M. McCarter, J. Wakeley, and J. Hey. 2000. The population genetics of the origin and divergence of the *Drosophila simulans* complex species. *Genetics* 156: 1913–1931.

Lambert, D. M. and H. E. H. Paterson. 1984. "On bridging the gap between race and species": the isolation concept and an alternative. *Proc. Linn. Soc. NSW* 107: 501–514.

Lambert, D. M., B. Michaux, and C. S. White. 1987. Are species self-defining? *Syst. Zool.* 36: 196–205.

Leaché, A. D. and T. W. Reeder. 2002. Molecular systematics of the eastern fence lizard (*Sceloporus undulatus*): a comparison of parsimony, likelihood, and Bayesian approaches. *Syst. Biol.* 51: 44–68.

Lewontin, R. C. 1983. The organism as the subject and object of evolution. *Scientia* 118: 63–82.

Mallet, J. 1995. A species definition for the Modern Synthesis. *Trends Ecol. Evol.* 10: 294–299.

Martinsen, G. D., T. G. Whitham, R. J. Turek, and P. Keim. 2001. Hybrid populations selectively filter gene introgression between species. *Evolution* 55: 1325–1335.

Masters, J. C. and H. G. Spencer. 1989. Why we need a new genetic species concept. *Syst. Zool.* 38: 270–279.

Masters, J. C., R. J. Rayner, I. J. McKay, A. D. Potts, D. Nails, J. W. Ferguson, B. K. Weissenbacher, M. Alsopp, and M. L. Anderson. 1987. The concept of species: Recognition versus isolation. *S. Afr. J. Sci.* 83: 534–537.

Mayr, E. 1942. *Systematics and the Origin of Species*. Columbia University Press, New York.

Mayr, E. 1963. *Animal Species and Evolution*. Belknap Press, Cambridge, MA.

Melnick, D. J., G. A. Hoelzer, R. Absher, and M. U. Ashley. 1993. mtDNA diversity in Rhesus monkeys reveals overestimates of divergence time and paraphyly with neighboring species. *Mol. Biol. Evol.* 10: 282–295.

McNeilly, T. and J. Autonovics. 1968. Evolution in closely adjacent plant populations. IV Barriers to gene flow. *Heredity* 23: 205–218.

Mishler, B. D. and R. N. Brandon. 1987. Individuality, pluralism, and the phylogenetic species concept. *Biol. Philos.* 2: 397–414.

Nelson, G. 1989. Species and taxa: Systematics and evolution. Pp. 60–81 *in* D. Otte and J. A. Endler (eds.) *Speciation and Its Consequences*. Sinauer Associates, Sunderland, MA.

Nosil, P., B. J. Crespi, and C. P. Sandoval. 2002. Host-plant adaptation drives the parallel evolution of reproductive isolation. *Nature* 417: 440–443.

Omland, K. E. 1997. Examining two standard assumptions of ancestral reconstructions: Repeated losses of dichromatism in dabbling ducks (*Anatini*). *Evolution* 51: 1636–1646.

Omland, K. E., C. L. Tarr, W. I. Boarman, J. M. Marzluff, and R. C. Fleischeer. 2000. Cryptic genetic variation and paraphyly in ravens. *Proc. R. Soc. Lond. B* 267: 2475–2482.

Paterson, H. E. H. 1985. The recognition concept of species. Pp. 21–29 *in* E. S. Vrba (ed.) *Species and Speciation*. Transvaal Museum Monograph No. 4, Pretoria.

Patton, J. L. and M. F. Smith. 1994. Paraphyly, polyphyly, and the nature of species boundaries in pocket gophers (Genus *Thomomys*). *Syst. Biol.* 43: 11–26.

Raubenheimer, D. and T. M. Crowe. 1987. The Recognition Species Concept: Is it really an alternative? *S. Afr. J. Sci.* 83: 530–534.

Rosen, D. E. 1979. Fishes from the uplands and intermontane basins of Guatemala: Revisionary studies and comparative geography. *Bull. Am. Mus. Nat. Hist.* 162: 267–376.

Rundle, H. D., L. Nagel, J. W. Boughman, and D. Schluter. 2000. Natural selection and parallel speciation in sympatric sticklebacks. *Science* 287: 306–308.

Shaw, K. L. 1996. Polygenic inheritance of a behavioral phenotype: Interspecific genetics of song in the Hawaiian cricket genus *Laupala*. *Evolution* 50: 256–266.

Shaw, K. L. 1998. Species and the diversity of natural groups. Pp. 44–56 *in* D. J. Howard and S. J. Berlocher (eds.) *Endless Forms: Species and Speciation*. Oxford University Press, Oxford.

Shaw, K. L. 2001. The genealogical view of speciation. *J. Evol. Biol.* 14: 880–882.

Shaw, K. L. 2002. Conflict between nuclear and mitochondrial DNA phylogenies of a recent species radiation: What mitochondrial DNA reveals and conceals about modes of speciation in Hawaiian crickets. *Proc. Natl. Acad. Sci. USA* 99: 16122–16129.

Simpson, G. G. 1961. *Principles of Animal Taxonomy*. Columbia University Press, New York.

Smith, T. B. 1987. Bill size polymorphism and inter-specific niche utilization in an African finch. *Nature* 329: 717–719.

Sokal, R. R. and T. J. Crovello. 1970. The Biological Species Concept: A critical evaluation. *Am. Nat.* 104: 127–153.

Spencer, H. G., D. M. Lambert, and B. H. McArdle. 1987. Reinforcement, species, and speciation: A reply to Butlin. *Am. Nat.* 130: 958–962.

Taylor, E. B. and J. D. McPhail. 2000. Historical contingency and ecological determinism interact to prime speciation in sticklebacks, *Gasterosteus*. *Proc. R. Soc. Lond. B* 267: 2375–2384.

Templeton, A. R. 1989. The meaning of species and speciation: A genetic perspective. Pp. 3–27 *in* D. Otte and J. A. Endler (eds.) *Speciation and Its Consequences*. Sinauer Associates, Sunderland, MA.

Ueshima, R. and T. Asami. 2003. Single-gene speciation by left-right reversal. *Nature* 425: 679.

Uyenoyama, M. 1995. A generalized least-squares estimate for the origin of sporophytic self-incompatibility. *Genetics* 139: 975–992.

Van Valen, L. 1976. Ecological species, multispecies, and oaks. *Taxon* 25: 233–239.

Wheeler, Q. D. and K. C. Nixon. 1990. Another way of *looking at* the species problem: A reply to de Queiroz and Donoghue. *Cladistics* 6: 77–81.

Wiley, E. O. 1978. The evolutionary species concept reconsidered. *Syst. Zool.* 27: 17–26.

Wilson, D. S. and M. Turelli. 1986. Stable under-dominance and the evolutionary invasion of empty niches. *Am. Nat.* 127: 835–850.

Young, J. E. and B. I. Crother. 2001. Allozyme evidence for the separation of *Rana areolata* and *Rana capita* and for the resurrection of *Rana sevosa*. *Copeia* 2001: 382–388.

PART IX

THE UNITS OF SELECTION DEBATE

Introduction

The debate about whether natural selection acts at only one level of organization or at many different levels at the same or different times is a concrete biological case of the broader debate about reductionism in the natural sciences. If natural selection is a process that operates at the level of the gene, for example, instead of the genotype or the individual, then evolutionary processes obtaining at the level of the individual, lineage, population, or species, will all have to be explained by the operation of natural selection at the level of the gene. Or, more radically, they may all have to be treated as just symptoms – by-products or side-effects – of a process that really does not take place anywhere else in the biological realm than at the level of the DNA.

Richard Dawkins argued for the more radical view in his famous work, *The Selfish Gene* (1976), and extended his arguments for the claim in other influential work. Lewontin and Sober (chapter 20) offered one of the first knowledgeable critiques of this view. Their argument, if correct, provides a further objection to reductionism in biology to go along with Kitcher's (see Part VII), or perhaps an example of the sort of irreducibility that Kitcher identifies in biology. Independently, it advanced a serious challenge to the "gene's eye' point of view" which, since the publication of *The Selfish Gene*, had become so influential in biology.

It is somewhat ironic that (Chapter 21) Kitcher should have produced, together with Kim Sterelny, a powerful argument against Lewontin and Sober's antireductionist claims. They do not argue for the strongest version of Dawkins' theory – that it is only at the level of the gene that selection takes place. But they may do the next best thing, arguing that wherever selection is taking place in the biological realm – at the level of the population, the lineage, the individual, the genotype, and so on – it is also always taking place at the level of the gene! Genic reductionists will be pardoned if they consider Kitcher and Sterelny's conclusion a minor concession to antireductionists about the levels of selection.

In the years since the debate between genic and genotypic selectionists began, the problem of the levels of selection has become far more consequential. The genic selectionist is faced by the mystery of explaining why selfish genes selected for fitness maximization should merge their fates with other genes and chromosomes, why they should allow half their copies to be lost in sexual reproduction, and why they should become somatic genes with no long-term reproductive prospects. This mysterious outcome poses a problem not just for genes, but one that arises repeatedly at "the major transitions of evolution" from genes to chromosomes, to multicellularity, to tissue specialization, to individual organisms, and to groups of them. Samir Okasha's review paper (chapter 22) shows how the initial debate revealed the tip of a biological iceberg.

Further Reading

Brandon, R., & Burian, R. (eds.) (1984). *Genes, organisms, population: Controversies over the units of selection*. Cambridge MA: MIT Press.

Craig, D. (1982). Group selection vs. individual selection: An experimental analysis. *Evolution*, 36, 271–282.

Dawkins, R. (1976). *The selfish gene*. Oxford: Oxford University Press.

Dawkins, R. (1978). Replicator selection and the extended phenotype. *Zeitschrift fur Tierpsychologie*, 47, 61–76.

Hull, D. (1980). Individuality and Selection. *Annual Review of Ecology and Systematics*, 11, 311–332.

Lloyd, E. (2005). Why the gene will not return. *Philosophy of Science*, 72, 287–310.

Lloyd, E., & Gould, S. (1993). Species selection on variability. *Proceedings of the National Academy of Sciences (USA)*, 90, 595–599.

Maynard Smith, J. (1998). *Evolutionary genetics*. Oxford: Oxford University Press.

Okasha, S. (2001). Why won't the group selection controversy go away? *British Journal for the Philosophy of Science*, 52, 25–50.

Okasha, S. (2003). The concept of group heritability. *Biology and Philosophy*, 18, 445–461.

Okasha, S. (2004). Multi-level selection, covariance and contextual analysis. *British Journal for the Philosophy of Science*, 55, 481–504.

Rosenberg, A., & McShea, D. (2008). *Philosophy of biology: A contemporary introduction* (Chapter 6). London: Routledge.

Shanahan, T. (1997). Pluralism, antirealism, and the units of selection. *Acta Biotheoretica*, 45, 117–126.

Sober, E. (1984). *The nature of selection: Evolutionary theory in philosophical focus*. Chicago: University of Chicago Press.

Sober, E., & Wilson, D. (1994). A critical review of philosophical work on the units of selection problem. *Philosophy of Science*, 61, 534–555.

Sober, E., & Wilson, D. (1998). *Unto others: The evolution and psychology of unselfish behavior*. Cambridge, MA: Harvard University Press.

Waters, C. (2005). Why genic and multilevel selection theories are here to stay. *Philosophy of Science*, 72, 311–333.

Wilson, R. (2003). Pluralism, entwinement, and the levels of selection. *Philosophy of Science*, 70, 531–552.

20

Artifact, Cause and Genic Selection

Elliott Sober and Richard C. Lewontin

Several evolutionary biologists have used a parsimony argument to argue that the single gene is the unit of selection. Since all evolution by natural selection can be represented in terms of selection coefficients attaching to single genes, it is, they say, "more parsimonious" to think that all selection is selection for or against single genes. We examine the limitations of this genic point of view, and then relate our criticisms to a broader view of the role of causal concepts and the dangers of reification in science.

Introduction

Although predicting an event and saying what brought it about are different, a science may yet hope that its theories will do double duty. Ideally, the laws will provide a set of parameters which facilitate computation and pinpoint causes; later states of a system can be predicted from its earlier parameter values, where these earlier parameter values are the ones which cause the system to enter its subsequent state.

In this paper, we argue that these twin goals are not jointly attainable by some standard ideas used in evolutionary theory. The idea that natural selection is always, or for the most part, selection for and against single genes has been vigorously defended by George C. Williams (*Adaptation and Natural Selection*) and Richard

Dawkins (*The Selfish Gene*). Although models of evolutionary processes conforming to this view of genic selection may permit computation, they often misrepresent the causes of evolution. The reason is that genic selection coefficients are *artifacts*, not causes, of population dynamics. Since the gene's eye point of view exerts such a powerful influence both within biology and in popular discussions of sociobiology, it is important to show how limited it is. Our discussion will not focus on cultural evolution or on group selection, but rather will be restricted to genetic cases of selection in a single population. The selfish gene fails to do justice to standard textbook examples of Darwinian selection.

The philosophical implications and presuppositions of our critique are various. First, it will be clear that we reject a narrowly instrumentalist

Elliott Sober and Richard Lewontin, "Artifact, Cause, and Genic Selection," *Philosophy of Science* (Philosophy of Science Association, 1982) 49, 157–180. Reprinted by permission of the publisher, the University of Chicago Press.

This paper was written while the authors held grants, respectively, from the University of Wisconsin Graduate School and the John Simon Guggenheim Foundation and from the Department of Energy (DE-AS02-76EV02472). We thank John Beatty, James Crow, and Steven Orzack for helpful suggestions.

interpretation of scientific theories; models of evolutionary processes must do more than correctly predict changes in gene frequencies. In addition, our arguments go contrary to certain regularity and counterfactual interpretations of the concepts of causality and force. To say that *a* caused *b* is to say more than just that any event that is relevantly similar to *a* would be followed by an event that is relevantly similar to *b* (we ignore issues concerning indeterministic causation); and to say that a system of objects is subject to certain forces is to say more than just that they will change in various ways, as long as nothing interferes. And lastly, our account of what is wrong with genic selection coefficients points to a characterization of the conditions under which a predicate will pick out a real property. Selfish genes and grue emeralds bear a remarkable similarity.

1. The "Canonical Objects" of Evolutionary Theory

The Modern Synthesis received from Mendel a workable conception of the mechanism of heredity. But as important as this contribution was, the role of Mendelian "factors" was more profound. Not only did Mendelism succeed in filling in a missing link in the three-part structure of variation, selection, and transmission; it also provided a canonical form in which *all* evolutionary processes could be characterized. Evolutionary models must describe the interactions of diverse forces and phenomena. To characterize selection, inbreeding, mutation, migration, and sampling error in a single predictive theoretical structure, it is necessary to describe their respective effects in a common currency. Change in gene frequencies is the "normal form" in which all these aspects are to be represented, and so genes might be termed the canonical objects of evolutionary theory.

Evolutionary phenomena can be distilled into a tractable mathematical form by treating them as preeminently genetic. It by no means follows from this that the normal form characterization captures everything that is biologically significant. In particular, the computational adequacy of genetic models leaves open the question of whether they also correctly identify the causes

of evolution. The canonical form of the models has encouraged many biologists to think of all natural selection as genic selection, but there has always been a tradition within the Modern Synthesis which thinks of natural selection differently and holds this gene's eye view to be fundamentally distorted.

Ernst Mayr perhaps typifies this perspective. Although it is clear that selection has an *effect* on gene frequencies, it is not so clear that natural selection is always selection for or against particular genes. Mayr has given two reasons for thinking that the idea of genic selection is wrong. One of the interesting things about his criticisms is their simplicity; they do not report any recondite facts about evolutionary processes but merely remind evolutionary theorists of what they already know (although perhaps lose sight of at times). As we will see, genic selectionists have ready replies for these criticisms.

The first elementary observation is that "natural selection favors (or discriminates against) phenotypes, not genes or genotypes" (1963, p. 184). Protective coloration and immunity from DDT are phenotypic traits. Organisms differ in their reproductive success under natural selection because of their phenotypes. If those phenotypes are heritable, then natural selection will produce evolutionary change (*ceteris paribus*, of course). But genes are affected by natural selection only indirectly. So the gene's eye view, says Mayr, may have its uses, but it does not correctly represent how natural selection works.

Mayr calls his second point *the genetic theory of relativity* (1963, p. 296). This principle says that "no gene has a fixed selective value, the same gene may confer high fitness on one genetic background and be virtually lethal on another." Should we conclude from this remark that there is never selection for single genes or that a single gene simultaneously experiences different selection pressures in different genetic backgrounds? In either case, the lesson here seems to be quite different from that provided by Mayr's first point – which was that phenotypes, not genotypes, are selected for. In this case, however, it seems to be gene complexes, rather than single genes, which are the objects of selection.

Mayr's first point about phenotypes and genotypes raises the following question: if we grant that selection acts "directly" on phenotypes and only "indirectly" on genotypes, why should it follow that natural selection is not selection for genetic attributes? Natural selection is a causal process; to say that there is selection for some (genotypic or phenotypic) trait X is to say that having X causes differential reproductive success (*ceteris paribus*).[1] So, if there is selection for protective coloration, this just means that protective coloration generates a reproductive advantage. But suppose that this phenotype is itself caused by one or more genes. Then having those genes causes a reproductive advantage as well. Thus, if selection is a causal process, in acting on phenotypes it also acts on the underlying genotypes. Whether this is "direct" or not may be important, but it doesn't bear on the question of what is and what is not selected for. Selection, in virtue of its causal character and on the assumption that causality is transitive, seems to block the sort of asymmetry that Mayr demands. Asking whether phenotypes or genotypes are selected for seems to resemble asking whether a person's death was caused by the entry of the bullet or by the pulling of the trigger.

Mayr's second point – his genetic principle of relativity – is independent of the alleged asymmetry between phenotype and genotype. It is, of course, not in dispute that a gene's fitness depends on its genetic (as well as its extrasomatic) environment. But does this fact show that there is selection for gene complexes and not for single genes? Advocates of genic selection tend to acknowledge the relativity but to deny the conclusion that Mayr draws. Williams (1966, pp. 56–7) gives clear expression to this common reaction when he writes:

Obviously it is unrealistic to believe that a gene actually exists in its own world with no complications other than abstract selection coefficients and mutation rates. The unity of the genotype and the functional subordination of the individual genes to each other and to their surroundings would seem at first sight, to invalidate the one-locus model of natural selection. Actually these considerations do not bear on the basic postulates of the theory. No matter how functionally dependent a gene may be, and no matter how complicated its interactions with other genes and environmental factors, it must always be true that a given gene substitution will have an arithmetic mean effect on fitness in any population. One allele can always be regarded as having a certain selection coefficient relative to another at the same locus at any given point in time. Such coefficients are numbers that can be treated algebraically, and conclusions inferred for one locus can be iterated over all loci. Adaptation can thus be attributed to the effect of selection acting independently at each locus.

Dawkins (1976, p. 40) considers the same problem: how can single genes be selected for, if genes build organisms only in elaborate collaboration with each other and with the environment? He answers by way of an analogy:

One oarsman on his own cannot win the Oxford and Cambridge boat race. He needs eight colleagues. Each one is a specialist who always sits in a particular part of the boat – bow or stroke or cox, etc. Rowing the boat is a cooperative venture, but some men are nevertheless better at it than others. Suppose a coach has to choose his ideal crew from a pool of candidates, some specializing in the bow position, others specializing as cox, and so on. Suppose that he makes his selection as follows. Every day he puts together three new trial crews, by random shuffling of the candidates, for each position, and he makes the three crews race against each other. After some weeks of this it will start to emerge that the winning boat often tends to contain the same individual men. These are marked up as good oarsmen. Other individuals seem consistently to be found in slower crews, and these are eventually rejected. But even an outstandingly good oarsman might sometimes be a member of a slow crew, either because of the inferiority of the other members, or because of bad luck – say a strong adverse wind. It is only *on average* that the best men tend to be in the winning boat.

The oarsmen are genes. The rivals for each seat in the boat are alleles potentially capable of occupying the same slot along the length of a chromosome. Rowing fast corresponds to building a body which is successful at surviving. The wind is the external environment. The pool of alternative candidates is the gene pool. As far as the survival of any one body is concerned, all its genes are in the same boat. Many a good gene gets into bad company, and finds itself sharing a body with a lethal gene, which kills the body off in childhood. Then the good gene is destroyed along with the rest. But this is only one body, and replicas of the same good gene live on in other bodies which lack the lethal gene. Many copies of good genes are dragged under because they happen to share a body with bad genes, and many perish through other forms of ill luck, say when their body is struck by lightning. But by definition luck, good and bad, strikes at random, and a gene which is consistently on the losing side is not unlucky; it is a bad gene.

Notice that this passage imagines that oarsmen (genes) are good and bad pretty much *independently* of their context. But even when fitness is heavily influenced by context, Dawkins still feels that selection functions at the level of the single gene. Later in the book (pp. 91–2), he considers what would happen if a team's performance were improved by having the members communicate with each other. Suppose that half of the oarsmen spoke only English and the other half spoke only German:

> What will emerge as the overall best crew will be one of the two stable states – pure English or pure German, but not mixed. Superficially it looks as though the coach is selecting whole language groups *as units*. This is not what he is doing. He is selecting individual oarsmen for their apparent ability to win races. It so happens that the tendency for an individual to win races depends on which other individuals are present in the pool of candidates.

Thus, Dawkins follows Williams in thinking that genic selectionism is quite compatible with the fact that a gene's fitness depends on context.

Right after the passage just quoted, Dawkins says that he favors the perspective of genic selectionism because it is more "parsimonious". Here, too, he is at one with Williams (1966), who uses parsimony as one of two main lines of attack against hypotheses of group selection. The appeal to simplicity may confirm a suspicion that already arises in this context: perhaps it is a matter of taste whether one prefers the single gene perspective or the view of selection processes as functioning at a higher level of organization. As long as we agree that genic fitnesses depend on context, what difference does it make how we tell the story? As natural as this suspicion is in the light of Dawkins' rowing analogy, it is mistaken. Hypotheses of group selection can be genuinely incompatible with hypotheses of organismic selection (Sober 1980), and, as we will see in what follows, claims of single gene selection are at times incompatible with claims that gene complexes are selected for and against. Regardless of one's aesthetic inclinations and regardless of whether one thinks of parsimony as a "real" reason for hypothesis choice, the general perspective of genic selectionism is mistaken for biological reasons.[2]

Before stating our objections to genic selectionism, we want to make clear one defect that this perspective does *not* embody. A quantitative genetic model that is, given at any level can be recast in terms of parameters that attach to genes. This genic representation will correctly trace the trajectory of the population as its gene frequencies change. In a minimal sense (to be made clear in what follows), it will be "descriptively adequate". Since the parameters encapsulate information about the environment, both somatic and extrasomatic, genic selectionism cannot be accused of ignoring the complications of linkage or of thinking that genes exist in a vacuum. The defects of genic selectionism concern its distortion of causal processes, not whether its models allow one to predict future states of the population.[3]

The causal considerations which will play a preeminent role in what follows are not being imposed from without, but already figure centrally in evolutionary theory. We have already mentioned how we understand the idea of *selection for X*. Our causal construal is natural in view of how the phenomena of linkage and pleiotropy are understood (see Sober 1981). Two genes may be linked together on the same chromosome, and so selection for one may cause them both to increase in frequency. Yet the linked gene – the "free rider" – may be neutral or even deleterious; there was no selection *for it*. In describing pleiotropy, the same distinction is made. Two phenotypic traits may be caused by the same underlying gene complex, so that selection for one leads to a proliferation of both. But, again, there was no selection for the free rider. So it is a familiar idea that two traits can attach to exactly the same organisms and yet differ in their causal roles in a selection process. What is perhaps less familiar is that two sets of selection coefficients may both attach to the same population and yet differ in their causal roles – the one causing change in frequencies, the other merely reflecting the changes that ensue.

2. Averaging and Reification

Perhaps the simplest model exhibiting the strategy of averaging recommended by Williams and Dawkins is used in describing heterozygote superiority. In organisms whose chromosomes come in pairs, individuals with different genes (or alleles) at the same location on two homologous chromosomes are called heterozygotes. When a population has only two alleles at a locus, there will be one heterozygote form (Aa) and two homozygotes (AA and aa). If the heterozygote is superior in fitness to both homozygotes, then natural selection may modify the frequencies of the two alleles A and a, but will not drive either to fixation (i.e., 100%), since reproduction by heterozygotes will inevitably replenish the supply of homozygotes, even when homozygotes are severely selected against. A textbook example of this phenomenon is the sickle cell trait in human beings. Homozygotes for the allele controlling the trait develop severe anemia that is often fatal in childhood. Heterozygotes, however, suffer no deleterious effects, but enjoy a greater than average resistance to malaria. Homozygotes for the other allele have neither the anemia nor the immunity, and so are intermediate in fitness. Human populations with both alleles that live in malarial areas have remained polymorphic, but with the eradication of malaria, the sickle cell allele has been eliminated.

Population genetics provides a simple model of the selection process that results from heterozygotes' having greater viability than either of the homozygotes (Li 1955). Let p be the frequency of A and q be the frequency of a (where $p + q = 1$). Usually, the maximal fitness, of Aa is normalized and set equal to 1. But for clarity of exposition we will let w_1 be the fitness of AA, w_2 be the fitness of Aa, and w_3 be the fitness of aa. These genotypic fitness values play the mathematical role of transforming genotype frequencies before selection into genotype frequencies after selection:

	AA	Aa	aa
Proportion before selection	p^2	$2pq$	q^2
Fitness	w_1	w_2	w_3
Proportion after selection	$\dfrac{p^2 w_1}{\bar{W}}$	$\dfrac{2pq w_2}{\bar{W}}$	$\dfrac{q^2 w_3}{\bar{W}}$

Here, $\bar{W}$, the average fitness of the population, is $p^2 w_1 + 2pq w_2 + q^2 w_3$. Assuming random mating, the population will move towards a stable equilibrium frequency $\hat{p}$ where

$$\hat{p} = \frac{w_3 - w_2}{(w_1 - w_2) + (w_3 - w_2)}.$$

It is important to see that this model attributes fitness values and selection coefficients to diploid genotypes and not to the single genes A and a. But, as genic selectionists are quick to emphasize, one can always define the required parameters. Let us do so.

We want to define W_A, which is the fitness of A. If we mimic the mathematical role of genotype fitness values in the previous model, we will require that W_A obey the following condition:

W_A × frequency of A before selection = frequency of A after selection × $\bar{W}$.

Since the frequency of A before selection is p and the frequency of A after selection is

$$\frac{w_1 p^2 - w_{21} pq}{\bar{W}},$$

it follows that

$$W_A = w_1 p + w_2 q.$$

By parity of reasoning,

$$W_a = w_3 q + w_2 p.$$

Notice that the fitness values of single genes are just weighted averages of the fitness values of the diploid genotypes in which they appear. The weighting is provided by their frequency of occurrence in the genotypes in question. The genotypic fitnesses specified in the first model are *constants*; as a population moves toward its equilibrium frequency, the selection coefficients attaching to the three diploid genotypes do not change. In contrast, the expression we have derived for allelic fitnesses says that allelic fitnesses change as a function of their own frequencies; as the population moves toward equilibrium, the fitnesses of the alleles must constantly be recomputed.

Heterozygote superiority illustrates the principle of genetic relativity. The gene *a* is maximally fit in one context (namely, when accompanied by *A*) but is inferior when it occurs in another (namely, when it is accompanied by another copy of itself). In spite of this, we can average over the two different contexts and provide the required representation in terms of genic fitness and genic selection.

In the diploid model discussed first, we represented the fitness of the three genotypes in terms of their *viability*, that is, in terms of the proportion of individuals surviving from egg to adult. It is assumed that the actual survivorship of a class of organisms sharing the same genotype precisely represents the fitness of that shared genotype. This assumes that random drift is playing no role. Ordinarily, fitness *cannot* be identified with actual reproductive success (Brandon 1978; Mills and Beatty 1979; Sober 1981). The same point holds true, of course, for the fitness coefficients we defined for the single genes.[4]

Of the two descriptions we have constructed of heterozygote superiority, the first model is the standard one; in it, *pairs* of genes are the bearers of fitness values and selection coefficients. In contrast to this diploid model, our second formulation adheres strictly to the dictates of genic selectionism, according to which it is *single genes* which are the bearers of the relevant evolutionary properties. We now want to describe what each of these models will say about a population that is at its equilibrium frequency.

Let's discuss this situation by way of an example. Suppose that both homozygotes are lethal. In that case, the equilibrium frequency is .5 for each of the alleles. Before selection, the three genotypes will be represented in proportions $1/4$, $1/2$, $1/4$, but after selection the frequencies will shift to 0, 1, 0. When the surviving heterozygotes reproduce, Mendelism will return the population to its initial $1/4$, $1/2$, $1/4$ configuration, and the population will continue to zig-zag

between these two genotype configurations, all the while maintaining each allele at .5. According to the second, single gene, model, at equilibrium the fitnesses of the two genes are both equal to 1 and the selection coefficients are therefore equal to zero. At equilibrium, no selection occurs, on this view. Why the population's *genotypic configuration* persists in zig-zagging, the gene's eye point of view is blind to see; it must be equally puzzling why $\bar{W}$, the average fitness of the population, also zig-zags. However, the standard diploid model yields the result that selection occurs when the population is at equilibrium, just as it does at other frequencies, favoring the heterozygote at the expense of the homozygotes, Mendelism *and selection* are the causes of the zig-zag. Although the models are computationally equivalent in their prediction of gene frequencies, they are not equivalent when it comes to saying whether or not selection is occurring.

It is hard to see how the adequacy of the single gene model can be defended in this case. The biological term for the phenomenon being described is apt. We are talking here about *heterozygote superiority*, and both terms of this label deserve emphasis. The heterozygote – i.e., the diploid genotype (not a single gene) – is superior *in fitness* and, therefore, enjoys a selective advantage. To insist that the single gene is always the level at which selection occurs obscures this and, in fact, generates precisely the wrong answer to the question of what is happening at equilibrium. Although the mathematical calculations can be carried out in the single gene model just as they can in the diploid genotypic model, the phenomenon of heterozygote superiority cannot be adequately "represented" in terms of single genes. This model does not tell us what is patently obvious about this case: even at equilibrium, what happens to gene frequencies is an artifact of selection acting on diploid genotypes.

One might be tempted to argue that in the heterozygote superiority case, the kind of averaging we have criticized is just an example of frequency dependent selection and that theories of frequency dependent selection are biologically plausible and also compatible with the dictates of genic selectionism. To see where this objection goes wrong, one must distinguish genuine from spurious cases of frequency dependent selection.

The former occurs when the frequency of an allele has some *biological impact* on its fitness; an example would be the phenomenon of mimicry in which the rarity of a mimic enhances its fitness. Here one can tell a biological story explaining why the fitness values have the mathematical form they do. The case of heterozygote superiority is altogether different; here frequencies are taken into account simply as a mathematical contrivance, the only point being to get the parameters to multiply out in the right way.

The diploid model is, in a sense, more contentful and informative than the single gene model. We noted before that from the *constant* fitness values of the three genotypes we could obtain a formula for calculating the fitnesses of the two alleles. Allelic fitnesses are implied by genotype fitness values and allelic frequencies; since allelic frequencies change as the population moves toward equilibrium, allelic fitnesses must constantly be recomputed. However, the derivation in the opposite direction cannot be made.[5] One cannot deduce the fitnesses of the genotypes from allelic fitnesses and frequencies. This is especially evident when the population is at equilibrium. At equilibrium, the allelic fitnesses are identical. From this information alone, we cannot tell whether there is no selection at all or whether some higher level selection process is taking place. Allelic frequencies plus genotypic fitness imply allelic fitness values, but allelic frequencies plus allelic fitness values do not imply genotypic fitness values. This derivational asymmetry suggests that the genotypic description is more informative.

Discussions of reductionism often suggest that theories at lower levels of organization will be more detailed and informative than ones at higher levels. However, here, the more contentful, constraining model is provided at the higher level. The idea that genic selection models are "deeper" and describe the fundamental level at which selection "really" occurs is simply not universally correct.

The strategy of averaging fosters the illusion that selection is acting at a lower level of organization than it in fact does. Far from being an idiosyncratic property of the genic model of heterozygote superiority just discussed, averaging is a standard technique in modelling a variety of selection processes. We will now describe another example in which this technique of representation is used. The example of heterozygote

superiority focused on differences in genotypic *viabilities*. Let us now consider the way differential fertilities can be modelled for one locus with two alleles. In the fully general case, fertility is a property of a mating pair, not of an individual. It may be true that a cross between an *AA* male and *aa* female has an expected number of offspring different from a cross between an *AA* female and an *aa* male. If fitnesses are a unique function of the pair, the model must represent nine possible fitnesses, one for each mating pair. Several special cases permit a reduction in dimensionality. If the sex of a genotype does not affect its fertility, then only six fitnesses need be given; and if fertility depends only on one of the sexes, say the females, the three female genotypes may be assigned values which fix the fertilities of all mating pairs.

But even when these special cases fail to obtain, the technique of averaging over contexts can nevertheless provide us with a fitness value for each genotype. Perhaps an *aa* female is highly fertile when mated with an *Aa* male but is much less so when mated with an *AA* male; perhaps *aa* females are quite fertile on average, but *aa* males are uniformly sterile. No matter – we can merely average over all contexts and find the average effect of the *aa* genotype. This number will fluctuate with the frequency distributions of the different mating pairs. Again, the model appears to locate selection at a level lower than what might first appear to be the case. Rather than assigning fertilities to mating pairs, we now seem to be assigning them to genotypes. This mathematical contrivance is harmless as long as it does not lead us to think that selection really acts at this lower level of organization.[6]

Our criticism of genic selectionism has so far focused on two forms of selection at a single locus. We now need to take account of how a multilocus theory can imply that selection is not at the level of the selfish gene. The pattern of argument is the same. Even though the fitness of a pair of genes at one locus may depend on what genes are found at other loci, the technique of averaging may still be pressed into service. But the selection values thereby assigned to the three genotypes at a single locus will be artifacts of the fitnesses of the nine genotype complexes that exist at the two loci. As in the examples we already described, the lower-level selection coefficients will change as a function of genotype frequencies, whereas the higher-level selection coefficients will remain constant. An example of this is provided by the work of Lewontin and White (reported in Lewontin 1974) on the interaction of two chromosome inversions found in the grasshopper *Moraba scura*. On each of the chromosomes of the EF pair, Standard (ST) and Tidbinbilla (TD) may be found. On the CD chromosome pair, Standard (ST) and Blundell (BL) are the two alternatives. The fitness values of the nine possible genotypes were estimated from nature as follows:

Chromosome EF	Chromosome CD		
	ST/ST	ST/BL	BL/BL
ST/ST	0.791	1.000	0.834
ST/TD	0.670	1.006	0.901
TD/TD	0.657	0.657	1.067

Notice that there is heterozygote superiority on the CD chromosome if the EF chromosome is either ST/ST or ST/TD, but that BL/BL dominance ensues when the EF chromosome is homozygous for TD. Moreover, TD/TD is superior when in the context BL/BL but is inferior in the other contexts provided by the CD pair. These fitness values represent differences in viability, and again the inference seems clear that selection acts on multilocus genotypic configurations and not on the genotype at a single locus, let alone on the separate genes at that locus.

3.   Individuating Selection Processes

The examples in the previous section have a common structure. We noted that the fitness of an object (a gene, a genotype) varied significantly from context to context. We concluded that selection was operating at a level higher than the one posited by the model – at the level of genotypes in the case of heterozygote superiority, at the level of the mating pair in the fertility model, and at the level of pairs of chromosome inversions in the *Moraba scura* example. These analyses suggest the following principle: *if the fitness of X is context sensitive, then there is not selection for X;*

rather, there is selection at a level of organization higher than X.

We believe that this principle requires qualification. To see why context sensitivity is not a *sufficient* condition for higher level selection, consider the following example. Imagine a dominant lethal gene; it kills any organisms in which it is found unless the organism also has a suppressor gene at another locus. Let's consider two populations. In the first population, each organism is homozygous for a suppressor gene which prevents copies of the lethal gene from having any effect. In the second population, no organism has a suppressor, so, whenever the lethal gene occurs, it is selected against. A natural way of describing this situation is that there is selection against the lethal gene in one population, but, in the other, there is no selection going on at all. It would be a mistake (of the kind we have already examined) to think that there is a single selection process at work here against the lethal gene, whose magnitude we calculate by averaging over the two populations. However, we do not conclude from this that there is a selection process at work at some higher level of organization than the single gene. Rather, we conclude that there are *two* populations; in one, *genic* selection occurs, and in the other *nothing* occurs. So the context sensitivity of fitness is an ambiguous clue. If the fitness of X depends on genetic context, this may mean that there is a single selection process at some higher level, *or* it may mean that there are several different selection processes at the level of X. Context sensitivity does not suffice for there to be selection at a higher level.[7]

Thus, the fitness of an object can be sensitive to genetic context for at least two reasons. How are they to be distinguished? This question leads to an issue at the foundation of *all* evolutionary models. What unites a set of objects as all being subject to a single selection process? Biological modelling of evolution by natural selection is based on three necessary and sufficient conditions (Lewontin 1970): a given set of objects must exhibit variation; some individuals must be fitter than others; and there must be correlation between the fitness of parents and the fitness of offspring. Here, as before, we will identify fitness with actual reproductive success, subject to the proviso that these will coincide only in special cases. Hence, evolution by natural selection exists when and only when there is heritable variation in fitness.

Using these conditions presupposes that some antecedent decision has been made about which objects can appropriately be lumped together as participating in a single selection process (or, put differently, the conditions are not sufficient after all). Biologists do not talk about a *single* selection process subsuming widely scattered organisms of different species which are each subject to quite different local conditions. Yet, such a gerrymandered assemblage of objects may well exhibit heritable variation in fitness. And even within the same species, it would be artificial to think of two local populations as participating in the same selection process because one encounters a disease and the other experiences a food shortage as its principal selection pressure. Admittedly, the gene frequencies can be tabulated and pooled, but in some sense the relation of organisms to environments is too heterogeneous for this kind of averaging to be mote than a mathematical contrivance.

It is very difficult to spell out necessary and sufficient conditions for when a set of organisms experience "the same" selection pressure. They need not compete with each other. To paraphrase Darwin, two plants may struggle for life at the edge of a desert, and selection may favor the one more suited to the stressful conditions. But it needn't be the case that some resource is in short supply, so that the amount expropriated by one reduces the amount available to the other. Nor need it be true that the two organisms be present in the same geographical locale; organisms in the semi-isolated local populations of a species may experience the same selection pressures. What seems to be required, roughly, is that some common causal influence impinge on the organisms. This sameness of causal influence is as much determined by the biology of the organisms as it is by the physical characteristics of the environment. Although two organisms may experience the same temperature fluctuations, there may be no selective force acting on both. Similarly, two organisms may experience the same selection pressure (for greater temperature tolerance, say) even though the one is in a cold environment and the other is in a hot one. Sameness of causal influence needs to be understood biologically.

For all the vagueness of this requirement, let us assume that we have managed to single out the class of objects which may properly be viewed as participating in a single selection process. To simplify matters, let us suppose that they are all organisms within the same breeding population. What, then, will tell us whether selection is at the level of the single gene or at the level of gene complexes? To talk about either of these forms of selection is, in a certain important but nonstandard sense, to talk about "group selection". Models of selection do not concern single organisms or the individual physical copies of genes (i.e., geno*tokens*) that they contain. Rather, such theories are about groups of organisms which have in common certain geno*types*. To talk about selection fox X, where X is some single gene or gene cluster, is to say something about the effect of having X and of lacking X on the relevant subgroups of the breeding population. If there is selection for X, every object which has X has its reproductive chances augmented by its possessing X. This does not mean that every organism which has X has precisely the same overall fitness, nor does it mean that every organism must be affected in precisely the same way (down to the minutest details of developmental pathways). Rather, what is required is that the effect of X on each organism be in the same direction as far as its overall fitness is concerned. Perhaps this characterization is best viewed as a limiting ideal. To the degree that the population conforms to this requirement, it will be appropriate to talk about genic selection. To the degree that the population falls short of this, it will be a contrivance to represent matters in terms of genic selection.[8]

It is important to be clear on why the context sensitivity of a gene's effect on organismic fitness is crucial to the question of genic selection. Selection theories deal with groups of single organisms and not with organisms taken one at a time. It is no news that the way a gene inside of a single organism will affect that organism's phenotype and its fitness depends on the way it is situated in a context of background conditions. But to grant this fact of context sensitivity does not impugn the claim of causation; striking the match caused it to light, even though the match had to be dry and in the presence of oxygen for the cause to produce the effect.

Selection theory is about geno*types* not geno-*tokens*. We are concerned with what properties are selected for and against in a population. We do not describe single organisms and their physical constituents one by one. It is for this reason that the question of context sensitivity becomes crucial. If we wish to talk about selection for a single gene, then there must be such a thing as *the* causal upshot of possessing that gene. A gene which is beneficial in some contexts and deleterious in others will have many *organismic* effects. But at *the population level*, there will be no selection for or against that gene.

It is not simply the averaging over contexts which reveals the fact that genic selection coefficients are pseudoparameters; the fact that such parameters *change* in value as the population evolves while the biological relations stay fixed also points to their being artifacts. In the case of heterozygote superiority, genotypic fitnesses remain constant, mirroring the fact that the three genotypes have a uniform effect on the viability of the organisms in which they occur. The population is thereby driven to its equilibrium value while genic fitness values are constantly modified. A fixed set of biological relationships fuels both of these changes; the evolution of genic fitness values is effect, not cause.[9]

Are there real cases of genic selection? A dominant lethal – a gene which causes the individual to die regardless of the context in which it occurs – would be selected against. And selection for or against a phenotypic trait controlled by a single locus having two alleles might also be describable in terms of genic selection, provided that the heterozygote is intermediate in fitness between the two homozygotes. In addition, meiotic drive, such as is found in the house mouse *Mus musculus*, similarly seems to involve genic selection (Lewontin and Dunn 1960). Among heterozygote males, the proportion of t-alleles in the sperm pool is greater than $^1/_2$. Chromosomes with the t-allele have enhanced chances of representation in the gamete pool, and this directional effect seems to hold true regardless of what other genes are present at other loci.[10] At this level, but not at the others at which the t-allele affects the population, it is appropriate to talk about genic selection.

We so far have construed genic selection in terms of the way that having or lacking a gene can affect the reproductive chances of organisms. But there is another possibility – namely, that genes differentially proliferate even though they have *no* effect on the phenotypes of organisms. A considerable quantity of DNA has no known function; Orgel and Crick (1980) and Doolittle and Sapienza (1980) have suggest that this DNA may in fact be "junk". Such "selfish DNA", as they call it, could nonetheless undergo a selection process, provided that some segments are better replicators than others. Although these authors associate their ideas with Dawkins' selfish gene, their conception is far more restrictive. For Dawkins, *all* selection is genic selection, whereas for these authors, selfish DNA is possible only when the differential replication of genes is not exhaustively accounted for by the differential reproductive success of organisms.

Standard ways of understanding natural selection rule out rather than substantiate the operation of genic selection. It is often supposed that much of natural selection is *stabilizing selection*, in which an intermediate phenotype is optimal (e.g., birth weight in human beings). Although the exact genetic bases of such phenotypes are frequently unknown, biologists often model this selection process as follows. It is hypothesized that the phenotypic value is a monotone increasing function of the number of "plus alleles" found at a number of loci. Whether selection favors the presence of plus genes at one locus depends on how many such genes exist at other loci. Although this model does not view heterozygote superiority as the most common fitness relation *at a locus*, it nevertheless implies that a *heterogeneous genome* is superior in fitness. Exceptions to this intermediate optimum model exist, and the exact extent of its applicability is still an open question. Still, it appears to be widely applicable. If it is generally correct, we must conclude that the conditions in which genic selection exists are extremely narrow. Genic selection is not impossible, but the biological constraints on its operation are extremely demanding.

Although it is just barely conceivable that a critique of a scientific habit of thought might be devoid of philosophical presuppositions, our strictures against genic selectionism are not a case in point. We have described selection processes in which genic selection coefficients are *reifications*; they are artifacts, not causes, of evolution. For this to count as a criticism, one must abandon a narrowly instrumentalist view of scientific theories; this we gladly do, in that we assume that selection theory ought to pinpoint causes as well as facilitate predictions.

But even assuming this broadly noninstrumentalist outlook, our criticisms are philosophically partisan in additional ways. In that we have argued that genic selection coefficients are often "pseudoproperties" of genes, our criticisms of the gene's eye point of view are connected with more general metaphysical questions about the ontological status of properties. Some of these we take up in the following section. And in that we have understood "selection for" as a causal locution, it turns out that our account goes contrary to certain regularity analyses of causation. In populations in which selection generated by heterozygote superiority is the only evolutionary force, it is true that gene frequencies will move to a stable equilibrium. But this law-like regularity does not imply that there is selection for or against any individual gene. To say that "the gene's fitness value caused it to increase in frequency" is not simply to say that "any gene with that fitness value (in a relevantly similar population) would increase in frequency", since the former is false and the latter is true. Because we take natural selection to be a force of evolution, these remarks about causation have implications (explored in section 5) for how the concept of force is to be understood.

4. Properties

The properties, theoretical magnitudes, and natural kinds investigated by science ought not to be identified with the meanings that terms in scientific language possess. Nonsynonymous predicates (like "temperature" and "mean kinetic energy" and like "water" and "H_2O") may pick out the same property, and predicates which are quite meaningful (like "phlogiston" and "classical mass") may fail to pick out a property at all. Several recent writers have explored the idea that properties are to be individuated by their potential causal efficacy (Achinstein 1974; Armstrong 1978; Shoemaker 1980; and Sober 1982b). Besides capturing much of the intuitive content of our informal talk of properties, this view also helps

explicate the role of property-talk in science (Sober 1981). In this section, we will connect our discussion of genic selectionism with this metaphysical problem.

The definitional power of ordinary and scientific language allows us to take predicates which each pick out properties and to construct logically from these components a predicate which evidently does not pick out a property at all. An example of this is that old philosophical chestnut, the predicate "grue". We will say that an object is grue at a given time if it is green and the time is before the year 2000, or it is blue and the time is not before the year 2000. The predicate "grue" is defined from the predicates "green", "blue", and "time", each of which, we may assume for the purposes of the example, picks out a "real" property. Yet "grue" does not. A theory of properties should explain the basis of this distinction.

The difference between real and pseudoproperty is not captured by the ideas that animate the metaphysical issues usually associated with doctrines of realism, idealism, and conventionalism. Suppose that one adopts a "realist" position toward color and time, holding that things have the colors and temporal properties they do independently of human thought and language. This typical realist declaration of independence (Sober 1982a) will then imply that objects which are grue are so independently of human thought and language as well. In this sense, the "reality" of grulers is insured by the "reality" of colors and time. The distinction between real properties and pseudoproperties must be sought elsewhere.

Another suggestion is that properties can be distinguished from non-properties by appeal to the idea of *similarity* or of *predictive power*. One might guess that green things are more similar to each other than grue things are to each other, or that the fact that a thing is green is a better predictor of its further characteristics than the fact that it is grue. The standard criticism of these suggestions is that they are circular. We understand the idea of similarity in terms of shared *properties*, and the idea of predictive power in terms of the capacity to facilitate inference of further *properties*. However, a more fundamental difficulty with these suggestions presents itself: even if grue things happened to be very similar to each other, this would not make grue a real property. If there were no blue things after the year 2000, then the class of grue things would simply be the class of green things before the year 2000. The idea of similarity and the idea of predictive power fail to pinpoint the *intrinsic* defects of nonproperties like grue. Instead, they focus on somewhat accidental facts about the objects which happen to exist.

Grue is not a property for the same reason that genic selection coefficients are pseudoparameters in models of heterozygote superiority. The key idea is not that nonproperties are mind-dependent or are impoverished predictors; rather, they cannot be causally efficacious. To develop this idea, let's note a certain similarity between grue and genic selection coefficients. We pointed out before that genotype fitnesses plus initial genotype frequencies in the population causally determine the gene frequencies after selection. These same parameters also permit the mathematical derivation of genie fitness values, but, we asserted, these genie fitness values are artifacts; they do not cause the subsequent alteration in gene frequencies. The structure of these relationships is as follows.

$$\begin{array}{c} \text{genotype fitness values and} \rightarrow \text{genic frequencies} \\ \text{frequencies at time } t \qquad \text{at } t+1 \\ \downarrow \\ \text{genic fitness values at time } t \end{array}$$

Note that there are two different kinds of determination at work here. Genic fitness values at a given time are not *caused* by the genotypic fitness values at the same time. We assume that causal relations do not obtain between simultaneous events; rather, the relationship is one of logical or mathematical deducibility (symbolized by a broken line). On the other hand, the relation of initial genotype fitnesses and frequencies and subsequent gene frequencies is one of causal determination (represented by a solid line).

Now let's sketch the causal relations involved in a situation in which an object's being green produces some effect. Let the object be a grasshopper, Suppose that it matches its grassy background and that this protective coloration hides it from a hungry predator nearby. The relationships involved might be represented as follows.

$$\begin{array}{c} \text{the grasshopper is} \rightarrow \text{the grasshopper evades} \\ \text{green at time } t \qquad \text{the predator at time } t+1 \\ \downarrow \\ \text{the grasshopper is grue at time } t \end{array}$$

Just as in the above case, the object's color at the time *logically implies* that it is grue at that time but is the *cause* of its evading the predator at a subsequent time. And just as genic fitness values do not cause changes in gene frequencies, so the grasshopper's being grue does not cause it to have evaded its predator.

Our assessment of genic selectionism was not that genic fitness values are *always* artifactual. In cases other than that of heterozygote superiority – say, in the analysis of the *t*-allele – it may be perfectly correct to attribute causal efficacy to genic selection coefficients. So a predicate can pick out a real (causally efficacious) property in one context and fail to do so in another. This does not rule out the possibility, of course, that a predicate like "grue" is *globally artifactual*. But this consequence should not be thought to follow from a demonstration that grue is artifactual in a single kind of causal process.

The comparison of grue with genie selection is not meant to solve the epistemological problems of induction that led Goodman (1965) to formulate the example. Nor does the discussion provide any *a priori* grounds for distinguishing properties from nonproperties. Nor is it even a straightforward and automatic consequence of the truth of any scientific model that grue is artifactual, or that the idea of causal efficacy captures the metaphysical distinction at issue. Instead, the point is that a certain natural interpretation of a biological phenomenon helps to indicate how we ought to understand a rather abstract metaphysical issue.[11]

5.   Forces

Our arguments against genic selectionism contradict a standard positivist view of the concept of force. Positivists have often alleged that Newtonian mechanics tells us that forces are not "things", but that claims about forces are simply to be understood as claims about how objects actually behave, or would behave, if nothing else gets in the way. An exhaustive catalog of the forces acting on a system is to be understood as simply specifying a set of counterfactuals that describe objects.[12]

A Newtonian theory of forces will characterize each force in its domain in terms of the changes it would produce, were it the only force at work. The theory will take pair-wise combinations of forces and describe the joint effects that the two forces would have were they the only ones acting on a system. Then the forces would be taken three at a time, and so on, until a fully realistic model is constructed, one which tells us how real objects, which after all are subject to many forces, can be expected to behave. Each step in this program may face major theoretical difficulties, as the recent history of physics reveals (Cartwright 1980b; Joseph 1980).

This Newtonian paradigm is a hospitable home for the modelling of evolutionary forces provided in population genetics. The Hardy–Weinberg Law says what happens to gene frequencies when no evolutionary forces are at work. Mutation, migration, selection, and random drift are taken up one at a time, and models are provided for their effects on gene frequencies when no other forces are at work. Then these (and other) factors are taken up in combination. Each of these steps increases the model's realism. The culmination of this project would be a model that simultaneously represents the interactions of all evolutionary forces.

Both in physics and in population genetics, it is useful to conceive of forces in terms of their *ceteris paribus* effects, But there is more to a force than the truth of counterfactuals concerning change in velocity, or change in gene frequencies. The laws of motion describe the *effects* of forces, but they are supplemented by source laws which describe their *causes*. The standard genotypic model of heterozygote superiority not only says what will happen to a population, but also tells us what makes the population change.

It is quite true that when a population moves to an equilibrium value, due to the selection pressures generated by heterozygote superiority, the alleles are "disposed" to change in frequency in certain ways.[13] That is, the frequencies *will* change in certain ways, as long as no other evolutionary forces impinge. Yet, there is no force of genic selection at work here. If this is right, then the claim that genic selection is occurring must involve more than the unproblematic observations that gene frequencies are disposed to change in certain ways.

There is something more to the concept of force because it involves the idea of *causality*, and there is more to the idea of causality than is spelled out by such counterfactuals as the ones cited above. Suppose that something pushes (i.e., causally interacts with in a certain way) a billiard ball due north, and something else pushes it due west. Assuming that nothing else gets in the way, the ball will move northwest. There are two "component" forces at work here, and, as we like to say, one "net" force. However, there is a difference between the components and the resultant. Although something pushes the ball due north and something else pushes it due west, nothing pushes it northwest. In a sense, the resultant force is not a force at all, if by force we mean a causal agency. The resultant force is an artifact of the forces at work in the system. For mathematical purposes this distinction may make no difference. But if we want to understand why the ball moves the way it does, there is all the difference in the world between component and net.[14]

The "force" of genic selection in the evolutionary process propelled by heterozygote superiority is no more acceptable than the resultant "force" which is in the northwesterly direction. In fact, it is much worse. The resultant force, at least, is defined from the same conceptual building blocks as the component forces are. Genic selection coefficients, however, are gerrymandered hodgepodges, conceptually and dynamically quite unlike the genotypic selection coefficients that go into their construction. For genic selection coefficients are defined in terms of genotypic selection coefficients *and* gene frequencies. As noted before, they vary as the population changes in gene frequency, whereas the genotypic coefficients remain constant. And if their uniform zero value at equilibrium is interpreted as meaning that no selection is going on, one obtains a series of false assertions about the character of the population.

The concept of force is richer than that of disposition. The array of forces that act on a system uniquely determine the disposition of that system to change, but not conversely. If natural selection is a force and fitness is a disposition (to be reproductively successful), then the concept of selection is richer than that of fitness. To say that objects differ in fitness is not yet to say *why* they do so. The possible causes of such differences may be various, in that many different combinations of selection pressures acting at different levels of organization can have the same instantaneous effect on gene frequencies. Although selection coefficients and fitness values are interdefinable mathematically (so that, typically, $s = 1 - w$), they play different conceptual roles in evolutionary theory (Sober 1980).

Notes

1 The "*ceteris paribus*" is intended to convey the fact that selection for X can fail to bring about greater reproductive success for objects that have X, if countervailing forces act. Selection for X, against Y, and so on, are component forces that combine vectorially to determine the dynamics of the population.

2 In the passages quoted, Williams and Dawkins adopt a very bold position: any selection process which *can* be represented as genic selection *is* genic selection. Dawkins never draws back from this monolithic view, although Williams' more detailed argumentation leads him to hedge. Williams allows that group selection (clearly understood to be an alternative to genic selection) *is* possible and has actually been documented once (see his discussion of the t-allele). But *all* selection processes – including group selection – can be "represented" in terms of selection coefficients attaching to single genes. This means that the representation argument proves far too much.

3 Wimwatt (1980) criticizes genic selectionist models for being computationally inadequate and for at best providing a kind of "genetic bookkeeping" rather than a "theory of evolutionary change". Although we dissent from the first criticism, our discussion in what follows supports Wimsatt's second point.

4 We see from this that Dawkins' remark, that a gene that is "consistently on the losing side is not unlucky; it's a bad gene" is not quite right. Just at a single genotoken (and the organism in which it is housed) may enjoy a degree of reproductive success that is not an accurate representation of its fitness, so a set of genotokens (which are tokens of the same genotype) may encounter the same fate. Fitness and actual reproductive success are guaranteed to be identical only in models which

ignore random drift and thereby presuppose an infinite population.

5 If the heterozygote fitness is set equal to 1, the derivation is possible for the one locus two allele case considered. But if more than two alleles are considered, the asymmetry exists even in the face of normalization.

6 The averaging of effects can also be used to foster the illusion that a group selection process is really just a case of individual selection. But since this seems to be a relatively infrequent source of abuse, we will not take the space to spell out an example.

7 The argument given here has the same form as one presented in Sober (1980) which showed that the following is not a sufficient condition for group selection: there is heritable variation in the fitness of groups in which the fitness of an organism depends on the character of the group it is in.

8 The definition of genic selection just offered is structurally similar to the definition of group selection offered in Sober (1980). There, the requirement was that for there to be selection for groups which are X, it must be the case that every organism in a group that is X has one component of its fitness determined by the fact that it is in a group which is X. In group selection, organisms within the same group are bound together by a common group characteristic just as in genic selection organisms with the same gene are influenced in the same way by their shared characteristic.

9 In our earlier discussion of Mayr's ideas, we granted that selection usually acts "directly" on phenotypes and only "indirectly" on genotypes. But given the transitivity of causality, we argued that this fact is perfectly compatible with the existence of genotypic selection. However, our present discussion provides a characterization of when phenotypic selection can exist without there being any selection at the genotypic level. Suppose that individuals with the same genotype in a population end up with different phenotypes, because of the different microenvironments in which they develop. Selection for a given phenotype may then cross-classify the genotypes, and by our argument above, there will be no such thing as *the* causal upshot of a genotype. Averaging over effects will be possible, as always, but this will not imply genotypic selection. It is important to notice that this situation can allow evolution by natural selection to occur, gene frequencies can change in the face of phenotypic selection that is not accompanied by any sort of genotypic selection. Without this possibility, the idea of phenotypic selection is deprived of its main interest. There is no reason to deny that there can

be selection for phenotypic differences that have no underlying genetic differences, but this process will not produce any change in the population (ignoring cultural evolution and the like).

10 Genes at other loci which modify the intensity of segregator distortion are known to exist in *Drosophila*; the situation in the house mouse is not well understood. Note that the existence of such modifiers is consistent with genic selection, as long as they do not affect the *direction* of selection.

11 Another consequence of this analogy is that one standard diagnosis of what is wrong with "grue" fails to get to the heart of the matter. Carnap (1947) alleged that "green", unlike "grue", is purely qualitative, in that it makes no essential reference to particular places, individuals, or times. Goodman (1965) responded by pointing out that *both* predicates can be defined with reference to the year 2000. But a more fundamental problem arises: even if "grue" were, in some sense, not purely qualitative, this would not provide a fully general characterization of when a predicate fails to pick out a real property. Genic selection coefficients are "purely qualitative" if genotypic coefficients are, yet their logical relationship to each other exactly parallels that of "grue" to "green". Predicates picking out real properties can be "gruified" in a purely qualitative way: Let F and G be purely qualitative and be true of all the objects sampled (the emeralds, say). The predicate "(F and G) or ($-F$ and $-G$)" is a gruification of F and poses the same set of problems as Goodman's "grue".

12 Joseph (1980) has argued that this position, in treating the distribution of objects as given and then raising epistemological problems about the existence of forces, is committed to the existence of an asymmetry between attributions of quantities of *mass* to points in space-time and attributions of quantities of *energy* thereto. He argues that this idea, implicit in Reichenbach's (1958) classic argument for the conventionality of geometry, contradicts the relativistic equivalence of mass and energy. If this is right, then the positivistic view of force just described, far from falling out of received physical theory, in fact contradicts it.

13 For the purpose of this discussion, we will assume that attributions of dispositions and subjunctive conditionals of certain kinds are equivalent. That is, we will assume that to say that x is disposed to F is merely to say that if conditions were such-and-such, x would F.

14 This position is precisely the opposite of that taken by Cartwright (1980a), who argues that net forces, rather than component forces, are the items which realty exist. Cartwright argues this

by pointing out that the billiard ball moves northwesterly and not due north or due west. However, this appears to conflate the *effect* of a force with the force or forces actually at work.

References

Achinstein, P. (1974), "The Identity of Properties", *American Philosophical Quarterly* 11, 4: 257–75.

Armstrong, D. (1978), *Universals and Scientific Realism*. Cambridge: Cambridge University Press.

Brandon, R. (1978), "Evolution", *Philosophy of Science* 45: 96–109.

Carnap, R. (1947), "On the Application of Inductive Logic", *Philosophy and Phenomenological Research* 8: 133–47.

Cartwright, N. (1980a), "Do the Laws of Nature State the Facts?", *Pacific Philosophical Quarterly* 61, 1: 75–84.

Cartwright, N. (1980b), "The Truth Doesn't Explain Much", *American Philosophical Quarterly* 17, 2: 159–63.

Dawkins, R. (1976), *The Selfish Gene*. Oxford: Oxford University Press.

Doolittle, W. and Sapienza, C. (1980), "Selfish Genes, The Phenotype Paradigm, and Genome Evolution", *Nature* 284: 601–3.

Fisher, R. (1930), *The Genetical Theory of Natural Selection*. New York: Dover.

Goodman, N. (1965), *Fact, Fiction, and Forecast*. Indianapolis: Bobbs Merrill.

Joseph, G. (1979), "Riemannian Geometry and Philosophical Conventionalism", *Australasian Journal of Philosophy* 57, 3: 225–36.

Joseph, G. (1980), "The Many Sciences and the One World", *Journal of Philosophy* LXXVII, 12: 773–90.

Lewontin, R. (1970), "The Units of Selection", *Annual Review of Ecology and Systematics* 1, 1: 1–14.

Lewontin, R. (1974), *The Genetic Basis of Evolutionary Change*. New York: Columbia University Press.

Lewontin, R. and Dunn, L. (1960), "The Evolutionary Dynamics of a Polymorphism in the House Mouse", *Genetics* 45: 705–22.

Li, C. (1955), *Population Genetics*. Chicago: University of Chicago Press.

Mayr, E. (1963), *Animal Species and Evolution*. Cambridge: Harvard University Press.

Mills, S. and Beatty, J. (1979), "The Propensity Interpretation of Fitness", *Philosophy of Science* 46: 263–86.

Orgel, L. and Crick, F. (1980), "Selfish DNA: The Ultimate Parasite", *Nature* 284: 604–7.

Reichenbach, H. (1958), *The Philosophy of Space and Time*. New York: Dover.

Shoemaker, S. (1980), "Causality and Properties", in P. van Inwagen (ed.), *Essays in Honor of Richard Taylor*. Dordrecht: Reidel.

Sober, E. (1980), "Holism, Individualism, and the Units of Selection", in P. Asquith and R. Glere (eds.) *PSA 1980*, vol. 2, Proceedings of the 1980 Biennial Meeting of the Philosophy of Science Association: East Lansing, Michigan.

Sober, E. (1981), "Evolutionary Theory and the Ontological Status of Properties", *Philosophical Studies* 40: 147–76.

Sober, E. (1982a), "Realism and Independence", *Noûs* 16: 369–85.

Sober, E. (1982b), "Why Logically Equivalent Predicates May Pick Out Different Properties", *American Philosophical Quarterly* 19: 183–9.

Williams, G. (1966), *Adaptation and Natural Selection*. Princeton: Princeton University Press.

Wirnsatt, W. (1980), "Reductionlstic Research Strategies and Their Biases in the Units of Selection Controversy", in T. Nickles (ed.), *Scientific Discovery*, vol. 2, *Case Studies*. Dordrecht: Reidel.

21

The Return of the Gene

Kim Sterelny and Philip Kitcher

We have two images of natural selection. The orthodox story is told in terms of individuals. More organisms of any given kind are produced than can survive and reproduce to their full potential. Although these organisms are of a kind, they are not identical. Some of the differences among them make a difference to their prospects for survival or reproduction, and hence, on the average, to their actual reproduction. Some of the differences which are relevant to survival and reproduction are (at least partly) heritable. The result is evolution under natural selection, a process in which, barring complications, the average fitness of the organisms within a kind can be expected to increase with time.

There is an alternative story. Richard Dawkins[1] claims that the "unit of selection" is the gene. By this he means not just that the result of selection is (almost always) an increase in frequency of some gene in the gene pool. That is uncontroversial. On Dawkins's conception, we should think of genes as differing with respect to properties that affect their abilities to leave copies of themselves. More genes appear in each generation than can copy themselves up to their full potential. Some of the differences among them make a difference to their prospects for successful copying and hence to the number of actual copies that appear in the next generation. Evolution under natural selection is thus a process in which, barring complication, the average ability of the genes in the gene pool to leave copies of themselves increases with time.

Dawkins's story can be formulated succinctly by introducing some of his terminology. Genes are *replicators* and selection is the struggle among *active germ-line* replicators. Replicators are entities that can be copied. Active replicators are those whose properties influence their chances of being copied. Germ-line replicators are those which have the potential to leave infinitely many descendants. Early in the history of life, coalitions of replicators began to construct *vehicles* through which they spread copies of themselves. Better replicators build better vehicles, and hence are copied more often. Derivatively, the vehicles associated with them become more common too. The orthodox story focuses on the successes

Kim Sterelny and Philip Kitcher, "The Return of the Gene," *Journal of Philosophy* (1988) 85, 339–361. Reprinted by permission of the *Journal of Philosophy* and by Philip Kitcher.

We are equally responsible for this paper which was written when we discovered that we were writing it independently. We would like to thank those who have offered helpful suggestions to one or both of us, particularly Patrick Bateson, Robert Brandon, Peter Godfrey-Smith, David Hull, Richard Lewontin, Lisa Lloyd, Philip Pettit, David Scheel, and Elliott Sober.

of prominent vehicles – individual organisms. Dawkins claims to expose an underlying struggle among the replicators.

We believe that a lot of unnecessary dust has been kicked up in discussing the merits of the two stories. Philosophers have suggested that there are important connections to certain issues in the philosophy of science: reductionism, views on causation and natural kinds, the role of appeals to parsimony. We are unconvinced. Nor do we think that a willingness to talk about selection in Dawkinspeak brings any commitment to the adaptationist claims which Dawkins also holds. After all, adopting a particular perspective on selection is logically independent from claiming that selection is omnipresent in evolution.

In our judgment, the relative worth of the two images turns on two theoretical claims in evolutionary biology.

1. Candidate units of selection must have systematic causal consequences. If Xs are selected for, then X must have a systematic effect on its expected representation in future generations.
2. Dawkins's gene selectionism offers a *more general theory* of evolution. It can also handle those phenomena which are grist to the mill of individual selection, but there are evolutionary phenomena which fit the picture of individual selection ill or not at all, yet which can be accommodated naturally by the gene selection model.

Those skeptical of Dawkins's picture – in particular, Elliott Sober, Richard Lewontin, and Stephen Jay Gould – doubt whether genes can meet the condition demanded in (1). In their view, the phenomena of epigenesis and the extreme sensitivity of the phenotype to gene combinations and environmental effects undercut genic selectionism. Although we believe that these critics have offered valuable insights into the character of sophisticated evolutionary modeling, we shall try to show that these insights do not conflict with Dawkins's story of the workings of natural selection. We shall endeavor to free the thesis of genic selectionism from some of the troublesome excrescences which have attached themselves to an interesting story.

I. Gene Selection and Bean-bag Genetics

Sober and Lewontin[2] argue against the thesis that all selection is genic selection by contending that many instances of selection do not involve selection for properties of individual alleles. Stated rather loosely, the claim is that, in some populations, properties of individual alleles are not positive causal factors in the survival and reproductive success of the relevant organisms. Instead of simply resting this claim on an appeal to our intuitive ideas about causality, Sober has recently provided an account of causal discourse which is intended to yield the conclusion he favors, thus rebutting the proposals of those (like Dawkins) who think that properties of individual alleles can be causally efficacious.[3]

The general problem arises because replicators (genes) combine to build vehicles (organisms) and the effect of a gene is critically dependent on the company it keeps. However, recognizing the general problem, Dawkins seeks to disentangle the various contributions of the members of the coalition of replicators (the genome). To this end, he offers an analogy with a process of competition among rowers for seats in a boat. The coach may scrutinize the relative times of different teams but the competition can be analyzed by investigating the contributions of individual rowers in different contexts (SG 40/1 91/2, EP 239).

Sober's case

At the general level, we are left trading general intuitions and persuasive analogies. But Sober (and, earlier, Sober and Lewontin) attempted to clarify the case through a particular example. Sober argues that *heterozygote superiority* is a phenomenon that cannot be understood from Dawkins's standpoint. We shall discuss Sober's example in detail; our strategy is as follows. We first set out Sober's case: heterozygote superiority cannot be understood as a gene-level phenomenon, because only pairs of genes can be, or fail to be, heterozygous. Yet being heterozygous can be causally salient in the selective process. Against Sober, we first offer an analogy to show that there must be something wrong with his line of thought: from the gene's eye view, heterozygote superiority is an instance of a standard selective phenomenon, namely *frequency-dependent* selection. The advantage (or disadvantage) of a trait can depend on the frequency of that trait in other members of the relevant population.

Having claimed that there is something wrong with Sober's argument, we then try to say what is wrong. We identify two principles on which the reasoning depends. First is a general claim about causal uniformity. Sober thinks that there can be selection for a property only if that property has a positive uniform effect on reproductive success. Second, and more specifically, in cases where the heterozygote is fitter, the individuals have no uniform causal effect. We shall try to undermine both principles, but the bulk of our criticism will be directed against the first.

Heterozygote superiority occurs when a heterozygote (with genotype *Aa*, say) is fitter than either homozygote (*AA* or *aa*). The classic example is human sickle-cell anemia: homozygotes for the normal allele in African populations produce functional hemoglobin but are vulnerable to malaria, homozygotes for the mutant ("sickling") allele suffer anemia (usually fatal), and heterozygotes avoid anemia while also having resistance to malaria. The effect of each allele varies with context, and the contexts across which variation occurs are causally relevant. Sober writes:

> In this case, the *a* allele does not have a unique causal role. Whether the gene *a* will be a positive or a negative causal factor in the survival and reproductive success of an organism depends on the genetic context. If it is placed next to a copy of A, *a* will mean an increase in fitness. If it is placed next to a copy of itself, the gene will mean a decrement in fitness (NS 303).

The argument against Dawkins expressed here seems to come in two parts. Sober relies on the principle

(A) There is selection for property *P* only if in all causally relevant background conditions *P* has a positive effect on survival and reproduction.

He also adduces a claim about the particular case of heterozygote superiority.

(B) Although we can understand the situation by noting that the heterozygote has a uniform effect on survival and reproduction, the property of having the *A* allele and the property of

having the *a* allele cannot be seen as having uniform effects on survival and reproduction.

We shall argue that both (A) and (B) are problematic.

Let us start with the obvious reply to Sober's argument. It seems that the heterozygote superiority case is akin to a familiar type of frequency-dependent selection. If the population consists just of *AA*s and a mutation arises, the *a*-allele, then, initially *a* is favored by selection. Even though it is very bad to be *aa*, *a* alleles are initially likely to turn up in the company of *A* alleles. So they are likely to spread, and, as they spread, they find themselves alongside other *a* alleles, with the consequence that selection tells against them. The scenario is very similar to a story we might tell about interactions among individual organisms. If some animals resolve conflicts by playing hawk and others play dove, then, if a population is initially composed of hawks (and if the costs of bloody battle outweigh the benefits of gaining a single resource), doves will initially be favored by selection.[4] For they will typically interact with hawks, and, despite the fact that their expected gains from these interactions are zero, they will still fare better than their rivals whose expected gains from interactions are negative. But, as doves spread in the population, hawks will meet them more frequently, with the result that the expected payoffs to hawks from interactions will increase. Because they increase more rapidly than the expected payoffs to the doves, there will be a point at which hawks become favored by selection, so that the incursion of doves into the population is halted.

We believe that the analogy between the case of heterozygote superiority and the hawk-dove case reveals that there is something troublesome about Sober's argument. The challenge is to say exactly what has gone wrong.

Causal uniformity

Start with principle (A). Sober conceives of selection as a *force*, and he is concerned to make plain the effects of component forces in situations where different forces combine. Thus, he invites us to think of the heterozygote superiority case by analogy with situations in which a physical object remains at rest because equal and opposite forces are exerted on it. Considering the situation only in terms of net forces will conceal the causal structure of the situation. Hence, Sober concludes, our ideas about units of selection should penetrate beyond what occurs on the average, and we should attempt to isolate those properties which positively affect survival and reproduction in every causally relevant context.

Although Sober rejects determinism, principle (A) seems to hanker after something like the uniform association of effects with causes that deterministic accounts of causality provide. We believe that the principle cannot be satisfied without doing violence to ordinary ways of thinking about natural selection, and, once the violence has been exposed, it is not obvious that there is any way to reconstruct ideas about selection that will fit Sober's requirement.

Consider *the* example of natural selection, the case of industrial melanism.[5] We are inclined to say that the moths in a Cheshire wood, where lichens on many trees have been destroyed by industrial pollutants, have been subjected to selection pressure and that there has been selection for the property of being melanic. But a moment's reflection should reveal that this description is at odds with Sober's principle. For the wood is divisible into patches, among which are clumps of trees that have been shielded from the effects of industrialization. Moths who spend most of their lives in these areas are at a disadvantage if they are melanic. Hence, in the population comprising all the moths in the wood, there is no uniform effect on survival and reproduction: in some causally relevant contexts (for moths who have the property of living in regions where most of the trees are contaminated), the trait of being melanic has a positive effect on survival and reproduction, but there are other contexts in which the effect of the trait is negative.

The obvious way to defend principle (A) is to split the population into subpopulations and identify different selection processes as operative in different subgroups. This is a revisionary proposal, for our usual approach to examples of industrial melanism is to take a coarse-grained perspective on the environments, regarding the existence of isolated clumps of uncontaminated trees as a perturbation of the overall selective process. Nonetheless, we might be led to make the revision, not in the interest of honoring a

philosophical prejudice, but simply because our general views about selection are consonant with principle (A), so that the reform would bring our treatment of examples into line with our most fundamental beliefs about selection.

In our judgment, a defense of this kind fails for two connected reasons. First, the process of splitting populations may have to continue much further – perhaps even to the extent that we ultimately conceive of individual organisms as making up populations in which a particular type of selection occurs. For, even in contaminated patches, there may be variations in the camouflaging properties of the tree trunks and these variations may combine with propensities of the moths to cause local disadvantages for melanic moths. Second, as many writers have emphasized, evolutionary theory is a statistical theory, not only in its recognition of drift as a factor in evolution but also in its use of fitness coefficients to represent the expected survivorship and reproductive success of organisms. The envisaged splitting of populations to discover some partition in which principle (A) can be maintained is at odds with the strategy of abstracting from the thousand natural shocks that organisms in natural populations are heir to. In principle, we could relate the biography of each organism in the population, explaining in full detail how it developed, reproduced, and survived, just as we could track the motion of each molecule of a sample of gas. But evolutionary theory, like statistical mechanics, has no use for such a fine grain of description: the aim is to make clear the central tendencies in the history of evolving populations, and, to this end, the strategy of averaging, which Sober decries, is entirely appropriate. We conclude that there is no basis for any revision that would eliminate those descriptions which run counter to principle (A).

At this point, we can respond to the complaints about the gene's eye view representation of cases of heterozygote superiority. Just as we can give sense to the idea that the trait of being melanic has a unique environment-dependent effect on survival and reproduction, so too we can explicate the view that a property of alleles, to wit, the property of directing the formation of a particular kind of hemoglobin, has a unique environment-dependent effect on survival and reproduction. The alleles form parts of one another's environments, and, in an environment in which a copy of the *A* allele is present, the typical trait of the *S* allele (namely, directing the formation of deviant hemoglobin) will usually have a positive effect on the chances that copies of that allele will be left in the next generation. (Notice that the effect will not be invariable, for there are other parts of the genomic environment which could wreak havoc with it). If someone protests that the incorporation of alleles as themselves part of the environment is suspect, then the immediate rejoinder is that, in cases of behavioral interactions, we are compelled to treat organisms as parts of one another's environments.[6] The effects of playing hawk depend on the nature of the environment, specifically on the frequency of doves in the vicinity.[7]

The causal powers of alleles

We have tried to develop our complaints about principle (A) into a positive account of how cases of heterozygote superiority might look from the gene's eye view. We now want to focus more briefly on (B). Is it impossible to reinterpret the examples of heterozygote superiority so as to ascribe uniform effects on survival and reproduction to allelic properties? The first point to note is that Sober's approach formulates the Dawkinsian point of view in the wrong way: the emphasis should be on the effects of properties of alleles, not on allelic properties of organisms (like the property of having an *A* allele) and the accounting ought to be done in terms of allele copies. Second, although we argued above that the strategy of splitting populations was at odds with the character of evolutionary theory, it is worth noting that the same strategy will be available in the heterozygote superiority case.

Consider the following division of the original population: let P_1 be the collection of all those allele copies which occur next to an *S* allele, and let P_2 consist of all those allele copies which occur next to an *A* allele. Then the property of being *A* (or of directing the production of normal hemoglobin) has a positive effect on the production of copies in the next generation in P_1, and conversely in P_2. In this way, we are able to partition the population and to achieve a Dawkinsian redescription that meets Sober's principle (A) – just in the way that we might try to do so if we wanted to satisfy (A) in understanding the

operation of selection on melanism in a Cheshire wood or on fighting strategies in a population containing a mixture of hawks and doves.

Objection: the "populations" just defined are highly unnatural, and this can be seen once we recognize that, in some cases, allele copies in the same organisms (the heterozygotes) belong to different "populations." Reply: so what? From the allele's point of view, the copy next door is just a critical part of the environment. The populations P_1 and P_2 simply pick out the alleles that share the same environment. There would be an analogous partition of a population of competing organisms which occurred locally in pairs such that some organisms played dove and some hawk. (Here, mixed pairs would correspond to heterozygotes).

So the genic picture survives an important initial challenge. The moral of our story so far is that the picture must be applied consistently. Just as paradoxical conclusions will result if one offers a partial translation of geometry into arithmetic, it is possible to generate perplexities by failing to recognize that the Dawkinsian *Weltanschauung* leads to new conceptions of environment and of population. We now turn to a different worry, the objection that genes are not "visible" to selection.

II. Epigenesis and Visibility

In a lucid discussion of Dawkins's early views, Gould claims to find a "fatal flaw" in the genic approach to selection. According to Gould, Dawkins is unable to give genes "direct visibility to natural selection."[8] Bodies must play intermediary roles in the process of selection, and, since the properties of genes do not map in one-one fashion onto the properties of bodies, we cannot attribute selective advantages to individual alleles. We believe that Gould's concerns raise two important kinds of issues for the genic picture: (i) Can Dawkins sensibly talk of the effect of an individual allele on its expected copying frequency? (ii) Can Dawkins meet the charge that it is the phenotype that makes the difference to the copying of the underlying alleles, so that, whatever the causal basis of an advantageous trait, the associated allele copies will have enhanced chances of being replicated? We shall take up these questions in order.

Do alleles have effects?

Dawkins and Gould agree on the facts of embryology which subvert the simple Mendelian association of one gene with one character. But the salience of these facts to the debate is up for grabs. Dawkins regards Gould as conflating the demands of embryology with the demands of the theory of evolution. While genes' effects blend in embryological development, and while they have phenotypic effects only in concert with their gene-mates, genes "do not blend as they replicate and recombine down the generations. It is this that matters for the geneticist, and it is also this that matters for the student of units of selection" (EP 117).

Is Dawkins right? Chapter 2 of EP is an explicit defense of the meaningfulness of talk of "genes for" indefinitely complex morphological and behavioral traits. In this, we believe, Dawkins is faithful to the practice of classical geneticists. Consider the vast number of loci in *Drosophila melanogaster* which are labeled for eye-color traits – white, eosin, vermilion, raspberry, and so forth. Nobody who subscribes to this practice of labeling believes that a pair of appropriately chosen stretches of DNA, cultured in splendid isolation, would produce a detached eye of the pertinent color. Rather, the intent is to indicate the effect that certain changes at a locus would make against the background of the rest of the genome.

Dawkins's project here is important not just in conforming to traditions of nomenclature. Remember: Dawkins needs to show that we can sensibly speak of alleles having (environment-sensitive) effects, effects in virtue of which they are selected for or selected against. If we can talk of a gene for X, where X is a selectively important phenotypic characteristic, we can sensibly talk of the effect of an allele on its expected copying frequency, even if the effects are always indirect, via the characteristics of some vehicle.

What follows is a rather technical reconstruction of the relevant notion. The precision is needed to allow for the extreme environmental sensitivity of allelic causation. But the intuitive idea is simple: we can speak of genes for X if substitutions on a chromosome would lead, in the relevant environments, to a difference in the X-ishness of the phenotype.

Consider a species S and an arbitrary locus L in the genome of members of S. We want to give sense to the locution 'L is a locus affecting P' and derivatively to the phrase 'G is a gene for P^*' (where, typically, P will be a determinable and P^* a determinate form of P). Start by taking an *environment* for a locus to be an aggregate of DNA segments that would complement L to form the genome of a member of S together with a set of extra-organismic factors (those aspects of the world external to the organism which we would normally count as part of the organism's environment). Let a set of variants for L be any collection of DNA segments, none of which is debarred, on physico-chemical grounds, from occupying L. (This is obviously a very weak constraint, intended only to rule out those segments which are too long or which have peculiar physico-chemical properties). Now, we say that L is a locus affecting P in S relative to an environment E and a set of variants V just in case there are segments s, s^*, and s^{**} in V such that the substitution of s^{**} for s^* in an organism having s and s^* at L would cause a difference in the form of P, against the background of E. In other words, given the environment E, organisms who are ss^* at L differ in the form of P from organisms who are ss^{**} at L and the cause of the difference is the presence of s^* rather than s^{**}. (A minor clarification: while s^* and s^{**} are distinct, we do not assume that they are both different from s.)

L is a locus affecting P in S just in case L is a locus affecting P in S relative to any standard environment and a feasible set of variants. Intuitively, the geneticist's practice of labeling loci focuses on the "typical" character of the complementary part of the genome in the species, the "usual" extra-organismic environment, and the variant DNA segments which have arisen in the past by mutation or which "are likely to arise" by mutation. Can these vague ideas about standard conditions be made more precise? We think so. Consider first the genomic part of the environment. There will be numerous alternative combinations of genes at the loci other than L present in the species S. Given most of these gene combinations, we expect modifications at L to produce modifications in the form of P. But there are likely to be some exceptions, cases in which the presence of a rare allele at another locus or a rare combination of alleles produces a phenotypic effect that dominates any effect on P. We can either dismiss the exceptional cases as nonstandard because they are infrequent or we can give a more refined analysis, proposing that each of the nonstandard cases involves either (a) a rare allele at a locus L' or (b) a rare combination of alleles at loci L', L'' . . . such that that locus (a) or those loci jointly (b) affect some phenotypic trait Q that dominates P in the sense that there are modifications of Q which prevent the expression of any modifications of P. As a concrete example, consider the fact that there are modifications at some loci in *Drosophila* which produce embryos that fail to develop heads; given such modifications elsewhere in the genome, alleles affecting eye color do not produce their standard effects!

We can approach standard extra-genomic environments in the same way. If L affects the form of P in organisms with a typical gene complement, except for those organisms which encounter certain rare combinations of external factors, then we may count those combinations as nonstandard simply because of their infrequency. Alternatively, we may allow rare combinations of external factors to count provided that they do not produce some gross interference with the organism's development, and we can render the last notion more precise by taking nonstandard environments to be those in which the population mean fitness of organisms in S would be reduced by some arbitrarily chosen factor (say, $^1/_2$).

Finally, the feasible variants are those which actually occur at L in members of S, together with those which have occurred at L in past members of S and those which are easily attainable from segments that actually occur at L in members of S by means of insertion, deletion, substitution, or transposition. Here the criteria for ease of attainment are given by the details of molecular biology. If an allele is prevalent at L in S, then modifications at sites where the molecular structure favors insertions, deletions, substitutions, or transpositions (so-called "hot spots") should count as easily attainable even if some of these modifications do not actually occur.

Obviously, these concepts of "standard conditions" could be articulated in more detail, and we believe that it is possible to generate a variety of explications, agreeing on the core of central cases but adjusting the boundaries of the concepts in different ways. If we now assess the labeling practices of geneticists, we expect to find that virtually all of their claims about loci affecting a phenotypic trait are sanctioned by all of the explications. Thus, the challenge that there is no way to honor the facts of epigenesis while speaking of loci that affect certain traits would be turned back.

Once we have come this far, it is easy to take the final step. An allele A at a locus L in a species S is for the trait P^* (assumed to be a determinate form of the determinable characteristic P) relative to a local allele B and an environment E just in case (a) L affects the form of P in S, (b) E is a standard environment, and (c) in E organisms that are AB have phenotype P^*. The relativization to a local allele is necessary, of course, because, when we focus on a target allele rather than a locus, we have to extend the notion of the environment – as we saw in the last section, corresponding alleles are potentially important parts of one another's environments. If we say that A is for P^* (period), we are claiming that A is for P^* relative to standard environments and common local alleles or that A is for P^* relative to standard environments and itself.

Now, let us return to Dawkins and to the apparently outré claim that we can talk about genes for reading. Reading is an extraordinarily complex behavior pattern and surely no adaptation. Further, many genes must be present and the extra-organismic environment must be right for a human being to be able to acquire the ability to read. Dyslexia might result from the substitution of an unusual mutant allele at one of the loci, however. Given our account, it will be correct to say that the mutant allele is a gene for dyslexia and also that the more typical alleles at the locus are alleles for reading. Moreover, if the locus also affects some other (determinable) trait, say, the capacity to factor numbers into primes, then it may turn out that the mutant allele is also an allele for rapid factorization skill and that the typical allele is an allele for factorization disability. To say that A is an allele for P^* does not preclude saying that A is an allele for Q^*, nor does it commit us to supposing that the phenotypic properties in question are either both skills or both disabilities. Finally, because substitutions at many loci may produce (possibly different types of) dyslexia, there may be many genes for dyslexia and many genes for reading. Our reconstruction of the geneticists' idiom, the idiom which Dawkins wants to use, is innocent of any Mendelian theses about one-one mappings between genes and phenotypic traits.

Visibility

So we can defend Dawkins's thesis that alleles have properties that influence their chances of leaving copies in later generations by suggesting that, in concert with their environments (including their genetic environments), those alleles cause the presence of certain properties in vehicles (such as organisms) and that the properties of the vehicles are causally relevant to the spreading of copies of the alleles. But our answer to question (i) leads naturally to concerns about question (ii). Granting that an allele is for a phenotypic trait P^* and that the presence of P^* rather than alternative forms of the determinable trait P enhances the chances that an organism will survive and reproduce and thus transmit copies of the underlying allele, is it not P^* and its competition which are directly involved in the selection process? What selection "sees" are the phenotypic properties. When this vague, but suggestive, line of thought has been made precise, we think that there is an adequate Dawkinsian reply to it.

The idea that selection acts directly on phenotypes, expressed in metaphorical terms by Gould (and earlier by Ernst Mayr), has been explored in an interesting essay by Robert Brandon.[9] Brandon proposes that phenotypic traits screen off genotypic traits (in the sense of Wesley Salmon[10]):

$$\Pr(O_n/G\&P) = \Pr(O_n/P) \neq \Pr(O_n/G)$$

where $\Pr(O_n/G\&P)$ is the probability that an organism will produce n offspring given that it has both a phenotypic trait and the usual genetic basis for that trait, $\Pr(O_n/P)$ is the probability that an organism will produce n offspring given that it has the phenotypic trait, and $\Pr(O_n/G)$ is the probability that it will produce n offspring given that it has the usual genetic basis. So fitness seems to vary more directly with the phenotype and less directly with the underlying genotype.

Why is this? The root idea is that the successful phenotype may occur in the presence of the wrong allele as a result of judicious tampering, and, conversely, the typical effect of a "good" allele

may be subverted. If we treat moth larvae with appropriate injections, we can produce pseudomelanics that have the allele which normally gives rise to the speckled form and we can produce moths, foiled melanics, that carry the allele for melanin in which the developmental pathway to the emergence of black wings is blocked. The pseudomelanics will enjoy enhanced reproductive success in polluted woods and the foiled melanics will be at a disadvantage. Recognizing this type of possibility, Brandon concludes that selection acts at the level of the phenotype.[11]

Once again, there is no dispute about the facts. But our earlier discussion of epigenesis should reveal how genic selectionists will want to tell a different story. The interfering conditions that affect the phenotype of the vehicle are understood as parts of the allelic environment. In effect, Brandon, Gould, and Mayr contend that, in a polluted wood, there is selection for being dark colored rather than for the allelic property of directing the production of melanin, because it would be possible to have the reproductive advantage associated with the phenotype without having the allele (and conversely it would be possible to lack the advantage while possessing the allele). Champions of the gene's eye view will maintain that tampering with the phenotype reverses the typical effect of an allele by changing the environment. For these cases involve modification of the allelic environment and give rise to new selection processes in which allelic properties currently in favor prove detrimental. The fact that selection goes differently in the two environments is no more relevant than the fact that selection for melanic coloration may go differently in Cheshire and in Dorset.

If we do not relativize to a fixed environment, then Brandon's claims about screening off will not generally be true.[12] We suppose that Brandon intends to relativize to a fixed environment. But now he has effectively begged the question against the genic selectionist by deploying the orthodox conception of environment. Genic selectionists will also want to relativize to the environment, but they should resist the orthodox conception of it. On their view, the probability relations derived by Brandon involve an illicit averaging over environments (see note 12). Instead, genic selectionists should propose that the probability of an allele's leaving n copies of itself should be understood relative to the total allelic environment, and that the specification of the total environment ensures that there is no screening off of allelic properties by phenotypic properties. The probability of producing *n* copies of the allele for melanin in a total allelic environment is invariant under conditionalization on phenotype.

Here too the moral of our story is that Dawkinspeak must be undertaken consistently. Mixing orthodox concepts of the environment with ideas about genic selection is a recipe for trouble, but we have tried to show how the genic approach can be thoroughly articulated so as to meet major objections. But what is the point of doing so? We shall close with a survey of some advantages and potential drawbacks.

III. Genes and Generality

Relatively little fossicking is needed to uncover an extended defense of the view that gene selectionism offers a more general and unified picture of selective processes than can be had from its alternatives. Phenomena anomalous for the orthodox story of evolution by individual selection fall naturally into place from Dawkins' viewpoint. He offers a revision of the "central theorem" of Darwinism. Instead of expecting individuals to act in their best interests, we should expect an animal's behavior "to maximize the survival of genes 'for' that behavior, whether or not those genes happen to be in the body of that particular animal performing it" (EP 223).

The cases that Dawkins uses to illustrate the superiority of his own approach are a somewhat motley collection. They seem to fall into two general categories. First are outlaw and quasi–outlaw examples. Here there is competition among genes which cannot be translated into talk of vehicle fitness because the competition is among co-builders of a single vehicle. The second group comprises "extended phenotype" cases, instances in which a gene (or combination of genes) has selectively relevant phenotypic consequences which are not traits of the vehicle that it has helped build. Again the replication potential of the gene cannot be translated into talk of the adaptedness of its vehicle.

We shall begin with outlaws and quasi outlaws. From the perspective of the orthodox story of individual selection, "replicators at different loci within the same body can be expected to 'cooperate'." The allele surviving at any given locus tends to be one best (subject to all the constraints) for the whole genome. By and large this is a reasonable assumption. Whereas individual outlaw organisms are perfectly possible in groups and subvert the chances for groups to act as vehicles, outlaw genes seem problematic. Replication of any gene in the genome requires the organism to survive and reproduce, so genes share a substantial common interest. This is true of asexual reproduction, and, granting the fairness of meiosis, of sexual reproduction too.

But there is the rub. Outlaw genes are genes which subvert meiosis to give them a better than even chance of making it to the gamete, typically by sabotaging their corresponding allele (EP 136). Such genes are *segregation distorters* or *meiotic drive* genes. Usually, they are enemies not only of their alleles but of other parts of the genome, because they reduce the individual fitness of the organism they inhabit. Segregation distorters thrive, when they do, because they exercise their phenotypic power to beat the meiotic lottery. Selection for such genes cannot be selection for traits that make organisms more likely to survive and reproduce. They provide uncontroversial cases of selective processes in which the individualistic story cannot be told.

There are also related examples. Altruistic genes can be outlawlike, discriminating against their genome mates in favor of the inhabitants of other vehicles, vehicles that contain copies of themselves. Start with a hypothetical case, the so-called "green beard" effect. Consider a gene *Q* with two phenotypic effects. *Q* causes its vehicle to grow a green beard and to behave altruistically toward green-bearded conspecifics. *Q*'s replication prospects thus improve, but the particular vehicle that *Q* helped build does not have its prospects for survival and reproduction enhanced. Is *Q* an outlaw not just with respect to the vehicle but with respect to the vehicle builders? Will there be selection for alleles that suppress *Q*'s effect? How the selection process goes will depend on the probability that *Q*'s cobuilders are beneficiaries as well. If *Q* is reliably associated with other gene kinds, those kinds will reap a net benefit from *Q*'s outlawry.

So altruistic genes are sometimes outlaws. Whether coalitions of other genes act to suppress them depends on the degree to which they benefit only themselves. Let us now move from a hypothetical example to the parade case.

Classical fitness, an organism's propensity to leave descendants in the next generation, seems a relatively straightforward notion. Once it was recognized that Darwinian processes do not necessarily favor organisms with high classical fitness, because classical fitness ignores indirect effects of costs and benefits to relatives, a variety of alternative measures entered the literature. The simplest of these would be to add to the classical fitness of an organism contributions from the classical fitness of relatives (weighted in each case by the coefficient of relatedness). Although accounting of this sort is prevalent, Dawkins (rightly) regards it as just wrong, for it involves double bookkeeping and, in consequence, there is no guarantee that populations will move to local maxima of the defined quantity. This measure and measures akin to it, however, are prompted by Hamilton's rigorous development of the theory of inclusive fitness (in which it is shown that populations will tend toward local maxima of inclusive fitness).[13] In the misunderstanding and misformulation of Hamilton's ideas, Dawkins sees an important moral.

Hamilton, he suggests, appreciated the gene selectionist insight that natural selection will favor "organs and behavior that cause the individual's genes to be passed on, whether or not the individual is an ancestor" (EP 185). But Hamilton's own complex (and much misunderstood) notion of inclusive fitness was, for all its theoretical importance, a dodge, a "brilliant last-ditch rescue attempt to save the individual organism as the level at which we think about natural selection" (EP 187). More concretely, Dawkins is urging two claims: first, that the uses of the concept of inclusive fitness in practice are difficult, so that scientists often make mistakes; second, that such uses are conceptually misleading. The first point is defended by identifying examples from the literature in which good researchers have made errors, errors which become obvious once we adopt the gene selectionist perspective. Moreover, even when the inclusive fitness calculations make the right predictions, they often seem to mystify the selective process involved (thus buttressing

Dawkins's second thesis). Even those who are not convinced of the virtues of gene selectionism should admit that it is very hard to see the reproductive output of an organism's relatives as a property of that organism.

Let us now turn to the other family of examples, the "extended phenotype" cases. Dawkins gives three sorts of "extended" phenotypic effects: effects of genes – indeed key weapons in the competitive struggle to replicate – which are not traits of the vehicle the genes inhabit. The examples are of artifacts, of parasitic effects on host bodies and behaviors, and of "manipulation" (the subversion of an organism's normal patterns of behavior by the genes of another organism via the manipulated organism's nervous system).

Among many vivid, even haunting, examples of parasitic behavior, Dawkins describes cases in which parasites synthesize special hormones with the consequence that their hosts take on phenotypic traits that decrease their own prospects for reproduction but enhance those of the parasites (see, for a striking instance, EP 215). There are equally forceful cases of manipulation: cuckoo fledglings subverting their host's parental program, parasitic queens taking over a hive and having its members work for her. Dawkins suggests that the traits in question should be viewed as adaptations – properties for which selection has occurred – even though they cannot be seen as adaptations of the individuals whose reproductive success they promote, for those individuals do not possess the relevant traits. Instead, we are to think in terms of selectively advantageous characteristics of alleles which orchestrate the behavior of several different vehicles, some of which do not include them.

At this point there is an obvious objection. Can we not understand the selective processes that are at work by focusing not on the traits that are external to the vehicle that carries the genes, but on the behavior that the vehicle performs which brings those traits about? Consider a spider's web. Dawkins wants to talk of a gene for a web. A web, of course, is not a characteristic of a spider. Apparently, however, we could talk of a gene for web building. Web building is a trait of spiders, and, if we choose to redescribe the phenomena in these terms, the extended phenotype is brought closer to home. We now have a trait of the vehicle in which the genes reside, and we can tell an orthodox story about natural selection for this trait.

It would be tempting to reply to this objection by stressing that the selective force acts through the artifact. The causal chain from the gene to the web is complex and indirect; the behavior is only a part of it. Only one element of the chain is distinguished, the endpoint, the web itself, and that is because, independently of what has gone on earlier, provided that the web is in place, the enhancement of the replication chances of the underlying allele will ensue. But this reply is exactly parallel to the Mayr–Could–Brandon argument discussed in the last section, and it should be rejected for exactly parallel reasons.

The correct response, we believe, is to take Dawkins at his word when he insists on the possibility of a number of different ways of looking at the same selective processes. Dawkins's two main treatments of natural selection, SG and EP, offer distinct versions of the thesis of genic selectionism. In the earlier discussion (and occasionally in the later) the thesis is that, for any selection process, there is a uniquely correct representation of that process, a representation which captures the causal structure of the process, and this representation attributes causal efficacy to genic properties. In EP, especially in chapters 1 and 13, Dawkins proposes a weaker version of the thesis, to the effect that there are often alternative, equally adequate representations of selection processes and that, for any selection process, there is a maximally adequate representation which attributes causal efficacy to genic properties. We shall call the strong (early) version *monist genic selectionism* and the weak (later) version *pluralist genic selectionism*. We believe that the monist version is faulty but that the pluralist thesis is defensible.

In presenting the "extended phenotype" cases, Dawkins is offering an alternative representation of processes that individualists can redescribe in their own preferred terms by adopting the strategy illustrated in our discussion of spider webs. Instead of talking of genes for webs and their selective advantages, it is possible to discuss the case in terms of the benefits that accrue to spiders who have a disposition to engage in web building. There is no privileged way to segment the causal chain and isolate the (really) real causal story. As we noted two paragraphs back, the analog

of the Mayr–Gould–Brandon argument for the priority of those properties which are most directly connected with survival and reproduction – here the webs themselves – is fallacious. Equally, it is fallacious to insist that the causal story must be told by focusing on traits of individuals which contribute to the reproductive success of those individuals. We are left with the general thesis of pluralism: there are alternative, maximally adequate representations of the causal structure of the selection process. Add to this Dawkins's claim that one can always find a way to achieve a representation in terms of the causal efficacy of genic properties, and we have pluralist genic selectionism.

Pluralism of the kind we espouse has affinities with some traditional views in the philosophy of science. Specifically, our approach is instrumentalist, not of course in denying the existence of entities like genes, but in opposing the idea that natural selection is a force that acts on some determinate target, such as the genotype or the phenotype. Monists err, we believe, in claiming that selection processes must be described in a particular way, and their error involves them in positing entities, "targets of selection," that do not exist.

Another way to understand our pluralism is to connect it with conventionalist approaches to space-time theories. Just as conventionalists have insisted that there are alternative accounts of the phenomena which meet all our methodological desiderata, so too we maintain that selection processes can usually be treated, equally adequately, from more than one point of view. The virtue of the genic point of view, on the pluralist account, is not that it alone gets the causal structure right but that it is always available.

What is the rival position? Well, it cannot be the thesis that the only adequate representations are those in terms of individual traits which promote the reproductive success of their bearers, because there are instances in which no such representation is available (outlaws) and instances in which the representation is (at best) heuristically misleading (quasi-outlaws, altruism). The sensible rival position is that there is a hierarchy of selection processes: some cases are aptly represented in terms of genic selection, some in terms of individual selection, some in terms of group selection, and some (maybe) in terms of species selection. Hierarchical monism claims that, for any selection process, there is a unique level of the hierarchy such that only representations that depict selection as acting at that level are maximally adequate. (Intuitively, representations that see selection as acting at other levels get the causal structure wrong.) Hierarchical monism differs from pluralist genic selectionism in an interesting way: whereas the pluralist insists that, for any process, there are many adequate representations, one of which will always be a genic representation, the hierarchical monist maintains that for each process there is just one kind of adequate representation, but that processes are diverse in the kinds of representation they demand.[14]

Just as the simple orthodoxy of individualism is ambushed by outlaws and their kin, so too hierarchical monism is entangled in spider webs. In the "extended phenotype" cases, Dawkins shows that there are genic representations of selection processes which can be no more adequately illuminated from alternative perspectives. Since we believe that there is no compelling reason to deny the legitimacy of the individualist redescription in terms of web-building behavior (or dispositions to such behavior), we conclude that Dawkins should be taken at face value: just as we can adopt different perspectives on a Necker cube, so too we can look at the workings of selection in different ways (EP ch. 1).

In previous sections, we have tried to show how genic representations are available in cases that have previously been viewed as troublesome. To complete the defense of genic selectionism, we would need to extend our survey of problematic examples. But the general strategy should be evident. Faced with processes that others see in terms of group selection or species selection, genic selectionists will first try to achieve an individualist representation and then apply the ideas we have developed from Dawkins to make the translation to genic terms.

Pluralist genic selectionists recommend that practicing biologists take advantage of the full range of strategies for representing the workings of selection. The chief merit of Dawkinspeak is its generality. Whereas the individualist perspective may sometimes break down, the gene's eye view is apparently always available. Moreover, as illustrated by the treatment of inclusive fitness,

adopting it may sometimes help us to avoid errors and confusions. Thinking of selection in terms of the devices, sometimes highly indirect, through which genes lever themselves into future generations may also suggest new approaches to familiar problems.

But are there drawbacks? Yes. The principal purpose of the early sections of this paper was to extend some of the ideas of genic selectionism to respond to concerns that are deep and important. Without an adequate rethinking of the concepts of population and of environment, genic representations will fail to capture processes that involve genic interactions or epigenetic constraints. Genic selectionism can easily slide into naive adaptationism as one comes to credit the individual alleles with powers that enable them to operate independently of one another. The move from the "genes for P" locution to the claim that selection can fashion P independently of other traits of the organism is perennially tempting.[15] But, in our version, genic represetations must be constructed in full recognition of the possibilities for constraints in gene-environment coevolution. The dangers of genic selectionism, illustrated in some of Dawkins's own writings, are that the commitment to the complexity of the allelic environment is forgotten in practice. In defending the genic approach against important objections, we have been trying to make this commitment explicit, and thus to exhibit both the potential and the demands of correct Dawkinspeak. The return of the gene should not mean the exile of the organism.[16]

Notes

1 The claim is made in *The Selfish Gene* (New York: Oxford University Press, 1976); and, in a somewhat modified form, in *The Extended Phenotype* (San Francisco: Freeman, 1962). We shall discuss the difference between the two versions in the final section of this paper, and our reconstruction will be primarily concerned with the later version of Dawkins's thesis. We shall henceforth refer to *The Selfish Gene* as SG, and to *The Extended Phenotype* as EP. To forestall any possible confusion, our reconstruction of Dawkins's position does not commit us to the provocative claims about altruism and selfishness on which many early critics of SG fastened.

2 "Artifact, Cause and Genic Selection," *Philosophy of Science*, XLIX (1982): 157–180.

3 See Sober, *The Nature of Selection* (Cambridge MA: MIT, 1984), chs. 7–9, especially 302–314. We shall henceforth refer to this book as NS.

4 For details, see John Maynard Smith, *Evolution and the Theory of Games* (New York: Cambridge, 1982); and, for a capsule presentation, Philip Kitcher, *Vaulting Ambition: Sociobiology and the Quest for Human Nature* (Cambridge University Press: MIT, 1985), pp. 88–97.

5 The *locus classicus* for discussion of this example is H. B. D. Kettlewell, *The Evolution of Melanism* (New York: Oxford University Press, 1973).

6 In the spirit of Sober's original argument, one might press further. Genic selectionists contend that an *A* allele can find itself in two different environments, one in which the effect of directing the formation of a normal globin chain is positive and one in which that effect is negative. Should we not be alarmed by the fact that the distribution of environments in which alleles are selected is itself a function of the frequency of the alleles whose selection we are following? No. The phenomenon is thoroughly familiar from studies of behavioral interactions – in the hawk–dove case we treat the frequency of hawks both as the variable we are tracking and as a facet of the environment in which selection occurs. Maynard Smith makes the parallel fully explicit in his paper "How To Model Evolution," in John Dupre, ed., *The Latest on the Best: Essays on Optimality and Evolution* (Cambridge, MA: MIT, 1987), pp. 119–131. especially pp. 125/6.

7 Moreover, we can explicitly recognize the co-evolution of alleles with allelic environments. A fully detailed general approach to population genetics from the Dawkinsian point of view will involve equations that represent the functional dependence of the distribution of environments on the frequency of alleles, and equations that represent the fitnesses of individual alleles in different environments. In fact, this is just another way of looking at the standard population genetics equations. Instead of thinking of W_{AA} as the expected contribution to survival and reproduction of (an organism with) an allelic pair, we think of it as the expected contribution of copies of itself of the allele *A* in environment *A*. We now see W_{AS} as the expected contribution of *A* in environment *S* and also at the expected contribution of *S* in environment *A*. The frequencies *p*, *q* are not only the frequencies of the alleles, but also the frequencies with which certain environments occur. The standard definitions of the overall (net) fitnesses of the alleles are obtained by weighting the fitnesses in the different environments by the frequencies with which the environments occur.

Lewontin has suggested to us that problems may arise with this scheme of interpretation if the population should suddenly start to reproduce asexually. But this hypothetical change could be handled from the genic point of view by recognizing an alteration of the coevolutionary process between alleles and their environments: whereas certain alleles used to have descendants that would encounter a variety of environments, their descendants are now found only in one allelic environment. Once the algebra has been formulated, it is relatively straightforward to extend the reinterpretation to this case.

8 "Caring Groups and Selfish Genes," in *The Panda's Thumb* (New York: Norton, 1980), p. 90. There is a valuable discussion of Gould's claims in Sober, NS 287 ff.

9 Gould, *op.cit.*; Mayr, *Animal Species and Evolution* (Cambridge MA: Harvard, 1963), p. 184; and Brandon, "The Levels of Selection," in Brandon and Richard Burian, eds., *Genes, Organisms, Populations* (Cambridge MA: MIT, 1984), pp. 133–141.

10 Brandon refers to Salmon's "Statistical Explanation," in Salmon, ed., *Statistical Explanation and Statistical Relevance* (Pittsburgh: Pittsburgh University Press, 1971). It is now widely agreed that statistical relevance misses some distinctions which are important in explicating causal relevance. See, for example, Nancy Cartwright, "Causal Laws and Effective Strategies," *Noûs*, XIII (1979): 419–437; Sober, NS ch. 8; and Salmon, *Scientific Explanation and the Causal Structure of the World* (Princeton NJ: Princeton University Press, 1984).

11 Unless the treatments are repeated in each generation, the presence of the genetic basis for melanic coloration will be correlated with an increased frequency of grandoffspring, or of great-grandoffspring, or of descendants in some further generation. Thus, analogs of Brandon's probabilistic relations will hold only if the progeny of foiled melanics are treated so as to become foiled melanics, and the progeny of pseudomelanics are treated so as to become pseudomelanics. This point reinforces the claims about the relativization to the environment that we make below. Brandon has suggested to us in correspondence that now his preferred strategy for tackling issues of the units of selection would be to formulate a principle for identifying genuine environments.

12 Intuitively, this will be because Brandon's identities depend on there being no correlation between O_n and G in any environment, except through the property P. Thus, ironically, the screening-off relations only obtain under the assumptions of

simple bean-bag genetical Sober seems to appreciate this point in a cryptic footnote (NS 229–230).

To see how it applies in detail, imagine that we have more than one environment and that the reproductive advantages of melanic coloration differ in the different environments. Specifically, suppose that E_1 contains m_1 organisms that have P (melanic coloration) and G (the normal genetic basis of melanic coloration), that E_2 contains m_2 organisms that have P and G, and that the probabilities $Pr(O_n/G\&P\&E_1)$ and $Pr(O_n/G\&P\&E_2)$ are different. Then, if we do not relativize to environments, we shall compute $Pr(O_n/G\&P)$ as a weighted average of the probabilities relative to the two environments.

$$Pr(O_n/G\&P) = Pr(E_1/G\&P) \cdot Pr(O_n/G\&P\&E_1) + \\ Pr(E_2/G\&P) \cdot Pr(O_n/G\&P\&E_2) \\ = m_1/(m_1 + m_2) \cdot Pr(O_n/G\&P\&E_1) + \\ m_2/(m_1 + m_2) \cdot Pr(O_n/G\&P\&E_2)$$

Now, suppose that tampering occurs in E_2 so that there are m_3 pseudomelanics in E_2. We can write $Pr(O_n/P)$ as a weighted average of the probabilities relative to the two environments.

$$Pr(O_n/P) = Pr(E_1/P) \cdot Pr(O_n/P\&E_1) + Pr(E_2/P) \cdot \\ Pr(O_n/P\&E_2).$$

By the argument that Brandon uses to motivate his claims about screening off, we can take $Pr(O_n/G\&P\&E_1) = Pr(O_n/P\&E_1)$ for $i = 1, 2$. However, $Pr(E_1/P) = m_1/(m_1 + m_2 + m_3)$ and $Pr(E_2/P) = (m_2 + m_3)/(m_1 + m_2 + m_3)$, so that $Pr(E_1/P) \neq Pr(E_1/G\&P)$. Thus, $Pr(O_n/G\&P) \neq Pr(O_n/P)$, and the claim about screening off fails.

Notice that, if environments are lumped in this way, then it will only be under fortuitous circumstances that the tampering makes the probabilistic relations come out as Brandon claims. Pseudomelanics would have to be added in both environments so that the weights remain exactly the same.

13 For Hamilton's original demonstration, see "The Genetical Evolution of Social Behavior I," In G. C. Williams, ed., *Group Selection* (Chicago IL: Aldine, 1971), pp. 23–43. For a brief presentation of Hamilton's ideas, see Kitcher, *op. cit.*, pp. 77–87; and for penetrating diagnoses of misunderstandings, see A. Grafen, "How Not to Measure Inclusive Fitness," *Nature*, CCXCVIII (1982): 425/8; and R. Michod, "The Theory of Kin Selection," in Brandon and Burian, *op. cit.*, pp. 203–237.

14 In defending pluralism, we are very close to the views expressed by Maynard Smith in "How To Model Evolution." Indeed, we would like to think that Maynard Smith's article and the present essay complement one another in a number of respects. In particular, as Maynard Smith explicitly notes, "recommending a plurality of models of the same process" contrasts with the view (defended by Gould and by Sober) of "emphasizing a plurality of processes." Gould's views are clearly expressed in "Is A New and General Theory of Evolution Emerging?" *Paleobiology*, VI (1980): 119–130: and Sober's ideas are presented in NS ch. 9.

15 At least one of us believes that the claims of the present paper are perfectly compatible with the critique of adaptationism developed in Gould and Lewontin, "The Spandrels of San Marco and the Panglossian Paradigm: A Critique of the Adaptationist Programme," in Sober, ed., *Conceptual Problems in Evolutionary Biology* (Cambridge MA: MIT, 1984). For discussion of the difficulties with adaptationism, see Kitcher, *Vaulting Ambition*, ch. 7; and "Why Not The Best?" in Dupré, *op. cit.*

16 As, we believe, Dawkins himself appreciates. See the last chapter of EP, especially his reaction to the claim that "Richard Dawkins has rediscovered the organism" (251).

22

The Levels of Selection Debate: Philosophical Issues

Samir Okasha

1. Introduction

For a number of years, the debate in evolutionary biology over the 'levels of selection' has attracted intense interest from philosophers of science. This is because the debate comprises an intriguing mix of empirical, conceptual and methodological questions which makes it ideally suited to, and much in need of, philosophical scrutiny. The main question concerns the *level of the biological hierarchy* at which natural selection occurs. Does selection act on organisms, genes, groups, colonies, demes, species, or some combination of these? According to traditional Darwinian theory the answer is the organism – it is the differential survival and reproduction of individual organisms that drives the evolutionary process. But there are alternative views too. Proponents of 'group selection' argue that groups of organisms, rather than individual organisms, may sometimes function as levels of selection, while 'genic selectionists' argue that the true level of selection is in fact the gene, for genes alone are the 'ultimate beneficiaries' of the selection process.

Philosophers of science first turned their attentions to these issues in the 1980s; notable works from this period include Hull (1980), Sober (1984), Sober and Lewontin (1982), Brandon (1982), Wimsatt (1980), Lloyd (1988) and Sterelny and Kitcher (1988). For the most part, these early philosophical forays aimed at clarifying the key concepts in the debate (such as 'level of selection'), examining the logic of the arguments used for and against various different positions, and trying to separate out the empirical from the conceptual issues. This led to a number of important advances, including Hull's recognition that the expression 'unit of selection' was often ambiguous between 'replicators' and 'interactors', Sober's recognition that different biologists have often used different criteria for group selection, and Sterelny and Kitcher's articulation of 'pluralism', which says that the choice between different levels of selection is often a matter of perspective, not empirical fact (see below). Interestingly, and in sharp contrast to other areas of philosophy of science, these philosophical contributions attracted considerable attention from the biologists whose work was under scrutiny. It was not for nothing that Daniel Dennett (1995) described the levels of selection debate as one of the "brightest areas" of the philosophy of science.

The flurry of attention paid to the levels of selection question tailed off somewhat in the 1990s, leading to a widespread perception, among both biologists and philosophers, that the debate had

Samir Okasha, "The Levels of Selection Debate: Philosophical Issues," *Philosophy Compass* (2006) 1, 74–85. Reproduced with permission of Blackwell Publishing Ltd.

run its course. Despite this perception, recent years have in fact seen interesting and important new work on the levels of selection, some of which has significantly redefined the terms of the traditional debate. This paper aims to introduce the reader to these new developments.

2. Multi-Level Selection and the Major Transitions in Evolution

The body of ideas known loosely as 'multi-level selection theory' takes as its starting point the notion that natural selection can operate *simultaneously* at different levels of the biological hierarchy. So the evolution of a given trait can be affected by selection at more than one level. This means that it is a mistake to ask what *the* level of selection is in a given scenario, or in general – there need be no single answer. Hence to oppose 'genic selection' to 'individual selection' or to 'group selection', as authors such as Dawkins (1976) and Williams (1966) did, is to commit a conceptual mistake: selection can operate at all of these levels, and others. Another central theme of multi-level selection theory is the idea that the *direction* of selection may be different at different hierarchical levels; for example, a trait may be selectively disadvantageous at the individual level, but selectively advantageous at the group level. According to its proponents, a properly inclusive evolutionary theory, which seeks to understand all the forces affecting biotic evolution, must recognise the possibility of selection at multiple levels.

Though the label is new, the basic ideas behind multi-level selection theory have actually been with us for some time. Darwin's famous discussion of the evolution of self-sacrificial behaviour among early humans in *The Descent of Man* (1871) makes the point that a trait or behaviour may be favoured by group selection but disfavoured by individual selection; and as Gould (2002) has recently documented, August Weismann formulated very clearly the idea that selection can operate at multiple hierarchical levels, above and below that of the organism. Weismann (1903) wrote: "this extension of the principle of natural selection to all grades of vital units is the characteristic feature of my theories . . . this idea will endure even if everything else

in the book should prove transient", (quoted in Gould (2002) p. 223.)

Despite this impressive pedigree, it is only relatively recently that biologists have come to see multi-level selection as a potent explanatory principle. A number of prominent evolutionary theorists, including Williams (1992), Maynard Smith and Szathmary (1995), Michod (1999), Frank (1999), Sober and Wilson (1998), and Gould (2002) – some of whom were staunch *opponents* of higher-level selection in previous years – have recently endorsed versions of multi-level or 'hierarchical' selection theory, though each in slightly different ways and for different explanatory ends.

This growth of interest in multi-level selection is in some ways surprising, given that active discussion of the levels of selection has been going on since the early 1960s. What explains it? Part of the answer, I believe, stems from an increasing realisation that the traditional way of setting up the levels of selection question takes too much for granted. Traditionally, the question has been set up roughly as follows: "the biological world is hierarchically organised – genes are found on chromosomes, chromosomes in cells, cells in tissues, tissues in organs, organs in organisms, organisms in groups, groups in species etc. Moreover, the principle of natural selection can be formulated wholly abstractly – as Lewontin (1970) famously argued, selection will operate on any entities which exhibit 'heritable variation in fitness'. Entities at many hierarchical levels satisfy these three conditions, hence there is the potential for selection to operate at different levels."

The problem with this formulation is that it takes the existence of the biological hierarchy for granted, as if hierarchical organisation is simply an exogenously given fact about the organic world (Griesemer 2000, Okasha (2005a)). But of course, the biological hierarchy is *itself* the product of evolution – entities further up the hierarchy, such as multi-cellular organisms, have obviously not been there since the beginning of life on earth. The same is true of cells and chromosomes. So ideally, we would like an evolutionary theory which explains how lower-level entities became aggregated into higher-level entities, e.g. how independent genes joined up to form chromosomes, how organelles came to be incorporated into prokaryotic cells to form eukaryotic cells, how single-celled organisms gave rise to multi-cellular ones, how solitary insects came to form integrated colonies, and so on. (These are examples of what Maynard Smith and Szathmary (1995) call the 'major transitions' in evolution.) In short, we want to know how the biological hierarchy got there in the first place, rather than just treating it as a given.

Very probably, multi-level selection will have a role to play in explaining the transitions to new levels of hierarchical organisation. As Buss (1987), Michod (1999) and Maynard Smith and Szathmary (1995) have all stressed, we need to know why lower-level selection did not disrupt the formation of the higher-level entities, e.g. why intra-organismic selection at the cellular level did not disrupt the integrity of multi-cellular organisms. Clearly, selection on the higher-level entities themselves is one possible answer. If so, then we have a classic multi-level scenario: selection operates on lower-level entities, favouring those that survive best/replicate fastest without regard for the effect on the higher-level entity; selection *also* operates on the higher-level entities, ultimately leading to a high degree of functional integration and suppression of competition among the lower-level entities. So the levels of selection problem becomes, not just the problem of discovering at which hierarchical level or levels selection *now* acts, but the problem of figuring out how the various levels in the hierarchy evolved initially.

This new 'diachronic' perspective gives the levels of selection question a renewed sense of urgency. Some biologists were inclined to dismiss the traditional levels of selection debate as a storm in a tea-cup – arguing that in practice, selection on individual organisms is the only important selective force in evolution, whatever other theoretical possibilities claim. But as Michod (1999) stresses, multi-cellular organisms did not come from nowhere, and a complete evolutionary theory must surely try to explain how they evolved, rather than just taking their existence for granted. So levels of selection apart from that of the individual organism must have existed in the past, whether or not they still operate today. From this expanded point of view, the argument that individual selection is 'all that matters in practice' is clearly unsustainable. Michod's own

models of the evolution of multi-cellularity involve two levels of selection – the organismic and the cellular – in a scenario interestingly reminiscent of the classical group selection model for the evolution of altruism. 'Selfish' cells, which abandon somatic duties in favour of increased replication, are selected for at the cellular level; but at the level of the whole organism, there is selection against such cells, for they disrupt organismic function (Michod 1999). This particular model has been criticised, but the general principle of interacting levels of selection leading to the evolution of new hierarchical levels is widely accepted.

The contrast I have drawn between the modern 'diachronic' view of the levels of selection and the traditional 'synchronic' view should not be overdone. Even in the earlier discussions, there was always an awareness that entities at different levels of hierarchical organization form a temporal sequence, i.e. lower-level entities generally evolved before higher-level ones. But the importance of explaining the major transitions, and the need to invoke multi-level selection theory to do so, was not widely appreciated until Buss's seminal *The Evolution of Individuality* (1987). Consider for example Richard Dawkins' (1982) brief discussion of how independent replicating units may originally have come together to form chromosomes. Dawkins says that it is "easily understood" why independent genes might have gained an advantage by "ganging up together" into cells, because their biochemical effects might have complemented each other (p. 252). What Dawkins fails to realise is that his argument in effect invokes group selection! From the selective point of view, genes sacrificing their independence by combining to form higher-level functional units, e.g. chromosomes or cells, is strictly analogous to individuals combining themselves into higher-level functional units, e.g. groups. But Dawkins is an implacable *opponent* of group selection, insisting on the impotence of selection for group advantage as an evolutionary mechanism, compared with ordinary individual selection! Clearly, Dawkins has failed to realise that trying to explain the major transitions involves us in levels of selection issues closely analogous to those on which debate traditionally focused.

The surge of interest in multi-level selection among biologists has prompted many philosophers of biology to take another look at the levels of selection question, prompting a considerable body of new work. To some extent, this new philosophical work is continuous with work done in the 1980s by Sober, Brandon, Lloyd, Wimsatt, Sterelny, Kitcher and others; to some extent it reflects the new scientific developments. A summary of some of the main philosophical contributions is offered below.

3. Philosophical Issues in Multi-Level Selection Theory

One recurring theme in philosophical discussions of multi-level selection is the issue of realism versus 'pluralism' or 'conventionalism' about the levels of selection. Roughly speaking, realists maintain that there is always a 'fact of the matter' about the level or levels of selection operating in a given scenario. Pluralists hold that in at least some cases there is no such fact. So for example, a given selection process could equally well be viewed as group selection or as individual selection – we are faced with a choice of perspective, not fact, according to pluralists. Pluralism first raised its head in debates over 'genie selection' in the 1980s, where the main issue was whether the 'gene's eye' view of evolution, championed by Dawkins and G. C. Williams and others, was ultimately equivalent to the orthodox organismic viewpoint or not (cf Sterelny and Kitcher 1988, Waters 1994). On balance, most participants in this debate came down on the 'pluralist' side. Aided by Hull's replicator/interactor distinction, it was argued that to oppose genic selection to individual selection was to commit a category mistake, for genes are replicators while individual organisms are interactors, and entities of *both* sorts are involved in any selection process. This sort of pluralism simply stems from our freedom to focus on replicators or interactors, when describing natural selection.

In the 1990s, however, a somewhat different realist/pluralist dispute arose, that could not be resolved simply by distinguishing replicators from interactors, for the issue at stake was individuals versus *groups* – both of which are interactors, not replicators. The dispute hinged around a particular class of evolutionary models, often called 'trait-group' models after D. S. Wilson

(1975, 1980), or 'intra-demic' selection models. In these models the evolution of a trait, typically a social behaviour, is affected by population structure – individual organisms engage in fitness-affecting interactions with certain other members of the population (which form the individual's 'trait-group'), generating evolutionary outcomes that would not occur in a freely-mixing, unstructured population. The key question is: do such models involve a component of group selection or not? Some authors, including Sober and Wilson (1998), have insisted that the answer is 'yes' – since the trait-groups that make up the population typically exhibit differential productivity, there is selection between groups as well as selection between organisms within groups. Sober and Wilson thus favour a resolutely realistic line – it is a matter of fact, not convention, whether or not group selection is occurring in a trait-group scenario. However others theorists, including Dugatkin and Reeve (1994) and Sterelny (1996) have defended a pluralistic line. They argue that trait-group models *can* be construed as multi-level selection as per Sober and Wilson, but can equally be regarded as pure individual selection, simply by treating the organisms in a particular individual's trait-group as part of that individual's selective environment. There is no fact of the matter as to which is right, on this view.

One notable recent contribution to this debate comes from Kerr and Godfrey-Smith (2002a, b); see also the replies by Maynard Smith (2002), Sober and Wilson (2002), and Dugatkin (2002). Kerr and Godfrey-Smith offer a highly sophisticated defence of pluralism. They construct a simple evolutionary model of selection in a structured environment, and show that the model's dynamics can be fully described by two sets of parameter values, one of which ascribes fitness values only to individuals, the other of which ascribes fitnesses to groups and individuals. The former is called a 'contextual' parameterization, for the fitness of an individual depends on its group context, while the latter is called a 'multi-level' parameterization, for both individuals and groups are ascribed fitnesses. Kerr and Godfrey-Smith demonstrate that the two parameterizations are mathematically equivalent – each set of parameter values can be derived from the other. This does not *prove* that pluralism rather than realism is the correct position to take on trait-group selection – for it might be argued that that only one of the parameterizations correctly captures the causal facts, even though the two are mathematically interchangeable, hence computationally equivalent. But Kerr and Godfrey-Smith's work certainly makes a strong case for pluralism, as well as bringing a new degree of rigour to this ongoing debate.

One persistent source of philosophical concern in the levels of selection debate concerns the concept of causality. Virtually everybody agrees that the theory of natural selection is a causal theory – it aims to provide a causal–historical explanation for changes in gene/trait frequency over time (though see Walsh, Ariew and Matthen (2002) for a dissenting view). Therefore, where multiple levels of selection are in play, it follows that causes must be operating at more than one hierarchical level. Sober's seminal (1984) book contained a detailed attempt to use philosophical ideas about causality to help understand multi-level selection. Recent work by Okasha (2004a, 2004b, 2004c) also addresses the issue of causality, though from a somewhat different angle. Most conceptual/philosophical work on the levels of selection has addressed a purely *qualitative* question, namely, what are the level(s) of selection in a given situation? Okasha argues that this traditional focus fails to address an important *quantitative* question, namely, given the levels of selection that are in play, what fraction of the total evolutionary change can be attributed to each? For example, suppose we agree that individual and group-level selection are both operating in a given situation. How do we tell how *much* of the resulting evolutionary change is due to selection at each level? Okasha explores two different statistical techniques designed to address this question, and finds that they yield incompatible results – each technique decomposes the total change into different components, allegedly corresponding to distinct levels of selection. This raises an interesting, and as yet unresolved, philosophical issue: how do we choose between the two techniques? Or is there perhaps 'no fact of the matter' about which is correct? Focusing on the quantitative rather than just the qualitative question takes the realist/pluralist dispute into new and uncharted territory.

4. Further and Related Issues

The biological and philosophical work summarised in the previous two sections deals with what might be called the levels of selection question *sensu strictu*. However, there is a set of related issues sometimes included under the 'levels of selection' or 'units of selection' rubric, though they really concern the units of *inheritance* rather than selection, that have been the focus of considerable recent discussion. A very brief summary of some of this work is offered below.

The distinction between selection and inheritance is conceptually straightforward, or at least should be (though see Michod (1999) who argues for their inseparability). Selection concerns which variants survive best/reproduce the most, while inheritance concerns the transmission of genotypic and phenotypic characters across generations. Thus quantitative geneticists typically distinguish between selection itself and the evolutionary response to selection – where the latter depends on the heritability of the trait selected for. Nonetheless, issues about selection and inheritance were often run together in the traditional levels of selection debate, particularly by advocates of genic selection. Thus Dawkins, for example, used facts about *inheritance*, e.g. that genes are faithfully replicated across generations while whole genotypes and organismic characters are not, to privilege the gene as the unit of *selection*. In retrospect it is clear that arguments of this type wrongly conflate two distinct issues, and equivocate on the expression 'unit of selection'. One of the merits of Hull's replicator/interactor distinction was to make this equivocation clear. (It is partly for this reason that I used the expression 'level of selection' rather than 'unit of selection' in the previous sections; in Hull's terminology, the issues of the previous section concern the level of interaction, not replication.) Nonetheless, questions about the units of inheritance/replication, and the primacy or otherwise of genes in the evolutionary process, are interesting and important in their own right, even if they are orthogonal to questions about selection itself.

Advocates of the 'gene's eye' or 'replicator first' view of evolution, and many others, have tended to regard genes as somehow more important than the other causal determinants of biological form (such as the environment), at least from an evolutionary point of view. While no biologist would officially deny the importance of environmental factors in development, nor the importance of cytoplasmic as well as nuclear inheritance, genes are nonetheless often invested with a special significance. Dawkins emphasised the fidelity of DNA replication as a reason for thinking of genes as the 'ultimate beneficiaries' of evolution, for whom all adaptations are 'for the good of'. The use of informational vocabulary to characterise genes, enshrined in the (metaphorical) notion that genes constitute 'blueprints' for building organisms, is closely bound up with this privileged status accorded to genes. G. C. Williams (1966) emphasised the necessity of thinking of a gene as a store of information, rather than a physical DNA molecule, in order to fully appreciate the significance of the 'gene's eye view' of evolution. Whether or not the notion of genetic information is an essential aspect of the 'gene's eye' viewpoint, it is undoubtedly the case that many genic selectionists have emphasized that notion.

In recent years a number of theorists – biologists, philosophers and others – have subjected the hegemony of the gene, and the concept of genetic information, to intense critical scrutiny. In particular, advocates of 'developmental systems theory' (DST) such as Paul Griffiths, Russell Gray and Susan Oyama, have argued that genes are just one among many causal factors involved in development, and not uniquely responsible for the reliable transmission of biological form across generations (Griffiths, Gray and Oyama (2001)). Supporters of DST argue that in treating DNA as the master-molecule containing the 'information needed to build an organism', biologists have lost sight of the obvious fact that parents transmit far more to their offspring than nuclear DNA, and that many causal factors apart from genes are essential for normal development. There is no particular reason to single out genes as the prime determinants of organismic form, these theorists argue; from the logical point of view, all causal factors responsible for producing the normal adult phenotype are on a par. Population geneticists generally define evolution as 'change in gene frequency over time', a definition which has considerably influenced genic selectionists such as Dawkins, but from the DST viewpoint this is a seriously distorted conception.

Closely allied to this critique of the causal primacy of genes in development is a critique of the very notion of genetic information itself. (Moss (2003) offers a particularly sophisticated critique of both notions.) The historical significance of informational and 'coding' language for the genesis and development of molecular biology cannot be doubted, but many recent philosophers and biologists have wondered how seriously we should take the notion of genetic information. Is there any literal sense in which genes contain 'information' for building organisms, or even for producing proteins, in which *other* factors relevant to development, or to transcription/translation, do *not* contain information? Why do we speak about genetic information but not environmental information, for instance? Moss argues that there are in fact two quite different concepts of the gene in modern biology, and that the idea that genes contain the information needed to build an organism represents an illegitimate conflation of the two. An extended discussion of the notion of genetic information, and how if at all it should be understood, can be found in *Philosophy of Science* 2000, with contributions from Maynard Smith, Sarkar, Godfrey-Smith and Sterelny.

5. Conclusion

It may seem surprising that the levels of selection debate is still live today, given that it traces right back to Darwin. The reason lies partly in the difficulty of resolving the relevant empirical issues, and partly in the fact that the levels of selection question, like so much in evolutionary biology, involves a mixture of empirical and conceptual issues (cf. Sterelny and Griffiths (1999)); and conceptual issues are generally much harder to resolve definitively than empirical issues, where they admit to definitive resolution at all. I hope that the foregoing survey, incomplete though it is, conveys some sense of the direction in which the debate is currently moving, and why it continues to attract so much attention.

Bibliography

NB; This bibliography includes all works cited in the text, plus some other relevant recent work. It does not aim to be exhaustive.

Bonner, J. T. (1988) *The Evolution of Complexity*, Princeton NJ: Princeton University Press.

Brandon, R. (1982) 'The Levels of Selection', in P. Asquith and T. Nickles (eds.) *PSA 1*, Philosophy of Science Association, 315–322.

Brandon, R. (1990) *Organism and Environment*, Princeton NJ: Princeton University Press.

Brandon, R. and Burian, R. (1984) *Genes, Organisms and Populations*, Cambridge MA: MIT Press.

Buss, L. (1987) *The Evolution of Individuality*, Princeton NJ: Princeton University Press.

Damuth, J. and Heisler, I. L. (1988) 'Alternative Formulations of Multi-Level Selection', *Biology and Philosophy*, 3, 407–430.

Darwin, G. (1871) *The Descent of Man and Selection in Relation to Sex*, New York: Appleton.

Dawkins, R. (1976) *The Selfish Gene*, Oxford: Oxford University Press.

Dawkins, R. (1982) *The Extended Phenotype*, Oxford: Oxford University Press.

Dennett, D. (1995) *Danpin's Dangerous Idea*, London: Penguin.

Dugatkin, L. A. (2002) 'Will Peace Follow?', *Biology and Philosophy*, 17, 4, 519–522.

Dugatkin, L. A. and Reeve, H. K. (1994) 'Behavioural Ecology and Levels of Selection: Dissolving the Group Selection Controversy', *Advances in the Study of Behaviour*, 23, 101–133.

Frank, S. (1999) *Foundations of Social Evolution*, Princeton NJ: Princeton University Press.

Godfrey-Smith, P. (2000a) 'Information, Arbitrariness and Selection: Comments on Maynard Smith', *Philosophy of Science*, 67, 2, 202–207.

Godfrey-Smith, P. (2000b) 'The Replicator in Retrospect', *Biology and Philosophy*, 15, 403–423.

Gould, S. J. (2002) *The Structure of Evolutionary Theory*, Cambridge MA: Harvard University Press.

Griesemer, J. (1999) 'Materials for the Study of Evolutionary Transitions', *Biology and Philosophy*, 14, 127–142.

Griesemer, J. (2000) 'The Units of Evolutionary Transition', *Selection*, 1, 1–3, 67–80, http://www.akkrt.hu

Griffiths, P., Gray, P. and Oyama, S. (2001) *Cycles of Contingency: Developmental Systems and Evolution*, Cambridge MA: MIT Press.

Hamilton, W. D. (1975) 'Innate Social Aptitudes in Man: an Approach from Evolutionary Genetics', in R. Fox (ed.) *Biosocial Anthropology*, New York: Wiley.

Heisler, I. L. and Damuth, J. (1987) 'A Method for Analysing Selection in Hierarchically Structured Populations', *American Naturalist*, 130, 582–602.

Hull, D. (1980) 'Individuality and Selection', *Annual Review of Ecology and Systematics*, 11, 311–332.

Jablonka, E. and Lamb, M. (1995) *Epigenetic Inheritance and Evolution: the Lamarckian Dimension*, Oxford: Oxford University Press.

Keller, L. (ed.) (1999) *Levels of Selection in Evolution*, Princeton NJ: Princeton University Press.

Kerr, B. and Godfrey-Smith, P. (2002a) 'Individualist and Multi-level Perspectives on Selection in Structured Populations', *Biology and Philosophy*, 17, 4, 477–517.

Kerr, B. and Godfrey-Smith, P. (2002b) 'On Price's Equation and Average Fitness', *Biology and Philosophy*, 17, 4, 551–565.

Leigh, E. (1995) 'The Major Transitions of Evolution', *Evolution*, 49, 1302–1306.

Lewontin, R. (1970) 'The Units of Selection', *Annual Review of Ecology and Systematics*, 1, 1–18.

Lloyd, E. (1988) *The Structure and Confirmation of Evolutionary Theory*, New York: Greenwood Press.

Maynard Smith, J. (2000a) 'The Concept of Information in Biology', *Philosophy of Science*, 67, 2, 177–194.

Maynard Smith, J. (2000b) 'Reply to Commentaries', *Philosophy of Science*, 67, 2, 214–218.

Maynard Smith, J. (2002) 'Commentary on Kerr and Godfrey-Smith', *Biology and Philosophy*, 17, 4, 523–527.

Maynard Smith, J. and Szathmary, E. (1995) *The Major Transitions in Evolution*, New York: W. H. Freeman.

Michod, R. (1997) 'Cooperation and Conflict in the Evolution of Individuality I. Multilevel Selection of the Organism', *American Naturalist*, 149, 4, 607–645.

Michod, R. (1999) *Darwinian Dynamics*, Princeton NJ: Princeton University Press.

Moss, L. (2003) *What Genes Can't Do*, Cambridge MA: MIT Press.

Okasha, S. (2001) 'Why Won't the Group Selection Controversy Go Away?', *British Journal for the Philosophy of Science*, 51, 25–50.

Okasha, S. (2002) 'Genetic Relatedness and the Evolution of Altruism', *Philosophy of Science*, 69, 1, 138–149.

Okasha, S. (2003) 27. 'Does the Concept of "Clade Selection" Make Sense?', *Philosophy of Science*, 70, 739–751.

Okasha, S. (2004a) 'Multi-level Selection, Covariance and Contextual Analysis', *British Journal for the Philosophy of Science*, 55, 481–504.

Okasha, S. (2004b) 'Multi-level Selection and the Partitioning of Covariance: a Comparison of Three Approaches', *Evolution*, 58, 3, 486–494.

Okasha, S. (2004c) 'The "Averaging Fallacy" and the Levels of Selection', *Biology and Philosophy*, 19, 167–184.

Okasha, S. (2005a) 'Multi-level Selection and the Major Transitions in Evolution', *Philosophy of Science*, 72, 5, 1013–1025.

Okasha, S. (2005b) 'Group Selection, Altruism and Correlated Interaction', *British Journal for the Philosophy of Science*, 56, 4, 703–725.

Roze, D. and Michod, R. (2001) 'Mutation, Multilevel selection, and the Evolution of Propagule Size during the Origin of Multicellularity', *American Naturalist*, 158, 6, 638–654.

Sarkar, S. (2000) 'Information in Genetics and Developmental Biology: Comments on Maynard Smith', *Philosophy of Science*, 67, 2, 208–213.

Sober, E. and Lewontin, R. (1982) 'Artifact, Cause and Genic Selection', *Philosophy of Science*, 45, 157–180.

Sober, E. (1984) *The Nature of Selection*, Chicago: Chicago University Press.

Sober, E. and Wilson, D. S. (1998) *Unto Others: The Evolution and Psychology of Unselfish Behaviour*, Cambridge MA: Harvard University Press.

Sober, E. and Wilson, D. S. (2002) 'Perspectives and Parameterizations: Commentary on Benjamin Kerr and Peter Godfrey-Smith's "Individualist and Multi-Level Perspectives on Selection in Structured Populations"', *Biology and Philosophy*, 17, 4, 529–537.

Sterelny, K. (1996a) 'Explanatory Pluralism in Evolutionary Biology', *Biology and Philosophy*, 11, 193–214.

Sterelny, K. (1996b) 'The Return of the Group', *Philosophy of Science*, 63, 562–584.

Sterelny, K. (2000) 'The "Genetic Program" Program: A Commentary on Maynard Smith on Information in Biology', *Philosophy of Science*, 67, 2, 195–201.

Sterelny, K. and Griffiths, P. E. (1999) *Sex and Death: An Introduction to the Philosophy of Biology*, Chicago: University of Chicago Press.

Sterelny, K. and Kitcher, P. (1988) 'The Return of the Gene', *Journal of Philosophy*, 85, 339–360.

Walsh, D., Ariew, A. and Matthen, M. (2002) 'Trials of Life: Natural Selection and Random Drift', *Philosophy of Science*, 69, 429–446.

Waters, C. K. (1994) 'Tempered Realism about the Forces of Selection', *Philosophy of Science*, 58, 553–573.

Weismann, A. (1903) *The Evolution Theory*, London: Edward Arnold.

Williams, G. C. (1966) *Adaptation and Natural Selection*, Princeton NJ: Princeton University Press.

Williams, G. C. (1992) *Natural Selection: Domains, Levels, Challenges*, Oxford: Oxford University Press.

Wilson, D. S. (1975) 'A Theory of Group Selection', *Proceedings of the National Academy of Sciences USA*, 72, 143–146.

Wilson, D. S. (1980) *The Natural Selection of Populations and Communities*, Menlo Park CA, Benjamin Cummings.

Wilson, D. S. (1997) 'Altruism and Organism: Disentangling the Themes of Multilevel Selection Theory', *American Naturalist*, 150, S122–S134.

Wilson, D. S. and Sober, E. (1989) 'Reviving the Superorganism', *Journal of Theoretical Biology*, 16, 337–356.

Wilson, R. A. (2003) 'Pluralism, Entwinement and the Levels of Selection', *Philosophy of Science*, 70, 531–552.

Wimsatt, W. (1980) 'Reductionistic Research Strategies and their Biases in the Units of Selection Controversy' in T. Nickles (ed.) *Scientific Discovery*, vol. 2, Dordrecht: Reidel.

PART X
SOCIOBIOLOGY AND ETHICS

Introduction

The power of Darwin's theory to explain adaptation, complexity, and diversity in the biological realm makes the temptation to apply it to explain human affairs overwhelming. For human affairs show adaptation, complexity, and diversity as much as, or more than, other domains in biology. But for 100 years the extension of Darwinism to the social and behavioral sciences was obstructed by one obvious difficulty so great that hardly any social scientist paid attention to biology: the problem of *cooperation*. Human affairs are rife with altruism, reciprocation, trust, morality, and other forms of cooperation. It looked like these human traits could not have been the result of natural selection; that we had, in E. O. Wilson's terms, "slipped the leash of evolution."

It was in the late 1960s that evolutionary biologists – not social scientists – began both to realize that there is a lot of apparently fitness-reducing altruism and cooperation among animals, and that there was a powerful way of applying Darwinian theory to explaining it. It did not take them long to begin to see that the same theory could also be applied to explain cooperative human affairs. This was the origin of Sociobiology and Wilson was its most influential expositor. We anthologize here (chapter 23) extracts from two chapters of his path-breaking and highly controversial work, *Sociobiology: The New Synthesis*.

This is followed (chapter 24) by the product of a collaboration between one of the most important evolutionary theorists, W. D. Hamilton, and an influential political scientist, Robert Axelrod, showing how evolutionary game theory can be extended from the non-human to the human case to explain cooperation as an adaptational strategy.

It is worth noting that the incursion of Darwinism into the social sciences, and the program of explaining so much of interest in the domain of the human sciences as adaptations is a large part of what motivated Lewontin and Gould to write the "The Spandrels of San Marco and the Panglossian Paradigm: A Critique of the Adaptationist Programme." Enthusiasts about sociobiology and its successor movement, evolutionary psychology, are encouraged to go back to both Lewontin and Gould and Mayr's discussions of the adaptationalist research program above in Part IV.

Ever since Darwin, philosophers and biologists have sought both to explain human ethical norms as biological adaptations and also to somehow justify or ground morality on the process of natural selection. The first of these was Herbert Spencer, who had done Darwin the dubious honor of suggesting the slogan of "the survival of the fittest" as a catchy label for the theory of natural selection. Spencer went on to burden Darwinism with another idea that Darwin himself was never tempted by: the notion that whatever wins in the struggle for survival must be morally good, and that we should not interfere in that struggle lest we encourage the morally bad. This doctrine came wrongly to be called *social Darwinism*, and saddled the theory with an undeserved reputation for beastliness over a century or more.

Alex Rosenberg's chapter from the *Cambridge Companion to Darwin* (chapter 25 in this volume) shows the limits of Darwinian theory's relevance to ethics, and the mistakes made by the repeated attempts since Spencer to harness natural selection to the vindication of any ethical claims whatever.

Further Reading

Alexander, J., & Skyrms, B. (1999). Bargaining with neighbors: Is justice contagious? *Journal of Philosophy*, 588–598.

Bergstrom, T. (2002). Evolution of social behavior: Individual and group selection models. *Journal of Economic Perspectives*, 16, 231–238.

Binmore, K. (1994). *Playing fair: Game theory and the social contract*. Cambridge, MA: MIT Press.

Binmore, K., & Samuelson, L. (1994). An economic perspective on the evolution of norms. *Journal of Institutional and Theoretical Economics*, 150, 45–63.

Clayton, P., & Schloss, J. (eds.) (2004). *Evolution and ethics: Human morality in biological and religious perspective*. Grand Rapids, MI: Eerdmans.

Danielson, P. (1992). *Artificial morality: Virtuous robots for virtual games*. London: Routledge.

Danielson, P. (1998). Critical notice: *Evolution of the Social Contract. Canadian Journal of Philosophy*, 28, 627–652.

D'Arms, J. (2000). When evolutionary game theory explains morality, what does it explain? *Journal of Consciousness Studies*, 7, 296–299.

Darwin, C. (1871/1909). *The descent of man and selection in relation to sex: And selection in relation to sex*. London: John Murray.

Ellingsen, T. (1997). The evolution of bargaining behavior. *Quarterly Journal of Economics*, 7, 581–602.

Fogel, D. (1993). Evolving behaviours in the iterated prisoner's dilemma. *Evolutionary Computation*, 1, 77–97.

Frank, S. (1998). *Foundations of social evolution*. Princeton, NJ: Princeton University Press.

Harms, W., & Skyrms, B. (2008). Evolution of moral norms. In M. Ruse (ed.), *The Oxford handbook of philosophy of biology* (pp. 434–450). Oxford: Oxford University Press.

Joyce, J. (1999). *The foundations of causal decision theory*. Cambridge: Cambridge University Press.

Kuhn, S. (2004). Reflections on ethics and game theory. *Synthese*, 141, 1–44.

Maienschein, J., & Ruse, M. (eds.) (1999). *Biology and the foundation of ethics*. Cambridge: Cambridge University Press.

Maynard Smith, J. (1998). The origin of altruism. *Nature*, 393, 639–640.

Nitecki, M., & Nitecki, D. (eds.) (1993). *Evolutionary ethics*. Binghamton, NY: SUNY Press.

O'Hear, A. (1997). *Beyond evolution: Human nature and the limits of evolutionary explanation*. Oxford: Clarendon Press.

Poundstone, W. (1992). *Prisoner's dilemma*. New York: Doubleday.

Reijnders, L. (1978). On the applicability of game theory to evolution. *Journal of Theoretical Biology*, 75, 245–247.

Rolston, H. (ed.) (1995). *Biology, ethics, and the origins of life*. London: Jones and Bartlett.

Rosenberg, A. (1991). The biological justification of ethics: A best case scenario. *Social Policy and Philosophy*, 8, 86–101.

Rosenberg, A. (1992). Altruism: Theoretical contexts. In E. Keller & E. Lloyd (eds.), *Keywords in evolutionary biology* (pp. 19–28). Cambridge MA: Harvard University Press.

Rosenberg, A., & McShea, D. (2008). *Philosophy of biology: A contemporary introduction* (Chapter 7). London: Routledge.

Ruse, M. (1986). Evolutionary ethics: A phoenix arisen. *Zygon*, 21, 95–112.

Ruse, M. (1995a). Evolutionary ethics: A defense. In H. Rolston (ed.), *Biology, ethics, and the origins of life* (pp. 93–112). London: Jones and Bartlett.

Ruse, M. (1995b). *Evolutionary naturalism*. London: Routledge.

Teehan, J. & diCarlo, C. (2004). On the naturalistic fallacy: A conceptual basis for evolutionary ethics. *Evolutionary Psychology*, 2, 32–46.

Trivers, R. (1985). *Social evolution*. Menlo Park, CA: Benjamin/Cummings.

Walter, A. (2006). The anti-naturalistic fallacy: Evolutionary moral psychology and the insistence of brute facts. *Evolutionary Psychology*, 4, 33–48.

Wilson, D. (1992). On the relationship between evolutionary and psychological definitions of altruism and selfishness. *Biology and Philosophy*, 7, 61–68.

Woolcock, P. (1999). The case against evolutionary ethics today. In J. Maienschein & M. Ruse (eds.), *Biology and the foundation of ethics* (pp. 276–306). Cambridge: Cambridge University Press.

23

Sociobiology: The New Synthesis

Edward O. Wilson

The Morality of the Gene

Camus said that the only serious philosophical question is suicide. That is wrong even in the strict sense intended. The biologist, who is concerned with questions of physiology and evolutionary history, realizes that self-knowledge is constrained and shaped by the emotional control centers in the hypothalamus and limbic system of the brain. These centers flood our consciousness with all the emotions – hate, love, guilt, fear, and others – that are consulted by ethical philosophers who wish to intuit the standards of good and evil. What, we are then compelled to ask, made the hypothalamus and limbic system! They evolved by natural selection. That simple biological statement must be pursued to explain ethics and ethical philosophers, if not epistemology and epistemologists, at all depths. Self-existence, or the suicide that terminates it, is not the central question of philosophy. The hypothalamic-limbic complex automatically denies such logical reduction by countering it with feelings of guilt and altruism. In this one way the philosopher's own emotional control centers are wiser than his solipsist consciousness, "knowing" that in evolutionary time the individual organism counts for almost nothing. In a Darwinist sense the organism does not live for itself. Its primary function is not even to reproduce other organisms; it reproduces genes, and it serves as their temporary carrier. Each organism generated by sexual reproduction is a unique, accidental subset of all the genes constituting the species. Natural selection is the process whereby certain genes gain representation in the following generations superior to that of other genes located at the same chromosome positions. When new sex cells are manufactured in each generation, the winning genes are pulled apart and reassembled to manufacture new organisms that, on the average, contain a higher proportion of the same genes. But the individual organism is only their vehicle, part of an elaborate device to preserve and spread them with the least possible biochemical perturbation. Samuel Butler's famous aphorism, that the chicken is only an egg's way of making another egg, has been modernized: the organism is only DNA's way of making more DNA. More to the point, the hypothalamus and limbic system are engineered to perpetuate DNA.

In the process of natural selection, then, any device that can insert a higher proportion of certain genes into subsequent generations will come to characterize the species. One class of such devices promotes prolonged individual survival.

Another promotes superior mating performance and care of the resulting offspring. As more complex social behavior by the organism is added to the genes' techniques for replicating themselves, altruism becomes increasingly prevalent and eventually appears in exaggerated forms. This brings us to the central theoretical problem of sociobiology: how can altruism, which by definition reduces personal fitness, possibly evolve by natural selection? The answer is kinship: if the genes causing the altruism are shared by two organisms because of common descent, and if the altruistic act by one organism increases the joint contribution of these genes to the next generation, the propensity to altruism will spread through the gene pool. This occurs even though the altruist makes less of a solitary contribution to the gene pool as the price of its altruistic act.

To his own question, "Does the Absurd dictate death?" Camus replied that the struggle toward the heights is itself enough to fill a man's heart. This arid judgment is probably correct, but it makes little sense except when closely examined in the light of evolutionary theory. The hypothalamic-limbic complex of a highly social species, such as man, "knows," or more precisely it has been programmed to perform as if it knows, that its underlying genes will be proliferated maximally only if it orchestrates behavioral responses that bring into play an efficient mixture of personal survival, reproduction, and altruism. Consequently, the centers of the complex tax the conscious mind with ambivalences whenever the organisms encounter stressful situations. Love joins hate; aggression, fear; expansiveness, withdrawal; and so on; in blends designed not to promote the happiness and survival of the individual, but to favor the maximum transmission of the controlling genes.

The ambivalences stem from counteracting pressures on the units of natural selection. Their genetic consequences will be explored formally later. For the moment suffice it to note that what is good for the individual can be destructive to the family; what preserves the family can be harsh on both the individual and the tribe to which its family belongs; what promotes the tribe can weaken the family and destroy the individual; and so on upward through the permutations of levels of organization. Counter-acting selection on these different units will result in certain genes

being multiplied and fixed, others lost, and combinations of still others held in static proportions. According to the present theory, some of the genes will produce emotional states that reflect the balance of counteracting selection forces at the different levels.

I have raised a problem in ethical philosophy in order to characterize the essence of sociobiology. Sociobiology is defined as the systematic study of the biological basis of all social behavior. For the present it focuses on animal societies, their population structure castes, and communication, together with all of the physiology underlying the social adaptations. But the discipline is also concerned with the social behavior of early man and the adaptive features of organization in the more primitive contemporary human societies. Sociology *sensu stricto*, the study of human societies at all levels of complexity, still stands apart from sociobiology because of its largely structuralist and nongenetic approach. It attempts to explain human behavior primarily by empirical description of the outermost phenotypes and by unaided intuition, without reference to evolutionary explanations in the true genetic sense. It is most successful, in the way descriptive taxonomy and ecology have been most successful, when it provides a detailed description of particular phenomena and demonstrates first-order correlations with features of the environment. Taxonomy and ecology, however, have been reshaped entirely during the past forty years by integration into neo-Darwinist evolutionary theory – the "Modern Synthesis," as it is often called – in which each phenomenon is weighed for its adaptive significance and then related to the basic principles of population genetics. It may not be too much to say that sociology and the other social sciences, as well as the humanities, are the last branches of biology waiting to be included in the Modern Synthesis. One of the functions of sociobiology, then, is to reformulate the foundations of the social sciences in a way that draws these subjects into the Modern Synthesis. Whether the social sciences can be truly biologicized in this fashion remains to be seen.

This book makes an attempt to codify sociobiology into a branch of evolutionary biology and particularly of modern population biology. I believe that the subject has an adequate richness of detail and aggregate of self-sufficient concepts

EVOLUTIONARY AND ECOLOGICAL PARAMETERS

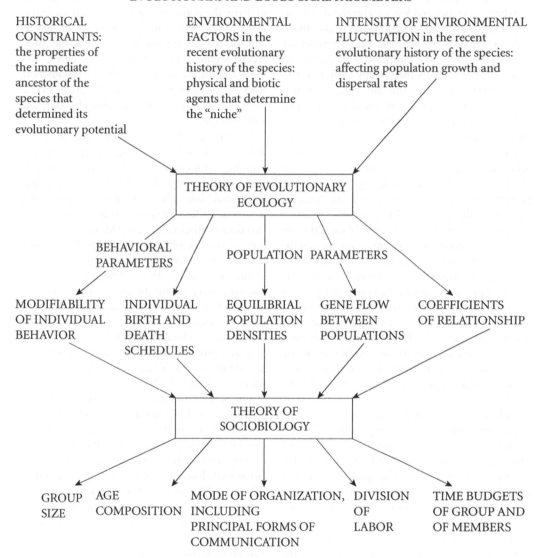

Figure 23.1 The connections that can be made between phylogenetic studies, ecology, and sociobiology.

to be ranked as coordinate with such disciplines as molecular biology and developmental biology. In the past its development has been slowed by too close an identification with ethology and behavioral physiology. In the view presented here, the new sociobiology should be compounded of roughly equal parts of invertebrate zoology, vertebrate zoology, and population biology. Figure 23.1 shows the schema with which I closed *The Insect Societies*, suggesting how the amalgam can be achieved. Biologists have always been intrigued by comparisons between societies of

invertebrates, especially insect societies, and those of vertebrates. They have dreamed of identifying the common properties of such disparate units in a way that would provide insight into all aspects of social evolution, including that of man. The goal can be expressed in modern terms as follows: when the same parameters and quantitative theory are used to analyze both termite colonies and troops of rhesus macaques, we will have a unified science of sociobiology. This may seem an impossibly difficult task. But as my own studies have advanced, I have been increasingly impressed with the functional similarities between invertebrate and vertebrate societies and less so with the structural differences that seem, at first glance, to constitute such an immense gulf between them. Consider for a moment termites and monkeys. Both are formed into cooperative groups that occupy territories. The group members communicate hunger, alarm, hostility, caste status or rank, and reproductive status among themselves by means of something on the order of 10 to 100 nonsyntactical signals. Individuals are intensely aware of the distinction between groupmates and nonmembers. Kinship plays an important role in group structure and probably served as a chief generative force of sociality in the first place. In both kinds of society there is a well-marked division of labor, although in the insect society there is a much stronger reproductive component. The details of organization have been evolved by an evolutionary optimization process of unknown precision, during which some measure of added fitness was given to individuals with cooperative tendencies – at least toward relatives. The fruits of cooperativeness depend upon the particular conditions of the environment and are available to only a minority of animal species during the course of their evolution.

This comparison may seem facile, but it is out of such deliberate oversimplification that the beginnings of a general theory are made. The formulation of a theory of sociobiology constitutes, in my opinion, one of the great manageable problems of biology for the next twenty or thirty years. The prolegomenon of Figure 23.1 guesses part of its future outline and some of the directions in which it is most likely to lead animal behavior research. Its central precept is that the evolution of social behavior can be fully comprehended only through an understanding, first, of demography, which yields the vital information concerning population growth and age structure, and, second, of the genetic structure of the populations, which tells us what we need to know about effective population size in the genetic sense, the coefficients of relationship within the societies, and the amounts of gene flow between them. The principal goal of a general theory of sociobiology should be an ability to predict features of social organization from a knowledge of these population parameters combined with information on the behavioral constraints imposed by the genetic constitution of the species. It will

be a chief task of evolutionary ecology, in turn, to derive the population parameters from a knowledge of the evolutionary history of the species and of the environment in which the most recent segment of that history unfolded. The most important feature of the prolegomenon, then, is the sequential relation between evolutionary studies, ecology, population biology, and sociobiology.

In stressing the tightness of this sequence, however, I do not wish to underrate the filial relationship that sociobiology has had in the past with the remainder of behavioral biology. Although behavioral biology is traditionally spoken of as if it were a unified subject, it is now emerging as two distinct disciplines centered on neurophysiology and on sociobiology, respectively. The conventional wisdom also speaks of ethology, which is the naturalistic study of whole patterns of animal behavior, and its companion enterprise, comparative psychology, as the central, unifying fields of behavioral biology. They are not, both are destined to be cannibalized by neurophysiology and sensory physiology from one end and sociobiology and behavioral ecology from the other (see Figure 23.2).

I hope not too many scholars in ethology and psychology will be offended by this vision of the future of behavioral biology. It seems to be indicated both by the extrapolation of current events and by consideration of the logical relationship behavioral biology holds with the remainder of science. The future, it seems clear, cannot be with the ad hoc terminology, crude models, and curve fitting that characterize most of contemporary ethology and comparative psychology. Whole patterns of animal behavior will inevitably be explained within the framework, first, of integrative neurophysiology, which classifies neurons and reconstructs their circuitry, and, second, of sensory physiology, which seeks to characterize the cellular transducers at the molecular level. Endocrinology will continue to play a peripheral role, since it is concerned with the cruder tuning devices of nervous activity. To pass from this level and reach the next really distinct discipline, we must travel all the way up to the society and the population. Not only are the phenomena best described by families of models different from those of cellular and molecular biology, but the explanations become largely evolutionary. There should be nothing surprising in this

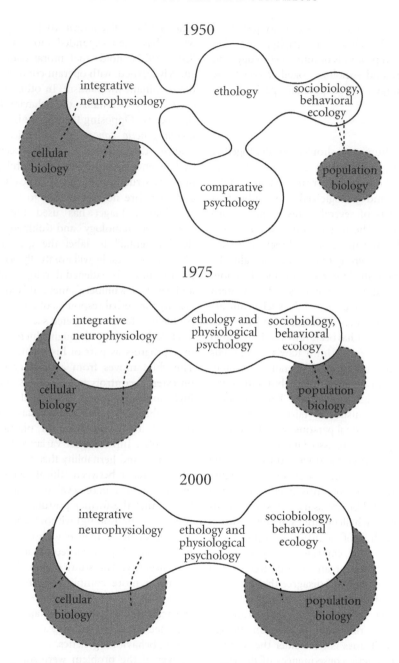

Figure 23.2 A subjective conception of the relative number of ideas in various disciplines in and adjacent to behavioral biology to the present time and as it might be in the future.

distinction. It is only a reflection of the larger division that separates the two greater domains of evolutionary biology and functional biology. As Lewontin (1972) has truly said: "Natural selection of the character states themselves is the essence of Darwinism. All else is molecular biology."

Man: From Sociobiology to Sociology

Let us now consider man in the free spirit of natural history, as though we were zoologists from another planet completing a catalog of social species on Earth. In this macroscopic view the humanities and social sciences shrink to specialized branches of biology; history, biography, and fiction are the research protocols of human ethology; and anthropology and sociology together constitute the sociobiology of single primate species.

Ethics

Scientists and humanists should consider together the possibility that the time has come for ethics to be removed temporarily from the hands of the philosophers and biologicized. The subject at present consists of several oddly disjunct conceptualizations. The first is *ethical intuitionism,* the belief that the mind has a direct awareness of true right and wrong that it can formalize by logic and translate into rules of social action. The purest guiding precept of secular Western thought has been the theory of the social contract as formulated by Locke, Roussean, and Kant. In our time the precept has been rewoven into a solid philosophical system by John Rawls (1971). His imperative is that justice should be not merely integral to a system of government but rather the object of the original contract. The principles called by Rawls "justice as fairness" are those which free and rational persons would choose if they were beginning an association from a position of equal advantage and wished to define the fundamental rules of the association. In judging the appropriateness of subsequent laws and behavior, it would be necessary to test their conformity to the unchallengeable starting position.

The Achilles heel of the intuitionist position is that it relies on the emotive judgment of the brain as though that organ must be treated as a black box. While few will disagree that justice as fairness is an ideal state for disembodied spirits, the conception is in no way explanatory or predictive with reference to human beings. Consequently, it does not consider the ultimate ecological or genetic consequences of the rigorous prosecution of its conclusions. Perhaps explanation and prediction will not be needed for the millennium. But this is unlikely – the human genotype and the ecosystem in which it evolved were fashioned out of extreme unfairness. In either case the full exploration of the neural machinery of ethical judgement is desirable and already in progress. One such effort, constituting the second mode of conceptualization, can be called *ethical behaviorism*. Its basic proposition, which has been expanded most fully by J. E. Scott (1971), holds that moral commitment is entirely learned, with operant conditioning being the dominant mechanism. In other words, children simply internalize the behavioral norms of the society. Opposing this theory is the *developmental-genetic conception* of ethical behavior. The best-documented version has been provided by Lawrence Kohlberg (1969). Kohlberg's viewpoint is structuralist and specifically Piagetian, and therefore not yet related to the remainder of biology. Piaget has used the expression "genetic epistemology" and Kohlberg "cognitive-developmental" to label the general concept. However, the results will eventually become incorporated into a broadened developmental biology and genetics. Kohlberg's method is to record and classify the verbal responses of children to moral problems. He has delineated six sequential stages of ethical reasoning through which an individual may progress as part of his mental maturation. The child moves from a primary dependence on external controls and sanctions to an increasingly sophisticated set of internalized standards (see Table 23.1). The analysis has not yet been directed to the question of plasticity in the basic rules. Intracultural variance has not been measured, and heritability therefore not assessed. The difference between ethical behaviorism and the current version of developmental-genetic analysis is that the former postulates a mechanism (operant conditioning) without evidence and the latter presents evidence without postulating a mechanism. No great conceptual difficulty underlies this disparity. The study of moral development is only a more complicated and less tractable version of the genetic variance problem. With the accretion of data the two approaches can be expected to merge to form a recognizable exercise in behavioral genetics.

Even if the problem were solved tomorrow, however, an important piece would still be missing. This is the *genetic evolution of ethics*. I argue that ethical philosophers intuit the deontological canons of morality by consulting the emotive centers of their own hypothalamic-limbic system. This is also true of the developmentalists, even when they are being their most severely objective. Only by interpreting the activity of the emotive centers as a biological adaptation can the meaning

Table 23.1 The classification of moral judgment into levels and stages of development. (Based on Kohlberg, 1969.)

Level	Basis of moral judgment	Stage of development	
I	Moral value is defined by punishment and reward	1.	Obedience to rules and authority to avoid punishment
		2.	Conformity to obtain rewards and to exchange favors
II	Moral value resides in filling the correct roles, in maintaining order and meeting the expectations of others	3.	Good-boy orientation: conformity to avoid dislike and rejection by others
		4.	Duty orientation: conformity to avoid censure by authority, disruption of order, and resulting guilt
III	Moral value resides in conformity to shared standards, rights, and duties	5.	Legalistic orientation: recognition of the value of contracts, some arbitrariness in rule formation to maintain the common good
		6.	Conscience or principle orientation: primary allegiance to principles of choice, which can overrule law in cases where the law is judged to do more harm than good

of the canons be deciphered. Some of the activity is likely to be outdated, a relic of adjustment to the most primitive form of tribal organization. Some of it may prove to be *in statu nascendi*, constituting new and quickly changing adaptations to agrarian and urban life. The resulting confusion will be reinforced by other factors. To the extent that unilaterally altruistic genes have been established in the population by group selection, they will be opposed by allelomorphs favored by individual selection. The conflict of impulses under their various controls is likely to be widespread in the population, since current theory predicts that the genes will be at best maintained in a state of balanced polymorphism. Moral ambivalency will be further intensified by the circumstance that a schedule of sex- and age-dependent ethics can impart higher genetic fitness than a single moral code which is applied uniformly to all sex-age groups. The argument for this statement is the special case of the Gadgil-Bossert distribution in which the contributions of social interactions to survivorship and fertility schedules are specified. Some of the differences in the Kohlberg stages could be explained in this manner. For example, it should be of selective advantage for young children to be self-centered and relatively disinclined to perform altruistic acts based on personal principle. Similarly, adolescents should be more tightly bound by age-peer bonds within their own sex and hence unusually sensitive to peer approval. The reason is that at this time greater advantage accrues to the formation of alliances and rise in status than later, when sexual and parental morality become the paramount determinants of fitness. Genetically programmed sexual and parent-offspring conflict of the kind predicted by the Trivers models are also likely to promote age differences in the kinds and degrees of moral commitment. Finally, the moral standards of individuals during early phases of colony growth should differ in many details from those of individuals at demographic equilibrium or during episodes of overpopulation. Metapopulations subject to high levels of r extinction will tend to diverge genetically from other kinds of populations in ethical behavior.

If there is any truth to this theory of innate moral pluralism, the requirement for an evolutionary approach to ethics is self-evident. It should also be clear that no single set of moral standards can be applied to all human populations, let alone all sex-age classes within each population. To impose a uniform code is therefore to create complex, intractable moral dilemmas – these, of course, are the current condition of mankind.

References

Kohlberg, L. (1969). "Stage and Sequences: The Cognitive-Developmental Approach to Socialization." In *The Handbook of Socialization Theory and Research*, ed. David A. Goslin. Chicago: Rand McNally (pp. 347–480).

Lewontin, R. (1972). "The Apportionment of Human Diversity." *Evolutionary Biology* 6: 381–398.

Rawls, J. (1971). *A Theory of Justice*. Cambridge, MA: Harvard University Press.

Scott, J. (1971). *Internalization of Norms: A Sociological Theory of Moral Commitment*. Englewood Cliffs, NJ: Prentice-Hall.

24

The Evolution of Cooperation

Robert Axelrod and William D. Hamilton

The theory of evolution is based on the struggle for life and the survival of the fittest. Yet cooperation is common between members of the same species and even between members of different species. Before about 1960, accounts of the evolutionary process largely dismissed cooperative phenomena as not requiring special attention. This position followed from a misreading of theory that assigned most adaptation to selection at the level of populations or whole species. As a result of such misreading, cooperation was always considered adaptive. Recent reviews of the evolutionary process, however, have shown no sound basis for a pervasive group-benefit view of selection; at the level of a species or a population, the processes of selection are weak. The original individualistic emphasis of Darwin's theory is more valid (1, 2).

To account for the manifest existence of cooperation and related group behavior, such as altruism and restraint in competition, evolutionary theory has recently acquired two kinds of extension. These extensions are, broadly, genetical kinship theory (3) and reciprocation theory (4, 5). Most of the recent activity, both in field work and in further developments of theory, has been on the side of kinship. Formal approaches have

varied, but kinship theory has increasingly taken a gene's-eye view of natural selection (6). A gene, in effect, looks beyond its mortal bearer to interests of the potentially immortal set of its replicas existing in other related individuals. If interactants are sufficiently closely related, altruism can benefit reproduction of the set, despite losses to the individual altruist. In accord with this theory's predictions, apart from the human species, almost all clear cases of altruism, and most observed cooperation, occur in contexts of high relatedness, usually between immediate family members. The evolution of the suicidal barbed sting of the honeybee worker could be taken as paradigm for this line of theory (7).

Conspicuous examples of cooperation (although almost never of ultimate self-sacrifice) also occur where relatedness is low or absent. Mutualistic symbioses offer striking examples such as these: the fungus and alga that compose a lichen; the ants and ant-acacias, where the trees house and feed the ants which, in turn, protect the trees (8); and the fig wasps and fig tree, where wasps, which are obligate parasites of fig flowers, serve as the tree's sole means of pollination and seed set (9). Usually the course of cooperation in such symbioses is smooth, but sometimes the partners

Robert Axelrod and William D. Hamilton, "The Evolution of Cooperation," *Science* (American Association for the Advancement of Science, 1981), 211, 1390–1396. Reprinted with permission from AAAS.

For helpful suggestions we thank Robert Boyd, Michael Cohen, and David Sloan Wilson.

show signs of antagonism, either spontaneous or elicited by particular treatments (*10*). Although kinship may be involved, as will be discussed later, symbioses mainly illustrate the other recent extension of evolutionary theory, the theory of reciprocation.

Cooperation per se has received comparatively little attention from biologists since the pioneer account of Trivers (*5*); but an associated issue, concerning restraint in conflict situations, has been developed theoretically. In this connection, a new concept, that of an evolutionarily stable strategy, has been formally developed (*6*, *11*). Cooperation in the more normal sense has remained clouded by certain difficulties, particularly those concerning initiation of cooperation from a previously asocial state (*12*) and its stable maintenance once established. A formal theory of cooperation is increasingly needed. The renewed emphasis on individualism has focused on the frequent ease of cheating in reciprocatory arrangements. This makes the stability of even mutualistic symbioses appear more questionable than under the old view of adaptation for species benefit. At the same time other cases that once appeared firmly in the domain of kinship theory now begin to reveal relatednesses of interactants that are too low for much nepotistic altruism to be expected. This applies both to cooperative breeding in birds (*13*) and to cooperative acts more generally in primate groups (*14*). Here either the appearances of cooperation are deceptive – they are cases of part-kin altruism and part cheating – or a larger part of the behavior is attributable to stable reciprocity. Previous accounts that already invoke reciprocity, however, underemphasize the stringency of its conditions (*15*).

Our contribution in this area is new in three ways.

1. In a biological context, our model is novel in its probabilistic treatment of the possibility that two individuals may interact again. This allows us to shed new light on certain specific biological processes such as aging and territoriality.
2. Our analysis of the evolution of cooperation considers not just the final stability of a given strategy, but also the initial viability of a strategy in an environment dominated by noncooperating individuals, as well as the robustness of a strategy in a variegated environment composed of other individuals using a variety of more or less sophisticated strategies. This allows a richer understanding of the full chronology of the evolution of cooperation than has previously been possible.
3. Our applications include behavioral interaction at the microbial level. This leads us to some speculative suggestions of rationales able to account for the existence of both chronic and acute phases in many diseases, and for a certain class of chromosomal nondisjunction, exemplified by Down's syndrome.

Strategies in the Prisoner's Dilemma

Many of the benefits sought by living things are disproportionally available to cooperating groups. While there are considerable differences in what is meant by the terms "benefits" and "sought," this statement, insofar as it is true, lays down a fundamental basis for all social life. The problem is that while an individual can benefit from mutual cooperation, each one can also do even better by exploiting the cooperative efforts of others. Over a period of time, the same individuals may interact again, allowing for complex patterns of strategic interactions. Game theory in general, and the Prisoner's Dilemma game in particular, allow a formalization of the strategic possibilities inherent in such situations.

The Prisoner's Dilemma game is an elegant embodiment of the problem of achieving mutual cooperation (*16*), and therefore provides the basis for our analysis. To keep the analysis tractable, we focus on the two-player version of the game, which describes situations that involve interactions between pairs of individuals. In the Prisoner's Dilemma game, two individuals can each either cooperate or defect. The payoff to a player is in terms of the effect on its fitness (survival and fecundity). No matter what the other does, the selfish choice of defection yields a higher payoff than cooperation. But if both defect, both do worse than if both had cooperated.

Figure 24.1 shows the payoff matrix of the Prisoner's Dilemma. If the other player cooperates, there is a choice between cooperation which yields R (the reward for mutual cooperation)

Player B

Player A	C Cooperation	D Defection
C Cooperation	R=3 Reward for mutual cooperation	S=0 Sucker's payoff
D Defection	T=5 Temptation to defect	P=1 Punishment for mutual defection

Figure 24.1 The Prisoner's Dilemma game. The payoff to player A is shown with illustrative numerical values. The game is defined by $T > R > P > S$ and $R > (S + T)/2$.

or defection which yields T (the temptation to defect). By assumption, $T > R$, so that it pays to defect if the other player cooperates. On the other hand, if the other player defects, there is a choice between cooperation which yields S (the sucker's payoff) or defection which yields P (the punishment for mutual defection). By assumption $P > S$, so it pays to defect if the other player defects. Thus, no matter what the other player does, it pays to defect. But, if both defect, both get P rather than the larger value of R that they both could have gotten had both cooperated. Hence the dilemma (17).

With two individuals destined never to meet again, the only strategy that can be called a solution to the game is to defect always despite the seemingly paradoxical outcome that both do worse than they could have had they cooperated.

Apart from being the solution in game theory, defection is also the solution in biological evolution (18). It is the outcome of inevitable evolutionary trends through mutation and natural selection: if the payoffs are in terms of fitness, and the interactions between pairs of individuals are random and not repeated, then any population with a mixture of heritable strategies evolves to a state where all individuals are defectors. Moreover, no single differing mutant strategy can do better than others when the population is using this strategy. In these respects the strategy of defection is stable.

This concept of stability is essential to the discussion of what follows and it is useful to state it more formally. A strategy is evolutionarily stable if a population of individuals using that strategy cannot be invaded by a rare mutant adopting a different strategy (11). In the case of the Prisoner's Dilemma played only once, no strategy can invade the strategy of pure defection. This is because no other strategy can do better with the defecting individuals than the P achieved by the defecting players who interact with each other. So in the single-shot Prisoner's Dilemma, to defect always is an evolutionarily stable strategy.

In many biological settings, the same two individuals may meet more than once. If an individual can recognize a previous interactant and remember some aspects of the prior outcomes, then the strategic situation becomes an iterated Prisoner's Dilemma with a much richer set of possibilities. A strategy would take the form of a decision rule which determined the probability of cooperation or defection as a function of the history of the interaction so far. But if there is a known number of interactions between a pair of individuals, to defect always is still evolutionarily stable and is still the only strategy which is. The reason is that defection on the last interaction would be optimal for both sides, and consequently so would defection on the next-to-last interaction, and so on back to the first interaction.

Our model is based on the more realistic assumption that the number of interactions is not fixed in advance. Instead, there is some probability, w, that after the current interaction the same two individuals will meet again. Factors that affect the magnitude of this probability of meeting again include the average lifespan, relative mobility, and health of the individuals. For any value of w, the strategy of unconditional defection (ALL D) is evolutionarily stable; if everyone is using this strategy, no mutant strategy can invade the population. But other strategies may be evolutionarily stable as well. In fact, when w is sufficiently great, there is no single best strategy regardless of the behavior of the others in the population (19). Just because there is no single best strategy, it does not follow that analysis is hopeless. On the contrary, we demonstrate not only the stability of a given strategy, but also its robustness and initial viability.

Before turning to the development of the theory, let us consider the range of biological reality that is encompassed by the game theoretic approach. To start with, an organism does not need a brain to employ a strategy. Bacteria, for example, have a basic capacity to play games in that (i) bacteria are highly responsive to selected aspects of their environment, especially their chemical environment; (ii) this implies that they can respond differentially to what other organisms around them are doing; (iii) these conditional strategies of behavior can certainly be inherited; and (iv) the behavior of a bacterium can affect the fitness of other organisms around it, just as the behavior of other organisms can affect the fitness of a bacterium.

While the strategies can easily include differential responsiveness to recent changes in the environment or to cumulative averages over time, in other ways their range of responsiveness is limited. Bacteria cannot "remember" or "interpret" a complex past sequence of changes, and they probably cannot distinguish alternative origins of adverse or beneficial changes. Some bacteria, for example, produce their own antibiotics, bacteriocins; those are harmless to bacteria of the producing strain, but destructive to others. A bacterium might easily have production of its own bacteriocin dependent on the perceived presence of like hostile products in its environment, but it could not aim the toxin produced toward an offending initiator. From existing evidence, so far from an individual level, discrimination seems to be by species rather even than variety. For example, a *Rhizobium* strain may occur in nodules which it causes on the roots of many species of leguminous plants, but it may fix nitrogen for the benefit of the plant in only a few of these species (*20*). Thus, in many legumes the *Rhizobium* seems to be a pure parasite. In the light of theory to follow, it would be interesting to know whether these parasitized legumes are perhaps less beneficial to free living *Rhizobium* in the surrounding soil than are those in which the full symbiosis is established. But the main point of concern here is that such discrimination by a *Rhizobium* seems not to be known even at the level of varieties within a species.

As one moves up the evolutionary ladder in neural complexity, game-playing behavior becomes richer. The intelligence of primates, including humans, allows a number of relevant improvements: a more complex memory, more complex processing of information to determine the next action as a function of the interaction so far, a better estimate of the probability of future interaction with the same individual, and a better ability to distinguish between different individuals. The discrimination of others may be among the most important of abilities because it allows one to handle interactions with many individuals without having to treat them all the same, thus making possible the rewarding of cooperation from one individual and the punishing of defection from another.

The model of the iterated Prisoner's Dilemma is much less restricted than it may at first appear. Not only can it apply to interactions between two bacteria or interactions between two primates, but it can also apply to the interactions between a colony of bacteria and, say, a primate serving as a host. There is no assumption of commensurability of payoffs between the two sides. Provided that the payoffs to each side satisfy the inequalities that define the Prisoner's Dilemma (Fig. 24.1), the results of the analysis will be applicable.

The model does assume that the choices are made simultaneously and with discrete time intervals. For most analytic purposes, this is equivalent to a continuous interaction over time, with the time period of the model corresponding to the minimum time between a change in behavior by one side and a response by the other. And while the model treats the choices as simultaneous, it would make little difference if they were treated as sequential (*21*).

Turning to the development of the theory, the evolution of cooperation can be conceptualized in terms of three separate questions:

1. *Robustness.* What type of strategy can thrive in a variegated environment composed of others using a wide variety of more or less sophisticated strategies?
2. *Stability.* Under what conditions can such a strategy, once fully established, resist invasion by mutant strategies?
3. *Initial viability.* Even if a strategy is robust and stable, how can it ever get a foothold in an environment which is predominantly noncooperative?

Robustness

To see what type of strategy can thrive in a variegated environment of more or less sophisticated strategies, one of us (R.A.) conducted a computer tournament for the Prisoner's Dilemma. The strategies were submitted by game theorists in economics, sociology, political science, and mathematics (22). The rules implied the payoff matrix shown in Fig. 24.1 and a game length of 200 moves. The 14 entries and a totally random strategy were paired with each other in a round robin tournament. Some of the strategies were quite intricate. An example is one which on each move models the behavior of the other player as a Markov process, and then uses Bayesian inference to select what seems the best choice for the long run. However, the result of the tournament was that the highest average score was attained by the simplest of all strategies submitted: TIT FOR TAT. This strategy is simply one of cooperating on the first move and then doing whatever the other player did on the preceding move. Thus TIT FOR TAT is a strategy of cooperation based on reciprocity.

The results of the first round were then circulated and entries for a second round were solicited. This time there were 62 entries from six countries (23). Most of the contestants were computer hobbyists, but there were also professors of evolutionary biology, physics, and computer science, as well as the five disciplines represented in the first round. TIT FOR TAT was again submitted by the winner of the first round, Professor Anatol Rapoport of the Institute for Advanced Study (Vienna). It won again. An analysis of the 3 million choices which were made in the second round identified the impressive robustness of TIT FOR TAT as dependent on three features: it was never the first to defect, it was provocable into retaliation by a defection of the other, and it was forgiving after just one act of retaliation (24).

The robustness of TIT FOR TAT was also manifest in an ecological analysis of a whole series of future tournaments. The ecological approach takes as given the varieties which are present and investigates how they do over time when interacting with each other. This analysis was based on what would happen if each of the strategies in the second round were submitted to a hypothetical next round in proportion to its success in the previous round. The process was then repeated to generate the time path of the distribution of strategies. The results showed that, as the less successful rules were displaced, TIT FOR TAT continued to do well with the rules which initially scored near the top. In the long run, TIT FOR TAT displaced all the other rules and went to fixation (24). This provides further evidence that TIT FOR TAT's cooperation based on reciprocity is a robust strategy that can thrive in a variegated environment.

Stability

Once a strategy has gone to fixation, the question of evolutionary stability deals with whether it can resist invasion by a mutant strategy. In fact, we will now show that once TIT FOR TAT is established, it can resist invasion by any possible mutant strategy provided that the individuals who interact have a sufficiently large probability, w, of meeting again. The proof is described in the next two paragraphs.

As a first step in the proof we note that since TIT FOR TAT "remembers" only one move back, one C by the other player in any round is sufficient to reset the situation as it was at the beginning of the game. Likewise, one D sets the situation to what it was at the second round after a D was played in the first. Since there is a fixed chance, w, of the interaction not ending at any given move, a strategy cannot be maximal in playing with TIT FOR TAT unless it does the same thing both at the first occurrence of a given state and at each resetting to that state. Thus, if a rule is maximal and begins with C, the second round has the same state as the first, and thus a maximal rule will continue with C and hence always cooperate with TIT FOR TAT. But such a rule will not do better than TIT FOR TAT does with another TIT FOR TAT, and hence it cannot invade. If, on the other hand, a rule begins with D, then this first D induces a switch in the state of TIT FOR TAT and there are two possibilities for continuation that could be maximal. If D follows the first D, then this being maximal at the start implies that it is everywhere maximal to follow D with D, making the strategy equivalent to ALL D. If C follows the initial D, the game is then reset as for the first move; so it must be maximal to repeat the sequence of DC indefinitely. These points show that the task of searching a seemingly infinite array of rules of behavior for one potentially capable of invading TIT FOR TAT is really easier than it seemed: if neither ALL D nor alternation of D and C can invade TIT FOR TAT, then no strategy can.

To see when these strategies can invade, we note that the probability that the nth interaction actually occurs is w^{n-1}. Therefore, the expression for the total payoff is easily found by applying the weights 1, w, w^2 ... to the payoff sequence and summing the resultant series. When TIT FOR TAT plays another TIT FOR TAT, it gets a payoff of R each move for a total of $R + wR + w^2R \ldots$, which is $R/(1 - w)$. ALL D playing with TIT FOR TAT gets T on the first move and P thereafter, so it cannot invade TIT FOR TAT if

$$R/(1 - w) \geq T + wP/(1 - w)$$

Similarly when alternation of D and C plays TIT FOR TAT, it gets a payoff of

$$T = wS + w^2T + s^3S \ldots$$
$$= (T + wS)/(1 - w^2)$$

Alternation of D and C thus cannot invade TIT FOR TAT if

$$R/(1 - w) \geq (T + wS)/(1 - w^2)$$

Hence, with reference to the magnitude of w, we find that neither of these two strategies (and hence no strategy at all) can invade TIT FOR TAT if and only if both

$$w \geq (T - R)/(T - P) \text{ and}$$
$$w \geq (T - R)/(R - S) \qquad (1)$$

This demonstrates that TIT FOR TAT is evolutionarily stable if and only if the interactions between the individuals have a sufficiently large probability of continuing (19).

Initial Viability

TIT FOR TAT is not the only strategy that can be evolutionary stable. In fact, ALL D is evolutionarily stable no matter what is the probability

of interaction continuing. This raises the problem of how an evolutionary trend to cooperative behavior could ever have started in the first place.

Genetic kinship theory suggests a plausible escape from the equilibrium of ALL D. Close relatedness of interactants permits true altruism – sacrifice of fitness by one individual for the benefit of another. True altruism can evolve when the conditions of cost, benefit, and relatedness yield net gains for the altruism-causing genes that are resident in the related individuals (25). Not defecting in a single-move Prisoner's Dilemma is altruism of a kind (the individual is foregoing proceeds that might have been taken) and so can evolve if the two interactants are sufficiently related (18). In effect, recalculation of the payoff matrix in such a way that an individual has a part interest in the partner's gain (that is, reckoning payoffs in terms of inclusive fitness) can often eliminate the inequalities $T > R$ and $P > S$, in which case cooperation becomes unconditionally favored (18, 26). Thus it is possible to imagine that the benefits of cooperation in Prisoner's Dilemma–like situations can begin to be harvested by groups of closely related individuals. Obviously, as regards pairs, a parent and its offspring or a pair of siblings would be especially promising, and in fact many examples of cooperation or restraint of selfishness in such pairs are known.

Once the genes for cooperation exist, selection will promote strategies that base cooperative behavior on cues in the environment (4). Such factors as promiscuous fatherhood (27) and events at ill-defined group margins will always lead to uncertain relatedness among potential interactants. The recognition of any improved correlates of relatedness and use of these cues to determine cooperative behavior will always permit advance in inclusive fitness (4). When a cooperative choice has been made, one cue to relatedness is simply the fact of reciprocation of the cooperation. Thus modifiers for more selfish behavior after a negative response from the other are advantageous whenever the degree of relatedness is low or in doubt. As such, conditionality is acquired, and cooperation can spread into circumstances of less and less relatedness. Finally, when the probability of two individuals meeting each other again is sufficiently high, cooperation based on reciprocity can thrive and be evolutionarily stable in a population with no relatedness at all.

A case of cooperation that fits this scenario, at least on first evidence, has been discovered in the spawning relationships in a sea bass (28). The fish, which are hermaphroditic, form pairs and roughly may be said to take turns at being the high investment partner (laying eggs) and low investment partner (providing sperm to fertilize eggs). Up to ten spawnings occur in a day and only a few eggs are provided each time. Pairs tend to break up if sex roles are not divided evenly. The system appears to allow the evolution of much economy in the size of testes, but Fischer (28) has suggested that the testis condition may have evolved when the species was more sparse and inclined to inbreed. Inbreeding would imply relatedness in the pairs and this initially may have transferred the system to attractance of tit-for-tat cooperation – that is, to cooperation unneedful of relatedness.

Another mechanism that can get cooperation started when virtually everyone is using ALL D is clustering. Suppose that a small group of individuals is using a strategy such as TIT FOR TAT and that a certain proportion, p, of the interactions of members of this cluster are with other members of the cluster. Then the average score attained by the members of the cluster in playing the TIT FOR TAT strategy is

$$p[R/(1 - w)] + (1 - p)[S + wP/(1 - w)]$$

If the members of the cluster provide a negligible proportion of the interactions for the other individuals, then the score attained by those using ALL D is still $P/(1 - w)$. When p and w are large enough, a cluster of TIT FOR TAT individuals can then become initially viable in an environment composed overwhelmingly of ALL D (19).

Clustering is often associated with kinship, and the two mechanisms can reinforce each other in promoting the initial viability of reciprocal cooperation. However, it is possible for clustering to be effective without kinship (3).

We have seen that TIT FOR TAT can intrude in a cluster on a population of ALL D, even though ALL D is evolutionarily stable. This is possible because a cluster of TIT FOR TATs gives each member a nontrivial probability of meeting another individual who will reciprocate the cooperation. While this suggests a mechanism for the initiation of cooperation, it also raises the question about whether the reverse could happen once a strategy like TIT FOR TAT became established itself. Actually, there is an interesting asymmetry here. Let us define a nice strategy as one, such as TIT FOR TAT, which will never be the first to defect. Obviously, when two nice strategies interact, they both receive R each move, which is the highest average score an individual can get when interacting with another individual using the same strategy. Therefore, if a strategy is nice and is evolutionarily stable, it cannot be intruded upon by a cluster. This is because the score achieved by the strategy that comes in a cluster is a weighted average of how it does with others of its kind and with the predominant strategy. Each of these components is less than or equal to the score achieved by the predominant, nice, evolutionarily stable strategy, and therefore the strategy arriving in a cluster cannot intrude on the nice, evolutionarily stable strategy (19). This means that when w is large enough to make TIT FOR TAT an evolutionarily stable strategy it can resist intrusion by any cluster of any other strategy. The gear wheels of social evolution have a ratchet.

The chronological story that emerges from this analysis is the following. ALL D is the primeval state and is evolutionarily stable. This means that it can resist the invasion of any strategy that has virtually all of its interactions with ALL D. But cooperation based on reciprocity can gain a foothold through two different mechanisms. First, there can be kinship between mutant strategies, giving the genes of the mutants some stake in each other's success, thereby altering the effective payoff matrix of the interaction when viewed from the perspective of the gene rather than the individual. A second mechanism to overcome total defection is for the mutant strategies to arrive in a cluster so that they provide a nontrivial proportion of the interactions each has, even if they are so few as to provide a negligible proportion of the interactions which the ALL D individuals have. Then the tournament approach demonstrates that once a variety of strategies is present, TIT FOR TAT is an extremely robust one. It does well in a wide range of circumstances and gradually displaces all other strategies in a simulation of a great variety of more or less sophisticated decision rules. And if the probability that interaction between two individuals will continue is great enough, then TIT FOR TAT is itself evolutionarily stable. Moreover, its stability is especially secure because it can resist the intrusion of whole clusters of mutant strategies. Thus cooperation based on reciprocity can get started in a predominantly noncooperative world, can thrive in a variegated environment, and can defend itself once fully established.

Applications

A variety of specific biological applications of our approach follows from two of the requirements for the evolution of cooperation. The basic idea is that an individual must not be able to get away with defecting without the other individuals being able to retaliate effectively (29). The response requires that the defecting individual not be lost in an anonymous sea of others. Higher organisms avoid this problem by their well-developed ability to recognize many different individuals of their species, but lower organisms must rely on mechanisms that drastically limit the number of different individuals or colonies with which they can interact effectively. The other important requirement to make retaliation effective is that the probability, w, of the same two individuals' meeting again must be sufficiently high.

When an organism is not able to recognize the individual with which it had a prior interaction, a substitute mechanism is to make sure that all of one's interactions are with the same interactant. This can be done by maintaining continuous contact with the other. This method is applied in most interspecies mutualism, whether a hermit crab and his sea-anemone partner, a cicada and the varied microorganismic colonies housed in its body, or a tree and its mycorrhizal fungi.

The ability of such partners to respond specifically to defection is not known but seems possible. A host insect that carries symbionts often carries several kinds (for example, yeasts and bacteria). Differences in the roles of these are almost wholly obscure (30). Perhaps roles are actually the same, and being host to more than

one increases the security of retaliation against a particular exploitative colony. Where host and colony are not permanently paired, a method for immediate drastic retaliation is sometimes apparent instead. This is so with fig wasps. By nature of their remarkable role in pollination, female fig wasps serve the fig tree as a motile aerial male gamete. Through the extreme protogyny and simultaneity in flowering, fig wasps cannot remain with a single tree. It turns out in many cases that if a fig wasp entering a young fig does not pollinate enough flowers for seeds and instead lays eggs in almost all, the tree cuts off the developing fig at an early stage. All progeny of the wasp then perish.

Another mechanism to avoid the need for recognition is to guarantee the uniqueness of the pairing of interactants by employing a fixed place of meeting. Consider, for example, cleaner mutualisms in which a small fish or a crustacean removes and eats ectoparasites from the body (or even from the inside of the mouth) of a larger fish which is its potential predator. These aquatic cleaner mutualisms occur in coastal and reef situations where animals live in fixed home ranges or territories (4, 5). They seem to be unknown in the free-mixing circumstances of the open sea.

Other mutualisms are also characteristic of situations where continued association is likely, and normally they involve quasi-permanent pairing of individuals or of endogamous or asexual stocks, or of individuals with such stocks (7, 31). Conversely, conditions of free-mixing and transitory pairing conditions where recognition is impossible are much more likely to result in exploitation – parasitism, disease, and the like. Thus, whereas ant colonies participate in many symbioses and are sometimes largely dependent on them, honeybee colonies, which are much less permanent in place of abode, have no known symbionts but many parasites (32). The small freshwater animal *Chlorohydra viridissima* has a permanent stable association with green algae that are always naturally found in its tissues and are very difficult to remove. In this species the alga is transmitted to new generations by way of the egg. *Hydra vulgaris* and *H. attenuata* also associate with algae but do not have egg transmission. In these species it is said that "infection is preceded by enfeeblement of the animals and is accompanied by pathological symptoms indicating a definite parasitism by the plant" (33). Again, it is seen that impermanence of association tends to destabilize symbiosis.

In species with a limited ability to discriminate between other members of the same species, reciprocal cooperation can be stable with the aid of a mechanism that reduces the amount of discrimination necessary. Philopatry in general and territoriality in particular can serve this purpose. The phrase stable territories means that there are two quite different kinds of interaction: those in neighboring territories where the probability of interaction is high, and strangers whose probability of future interaction is low. In the case of male territorial birds, songs are used to allow neighbors to recognize each other. Consistent with our theory, such male territorial birds show much more aggressive reactions when the song of an unfamiliar male rather than a neighbor is reproduced nearby (34).

Reciprocal cooperation can be stable with a larger range of individuals if discrimination can cover a wide variety of others with less reliance on supplementary cues such as location. In humans this ability is well developed, and is largely based on the recognition of faces. The extent to which this function has become specialized is revealed by a brain disorder called prosopagnosia. A normal person can name someone from facial features alone, even if the features have changed substantially over the years. People with prosopagnosia are not able to make this association, but have few other neurological symptoms other than a loss of some part of the visual field. The lesions responsible for prosopagnosia occur in an identifiable part of the brain: the underside of both occipital lobes, extending forward to the inner surface of the temporal lobes. This localization of cause, and specificity of effect, indicates that the recognition of individual faces has been an important enough task for a significant portion of the brain's resources to be devoted to it (35).

Just as the ability to recognize the other interactant is invaluable in extending the range of stable cooperation, the ability to monitor cues for the likelihood of continued interaction is helpful as an indication of when reciprocal cooperation is or is not stable. In particular, when the value of w falls below the threshold for stability given in condition (1), it will no longer

pay to reciprocate the other's cooperation. Illness in one partner leading to reduced viability would be one detectable sign of declining w. Both animals in a partnership would then be expected to become less cooperative. Aging of a partner would be very like disease in this respect, resulting in an incentive to defect so as to take a onetime gain when the probability of future interaction becomes small enough.

These mechanisms could operate even at the microbial level. Any symbiont that still has a transmission "horizontally" (that is, infective) as well as vertically (that is, transovarial, or more rarely through sperm, or both) would be expected to shift from mutualism to parasitism when the probability of continued interaction with the host lessened. In the more parasitic phase it could exploit the host more severely by producing more infective propagules. This phase would be expected when the host is severely injured, contracted some other wholly parasitic infection that threatened death, or when it manifested signs of age. In fact, bacteria that are normal and seemingly harmless or even beneficial in the gut can be found contributing to sepsis in the body when the gut is perforated (implying a severe wound) (36). And normal inhabitants of the body surface (like *Candida albicans*) can become invasive and dangerous in either sick or elderly persons.

It is possible also that this argument has some bearing on the etiology of cancer, insofar as it turns out to be due to viruses potentially latent in the genome (37). Cancers do tend to have their onset at ages when the chances of vertical transmission are rapidly declining (38). One oncogenic virus, that of Burkitt's lymphoma, does not have vertical transmission but may have alternatives of slow or fast production of infectious propagules. The slow form appears as a chronic mononucleosis, the fast as an acute mononucleosis or as a lymphoma (39). The point of interest is that, as some evidence suggests, lymphoma can be triggered by the host's contracting malaria. The lymphoma grows extremely fast and so can probably compete with malaria for transmission (possibly by mosquitoes) before death results. Considering other cases of simultaneous infection by two or more species of pathogen, or by two strains of the same one, our theory may have relevance more generally to whether a disease will follow a slow, joint-optimal exploitation course ("chronic" for the host) or a rapid severe exploitation ("acute" for the host). With single infection the slow course would be expected. With double infection, crash exploitation might, as dictated by implied payoff functions, begin immediately, or have onset later at an appropriate stage of senescence (*40*).

Our model (with symmetry of the two parties) could also be tentatively applied to the increase with maternal age of chromosomal nondisjunction during ovum formation (oogenesis) (*41*). This effect leads to various conditions of severely handicapped offspring, Down's syndrome (caused by an extra copy of chromosome 21) being the most familiar example. It depends almost entirely on failure of the normal separation of the paired chromosomes in the mother, and this suggests the possible connection with our story. Cell divisions of oogenesis, but not usually of spermatogenesis, are characteristically unsymmetrical, with rejection (as a so-called polar body) of chromosomes that go to the unlucky pole of the cell. It seems possible that, while homologous chromosomes generally stand to gain by steadily cooperating in a diploid organism, the situation in oogenesis is a Prisoner's Dilemma: a chromosome which can be "first to defect" can get itself into the egg nucleus rather than the polar body. We may hypothesize that such an action triggers similar attempts by the homolog in subsequent meioses, and when both members of a homologous pair try it at once, an extra chromosome in the offspring could be the occasional result. The fitness of the bearers of extra chromosomes is generally extremely low, but a chromosome which lets itself be sent to the polar body makes a fitness contribution of zero. Thus $P > S$ holds. For the model to work, an incident of "defection" in one developing egg would have to be perceptible by others still waiting. That this would occur is pure speculation, as is the feasibility of self-promoting behavior by chromosomes during a gametic cell division. But the effects do not seem inconceivable: a bacterium, after all, with its single chromosome, can do complex conditional things. Given such effects, our model would explain the much greater incidence of abnormal chromosome increase in eggs (and not sperm) with parental age.

Conclusion

Darwin's emphasis on individual advantage has been formalized in terms of game theory. This establishes conditions under which cooperation based on reciprocity can evolve.

References and Notes

1 G. C. Williams, *Adaptations and Natural Selection* (Princeton Univ. Press, Princeton, 1966); W. D. Hamilton, in *Bisocial Anthropology*, R. Fox, Ed. (Malaby, London, 1975), p. 133.

2 For the best recent case for effective selection at group levels and for altruism based on genetic correction of non-kin interactants see D. S. Wilson, *Natural Selection of Populations and Communities* (Benjamin/Cummings, Menlo Park, Calif., 1979).

3 W. D. Hamilton, *J. Theoret. Biol.* 7, 1 (1964).

4 R. Trivers, *Q. Rev. Biol.* 46, 35 (1971).

5 For additions to the theory of biological cooperation see I. D. Chase [*Am. Nat.* 115, 827 (1980)], R. M. Fagen [*ibid.*, p. 858 (1980)], and S. A. Boorman and P. R. Levitt [*The Genetics of Altruism* (Academic Press, New York, 1980)].

6 R. Dawkins, *The Selfish Gene* (Oxford Univ. Press, Oxford, 1976).

7 W. D. Hamilton, *Annu. Rev. Ecol. Syst.* 3, 193 (1972).

8 D. H. Janzen, *Evolution* 20, 249 (1966).

9 J. T. Wiebes, *Gard. Bull. (Singapore)* 29, 207 (1976); D. H. Janzen, *Annu. Rev. Ecol. Syst.* 10, 31 (1979).

10 M. Caullery, *Parasitism and Symbiosis* (Sidgwick and Jackson, London, 1952). This gives examples of antagonism in orchid-fungus and lichen symbioses. For the example of wasp-ant symbiosis, see (*7*).

11 J. Maynard Smith and G. R. Price, *Nature (London)* 246, 15 (1973); J. Maynard Smith and G. A. Parker, *Anim. Behav.* 24, 159 (1976); G. A. Parker, *Nature (London)* 274, 849 (1978).

12 J. Elster, *Ulysses and the Sirens* (Cambridge Univ. Press, London, 1979).

13 S. T. Emlen, in *Behavioral Ecology: An Evolutionary Approach*, J. Krebs and N. Davies, Eds. (Blackwell, Oxford, 1978), p. 245; P. B. Stacey, *Behav. Ecol. Sociobiol.* 6, 53 (1979).

14 A. H. Harcourt, *Z. Tierpsychol.* 48, 401 (1978); C. Packer, *Anim. Behav.* 27, 1 (1979); R. W. Wrangham, *Soc. Sci Info.* 18, 335 (1979).

15 J. D. Ligon and S. H. Ligon, *Nature (London)* 276, 496 (1978).

16 A. Rapoport and A. M. Chammah, *Prisoner's Dilemma* (Univ. of Michigan Press, Ann Arbor, 1965). There are many other patterns of interaction which allow gains for cooperation. See for example the model of intraspecific combat in J. Maynard Smith and G. R. Price, in (*11*).

17 The condition that $R > (S + T)/2$ is also part of the definition to rule out the possibility that alternating exploitation could be better for both than mutual cooperation.

18 W. D. Hamilton, in *Man and Beast: Comparative Social Behavior* (Smithsonian Press, Washington, 1971), p. 57. R. M. Fagen [in (*5*)] shows some conditions for single encounters where defection is not the solution.

19 For a formal proof, see R. Axelrod, *Am. Political Sci. Rev.*, in press. For related results on the potential stability of cooperative behavior see R. D. Luce and H. Raiffa, *Games and Decisions* (Wiley, New York, 1957), p. 102; M. Taylor, *Anarchy and Cooperation* (Wiley, New York, 1976); M. Kurz, in *Economic Progress, Private Values and Public Policy*, B. Balassa and R. Nelson, Eds. (North-Holland, Amsterdam, 1977), p. 177.

20 M. Alexander, *Microbial Ecology* (Wiley, New York, 1971).

21 In either case, cooperation on a tit-for-tat basis is evolutionarily stable if and only if w is sufficiently high. In the case of sequential moves, suppose there is a fixed chance, p, that a given interactant of the pair will be the next one to need help. The critical value of w can be shown to be the minimum of the two side's value of $A/p(A + B)$ where A is the cost of giving assistance, and B is the benefit of assistance when received. See also P. R. Thompson, *Soc. Sci. Info.* **19**, 341 (1980).

22 R. Axelrod, *J. Conflict Resolution* **24**, 3 (1980).

23 In the second round, the length of the games was uncertain, with an expected probability of 200 moves. This was achieved by setting the probability that a given move would not be the last at $w = .99654$. As in the first round, each pair was matched in five games (*24*).

24 R. Axelrod, *J. Conflict Resolution* **24**, 379 (1980).

25 R. A. Fisher, *The Genetical Theory of Natural Selection* (Oxford Univ. Press, Oxford, 1930); J. B. S. Haldane, *Nature (London) New Biol.* **18**, 34 (1955); W. D. Hamilton, *Am. Nat.* **97**, 354 (1963).

26 M. J. Wade and F. Breden, *Behav. Ecol. Sociobiol* **7**, (No. 3), 167 (1980).

27 R. D. Alexander, *Annu. Rev. Ecol. Syst.* **5**, 325 (1974).

28 E. Fischer, *Anim. Behav.* **28**, 620 (1980); E. G. Leigh, Jr., *Proc. Natl. Acad. Sci. U.S.A.* **74**, 4542 (1977).

29 For economic theory on this point see G. Akerlof, *Q. J. Econ.* **84**, 488 (1970); M. R. Darby and E. Karni, *J. Law Econ.* **16**, 67 (1973); O. E. Williamson, *Markets and Hierarchies* (Free Press, New York, 1975).

30 P. Buchner, *Endosymbiosis of Animals with Plant Microorganisms* (Interscience, New York, 1965).

31 W. D. Hamilton, in *Diversity of Insect Faunas*, L. A. Mound and N. Waloff, Eds. (Blackwell, Oxford, 1978).

32 E. O. Wilson, *The Insect Societies* (Bellknap, Cambridge, Mass., 1971); M. Treisman, *Anim. Behav.* **28**, 311 (1980).

33 C. M. Yonge [*Nature (London)* **134**, 12 (1979)] gives other examples of invertebrates with unicellular algae.

34 E. O. Wilson, *Sociobiology* (Harvard Univ. Press, Cambridge, Mass., 1975), p. 273.

35 N. Geschwind, *Sci. Am.* **241**, (No. 3), 180 (1979).

36 D. C. Savage, in *Microbial Ecology of the Gut*, R. T. J. Clarke and T. Bauchop, Eds. (Academic Press, New York, 1977), p. 300.

37 J. T. Manning, *J. Theoret. Biol.* **55**, 397 (1975); M. J. Orlove, *ibid.* **65**, 605 (1977).

38 W. D. Hamilton, *ibid.* **12**, 12 (1966).

39 W. Henle, G. Henle, E. T. Lenette, *Sci. Am.* **241** (No. 1), 48 (1979).

40 See also I. Eshel, *Theoret. Pop. Biol.* **11**, 410 (1977) for a related possible implication of multiclonal infection.

41 C. Stern, *Principles of Human Genetics* (Freeman, San Francisco, 1973).

25

Darwinism in Contemporary Moral Philosophy and Social Theory

Alex Rosenberg

1. Darwinism Characterized

Philosophical Darwinism is a species of naturalism. Among philosophers, naturalism is widely treated as the view that contemporary scientific theory is the source of solutions to philosophical problems. Thus, naturalists look to the theory of natural selection as the primary source in coming to solve philosophical problems raised by human affairs. For it combines more strongly than any other theory relevance to human affairs and scientific warrant. Other theories, especially in physics and chemistry, are more strongly confirmed, especially because their more precise predictions can be tested in real time. But these theories have little to tell us about human conduct and institutions. On the other hand, actual and possible theories, in the social and behavioral sciences, may in the future have more to tell us about humanity than Darwinian theory, but these theories do not as yet have anything like the degree of confirmation of Darwin's theory. Since Darwinism has important consequences for human affairs, the naturalist must look to Darwin's theory, above all others, in the search for philosophical understanding.

For present purposes, Darwinism is the thesis that the diversity, complexity and especially the adaptatedness which organic phenomena manifest is solely the result of successive rounds of random variation and natural selection. Both the notions of 'selection' and 'random' need to be understood in special ways. Properly understood, Darwin's theory undermines the place of purposes in nature. Natural "selection" is a metaphor. There is no foresight in the way mutation and recombination produce variations on which the environment acts, filtering out those organisms which lack fitness minimal for survival long enough to reproduce themselves. One may hold that the theory of natural selection rids the world of purposes by showing that the apparent purposes manifest in adaptations are not real, as adaptation is the result of causes in which no purposes are represented. Or one may hold that Darwin's theory naturalizes purposes, showing how a naturalist can accept descriptions of nature in terms of purpose ("the heart beats in order to circulate the blood") as reflecting an evolutionary etiology (that brought about hearts). The first alternative banishes purpose from the universe; the second reduces it to mechanistic forces that naturalism countenances. Both threaten the "higher" purposes morality is traditionally supposed to serve. Most naturalists have long denied this threat, and have in fact held that Darwinism can illuminate and underwrite human values and moral commitments.

Alex Rosenberg, "Darwinism in Moral Philosophy and Social Theory," in Jonathan Hodge & Gregory Radick (eds.) *The Cambridge Companion to Darwin* (Cambridge University Press, 2003), pp. 310–332. Reprinted with permission.

This chapter surveys contemporary strategies for proving a naturalistic understanding and vindication of morality, ethical norms, our conception of justice, and the cooperative human institutions which these norms and conceptions underlie. We will see that while the prospects for vindication of moral claims as true or well-founded remain clouded, those for explaining the normative dimension of human affairs by appeal to Darwinism appear to be improving. Moreover, the sort of evolutionary understanding of why human beings have been selected for being moral agents comes as close to a vindication of morality in human affairs as naturalism will allow.

2. Two Tasks for Darwinism in Ethics

The ubiquitous human practice of making judgements of right and wrong, moral goodness and badness, imposing standards of fairness and justice, attributing moral duties and responsibility, and according autonomy, constitutes one of the most difficult challenges Naturalism faces. For the truth of statements expressing these judgements, standards and assumptions does not appear to be dependent on facts about the world accessible to scientific discovery. Indeed, these statements appear to report non-natural facts which cannot be accommodated in naturalism's metaphysics, nor are they amenable to evidential support by the employment of scientific methods that naturalism countenances. Naturalism has therefore called upon Darwinian considerations to reconcile our commitment to such normative judgements with a purely scientific world view.

There are broadly two "programs" which attempt to discharge this duty with which naturalism burdens the theory of natural selection: One of these programs seeks to underwrite either received moral judgements or some successor to them as true or correct in the light not of special normative truthmakers (this option being ruled out by naturalism) but in the light of the history of variation and selection through which they emerged. The second of these programs seeks to explain or explain away moral judgments as reflecting the operation of natural selection on hereditary variation in human activities. This second alternatives, naturalists will argue, is a new twist on the enterprise of analyzing the meaning

of ethical claims which philosophers identify as metaethics. It is a new twist on traditional metaethics because it expresses naturalistic doubts about separating claims about meanings of ethical concepts from claims about the causes of ethical commitments expressed in these concepts. Thus, if naturalism can give an explanation of why we make the normative claims we do, it will claim to have provided as much of their meaning as can be provided. Call this project Darwinian metaethics. The first program is a compartment of substantive normative ethics, which identifies what is morally right and wrong, good and bad, just and unjust in terms of some evolutionary considerations. Call this project Darwinian morality.

Both Darwinian metaethics and morality must take account of the peculiar fact about moral judgments that they are supposed to motivate us to do certain things, and to enjoin certain actions, not just as prudentially (in)advisable, in the light of our interests, but as right (or wrong) in themselves. This is a feature of normative claims which philosophers have dubbed "ethical internalism". If we accept that moral claims have this feature then they cannot, for instance, be merely injunctions of prudence, matters of merely instrumental ends-means rationality. On such an account of morality as instrumentally rational, if we do not accept the ends to which moral judgments report the means, we may disregard these judgements. But at least some moral judgements seem to make claims on us that are not merely instrumental, but categorical: "thou shalt not commit adultery", not "If thou wish to avoid some bad end, or to attain some good one, thou should not commit adultery." Darwinian metaethics may explain away the internalism of moral judgements as an illusion, though perhaps an adaptive illusion. Darwinian morality must harness it to evolutionary values as the motive for moral conduct. It will have to identify some naturalistically accepted normative grounds, some commitment to the ends or objectives such as species perpetuation or ecological preservation that make Darwinian morality internally motivating.

According to most philosophers the trouble with Darwinian morality has been well known for almost a century: As a philosophical project it rests on a mistake: the so-called "naturalistic fallacy".

In *Principia Ethica*[1] G. E. Moore offered the so-called "open question" argument against any identification of a normative property, like goodness, with a non-normative or "natural" property, like pleasure, or happiness or for that matter the survival of the individual or the species or for that matter the eco-system, planet or universe. Of any property, say an emotion such as love or a virtue such as heroism or a generalized feeling of pleasure, which is exemplified by someone, it may sensibly be asked whether the virtue or emotion or feeling is good. Accordingly, the identification of any such natural property with goodness cannot be correct. For if it were, the question "is Jones' love for Smith, or for that matter of human-kind as a whole good?" would not be an "open question" to which a negative answer might be given. It would be a question like "Is Mr Jones' mother a woman?" This question is not open to a negative answer. But all questions about whether some natural fact has a normative property are decidedly open questions, to which a negative answer may be intelligibly given. Accordingly Moore argued all attempts to naturalize the normative are fallacious. His open question argument defines the "naturalistic fallacy". Its acceptance by philosophers has made Darwinian morality an unattractive option to most naturalists.[2]

3.   Darwinian Morality

The objection lodged against Darwinian morality may be illustrated by considering a philosophically sophisticated late 20th century version of this project: the attempt to establish certain normative principles as objective truths open to scientific discovery. The program took the name "moral realism" to echo the epistemological program of scientific realism, which argues that scientific theories about unobservable properties and entities should be treated as literally true descriptions of reality, and that the properties and entities to which they advert must exist in spite of the absence of direct empirical evidence for them. Similarly, latter day moral realism holds, we may know certain that certain favored moral properties – like goodness, in particular – exist, and that some social arrangements have these moral properties, on the basis of scientific theory

– in particular through considerations from a theory of the natural selection of moral norms. Peter Railton provides an excellent example of this school of Darwinism.[3] Railton's aim is to provide "descriptions and explanations of certain prominent features of the evolution of moral norms" (p. 203) that will establish their naturalistic foundations. If Darwin's name does not figure in his account it is because Railton recognizes that when it comes to the emergence of normatively right social institutions in the absence of ruling intentions to establish them, the only explanation can be Darwinian (see Railton, 1986, section III and IV, and especially footnote 21).[4]

According to Railton the morally good reflects what it is rational to want, not from an individual point of view, but from "the social point of view" (p. 180). What is rational from the social point of view is what would be rationally approved of were the objective interests of all potentially affected individuals counted equally. Railton holds that social arrangements depart from rationality when they significantly dis-count the interests of particular groups. When this happens there is "potential for dissatisfaction and unrest" which reduces the viability, i.e the fitness, of these social arrangements and of the whole society so arranged. On Railton's view, reduced viability of an arrangement – a norm, an institution, etc. – is reflected in "alienation, loss of morale, decline in the effectiveness of authority ... potential for unrest, ... a tendency towards religious or ideological doctrines, or towards certain forms of repressive apparatus, ..." etc. (p. 192).

On the other hand, social arrangements which are more rational, i.e. tend more fully to be in the interests of all individuals in the society counted equally, will be selected for. That is, the societies bearing these traits will be more viable, presumably because arrangements that enhance equality of treatment are more adapted to the environments in which societies find themselves. This environment is not just the physical, geographical location of a society, it also includes societies with which it is in competition for scarce resources, and the society's environment also includes the fact that the individuals composing it have been selected for fitness- (and thus utility-) maximizing by natural selection. In the long run, just as biological natural selection winnows for those available traits that best "match" organisms' local environments, similarly, the struggle for survival among societies with varying moral traits will eventually winnow for those moral traits – i.e. principles, norms, institutions – that best match societies environment, and these, according to Railton, will invariably be ones that foster equality of various kinds. This will be so, since egalitarian arrangements most nearly fulfil individual people's objective – scientifically determinable – interests.

One objection to this approach is its commitment to natural selection of groups, whole societies, as opposed to individuals. What if in a society more viable than others because of its more extensively egalitarian norms, individuals arise who free-ride on and float these norms when they can. In this case, within group selection for immorality (i.e. inequality in treatment of others) may be stronger than between group selection for morality. In this case, evolution will not proceed in the direction of greater egalitarianism. Of this more in section 5 below. Meanwhile, Railton's account requires the truth of substantive claims that social arrangements which treat society's members in more nearly equal ways will be more adaptive under any conditions, for the society as a whole than those which entrain, enhance or preserve inequalities. Even if this claim were right, Railton's moral realism would still be subject to Moore's objection. There is no reason to think that the survival of any particular social group, individual, or *Homo sapiens* in general for that matter, is intrinsically good or morally required. There is in a naturalistic world view no scope for grounding such claims of intrinsic value.

Suppose it is retorted that Railton's thesis is analysis of what moral goodness consists in, not a justificatory endorsement of it. If this is true, moral realism does not accomplish what it has set out to do for Darwinian morality. For then Darwinism does not motivate any commitment to the moral principles it singles out as true. In effect, so understood, naturalisms like Railton's would deny or ignore the internal normativity of moral judgements and treat them as implicit claims about instrumental rationality, that is rules justified by the success of those (individuals or groups) who (or which) employ them in attaining their non-normative objectives (Railton, 1986, p. 200). Railton may well view his normative claims as merely instrumentally useful, and

without internal moral force. He describes them as part of "the skeleton of a explanatory theory that uses the notion of what is . . . rational from a social point of view . . . that parallels in an obvious way . . . assessments of [instrumental] rationality . . . in explanations of individual behaviors.") In fact, Railton recommends we surrender "the idea that moral evaluations must have categorical force" [p. 204], This denial of the internal normativity of moral judgments has the prospect of reducing Darwinian morality into some versions of Darwinian metaethics. For now it turns out that moral judgements are really just disguised claims about means-ends "instrumental rationality" to which we attribute some purely prudential normative force. Note that non-naturalistic forms of moral realism are not similarly threatened with such reduction to metaethics. For they claim that the normativity of moral judgment reflects some factual condition in the world which our moral detection apparatus enables us to identify. Thus, it has sometimes been claimed that we have direct intuition of the moral qualities of an act and these normative qualities motivate our approval or disapproval of the act in question. Naturalists of all stripes find such moral qualities either non-existent or unintelligible. It is certain there is no room for them in a naturalistic metaphysics.

The naturalists' denial that a range of distinctive moral facts exist and make true moral judgements, together with the force of Moore's diagnosis of a naturalistic fallacy, make Darwinian metaethics a far more attractive project for naturalists than Darwinian morality. Once we deny the existence of a separate range of moral facts to be learned by some sort of interaction either with nature or an with an abstract Platonic realm of values, metaethics becomes a matter of urgency. Metaethics is in large part the study of the nature and meaning of moral judgements. Without truth-makers for moral judgements, ethical claims may be threatened with meaninglessness. If they are meaningless we need at least an explanation of why we and all *Homo sapiens* make these apparent "judgements." If they are not meaningless, but say, all false, we still need an explanation of why the error should persist time out of mind. If moral judgements are neither true nor false, but expressions of our emotions, we need an account of why this expression takes the form it does and why these expressions of our subjective states are coordinated in the way they are. And if moral judgements express the norms of conduct we embrace, we again need a theory to explain why we embrace these norms and not other ones. And in every case, an account needs to be provided of why we feel the commitment to an objective morality reflecting facts independent of us, and which motivate our conduct. About the only Naturalistic metaethical theory that can do any of these things is a Darwinian one.

4.   Darwinian Metaethics

Most of the metaethical theories by Darwinian considerations belong to a species of metaethical theories collectively called "noncognitivist" owing to the fact that they share agreement that moral judgements are neither true nor false reports about the world – they have no propositional or "cognitive content". Among the earliest non-cognitivist theories was the "emotivist" doctrine advanced by A. J. Ayer and C. L. Stevenson and associated with Logical Positivism.[5] This doctrine held that moral judgements expressed emotional states and attitudes of the utterer. Two virtues of this otherwise implausible theory are its ability to explain intransigent moral disagreement as expression of incompatible emotions, and its account of the apparent internalism of moral judgments: ethical judgments have motivational force derived from the emotional attitudes they express. But non-cognitivism will not account for the complex character of ethical reasoning characteristic of human life. More important, we often issue moral judgements on events distant in space/or time in such a cool and bloodless a way that they seem not to express emotions at all. Few latter day naturalists have been attracted by emotivism.

A more sophisticated version of non-cognitivist metaethics has been developed, with an eye to its place in a Darwinian framework, by Alan Gibbard.[6] This widely discussed theory avoids many of the traditional objections to non-cognitivism, while making as strong a positive case for moral objectivity as naturalism will allow. Moreover, Gibbard's theory of the nature of moral judgments seeks to show at least how the emergence of morality might have reflected

coordinated strategies that are adaptive for the individuals who employ them. As such Gibbard provides the philosophical foundation for an explosion of developments in evolutionary game theory and Darwinian political philosophy that we will explore below. Gibbard's theory is only one of a number of actual and possible Darwinian metaethics. The details of any such a theory will be important to philosophers anxious about the meaning of moral judgements. Biologists and others interested in the more general question of how moral judgements are possible within the Darwinian perspective will be more interested in how Gibbard develops the general strategy of a Darwinian metaethics. Before proceeding it is worth noting that the crucial difference between a moral realist like Railton's appeal to natural selection and Gibbard's is that the latter is not out to vindicate the norms which have in fact evolved as the morally right ones, only as the most adaptive ones.

The key to human moral nature lies in coordination broadly considered. [p. 26]

Organisms like *Homo sapiens* needed to coordinate their actions if they are to survive and flourish in competition with megafauna, and cooperative enterprises of proto-agriculture. The design-problem nature set for *Homo sapiens* of establishing and securing this coordination among them is accomplished in large measure by coordinated emotions (here Gibbard's noncognitivism shows its hand). Gibbard's objective is not to establish how institutions of morality or particular moral judgements emerged or might have emerged as a result of random variation and natural selection, but rather to give an analysis of the meaning of moral judgements which, *inter alia*, make such a derivation possible.

A moral judgement is not the expression of an emotion, but a judgement of what sort of emotion or feeling it is rational to have; an emotion is a rational one to have if it is permissible in light of the norms one accepts. The capacity to accept norms depends on language, because language is required to coordinate several agents' norms in ways that are mutually fitness-enhancing. The environment of early man presumably selected for emotional propensities which enhanced coordination, and for linguistic potential that enable norms governing the display of these emotions to do so as well. Gibbard identifies resentment and anger, guilt and shame, as central moral feelings. Norms describing when it is appropriate to feel these emotions, are coordinated with one another so as to encourage or reestablish cooperative conduct among moral agents. Thus, what a person does is morally wrong if it makes sense, in the light of norms she accepts, for her to feel guilty about it, and it makes sense for others to feel resentment about her conduct in the light of their norms. A's guilt meshes with B's anger, C's shame with D's disdain. If uncoordinated, these emotions can lead to escalating conflict; coordinated they make possible the acknowledgment of wrong-doing and reconciliation. What it makes sense to do, or to feel, in the light of norms a person accepts is what Gibbard defines as 'rational'. He rejects a purely instrumental account of rationality, both because of classical puzzle cases in the theory of decision, and more important, because 'rational' has an appraising or approving connotation (a reflection of the internalism of moral judgements), which analyses inspired by rational choice theory cannot capture. But to call an act or feeling rational is not to state a fact about it. It is to express ones acceptance of norms that permit the act or feeling.

Why is Gibbard's doctrine naturalistic and where is the special role for Darwinism in metaethics? The metaethics here is naturalistic because it requires no distinct range of independent moral facts to make true moral judgements. Our moral psychologies do not consist in systems which recognize and represent independent existing normative facts. Rather they are systems that coordinate what is in one agent's head with what is in other agents' heads. What is coordinated is the acceptance of norms in the light of which people's actions and emotions mesh to mutual advantage. The Darwinism emerges in the search for functions which these psychological mechanisms have, for a function, on Gibbard and most Darwinians' views, is as Larry Wright argued,[7] is what emerges from an etiology of variation and selection.

The "design problem" which our hominid ancestors faced was how to establish and ensure cooperation – acts of reciprocal altruism. Cooperation requires coordinated expectations of the sort that might emerge from a bargaining context. Gibbard suggests the emotions' function to enhance coordination must have been selected for by the same forces that made language adaptive for *Homo sapiens*. Gibbard's answer to the question of why did language emerge in the genus Homo has it that the capacity to be guided by words in action and emotion is indispensable to the acceptance of norms which produce cooperation. For such acceptance proceeds by discussion which tends towards consensus, consistency and similarity of motives.

But if moral judgement is not a matter of discerning truths but of expressing one's acceptance of norms that make sense of anger, resentment, guilt and shame, whence their apparent feeling of objectivity, of existence independent of us? Certainly not, Gibbard insists, from the existence of any Platonic range of moral facts or truths we can apprehend. The feeling of objectivity that accompanies these norms is a matter of how strongly we accept them. A norm is felt to be objective if one who holds it would consider it rational even if the holder did not accept the norm himself. And then there is a hierarchy of norms which agents accept. Judgements of objectivity will be a matter of derivation from these higher level norms. Gibbard is tempted by a parallel to the doctrine of secondary qualities. Color, it has long been argued by some empiricists, is a secondary property, that is a property of our experience, in us, and not in the objects we see as colored. But we mistakenly project this property on to objects in the world as an objective feature of it. Color is not a property of things out there in the world, but color attributions have considerable "objectivity". That is, a thing is red if and only if normal observers in normal conditions have red sensations when looking at it. Similarly, the objectivity of moral judgements is a matter of normal agents in normal circumstances accepting the same set of norms of anger, guilt, disdain and resentment.

So, in sum, a moral judgment is rational if it is in accordance with norms we embrace. These norms are ones selected for because they solve problems of cooperation, and their felt objectivity consists in their evolutionarily shaped ubiquity. The emotions that give these norms their internal motivational force are selected for because they coordinate and convey commitment to action in accordance with these norms.

Gibbard speculates that higher cognitive functions and language in *Homo sapiens* were selected for, owing to their role in the facilitation of social coordination. Language in particular enables agents to express norms and enhance their motivational power. The capacity to accept norms depends on language, and the discussions which language makes possible enhance mutual influence, consistency, and move people to act according to agreed-upon norms (whence their apparent internalist characters). The psychological state of accepting a norm, Gibbard holds, can at present only be identified as that psychological state which gives rise to the avowal of the norm and to governance by it. Thus metaethics turns out to give empirical promissory notes about the origins of cognition and language that only biological anthropology and evolutionary psychology can cash in. It also requires demonstrations that the sort of cooperation which characterize morality is in fact adaptive.

It is worth noting that independent of Gibbard, developments in biological anthropology were in fact substantiating several factual presuppositions of his Darwinian metaethic. Only a sketch of these considerations can be given. To begin with, there is evidence that our hominid ancestors were originally solitary and highly competitive, not members of extended family troops with strong kinship relation. Cooperation can be expected to emerge among kin groups through the maximization of inclusive fitness (which calculates individual fitness as a function of total off-spring gene-copies an organism's genes leaves). But cooperation among originally solitary unrelated hominids requires communication of strategies. Independently, the shift from forest to savannah environments selects for the shift of vocalization from limbic to neocortical control (uncontrolled reflex vocalization in the vicinity of predators of the sort arboreal apes display is maladaptive on the savannah where there are no trees to climb). Whatever selected for the hominid shift to the savannah, selected also for neocortical control of vocalization that is necessary for language.[8] The need for cooperation among unrelated individuals puts a further adaptive premium on language, as well as on the cognitive equipment required for recognizing cooperative strategies and non-cooperative ones. And this latter result is one evolutionary psychology has provided

some evidence for.[9] Finally, recent work on the theory of emotions provides further evidence that an adaptational account of anger especially as irrational precommitment to cooperative outcomes seems correct.[10]

5. Can Cooperation Evolve?

Most of all, what this account of moral judgment requires is a great deal of detailed explanation of how natural selection could have brought about the norms of cooperation of which Gibbard claims our moral judgements express acceptance. Without the detail, such a Darwinian metaethic is little more than what S. J. Gould has stigmatized as a "just so" story. This need Gibbard's theory shares with any Darwinian metaethic.

It is just this sort of detail which evolutionary biology, game theory and political philosophy altogether provide, thus freeing a Darwinian metaethic from the charge of being merely a just so story. The substantiation of a naturalistic theory like Gibbard's, begins ironically with a major evolutionary problem. As E. O. Wilson wrote in *Sociobiology*, cooperation and altruism constitute "the central theoretical problem of sociobiology: how can altruism, which by definition reduces personal fitness, possibly evolve by natural selection."[11] Natural selection relentlessly shapes organisms for individual fitness maximization: leaving the largest number of off-spring carrying the organism's genes. Call an act altruistic if it results in an increase in the fitness of another organism and a decrease in the fitness of the organism so acting. Now, the persistence of cooperation among organisms requires acts of reciprocated altruism so that the netpay offs to mutual cooperators is greater than the rewards of mutual non-cooperation. Other things being equal, natural selection blocks the building up of altruism among randomly chosen organisms because altruistic acts offers opportunities to free-ride, to decline to reciprocate, and natural selection drives organisms to maximize fitness by taking every opportunity to free ride. Since altruism, and cooperation characterizes several infrahuman species, and all *Homo sapiens* societies, it appears that evolutionary theory has little to tell us about human conduct. This is what led

Wilson to hold that the existence of altruism posed the gravest challenge to sociobiology.

Wrestling with this problem earlier in the 20th century some theorists, concluded that individual altruism is selected for because of the contribution it makes to the fitness of the group in which the individual finds itself. Group selection as an account of the evolution of altruism fell into great disfavor, however, for individual fitness maximization will swamp group selection. Suppose all members of a group are predisposed to cooperate, to engage in altruistic acts because their genes programmed them to act in this way. Suppose that through mutation, recombination, or immigration a new organism joins the group, lacking the gene for the propensity to cooperate. Instead it is genetically programmed to free-ride, cheat, slack off, shirk and take more than its share, whenever it can do so undetected. The free-rider has only to get away with free-riding some of the time to have a higher fitness level than the rest of the group. Its off-spring will in turn bear the free-riding gene, and will take advantage of altruists as their immediate ancestor did. And so on, generation after generation, until genetically encoded reciprocal altruism has been extirpated from this group, which now of course has lower average fitness than it had when composed of altruists. In Maynard Smith's terms, genetically programmed altruism in a group is not an "evolutionary stable strategy." Wilson' problem of reconciling the ubiquity of human cooperation with natural selection remains.

However, if fitness is measured in terms of the number of copies of itself a gene leaves, genetic selfishness must lead to one kind of organismal unselfishness. If an organism behaves altruistically towards its off-spring, enhancing their survival and reproductive opportunities, the result may be a decline in the altruistic parent's viability, but not a decline in its fitness or rather the fitness of its genes. This is kin-selection. But of course cooperation is far more widespread among *Homo sapiens* than selection for altruism towards kin. So, sociobiology is still faced with the problem that Wilson posed, of how altruism is possible. And Darwinian metaethics' claim that moral judgements are selected for coordinating behavior into cooperative exchanges remains ungrounded.

It was by exploring the economists puzzle of the prisoner's dilemma that evolutionary theorists were able to show how reciprocal altruism can be generated as the optimal strategy for fitness maximizing agents. Two agents, A and B are faced with mirror image choices of whether to cooperate with one another or to decline to do so, in other words, to defect. Payoffs to mutual defection are lower than payoffs to mutual cooperation, but defecting when the other party cooperates gives the highest payoff. The Prisoner's dilemma is a dilemma because the rational strategy for each player—defection – leads to an outcome neither prefers.

Something very much like the prisoner's dilemma situations occur frequently in real life. Every exchange across a store-counter, of money for goods represents what looks like such a problem: the customer would be best off if she grabbed the merchandise and left without paying, the sales-person would be best off if she could grab the money out of the customer's hands and with-hold the goods, the third best outcome for both is that the customer keeps the money while the sales-person keeps the goods, and yet almost always, both attain the second most preferred outcome for both of exchanging good for money. The parties to this exchange are not irrational, so we need to explain why they attained the cooperative outcome, why the situation is not a prisoner's dilemma.

The reason is that the store-counter exchange problem is part of a larger game, the iterated or repeated prisoner's dilemma in which the two agents play the game again and again whenever the customer comes to the store. What is the best strategy in an iterated prisoner's dilemma? In computer simulations famously carried out by R. Axelrod, the optimal strategy in most iterated prisoner's dilemma games of interest is one called "tit-for-tat": cooperate in game one, and then in each subsequent round do what the other player did in the previous round. In iterated prisoner's dilemmas among humans tit-for-tat is an effective strategy in part because it is clear – opponents don't need a great deal of cognitive skill to tell what strategy a player is using, it is nice – it starts out cooperatively, and it is forgiving – it retaliates only once for each attempt to free-ride on it. (It is important to bear in mind that tit-for-tat is an optimal strategy for maximizing the individual's pay-off (evolutionary or otherwise) only under certain conditions. See Axelrod, 1984.) When a group of players play tit-for-tat among themselves, the group and their strategy are not vulnerable to invasion by players using an always-free-ride-never cooperate strategy. Players who do not cooperate will do better on the first round with each of the tit-for-tat-ers, but will do worse on each subsequent round, and in the long run will be eliminated. Tit-for-tat is an evolutionarily stable strategy: if it gets enough of a foothold in a group it will expand until it is the dominant strategy, and once it is established it cannot be overwhelmed by another strategy.

We can expect that nature's relentless exploration of the space of adaptive strategies in cooperative situations will uncover tit-for-tat, that long before the appearance of *Homo sapiens*, this strategy will have been written in the genes, and with it the genetic predispositions that make cooperation actual. By the time we get to human beings, these dispositions will include the cognitive ability to detect the strategies others use, enough language to coordinate them, and the emotions that mesh sufficiently to reinforce cooperation. In other words, Darwinian selection for fitness maximizers will have provided the biological details that a Darwinian metaethic such as Gibbard's requires.

It may even do more. Once we have recognition of partners, and memory about how they played in previous iterations, we may even have sufficient cognitive resources so that the one-shot prisoner's dilemma can be solved cooperatively. This, at any rate, appears to be the conclusion of *Unto Others*, Sober and Wilson's revisionist argument that group selection for cooperation is after all possible, and that the adaptational conditions under which it is actual, may well have obtained in hominid and human evolution.[12] Their argument is disarmingly simple. Every one grants that kin-selection is not only possible but actual, as much evidence from infrahuman behavior demonstrates. It is also clear that kin-selection between one parent and one off-spring provides adaptational advantages to the two-membered "group" which they compose. In a one-shot prisoner's dilemma involving kin, both may be advantaged by cooperation regardless of the other's action, if the pay-off they are "designed" to maximize (reproductive fitness) satisfy the inequality, $r > b/c$, where r is the coefficient of relatedness ($\frac{1}{2}$ in the case of off-spring and siblings, 1 in the case of identical twins, $\frac{1}{4}$ in the case of cousins and nephews), b is the pay-off to mutual cooperation, and c is the cost of cooperation in the face of selfishness. If the group's fitness is a function of individual fitnesses, then groups of kin-related agents playing the cooperative (or "sucker's") strategy in a one-shot prisoner's dilemma will be fitter than groups composed of pairs of mutual free-riders playing the defector-strategy, or mixed groups of pairs of free-riders and suckers. The result generalizes to larger groups than pairs. But once players can recognize their degrees of relatedness, or for that matter what strategies they are genetically programmed to play in prisoner's dilemmas, they can preferentially aggregate into such fitter groups. When players seek out one on the basis of what strategy they play, the long term result is a "correlated equilibrium" of groups of cooperators only, the non-cooperating groups having been driven to extinction.

But recall the problem of invasion. Once these groups of cooperators get started, they are vulnerable to invasion or mutation that subverts from within, producing free-riders that take all other players in the group for suckers and increase in proportion from generation to generation until eventually selfishness becomes fixed in every erstwhile altruistic group. Wilson and Sober suggest that cooperating groups preserve themselves by means of secondary enforcement behaviors. Norms of cooperation are policed by norms of enforcement, and acting on these norms – shaming, reporting, confiscating – are far less costly to the enforcing individuals than are the norms of cooperation they preserve from break-down. Wilson and Sober argue that unrelated human cooperative groups attain stable equilibria (ones that cannot be invaded) through the en-forcement of social norms that lower the costs of cooperating and raise the costs of defecting.

6.   Is Justice Selected For?

So, evolutionary game theory seems capable of solving Wilson's problem of how to render human cooperation compatible with natural selection, and thus to help explain the emergence of the norms and emotions that underwrite them which Gibbard's Darwinian metaethical project needs. But evolutionary game theory may even be able to go further and identify the content of some of these norms. In *The Evolution of the Social Contract* Brian Skyrms shows how a Darwinian process can result in the fixation among humans of the norm of justice as fair division. The key to this demonstration is again the evolution of a "correlated equilibrium" among like strategies through a mechanism of random variation and natural selection. Consider the problem of divide the cake: two players bid independently on the size of the cake they want. If the bids add up to more than the whole cake, neither gets any cake. Otherwise, they get what they bid. Most people bid $\frac{1}{2}$, of course. This outcome is a equilibrium: neither can do better, no matter what strategy the other employs. There are indefinitely many other Nash equilibria: for example, I bid 90%, you bid 10%. But none of them is evolutionarily stable. A population whose members demand more than $\frac{1}{2}$ or less than $\frac{1}{2}$ of the cake will be invaded and swamped by pairs who demand $\frac{1}{2}$. Consider a bidding game in which random proportions of three strategies – bid $\frac{1}{3}$, bid $\frac{2}{3}$, bid $\frac{1}{2}$ – are represented to begin with. Skyrms has shown that in a computer simulation, in which strategies of lowest fitness are regularly removed, after 10,000 rounds, the fair division bid $\frac{1}{2}$ is the sole remaining strategy 62% of the time. When strategies correlate so that fair division plays against itself more frequently or with increasing frequency as the game proceeds, it almost always swamps any other strategy. Skyrms concludes, "In a finite population, in a finite time, where there is some random element in evolution, some reasonable amount of divisibility of the good and some correlation, we can say that it is likely that something close to share and share alike should evolve in dividing the cake situations. This is, perhaps, a beginning of an explanation of the origin of our concept of justice."[13]

Skyrms shows more than this: correlation among games enables selection of strategies for fitness to give rise to fair shares cooperation when cut-the-cake is played serially, instead of simultaneously (so that player one can demand more than $\frac{1}{2}$, forcing player 2 to choose between less than a fair share and nothing at all).

Correlation among strategies in the defense of territories can give rise to private property as a cooperative solution to an adaptational problem. And finally, as we shall see, Skyrms sketches a way in which correlated strategies can give rise to meaning. One of Skyrms larger aims is to show that these happy Nash-equilibrium outcomes are attainable when the choice of individual strategies is governed by natural selection for optimal outcomes. None are attainable, when the choice of individual strategies is governed by considerations of economic rational choice seeking maximal pay-off.

But how can we be confident that correlation required for the evolution of cooperation arose? Here is the problem, illustrated by one of Skyrms's results: In groups of kin-related individuals, for example vervet monkeys, signaling the presence of various threats – snake, leopard, eagle – can develop from correlated conventions about what noises consistently to make in the presence of different stimuli. Natural selection will prefer systems in which senders and receivers treat noises as bearing the same "news". It will also select for altruistic employment of signals to warn kin, even at signaler's expense. Note, this is a result that both Wilson and Sober, and Gibbard require. For norms of cooperation and enforcement require language; indeed, language is so important to the evolution of cooperation, that one might even argue that it emerged through selection for its impact on cooperation. But Skyrms's model for the evolution of language presupposes strong correlation. In the case of vervets, it is provided by the kin-structure of aboral monkeys. Hominid evolution most probably proceeded however through solitary individuals dispersed from their kin and roaming a savanna alone.[14] The cooperation they needed to establish to survive could not presuppose kin-based correlation. There seems no other source of correlation. But without correlation there is no basis in evolutionary game theory to be confident that cooperation, or its semantic prerequisites will arise. There is thus more work to do in developing plausible models of the evolution of cooperation among fitness maximizers like us.

But what has been done in evolutionary game theory certainly has begun to provide the empirical foundations that a Darwinian metaethic requires for its claims about meaning and foundations of moral judgement. And in some attenuated sense, the result may even satisfy the hopes for a Darwinian morality. Without vindicating the internalism of moral judgements as reflecting objective demands on our conduct, the Darwinian metaethic approaches the goals that one tradition in ethics since Hobbes has set for itself: the task of showing that it is rational to be moral. Cooperation makes us each better off than we would be in a state of nature. But this outcome is not attainable as a bargain among rational agents; rather it is the result of natural selection operating over blind variation. This is almost, but not quite Darwinian morality.

7. Broader Implications of Darwinism for Social Theory

Well before the developments reported above, Darwinism was guiding a research program in the empirical social sciences that took the name of sociobiology. Laterally, some sociobiologists have substituted to name "evolutionary psychology" for their program, in part to avoid the controversies which vexed sociobiology and in part to reflect a Darwinian commitment to individual selection, as opposed to group selection, as the force which shapes human behavior and social institutions. Sociobiology is controversial because it has been accused of adopting a Panglossian methodology that wrongly underwrites the status quo as inevitable and unchangeable. If social institutions – including the division of labor, both sexual and industrial, economic and racial inequality, vast power asymmetries, coercive violence, are the result of long term selection processes written into the genes, then they are no more subject to amelioration or change than eye-colour. And if this conclusion is derived from a method which simply finds some story about variation and selection that accords the status of an adaptation on extant institutions, without any empirical basis or even on the basis of a puerile misunderstanding of natural selection, then it will be no surprise that the research program is politically controversial. A sustained argument for this conclusion about Darwinism's baleful influence in the social sciences is offered in Lewontin, Kamin, and Rose's *Not in Our Genes*.[15] [1984, especially chapter 9.]

Some work carried on under the banner of Darwinian sociobiology may certainly warrant such criticism.[16] But not all of it can be so criticized. Reviewing this work would take us too far afield, but at least some of the criticism of the research program of Darwinism in the social sciences can be deflected by the developments in moral and political philosophy reported here. For if individual fitness maximization can result in the morality most of us share and in institutions of cooperation and justice, then it is not guilty of simply underwriting an unjust, non-egalitarian, sexist, racist status quo. There will have to be other factors at least in part responsible for these outcomes besides natural selection, and there will be environments – perhaps even attainable ones, in which natural selection will not inevitably lead to such nefarious outcomes.

Darwinian metaethics and evolutionary game theory have succeed, perhaps beyond the naturalist's hopes, in providing an account of how cooperative institutions are possible even where they not the result of conscious design or intention among any of their participants. Darwinism's success has also strongly encouraged other non-normative explanatory projects in the social sciences motivated by a search for stable equilibria that optimize some function without any participant intending or acting to attain such an outcome. In this respect Darwinism may in part vindicate the "invisible" or "hidden" hand strategy of the approach of Adam Smith and his market-oriented followers in economics. Smith's laissez-faire economic theory implies that self-seeking in free markets will lead as if by a hidden hand to unintended outcomes which advantage all. It is now well known that this is not the case. Rational choice behavior among economic agents leads to non-optimum outcomes in many different circumstances: in the provision of public goods, or when large companies can make things more cheaply than small ones (what economists call "positive returns to scale"), or there is a small numbers of traders, asymmetries of information, when transaction costs are great, or there is a difference in the interests of principals and agents. These "market failures" have led critics of the market both to deny that economic arrangements reflect the operation of an invisible hand optimizing welfare or satisfaction, and to deny that social institutions are the result of what Hayek called "spontaneous order."[17] What evolutionary approaches show are that a) when behavior is the result of natural selection for outcomes that enhance fitness, instead of rational choice of outcomes that enhance individual welfare, market failures can be avoided and optimal outcomes may after all be attainable, and b) these outcomes result from the aggregation of individual behaviors, not the selection of some properties of the group (beyond those correlated pairs Sober and Wilson's group selection countenances). Of course, if the maximization of welfare is among the ways in which fitness is often maximized, then natural selection for individual fitness maximization will bring individual welfare maximization along with it, thus substantiating Smith's laissez faire conclusions if not his reasoning. Thus, successful Darwinian explanations in the social sciences will substantiate both methodological individualism and invisible or hidden hand perspectives. But there is another potentially more promising adaptation of Darwinism in social science.

If genes and packages of genes can replicate and be selected for in virtue of the adaptational phenotypes they confer on organisms, why cannot beliefs, desires, and other cognitive states be selected for as a consequence of the benefits thinking of them confers on cognitive agents. Following Dawkins, call these cognitive states "memes" (mental "genes"), whose varying individually rewarding effects in behavior (phenotypes) result in their being differentially copied (reproduced) into the cognitive systems of other agents. Here again, the attractions of memetic natural selection are its freedom from assumptions about the conscious rational choices of individuals to adopt particular ideas, values, fashions, etc., and the availability of an invisible hand mechanism that explains how they spread, become fixed in a population, and often become less widespread as environmental change (or even frequency-dependent selection) makes them less adaptative.

It would be wrong to suppose that Darwinism vindicates the notion, sometimes attributed to Smith and his followers, that social interactions, and economic ones, are largely competitive ones in which there are inevitably losers made extinct by the competition. As we have already seen, in some environments – i.e. under some payoff distributions – individual selection makes cooperation the most adaptive strategy, not competition. That this is a possibility is something one might have inferred from Darwinian biology directly. For there are many natural cases in which selection fosters cooperation among organisms in different species, within the same species, whether closely related or not. And inference from Darwinism directly to a view of nature or society as "red in tooth and claw" is a mistake due to the neglect of the role of the environment which perhaps more often than not selects for competition and less often for cooperation. But Darwinism cannot deny the charge that non-competitive cooperation is in the end a strategy only locally adaptive, and adaptive for fundamentally "selfish genes" whose own fitness maximizing strategies organismal cooperation fosters.

8. Conclusion

Darwinian morality has been a recurrent goal among naturalists. But, if the present orthodoxy among philosophers holds, it will remain an unreachable goal. Darwinian metaethics, on the other hand, seems to be carried forward on a rising tide of research into human affairs that twentieth and twenty first century research in game theory, biological anthropology, and evolutionary psychology. Several philosophers have made the most of the results, theories and findings which these disciplines have offered to provide an account of the nature and significance of morality. They have shown how it may be expected to have emerged among fitness maximizing animals, and how nature may have selected for the cooperative norms and the emotions that express our commitment to them that give morality its universal content. The specificity and detail that these accounts seem already to have attained, should encourage opponents of naturalism to be modest in their claims about the long-term limits of a Darwinian understanding of human affairs.

Notes

1   Moore, G. E., *Principia Ethica*, London, Routledge, 1903, chapter one.

2   For further discussion see A. Rosenberg, "The biological justification of ethics: A best case scenario", in *Social Policy and Philosophy*, 8:1 1990, pp. 86–101, reprinted in Rosenberg, A., *Darwinism in Philosophy, Social Science and Policy*, Cambridge, Cambridge University Press, 2000, pp. 118–136.

3   Peter Railton, "Moral Realism," *Philosophical Review* 95 (1986): 163–207. Page references in this section of the chapter are to this to this paper.

4   See Railton, "Moral Realism", section III and IV, and especially footnote 21.

5   See Ayer, A. J., *Language, Truth and Logic*, London, Gollnaz, 1940, and Stevenson, C. L., *Ethics and Language*, New Haven, Yale University Press, 1944.

6   Gibbard, A., *Wise Choices, Apt Feelings*, Cambridge MA, Harvard University Press, 1992. Page references in this section of the chapter are to this to this paper.

7   Wright, L., *Teleological Explanation*, Berkeley, University of California Press, 1976.

8   Maryanski, A., and Turner, J., *The Social Cage*, Palo Alto, Stanford University Press, 1992.

9   Barkow, R., Tooby, J., and Cosmides, L., *The Adapted Mind*, Oxford, Oxford University Press, 1992.

10 Griffiths, Paul, *What Emotions Really Are*, Chicago, University of Chicago Press, 1997.

11 Wilson, E. O., *Sociobiology*, Cambridge MA, Harvard University Press, 1976.

12 Wilson, D. S., and Sober, E., *Unto Others*, Cambridge MA, Harvard University Press, 1998.

13 Skyrms, B., *Evolution of the Social Contract*, Cambridge, Cambridge University Press, 1996, p. 21.

14 See *Maryanski and Turner*, op. cit., note 6.

15 Lewontin, R., Kamin, L., and Rose, S., *Not in Our Genes*, New York, Pantheon Books, 1984, especially chapter nine.

16 See Kitcher, P., *Vaulting Ambition*, Cambridge MA, MIT Press, 1989, for examples of such work and criticism of them.

17 Hayek, F., *Law, Liberty and Legislation*, v.1, Rules and Order, Chicago, University of Chicago Press, 1981.

References

Axelrod, R., *The Evolution of Cooperation*, Ann Arbor, University of Michigan Press, 1984.

Ayer, A. J., *Language, Truth and Logic*, London, Gollancz, 1940.

Barkow, R., Tooby, J., and Cosmides, L., *The Adapted Mind*, Oxford, Oxford University Press, 1992.

Gibbard, A., *Wise Choices, Apt Feelings*, Cambridge MA, Harvard University Press, 1992.

Griffiths, Paul, *What Emotions Really Are*, Chicago, University of Chicago Press, 1997.

Hayek, F., *Law, Liberty and Legislation*, v.1, Rules and Order, Chicago, University of Chicago Press, 1981.

Kitcher, P., *Vaulting Ambition*, Cambridge, MIT Press, 1989.

Lewontin, R., Kamin, L., and Rose, S., *Not in Our Genes*, New York, Pantheon Books, 1984.

Maryanski, A., and Turner, J., *The Social Cage*, Palo Alto, Stanford University Press, 1992.

Moore, G. E., *Principia Ethica*, London, Routledge, 1903.

Peter Railton, "Moral Realism," *Philosophical Review*, 95 (1986): 163–207.

Rosenberg, A., *Darwinism in Philosophy, Social Science and Policy*, Cambridge, Cambridge University Press, 2000.

Skyrms, B., *Evolution of the Social Contract*, Cambridge, Cambridge University Press, 1996.

Stevenson, C. L., *Ethics and Language*, New Haven, Yale University Press, 1944.

Wilson, E. O., *Sociobiology*, Cambridge MA, Harvard University Press, 1976.

Wilson, D. S., and Sober, E., *Unto Others*, Cambridge MA, Harvard University Press, 1998.

Wright, L., *Teleological Explanation*, Berkeley, University of California Press, 1976.

PART XI
EVOLUTIONARY PSYCHOLOGY

Introduction

This section begins (chapter 26) with a selection from John Tooby and Leda Cosmides, two of the most famous advocates of what has come to be known as *evolutionary psychology*. All evolutionary psychologists posit that evolution is responsible not only for human physiology and anatomy, but also for certain human psychological and behavioral characteristics that evolved in our past to solve specific problems of survival. The time period that many evolutionary psychologists look to as being significant for the evolution of the mind is the Pleistocene Epoch (1.8 million years ago to 10,000 years ago), and this would make sense since mitochondrial DNA and fossil evidence indicate that modern *Homo sapiens* evolved during this time period. Evolutionary psychology has generated testable results and an entire research program all its own, and Cosmides and Tooby explain some of this in the material included in this section.

What about the uniquely human ability to solve problems creatively, which has led our species to construct novel tools, reason abstractly, and dominate the earth? After all, if it were not for our ability to be creative, we would *only* be able to deal with familiar, rather routine situations and respond to information in new situations in very limited ways. Creativity, then, seems to have played an essential role in our evolution. Can evolutionary psychologists account for such a creative problem-solving ability in humans? In the second selection in this section (chapter 27), Robert Arp notes that current theories of creative problem-solving in evolutionary psychology are deficient in explaining how our early hominin ancestors solved problems in wholly novel environments. Using an idea from archeologist Steven Mithen (1996) called *cognitive fluidity*, Arp attempts to rectify some of the deficiencies of the standard picture in evolutionary psychology and argues for the emergence of a mental capacity he calls *scenario visualization*, which he thinks better explains creativity in humans.

Further Reading

Arp, R. (2007). Resolving conflicts in evolutionary psychology with cognitive fluidity. *Southwest Philosophy Review*, 35, 91–101.

Arp, R. (2008). *Scenario visualization: An evolutionary account of creative problem solving.* Cambridge, MA: MIT Press.

Barkow, J., Cosmides, L., & Tooby, J. (eds.) (2002). *The adapted mind: Evolutionary psychology and the generation of culture.* Oxford: Oxford University Press.

Barrett, L., Dunbar, R., & Lycett, J. (2002). *Human evolutionary psychology.* Princeton, NJ: Princeton University Press.

Buller, D. (2005). *Adapting minds: Evolutionary psychology and the persistent quest for human nature.* Cambridge, MA: MIT Press.

Buss, D. (ed.). (2005). *The handbook of evolutionary psychology.* Chichester: Wiley.

Buss, D. (2007). *Evolutionary psychology: The new science of the mind.* New York: Allyn & Bacon.

Byrne, R., & Whiten, A. (eds.) (1988). *Machiavellian intelligence: Social expertise and the evolution of intellect in monkeys, apes and humans.* Oxford: Clarendon Press.

Cann, R., Stoneking, M., & Wilson, A. (1987). Mitochondrial DNA and human evolution. *Nature,* 325, 31–36.

Carruthers, P. (2002). Human creativity: Its evolution, its cognitive basis, and its connections with childhood pretence. *British Journal for the Philosophy of Science,* 53, 1–29.

Cosmides, L., & Tooby, J. (1992). The psychological foundations of culture. In J. Barkow, L. Cosmides, & J. Tooby (eds.), *The adapted mind: Evolutionary psychology and the generation of culture* (pp. 19–136). New York: Oxford University Press.

Cosmides, L., & Tooby, J. (1994). Origins of domain specificity: The evolution of functional organization. In L. Hirschfeld & S. Gelman (eds.), *Mapping the mind: Domain specificity in cognition and culture* (pp. 71–97). Cambridge: Cambridge University Press.

Donald, M. (1991). *Origins of the modern mind.* Cambridge, MA: Harvard University Press.

Donald, M. (1997). The mind considered from a historical perspective. In D. Johnson & C. Erneling (eds.), *The future of the cognitive revolution* (pp. 355–365). New York: Oxford University Press.

Dunbar, R., & Barrett, L. (eds.) (2007). *The Oxford handbook of evolutionary psychology.* Oxford: Oxford University Press.

Mameli, M. (2008). Sociobiology, evolutionary psychology, and cultural evolution. In M. Ruse (ed.), *The Oxford handbook of philosophy of biology* (pp. 410–433). Oxford: Oxford University Press.

Mithen, S. (1996). *The prehistory of the mind: The cognitive origins of art, religion and science.* London: Thames and Hudson.

Mithen, S. (ed.) (1998). *Creativity in human evolution and prehistory.* London: Routledge.

Oller, D., & Griebel, U. (eds.) (2008). *Evolution of communicative flexibility: Complexity, creativity, and adaptability in human and animal communication.* Cambridge, MA: MIT Press.

Palmer, J., & Palmer, L. (2001). *Evolutionary psychology: The ultimate origins of human behavior.* New York: Allyn & Bacon.

Pinker, S. (1997). *How the mind works.* New York: W. W. Norton.

Pinker, S. (2002). *The blank slate: The modern denial of human nature.* New York: Penguin Books.

Richardson, R. (2007). *Evolutionary psychology as maladapted psychology.* Cambridge, MA: MIT Press.

Rogers, A., & Jorde, L. (1995). Genetic evidence on modern human origins. *Human Biology,* 67, 1–36.

Rosenberg, A., & McShea, D. (2008). *Philosophy of biology: A contemporary introduction* (Chapter 7). London: Routledge.

Smith, S., Ward, T., & Finke, R. (eds.) (1992). *Creative cognition: Theory, research and applications.* Cambridge, MA: MIT Press.

Sternberg, R. (ed.) (1999). *Handbook of creativity.* Cambridge: Cambridge University Press.

Sternberg, R., & Kaufman, J. (eds.) (2002). *The evolution of intelligence.* Mahwah, NJ: Lawrence Erlbaum Associates.

Stringer, C. (2002). Modern human origins: Progress and prospects. *Philosophical Transactions of the Royal Society of London,* 357B, 563–579.

Stringer, C., & Andrews, P. (2005). *The complete world of human evolution.* London: Thames & Hudson.

Tattersall, I. (2001). How we came to be human. *Scientific American,* 285, 57–63.

Tattersall, I. (2002). *The monkey in the mirror: Essays on the science of what makes us human.* Oxford: Oxford University Press.

Templeton, A. (2002). Out of Africa again and again. *Nature,* 416, 45–51.

Wall, J., & Przeworski, M. (2000). When did the human population start increasing? *Genetics,* 155, 1865–1874.

26

Conceptual Foundations of Evolutionary Psychology

John Tooby and Leda Cosmides

The Emergence of Evolutionary Psychology: What Is at Stake?

The theory of evolution by natural selection has revolutionary implications for understanding the design of the human mind and brain, as Darwin himself was the first to recognize (Darwin, 1859). Indeed, a principled understanding of the network of causation that built the functional architecture of the human species offers the possibility of transforming the study of humanity into a natural science capable of precision and rapid progress. Yet, nearly a century and a half after *The Origin of Species* was published, the psychological, social, and behavioral sciences remain largely untouched by these implications, and many of these disciplines continue to be founded on assumptions evolutionarily informed researchers know to be false (Pinker, 2002; Tooby & Cosmides, 1992). Evolutionary psychology is the long-forestalled scientific attempt to assemble out of the disjointed, fragmentary, and mutually contradictory human disciplines a single, logically integrated research framework for the psychological, social, and behavioral sciences – a framework that not only incorporates the evolutionary sciences on a full and equal basis, but that systematically works out all of the revisions in existing belief and

research practice that such a synthesis requires (Tooby & Cosmides, 1992).

The long-term scientific goal toward which evolutionary psychologists are working is the mapping of our universal human nature. By this, we mean the construction of a set of empirically validated, high-resolution models of the evolved mechanisms that collectively constitute universal human nature. Because the evolved function of a psychological mechanism is computational – to regulate behavior and the body adaptively in response to informational inputs – such a model consists of a description of the functional circuit logic or information processing architecture of a mechanism (Cosmides & Tooby, 1987; Tooby & Cosmides, 1992). Eventually, these models should include the neural, developmental, and genetic bases of these mechanisms, and encompass the designs of other species as well.

A genuine, detailed specification of the circuit logic of human nature is expected to become the theoretical centerpiece of a newly reconstituted set of social sciences, because each model of an evolved psychological mechanism makes predictions about the psychological, behavioral, and social phenomena the circuits generate or influence. (For example, the evolutionarily specialized mechanisms underlying human alliance

From John Tooby and Leda Cosmides, "Conceptual Foundations of Evolutionary Psychology," in D. Buss (ed.) *The Handbook of Evolutionary Psychology* (Wiley, 2005), pp. 5–67.

help to explain phenomena such as racism and group dynamics; Kurzban, Tooby, & Cosmides, 2001.) A growing inventory of such models will catalyze the transformation of the social sciences from fields that are predominantly descriptive, soft, and particularistic into theoretically principled scientific disciplines with genuine predictive and explanatory power. Evolutionary psychology in the narrow sense is the scientific project of mapping our evolved psychological mechanisms; in the broad sense, it includes the project of reformulating and expanding the social sciences (and medical sciences) in light of the progressive mapping of our species' evolved architecture.

The resulting changes to the social sciences are expected to be dramatic and far-reaching because the traditional conceptual framework for the social and behavioral sciences – what we have called the *Standard Social Science Model* (SSSM) – was built from defective assumptions about the nature of the human psychological architecture (for an analysis of the SSSM, see Tooby & Cosmides, 1992). The most consequential assumption is that the human psychological architecture consists predominantly of learning and reasoning mechanisms that are general-purpose, content-independent, and equipotential (Pinker, 2002; Tooby & Cosmides, 1992). That is, the mind is blank-slate like, and lacks specialized circuits that were designed by natural selection to respond differentially to inputs by virtue of their evolved significance. This presumed psychology justifies a crucial foundational claim: Just as a blank piece of paper plays no causal role in determining the content that is inscribed on it, the blank-slate view of the mind rationalizes the belief that the evolved organization of the mind plays little causal role in generating the content of human social and mental life. The mind with its learning capacity absorbs its content and organization almost entirely from external sources. Hence, according to the standard model, the social and cultural phenomena studied by the social sciences are autonomous and disconnected from any nontrivial causal patterning originating in our evolved psychological mechanisms. Organization flows inward to the mind, but does not flow outward (Geertz, 1973; Sahlins, 1976).

Yet if – as evolutionary psychologists have been demonstrating – the blank-slate view of the mind is wrong, then the social science project of the past century is not only wrong but radically misconceived. The blank-slate assumption removes the central causal organizers of social phenomena – evolved psychological mechanisms – from the analysis of social events, rendering the social sciences powerless to understand the animating logic of the social world. Evolutionary psychology provokes so much reflexive opposition because the stakes for many social scientists, behavioral scientists, and humanists are so high: If evolutionary psychology turns out to be well-founded, then the existing superstructure of the social and behavioral sciences – the Standard Social Science Model – will have to be dismantled. Instead, a new social science framework will need to be assembled in its place that recognizes that models of psychological mechanisms are essential constituents of social theories (Boyer, 2001; Sperber, 1994, 1996; Tooby & Cosmides, 1992). Within such a framework, the circuit logic of each evolved mechanism contributes to the explanation of every social or cultural phenomenon it influences or helps to generate. For example, the nature of the social interactions between the sexes are partly rooted in the design features of evolved mechanisms for mate preference and acquisition (Buss, 1994, 2000; Daly & Wilson, 1988; Symons, 1979); the patterned incidence of violence is partly explained by our species' psychology of aggression, parenting, and sexuality (Daly & Wilson, 1988); the foundations of trade can be located in evolved cognitive specializations for social exchange (Cosmides & Tooby, 2005); both incest avoidance and love for family members are rooted in evolved mechanisms for kin recognition (Lieberman, Tooby, & Cosmides, 2003, 2005). Indeed, even though the field is in its infancy, evolutionary psychologists have already identified a large set of examples that touch almost every aspect of human life.

To summarize, evolutionary psychology's focus on psychological mechanisms as evolved programs was motivated by new development from a series of different fields:

[. . .]

Advance 1: The cognitive revolution was providing, for the first time in human history, a precise language for describing mental mechanisms as programs that process information. Galileo's discovery that mathematics provided a precise language for expressing the mechanical and physical relationships enabled the birth of modern physics. Analogously, cognitive scientists' discovery that computational-informational formalisms provide a precise language for describing the design, properties, regulatory architecture, and operation of psychological mechanisms enables a modern science of mind (and its physical basis). Computational language is not just a convenience for modeling anything with complex dynamics. The brain's evolved function is computational – to use information to adaptively regulate the body and behavior – so computational and informational formalisms are by their nature the most appropriate to capture the functional design of behavior regulation.

Advance 2: Advances in paleoanthropology, hunter-gatherer studies, and primatology were providing data about the adaptive problems our ancestors had to solve to survive and reproduce and the environments in which they did so.

Advance 3: Research in animal behavior, linguistics, and neuropsychology was showing that the mind is not a blank slate, passively recording the world. Organisms come "factory-equipped" with knowledge about the world, which allows them to learn some relationships easily and others only with great effort, if at all. Skinner's hypothesis – that there is one simple learning process governed by reward and punishment – was wrong.

Advance 4: Evolutionary biology was revolutionized by being placed on a more rigorous, formal foundation of replicator dynamics, leading to the derivation of a diversity of powerful selectionist theories, and the analytic tools to recognize and differentiate adaptations, from by-products and stochastically generated evolutionary noise (Williams, 1966).

Ethology had brought together advances 2 and 3, sociobiology had connected advances 2 and 4, sometimes with 3; nativist cognitive science connected advances 1 and 3, but neglected and still shrinks from advances 2 and 4. Cognitive neuroscience partially and erratically accepts 1 and 3, but omits 2 and 4. Outside of cognitive approaches, the rest of psychology lacks much of advance 1, most of advance 3, and all of advances 2 and 4. Evolutionary anthropology appreciates advances 2 and 4, but neglects 1 and 3. Social anthropology and sociology lack all four. So it goes. If one counts the adaptationist/computationalist

resolution of the nature-nurture issue as a critical advance, the situation is even bleaker.

We thought these new developments could be pieced together into an integrated framework that successfully addressed the difficulties that had plagued evolutionary and nonevolutionary approaches alike. The reason why the synthesis had not emerged earlier in the century was because the connections between the key concepts ran between fields rather than cleanly within them. Consequently, relatively few were in the fortunate position of being professionally equipped to see all the connections at once. This limited the field's initial appeal, because what seems self-evident from the synoptic vantage point seems esoteric, pedantic, or cultish from other vantage points. Nevertheless, we and those working along similar lines were confident that by bringing all four advances together, the evolutionary sciences could be united with the cognitive revolution in a way that provided a framework not only for psychology but for all of the social and behavioral sciences. To signal its distinctiveness from other approaches, the field was named *evolutionary psychology*.[1]

Evolutionary Psychology

Like cognitive scientists, when evolutionary psychologists refer to the *mind*, they mean the set of information processing devices, embodied in neural tissue, that is responsible for all conscious and nonconscious mental activity, that generates all behavior, and that regulates the body. Like other psychologists, evolutionary psychologists test hypotheses about the design of these computational devices using methods from, for example, cognitive psychology, social psychology, developmental psychology, experimental economics, cognitive neuroscience, genetics, pysiological psychology, and cross-cultural field work.

The primary tool that allows evolutionary psychologists to go beyond traditional psychologists in studying the mind is that they take full advantage in their research of an overlooked reality: The programs comprising the human mind were designed by natural selection to solve the adaptive problems regularly faced by our hunter-gatherer ancestors – problems such as finding a mate, cooperating with others, hunting, gathering, protecting children, navigating, avoiding pred- ators, avoiding exploitation, and so on. Knowing this allows evolutionary psychologists to approach the study of the mind like an engineer. You start by carefully specifying an adaptive information processing problem; then you do a task analysis of that problem. A task analysis consists of iden- tifying what properties a program would have to have to solve that problem well. This approach allows you to generate hypotheses about the struc- ture of the programs that comprise the mind, which can then be tested.

From this point of view, there are precise causal connections that link the four developments discussed earlier into a coherent framework for thinking about human nature and society (Tooby & Cosmides, 1992):

C-1:   Each organ in the body evolved to serve a function: The intestines digest, the heart pumps blood, and the liver detoxifies poisons. The brain's evolved function is to extract information from the environment and use that information to generate behavior and regulate physiology. Hence, the brain is not just like a computer. It is a computer – that is, a physical system that was designed to process information (Advance 1). Its programs were designed not by an engineer, but by natural selection, a causal process that retains and discards design features based on how well they solved adaptive problems in past environ- ments (Advance 4).

The fact that the brain processes information is not an accidental side effect of some metabolic process. The brain was designed by natural selec- tion *to be* a computer. Therefore, if you want to describe its operation in a way that captures its evolved function, you need to think of it as com- posed of programs that process information. The question then becomes: What programs are to be found in the human brain? What are the reliably developing, species-typical programs that, taken together, comprise the human mind?

C-2:   Individual behavior is generated by this evolved computer, in response to information that it extracts from the internal and external environ- ment (including the social environment, Advance 1). To understand an individual's behavior, there- fore, you need to know both the information that the person registered *and* the structure of the programs that generated his or her behavior.

C-3:   The programs that comprise the human brain were sculpted over evolutionary time by the ancestral environments and selection pressures experienced by the hunter-gatherers from whom we are descended (Advances 2 and 4). Each evolved program exists because it produced behavior that promoted the survival and reproduction of our ancestors better than alternative programs that arose during human evolutionary history. Evolutionary psychologists emphasize hunter- gatherer life because the evolutionary process is slow – it takes thousands of generations to build a program of any complexity. The industrial revolution – even the agricultural revolution – is too brief a period to have selected for complex new cognitive programs.[2]

C-4:   Although the behavior our evolved pro- grams generate would, on average, have been adaptive (reproduction promoting) in ancestral environments, there is no guarantee that it will be so now. Modern environments differ import- antly from ancestral ones, particularly when it comes to social behavior. We no longer live in small, face-to-face societies, in seminomadic bands of 20 to 100 people, many of whom were close relatives. Yet, our cognitive programs were designed for that social world.

C-5:   Perhaps most importantly, natural selection will ensure that the brain is composed of many different programs, many (or all) of which will be specialized for solving their own corresponding adaptive problems. That is, the evolutionary process will not produce a predominantly general- purpose, equipotential, domain-general archi- tecture (Advance 3).

In fact, this is a ubiquitous engineering outcome. The existence of recurrent computational prob- lems leads to functionally specialized application software. For example, the demand for effective word processing and good digital music play- back led to different application programs because many of the design features that make a program an effective word processing program are differ- ent from those that make a program a good dig- ital music player. Indeed, the greater the number of functionally specialized programs (or sub- routines) your computer has installed, the more intelligent your computer is, and the more things it can accomplish. The same is true for organisms. Armed with this insight, we can lay to rest the myth that the more evolved organization the human mind has, the more inflexible its response. Interpreting the emotional expressions of others, seeing beauty, learning language, loving your child – all these enhancements to human mental life are made possible by specialized neural programs built by natural selection.

To survive and reproduce reliably as a hunter-gatherer required the solution of a large and diverse array of adaptive information-processing problems. These ranged from predator vigilance and prey stalking to plant gathering, mate selection, childbirth, parental care, coalition formation, and disease avoidance. Design features that make a program good at choosing nutritious foods, for example, are ill suited for finding a fertile mate or recognizing free riders. Some sets of problems would have required differentiated computational solutions.

The demand for diverse computational designs can be clearly seen when results from evolutionary theory (Advance 4) are combined with data about ancestral environments (Advance 2) to model different ancestral computational problems. The design features necessary for solving one problem are usually markedly different from the features required to construct programs capable of solving another adaptive problem. For example, game theoretic analyses of conditional helping show that programs designed for logical reasoning would be poorly designed for detecting cheaters in social exchange and vice versa; this incommensurability selected for programs that are functionally specialized for reasoning about reciprocity or exchange (Cosmides & Tooby 2005).

C-6:   Finally, descriptions of the computational architecture of our evolved mechanisms allows a systematic understanding of cultural and social phenomena. The mind is not like a video camera, passively recording the world but imparting no content of its own. Domain-specific programs organize our experiences, create our inferences, inject certain recurrent concepts and motivations into our mental life, give us our passions, and provide cross-culturally universal frames of meaning that allow us to understand the actions and intentions of others. They invite us to think certain kinds of thoughts; they make certain ideas, feelings, and reactions seem reasonable, interesting, and memorable. Consequently, they play a key role in determining

which ideas and customs will easily spread from mind to mind and which will not (Boyer, 2001; Sperber, 1994, 1996; Tooby & Cosmides, 1992). That is, they play a crucial role in shaping human culture.

Instincts are often thought of as the opposite of reasoning, decision making, and learning. But the reasoning, decision-making, and learning programs that evolutionary psychologists have been discovering (1) are complexly specialized for solving an adaptive problem, (2) reliably develop in all normal human beings, (3) develop without any conscious effort and in the absence of formal instruction, (4) are applied without any awareness of their underlying logic, and (5) are distinct from more general abilities to process information or behave intelligently. In other words, they have all the hallmarks of what we usually think of as instinct (Pinker, 1994). In fact, we can think of these specialized circuits as instincts: *reasoning instincts, decision instincts, motivational instincts, and learning instincts.* They make certain kinds of inferences and decisions just as easy, effortless, and natural to us as humans as catching flies is to a frog or burrowing is to a mole.

Consider this example from the work of Simon Baron-Cohen (1995). Like adults, normal 4-year-olds easily and automatically note eye direction in others, and use it to make inferences about the mental states of the gazer. For example, 4-year-olds, like adults, infer that when presented with an array of candy, the gazer wants the particular candy he or she is looking at. Children with autism do not make this inference. Although children with this developmental disorder can compute eye direction correctly, they cannot use that information to infer what someone wants. Normal individuals know, spontaneously and with no mental effort, that the person wants the candy he or she is looking at. This is so obvious to us that it hardly seems to require an inference at all. It is just common sense. But "common sense" is caused: It is produced by cognitive mechanisms. To infer a mental state (wanting) from information about eye direction requires a computation. There is an inference circuit – a reasoning instinct – that produces this inference. When the circuit that does this computation is broken or fails to develop, the inference cannot be made. Those with autism fail this task because they lack this reasoning instinct, even though they often acquire very sophisticated competences of other sorts. If the mind consisted of a domain-general knowledge acquisition system, narrow impairments of this kind would not be possible.

Instincts are invisible to our intuitions, even as they generate them. They are no more accessible to consciousness than our retinal cells and line detectors but are just as important in manufacturing our perceptions of the world. As a species, we have been blind to the existence of these instincts, not because we lack them but precisely because they work so well. Because they process information so effortlessly and automatically, their operation disappears unnoticed into the background. Moreover, these instincts structure our thought and experience so powerfully we mistake their products for features of the external world: Color, beauty, status, friendship, charm – all are computed by the mind and then experienced as if they were objective properties of the objects they are attributed to. These mechanisms limit our sense of behavioral possibility to choices people commonly make, shielding us from seeing how complex and regulated the mechanics of choice is. Indeed, these mechanisms make it difficult to imagine how things could be otherwise. As a result, we take normal behavior for granted: We do not realize that normal behavior needs to be explained at all.

As behavioral scientists, we need corrective lenses to overcome our instinct blindness. The brain is fantastically complex, packed with programs, most of which are currently unknown to science. Theories of adaptive function can serve as corrective lenses for psychologists, allowing us to see computational problems that are invisible to human intuition. When carefully thought out, these functional theories can lead us to look for programs in the brain that no one had previously suspected.

[. . .]

Principles of Organic Design

Over evolutionary time, more and more design features accumulate to form an integrated structure or device that is well engineered to solve its particular adaptive problem. Such a structure or device is called an *adaptation*. Indeed, an organism can be thought of as a collection of

adaptations, together with the engineering by-products of adaptations, and evolutionary noise. The functional subcomponents of the ear, hand, intestines, uterus, or circulatory system are examples. Each of these adaptations exists in the human design now because it contributed to the process of direct and kin reproduction in the ancestral past. Adaptive problems are the only kind of problem that natural selection can design machinery for solving.

The environment of evolutionary adaptedness

One key to understanding the functional architecture of the mind is to remember that its programs were not selected for because they solved the problems faced by modern humans. Instead, they were shaped by how well they solved adaptive problems among our hunter-gatherer ancestors. The second key is to understand that the developmental processes that build each program, as well as each program in its mature state, evolved to use information and conditions that were reliably present in ancestral environments. The design of each adaptation assumes the presence of certain background conditions and operates as a successful problem solver only when those conditions are met. The *environment of evolutionary adaptedness* (EEA) refers jointly to the problems hunter-gatherers had to solve and the conditions under which they solved them (including their developmental environment).

Although the hominid line is thought to have originated on edges of the African savannahs, the EEA is not a particular place or time. The EEA for a given adaptation is the statistical composite of the enduring selection pressures or cause-and-effect relationships that pushed the alleles underlying an adaptation systematically upward in frequency until they became species-typical or reached a frequency-dependent equilibrium (most adaptations are species-typical; see Hagen, Chapter 5 in Buss, 2005). Because the coordinated fixation of alleles at different loci takes time, complex adaptations reflect enduring features of the ancestral world. The adaptation is the consequence of the EEA, and so the structure of the adaptation reflects the structure of the EEA. The adaptation evolved so that when it interacted with the stable features of the ancestral task environment, their interaction systematically promoted fitness (i.e., solves an adaptive problem). The concept of the EEA is essential to Darwinism, but its formalization was prompted by the evolutionary analysis of humans because human environments have changed more dramatically than the environments most other species occupy. The research problems faced by most biologists do not require them to distinguish the modern environment from a species' ancestral environment. Because adaptations evolved and assumed their modern form at different times and because different aspects of the environment were relevant to the design of each, the EEA for one adaptation may be somewhat different from the EEA for another. Conditions of terrestrial illumination, which form (part of) the EEA for the vertebrate eye, remained relatively constant for hundreds of millions of years – and can still be observed by turning off all artificial lights. In contrast, the social and foraging conditions that formed (part of) the EEA that selected for neural programs that cause human males to provision and care for their offspring (under certain conditions) is almost certainly less than two million years old.

When a program is operating outside the envelope of ancestral conditions that selected for its design, it may look like a poorly engineered problem solver. Efficient foraging, for example, requires good probability judgments, yet laboratory data suggested that people are poor intuitive statisticians, incapable of making simple inferences about conditional probabilities (Kahneman, Slovic, & Tversky, 1982). Evolutionary psychologists recognized that these findings were problematic, given that birds and insects solve similar problems with ease. The paradox evaporates when you consider the EEA for probability judgment. Behavioral ecologists presented birds and bees with information in ecologically valid formats; psychologists studying humans were not.

Being mindful of the EEA concept changes how research is designed and what is discovered. Giving people probability information in the form of absolute frequencies – an ecologically valid format for hunter-gatherers – reveals the presence of mechanisms that generate sound Bayesian inferences (Brase, Cosmides, & Tooby, 1998; Cosmides & Tooby, 1996; Gigerenzer, 1991; Gigerenzer, Todd, & the ABC Group, 1999).

Indeed, EEA-minded research on judgment under uncertainty is now showing that the human mind is equipped with a toolbox of "fast-and-frugal heuristics," each designed to make well-calibrated judgments quickly on the basis of limited information (Gigerenzer & Selten, 2002; Gigerenzer, Todd, & the ABC Group, 1999; Chapter 27). These procedures are *ecologically rational*, providing good solutions when operating in the task environments for which they evolved (Tooby & Cosmides in Buss, 2005).

Notes

1 We sometimes read that evolutionary psychology is simply sociobiology, with the name changed to avoid the bad political press that sociobiology had received. Although it is amusing (given the record) to be accused of ducking controversy, these claims are historically and substantively wrong. In the first place, evolutionary psychologists are generally admirers and defenders of sociobiology (or behavioral ecology, or evolutionary ecology). It has been the most useful and most sophisticated branch of modern evolutionary biology, and several have made contributions to this literature. Nonetheless, the lengthy and intense debates about how to apply evolution to behavior made it increasingly clear that markedly opposed views needed different labels if any theoretical and empirical project was to be clearly understood. In the 1980s, Martin Daly, Margo Wilson, Don Symons, John Tooby, Leda Cosmides, and David Buss had many discussions about what to call this new field, some at Daly and Wilson's kangaroo rat field site in Palar Desert, some in Santa Barbara, and some at the Center for Advanced Study in the Behavioral Sciences. Politics and the press did not enter these discussions, and we anticipated (correctly) that the same content-free ad hominem attacks would pursue us throughout our careers. What we *did* discuss was that this new field focused on psychology – on characterizing the adaptations comprising the psychological architecture – whereas sociobiology had not. Sociobiology had focused mostly on selectionist theories, with no consideration of the computational level and little interest in mapping psychological mechanisms. Both the subject matter of evolutionary psychology and the theoretical commitments were simply different from that of sociobiology, in the same way that sociobiology was quite different from the ethology that preceded it and cognitive psychology was different

from behaviorist psychology – necessitating a new name in each case.

2 Unidimensional traits, caused by quantitative genetic variation (e.g., taller, storter), can be adjusted in less time; see Tooby & Cosmides, 1990b.

References

Alcock, J. (2001). *The triumph of sociobiology*. Oxford: Oxford University Press.

Anderson, A., & Phelps, E. (2001). Lesions of the human amygdala impair enhanced perception of emotionally salient events. *Nature, 411*, 305–309.

Atran, S. (1990). *Cognitive foundations of natural history*. Cambridge: Cambridge University Press.

Atran, S. (1998). Folk biology and the anthropology of science: Cognitive unlversals and cultural particulars. *Behavioral and Brain Sciences, 21*, 547–611.

Baron-Cohen, S. (1995). *Mindblindness: An essay on antism and theory of mind*. Cambridge. MA: MIT Press.

Barrett, H. C. (1999). *From predator-prey adaptations to a general theory of understanding behavior*. Doctoral Dissertation, Department of Anthropology, University of Callfornin, Santa Barbara.

Barrett, H. C. (2005a). Adaptations to predators and prey. In D. M. Buss (Ed.), *Evolutionary psychology handbook*. New York: Wiley.

Barrett, H. C. (2005b). Enzymatic computation and cognitive modularity. *Mind and Language, 20*(3), 259–287.

Barrett, H. C., Cosmides, L., & Tooby, J. (in press). *By descent or by design? Evidence for two modes of biological reasoning*.

Barrett, H. C., Tooby, J., & Cosmides, L. (in press). *Children's understanding of predator-prey interactions: Cultural dissociations as tests of the impact of experience on evolved inference systems*.

Blurton Jones, N., & Konner, M. (1976). !Kung knowledge of animal behavior (or The proper study of mankind is animals). In R. Lee & I. DeVore (Eds.), *Kalahari hunter-gatherer: Studies of the !Kung San and their neighbors* (pp. 325–348). Cambridge, MA: Harvard.

Boole, G. (1848). The calculus of logic. *Cambridge and Dublin Mathematical Journal, III*, 183–198.

Boyer, P. (2001). *Religion explained: The evolutionary roots of religious thought*. New York: Basic Books.

Boyer, P., & Barrett, H. C. (in press). Domain-specificity and intuitive ontology. In D. M. Buss (Ed.), *Evolutionary psychology handbook*. New York: Wiley.

Brase, G., Cosmides, L., & Tooby, J. (1998). Individuation, counting, and statistical inference: The role of frequency and whole object representa-

tions in judgment under uncertainty. *Journal of Experimental Psychology: General, 127*, 1–19.

Braun, J. (2003). Natural scenes upset the visual applecart. *Trends in Cognitive Sciences, 7*(1), 7–9. (January).

Buss, D. M. (1994). *The evolution of desire*. New York: Basic Books.

Buss, D. M. (1999). *Evolutionary psychology: The new science of the mind*. Boston: Allyn & Bacon.

Buss, D. M. (2000). *The dangerous passion*. London: Bloomsbury Publishing.

Buss, D. M. (Ed.). (2005). *The handbook of evolutionary psychology*. New York: Wiley.

Campos, J., Bertenthal, B., & Kermolan, R. (1992). Early experience and emotional development: The emergence of wariness of heights. *Psychological Science, 3*, 61–64.

Cannon, W. (1929). *Bodily changes in pain, hunger, fear and rage*. Researches into the function of emotional excitement. New York: Harper & Row.

Caramazza, A. (2000). The organization of conceptual knowledge in the brain. In M. S. Gazzaniga (Ed.), *The new cognitive neurosciences* (2nd ed., pp. 1037–1046). Cambridge, MA: MIT Press.

Caramazza, A., & Shelton, J. (1998). Domain-specific knowledge systems in the brain: The animate–inanimate distinction. *Journal of Cognitive Neuroscience, 10*, 1–34.

Carruthers, P. (2006). The case for massively modular models of mind. In R. Stainton (Ed.), *Contemporary debates in cognitive science* (pp. 3–21). Oxford: Blackwell.

Cheney, D., Seyfarth, R., Smuts, R., & Wrangham, R. (Eds.). (1987). *Primate societies*. Chicago: University of Chicago Press.

Chomsky, N. (1959). A review of B. F. Skinner's verbal behavior. *Language, 35*(1), 26–58.

Chomsky, N. (1965). *Aspects of a theory of syntax*. Cambridge, MA: MIT Press.

Cosmides, L. (1985). *Deduction or Darwinian Algorithms? An explanation of the "elusive" content effect on the Wason selection task*. Doctoral dissertation, Harvard University. (UMI No. #86-02206).

Cosmides, L., & Tooby, J. (1981). Cytoplasmic inheritance and intragenomic conflict. *Journal of Theoretical Biology, 89*, 83–129.

Cosmides, L., & Tooby, J. (1987). From evolution to behavior: Evolutionary psychology as the missing link. In J. Dupré (Ed.), *The latest on the best: Essays on evolution and optimality*. Cambridge, MA: MIT Press.

Cosmides, L., & Tooby, J. (1992). The psychological foundations of culture. In J. Barkow, L. Cosmides & J. Tooby (Eds.), *The Adapted Mind* (pp. 19–136). New York: Oxford University Press.

Cosmides, L., & Tooby, J. (1994a). Beyond intuition and instinct blindness: The case for an evolutionarily rigorous cognitive science. *Cognition, 50*, 41–77.

Cosmides, L., & Tooby, J. (1994b). Origins of domain-specificity: The evolution of functional organization. In L. Hirschield & S. Gelman (Eds.), *Mapping the mind: Domain-specificity in cognition and culture.* New York: Cambridge University Press.

Cosmides, L., & Tooby, J. (1996). Are humans good intuitive statisticians after all?: Rethinking some conclusions of the literature on judgment under uncertainty. *Cognition, 58,* 1–73.

Cosmides, L., & Tooby, J. (2000a). Consider the source: The evolution of adaptations for decoupling and metarepresentation. In D. Sperber (Ed.), *Metarepresentations: A multidisciplinary perspective* (pp. 53–115). New York: Oxford University Press.

Cosmides, L. & Tooby, J. (2000b). Evolutionary psychology and the emotions. In M. Lewis & J. M. Haviland-Jones (Eds.), *Handbook of emotions* (2nd ed., pp. 91–115). New York: Guilford Press.

Cosmides, L., & Tooby, J. (2001). Unraveling the enigma of human intelligence: Evolutionary psychology and the multimodular mind. In R. J. Sternberg & J. C. Kaufman (Eds.), *The evolution of intelligence* (pp. 145–198). Hillsdale, NJ: Erlbaum.

Cosmides, L., & Tooby, J. (2005). Neurocognitive adaptations designed for social exchange. In D. M. Buss (Ed.), *The handbook of evolutionary psychology* (pp. 584–627). New York: Wiley.

Daly, M., & Wilson, M. (1988). *Homicide.* New York: Aldine.

Daly, M., Wilson, M., & Weghorst, S. J. (1982). Male sexual jealousy. *Ethology and Sociobiology, 3,* 11–27.

Darwin, C. (1859). *On the origin of species.* London: John Murray.

Dawkins, R. (1986). *The blind watchmaker.* New York: Norton.

Defeyter, M. A., & German, T. (2003). Acquiring an understanding of design: Evidence from children's insight problem solving. *Cognition, 89,* 133–155.

Dennett, D. (1987). *The intentional stance.* Cambridge, MA: MIT Press/Bradford.

DeSteno, D., Bartlett, M., Braverman, J., & Salovey, P. (2002). Sex differences in jealousy: Evolutionary mechanism or artifact of measurement. *Journal of Personality and Social Psychology, 83*(5): 1103–1116.

DeVore, I. (1965). *Primate behavior: Field studies of monkeys and apes.* New York: Holt, Rinehart & Winston.

Eaton, S. B., Shostak, M., & Konner, M. (1988). *The Paleolithic prescription: A program of diet, exercise and a design for living.* New York: Harper & Row.

Eibl-Ebesfeldt, I. (1970). *Ethology: The biology of behavior.* New York: Holt, Reinhart & Winston.

Ekman, P. (Ed.). (1982). *Emotion in the human face.* (2nd ed.). Cambridge, England: Cambridge University Press.

Fisher, R. A. (1930). *The genetical theory of natural selection.* Oxford: Clarendon Press.

Fodor, J. (1983). *The modularity of mind.* Cambridge, MA: MIT Press.

Fodor, J. (2000). *The mind doesn't work that may.* Cambridge, MA: MIT Press.

Frege, G. (1879). *Begriffsschrift ('concept notation'), cine der arithmetischen nachgebildete Formelsprnche des reinen denkens.* Halle A. S.

Frlesen, C., & Kingstone, A. (2003). Abrupt onsets and gaze direction cues trigger independent reflexive attentional effects. *Cognition, 87,* B1–B10.

Gallistel, C. R. (2000). The replacement of general-purpose learning models with adaptively specialized learning modules. In M. S. Gazzanlga (Eds.), *The new cognitive neurusciences* (pp. 1179–1191). Cambridge, MA: MIT Press.

Gallistel, C. R., & Gibbon, J. (2000). Time, rate and conditioning. *Psychological Review, 107,* 289–344.

Gallistel, C. R., Brown, A., Carey, S., Gelman, R., & Keil, F. (1991). Lessons from animal learning for the study of cognitive development. In S. Carey & R. Gelman (Eds.), *The epigenesis of mind.* Hillsdale, NJ: Erlbaum.

Geertz, C. (1973). *The interpretation of cultures.* New York: Basle Books.

German, T. P. (1999). Children's causal reasoning: Counterfactual thinking occurs for "negative" outcomes only. *Developmental Science, 2,* 442–447.

German, T. P., & Barrett, H. C. (in press). Functional fixedness in a technologically sparse culture. *Psychological Science 16*(1), 1–5.

Gigerenzer, G. (1991). How to make cognitive illusions disappear Beyond heuristics and biases. *European Review of Social Psychology, 2,* 83–115.

Gigerenzer, G., & Murray, D. (1987). *Cognition as intuitive statistics.* Hillsdale, NJ: Erlbaum.

Gigerenzer, G., & Selten, R. (Eds.). (2002). *Bounded rationality: The adaptive toolbox.* Cambridge, MA: MIT Press.

Gigerenzer, G., Todd, P., & the ABC Research Group. (1999). *Simple heuristics that make us smart.* New York: Oxford.

Gould, S. J., & Lewontin, R. C. (1979). The spandrels of San Marco and the panglossian paradigm: A critique of the adaptationist programme. *Proceedings of the Royal Society of London. Series B, Biological Sciences, 205,* 581–598.

Gray, H. (1918). *Gray's anatomy* (20th ed.). W. Lewis (Ed.), Philadelphia: Lea & Febiger.

Haidt, J. (2001). The emotional dog and its rational tail: A social intuitionist approach to moral judgment. *Psychological Review, 108*(4), 814–834.

Hamilton, W. D. (1964). The genetical evolution or social behavior. *Journal of Theoretical Biology, 7,* 1–52.

Haselton, M. G., & Buss, D. M. (2000). Error management theory: A new perspective on biases in cross-sex mind reading. *Journal of Personality and Social Psychology, 78,* 81–91.

Herrnstein, R. J. (1977). The evolution of behaviorism. *American Psychologist, 32,* 593–603.

Hirschfeld, L. A., & Gelman, S. A. (Eds.). (1994). *Mapping the mind: Domain specificity in cognition and culture.* Cambridge, England: Cambridge University Press.

Kahneman, D., Slovic, P., & Tversky, A. (Eds.). (1982). *Judgment under uncertainly: Heuristics and biases.* Cambridge, England: Cambridge University Press.

Kaplan, H., & Hill, K. (1985). Food sharing among Ache foragers: Tests of explanatory hypotheses. *Current Anthropology, 26*(2), 223–246.

Keil, F. (1989). *Concepts, kinds, and cognitive development.* Cambridge, MA: MIT Press.

Keil, F. C. (1994). The birth and nurturance of concepts by domains. The origins of concepts of living things. In L. A. Hirschfeld & S. A. Gelman (Eds.), *Mapping the mind: Domain specificity in cognition and culture.* Cambridge, England: Cambridge University Press.

Klein, S. (2005). The cognitive neuroscience of knowing one's self. In M. S. Gazzaniga (Ed.), *The cognitive neurosciences, III* (pp. 1077–1089). Cambridge, MA: MIT Press.

Klein, S., Cosmides, L., Tooby, J., & Chance, S. (2002). Decisions and the evolution of memory: Multiple systems, multiple functions. *Psychological Review, 109,* 306–329.

Klein, S., German, T., Cosmides, L., & Gabriel, R. (2004). A theory of autobiolographical memory: Necessary components and disorders resulting from their loss. *Social Cognition, 22*(5), 460–490.

Kurzban, R., Tooby, J., & Cosmides, L. (2001). Can race be erased? Coalitional computation and social categorization. *Proceedings of the National Academy of Sciences 98*(26), 15387–15392.

LeDoux, J. (1995). In search of an emotional system in the brain: Leaping from fear to emotion to consclousness. In M. S. Gazzaniga (Ed.), *The cognitive neurosciences* (pp. 1049–1061). Cambridge, MA: MIT Press.

Lee, R., & DeVore, I. (Eds.). (1968). *Man the hunter.* Chicago: Aldine.

Lee, R., & DeVore, I. (Eds.). (1976). *Kalahari hunter-gatherers: Studies of the !Kung San and their neighbors.* Cambridge, MA: Harvard.

Lenneberg, E. (1967). *Biological foundations of language.* New York: John Wiley & Sons.

Leslie, A. (1987). Pretense: and representation: The origins of "theory of mind." *Psychological Review, 94,* 412–426.

Leslie, A. M. (1994). ToMM, ToBy, and agency: Core architecture and domain specificity. In L. A. Hirachfeld & S. A. Gelman (Eds.), *Mapping the mind: Domain specificity in cognition and culture* (pp. 119–148). Cambridge, England: Cambridge University Press.

Leslie, A. M., & Thaiss, L. (1992). Domain specificity in conceptual development: Neuropsychological evidence from autism. *Cognition, 43,* 225–251.

Leslie, A. M., German, T. P., & Polizzi, P. (2005). Belief-desire reasoning as a process of selection. *Cognitive Psychology, 50,* 45–85.

Li, F. F., Van Rullen, R., Koch, C., & Perona, P. (2002). Rapid natural scene categorization in the near absence of attention. *Proceedings of the National Academy of Science, USA, 99,* 9596–9601.

Lieberman, D., Tooby, J., & Cosmides, L. (2000). The evolution of human incest avoidance mechanisms: An evolutionary psychological approach. In A. Wolf & J. P. Takala (Eds.), *Evolution and the moral emotions: Appreciating Edward Westermarck.* Stanford, CA: Stanford University Press.

Lieberman, D., Tooby, J., & Cosmides, L. (2003). Does morality have a biological basis? An empirical test of the factors governing moral sentiments relating to incest. *Proceedings of the Royal Society London (Biological Sciences), 270*(1517), 819–826.

Lieberman, D., Tooby, J., & Cosmides, L. (2007). The architecture of the human kin detection system. *Nature, 445,* 727–731.

López, A., Atran, S., Coley, J., Medin, D., & Smith, E. (1997). The tree of life: Universals of folkbiological taxonomies and inductions. *Cognitive Psychology, 32,* 251–295.

Lutz, C. A. (1988). *Unnatural emotions: Everyday sentiments on a Micronesian Atoll and their challenge to western theory.* Chicago: University of Chicago Press.

Mandler, J., & McDonough, L. (1998). Studies in inductive inference in infancy. *Cognitive Psychology, 37*(1), 60–96.

Markman, E. (1989). *Categorization and naming in children.* Cambridge, MA: MIT Press.

Marks, I. (1987). *Fears, phobias, and rituals.* New York: Oxford.

Maynard Smith, J. (1982). *Evolution and the theory of games.* Cambridge, England: Cambridge University Press.

Mineka, S., & Cook, M. (1993). Mechanisms involved in the observational conditioning of fear. *Journal of Experimental Psychology: General, 122,* 23–38.

Mineks, S., Davidson, M., Cook, M., & Keir, R. (1984). Observational conditioning of snake fear in rhesus monkeys. *Journal of Abnormal Psychology, 93,* 355–372.

New, J., Cosmides, L., & Tooby, J. (under review). Category-specific attention for animals reflects ancestral priorities not expertise. *Proceedings of the National Academy of the Sciences, 104*(42), 16598–16603.

Pinker, S. (1994). *The language instinct.* New York: Morrow.

Pinker, S. (2002). *The blank slate.* New York: Viking Press.

Pinker, S., & Bloom, P. (1990). Natural language and natural selection. *Behavioral and Brain Sciences* 13(4): 707–784.

Pitman, R., & Ott, S. (1995). Psychophysiology of emotional and memory networks in posttraumatic stress disorder. In J. McGaugh, N. Weinberger, & G. Lynch (Eds.), *Brain and memory: Modulation and mediation of neuroplasticity* (pp. 75–83). New York: Oxford.

Posner, M. (1978). *Chronometric explorations of mind.* New York: Oxford.

Rips, L. (1994). *The psychology of proof.* Cambridge, MA: MIT Press.

Ro, T., Russell, C., & Lavie, N. (2001). Changing faces: A detection advantage in the flicker paradigm. *Psychological Science,* 12, 94–99.

Rode, C., Cosmides, L., Hell, W., & Tooby, J. (1999). When and why do people avoid unknown probabilities in decisions under uncertainty? Testing some predictions from optimal foraging theory. *Cognition,* 72, 269–304.

Sahlins, M. (1976). *The use and abuse of biology: An anthropological critique of sociobiology.* Ann Arbor: University of Michigan Press.

Schacter, D., & Tulving, E. (Eds.). (1994). *Memory systems 1994.* Cambridge, MA: MIT Press.

Shannon, C. E. (1948). A mathematical theory of communication. *Bell System Technical Journal,* 27 379–423 & 623–656.

Shepard, R. N. (1984). Ecological constraints on internal representation: Resonant kinematics of perceiving, imagining, thinking, and dreaming. *Psychological Review,* 91, 417–447.

Shepard, R. N. (1987). Evolution of a mesh between principles of the mind and regularities of the world. In J. Dupré (Ed.), *The latest on the best: Essays on evolution and optimality* (pp. 251–275). Cambridge, MA: MIT Press.

Sherry, D., & Schacter, D. (1987). The evolution of multiple memory systems. *Psychological Review,* 94, 439–454.

Shostak, M. (1981). *Nisa: The life and words of a !Kung woman.* Cambridge, MA: Harvard.

Skinner, B. F. (1957). *Verbal behavior.* New York: Appleton-Century-Crofts.

Spelke, E. S. (1990). Principles of object perception. *Cognitive Science,* 14, 29–56.

Sperber, D. (1994). The modularity of thought and the epidemiology of representations. In L. A. Hirschfeld & S. A. Gelman (Eds.), *Mapping the mind: Domain specificity in cognition and culture.* Cambridge, England: Cambridge University Press.

Sperber, D. (1996). *Explaining culture: A naturalistic approach.* Oxford: Blackwell.

Sperber, D., & Wilson, D. (1995). *Relevance: Communication and cognition* (2nd ed.). Oxford, England: Blackwell.

Springer, K. (1992). Children's awareness of the implications of biological kinship. *Child Development,* 63, 950–959.

Steen, F., & Owens, S. (2001). Evolution's pedagogy: An adaptationist model of pretense and entertainment. *Journal of Cognition and Culture, 1*(4), 289–321.

Suarez, S. D., & Gallup, G. G. (1979). Tonic immobility as a response to rage in humans: A theoretical note. *Psychological Record,* 29, 315–320.

Symons, D. (1979). *The evolution of human sexuality.* New York: Oxford University Press.

Symons, D. (1987). If we're all Darwinians, what's the fuss about. In C. B. Crawford, M. F. Smith, & D. L. Krebs (Eds.), *Sociobiology and psychology* (pp. 121–146). Hillsdale, NJ: Erlbaum.

Symons, D. (1989). A critique of Darwinian anthropology. *Ethology and Sociobiology, 10,* 131–144.

Symons, D. (1992). On the use and misuse of Darwinism in the study of human behavior. In J. Barkow, L. Cosmides, & J. Tooby (Eds.), *The adapted mind: Evolutionary psychology and the generation of culture* (pp. 137–159). New York: Oxford University Press.

Tomaka, J., Blascovich, J., Kibler, J., & Ernst, J. (1997). Cognitive and physiological antecedents of threat and challenge appraisal. *Journal of Personality and Social Psychology,* 73, 63–72.

Tooby, J. (1982). Pathogens, polymorphism, and the evolution of sex. *Journal of Theoretical Biology, 97,* 557–576.

Tooby, J. (1985). The emergence of evolutionary psychology. In D. Pines (Ed.), *Emerging syntheses in science* (pp. 124–137). Santa Fe, NM: The Santa Fe Institute.

Tooby, J., & Cosmides, L. (1990a). The past explains the present. Emotional adaptations and the structure of ancestral environments. *Ethology and Sociobiology,* 11, 375–424.

Tooby, J., & Cosmides, L. (1990b). On the universality of human nature and the uniqueness of the individual: The role of genetics and adaptation. *Journal of Personality,* 58, 17–67.

Tooby, J., & Cosmides, L. (1992). The psychological foundations of culture. In J. Barkow, L. Cosmides, & J. Tooby (Eds.). *The adapted mind: Evolutionary psychology and the generation of culture* (pp. 19–136). New York: Oxford University Press.

Tooby, J., & Cosmides, L. (1996). Friendship and the banker's paradox: Other pathways to the evolution of adaptations for altruism. *Proceedings of the British Academy,* 88, 119–143.

Tooby, J., & Cosmides, L. (2001). Does beauty build adapted minds? Toward an evolutionary theory of aesthetics, fiction and the arts. *SubStance, 94/95*(1), 6–27.

Tooby, J., & Cosmides, L. (in press). Ecological rationality in a multimodular mind. In *Evolutionary psychology: Foundational papers*. Cambridge, MA: MIT Press.

Tooby, J., Cosmides, L., & Barrett, H. C. (2003). The second law of thermodynamics is the first law of psychology: Evolutionary developmental psychology and the theory of tandem, coordinated inheritances. *Psychological Bulletin, 129*(6), 858–865.

Tooby, J., Cosmides, L., & Barrett, H. C. (2005). Resolving the debate on innate ideas: Learnability constraints and the evolved interpenetration of motivational and conceptual functions. In P. Carruthers, S. Laurence, & S. Stich (Eds.), *The innate mind: Structure and content*. New York: Oxford University Press.

Tooby, J., & DeVore, I. (1987). The reconstruction of hominid behavioral evolution through strategic modeling. In W. Kinzey (Ed.), *Primate models of hominid behavior* (pp. 183–237). New York: SUNY Press.

Triesman, A. (2005). Psychological issues in selective attention. In M. S. Gazzaniga (Ed.), *The cognitive neurosciences, III* (pp. 529–544). Cambridge, MA: MIT Press.

Vining, D. R. (1986). Social versus reproductive success: The central theoretical problem of human sociobiology. *Behavioral and Brain Sciences, 9*, 167–216.

Walker, R., Hill, K., Kaplan, H., & McMillan, G. (2002). Age dependency of hunting ability among the Ache of eastern Paraguay. *Journal of Human Evolution, 42*, 639–657.

Wang, X. T. (2002). Risk as reproductive variance. *Evolution and Human Behavior, 23*, 35–57.

Weiner, N. (1948). *Cybernetics or control and communication in the animal and the machine*. Cambridge, MA: MIT Press.

Williams, G. C. (1966). *Adaptation and natural selection: A critique of some current evolutionary thought*. Princeton, NJ: Princeton University Press.

Wilson, E. O. (1975). *Sociobiology: The new synthesis*. Cambridge, MA: Belknap Press.

Wynn, K. (1998). Psychological foundations of number: Numerical competence in human infants. *Trends in Cognitive Sciences, 2*, 296–303.

Yerkes, R. M., & Yerkes, A. W. (1936). Nature and conditions of avoidance (fear) response in chimpanzee. *Journal of Comparative Psychology, 21*, 53–66.

27

The Environments of Our Hominin Ancestors, Tool-usage, and Scenario Visualization

Robert Arp

Introduction

All evolutionary psychologists posit that evolution is responsible not only for human physiology and anatomy, but also for certain human psychological and behavioral characteristics that evolved in our past to solve specific problems of survival (e.g., Cosmides and Tooby 1992, 1994; Gardner 1993, 1999; Pinker 1994, 1997, 2002; Mithen 1996, 1998, 2001; Wilson 2003; Scher and Rauscher 2003). However, two of the issues evolutionary psychologists debate about concern (a) the type and number of mental modules the human mind contains, as well as (b) the exact time-period or time-periods when these mental modules were solidified in the human psyche.

Recently, Scher and Rauscher (2003: XI) and Wilson (2003) have drawn a distinction between what they call *narrow* evolutionary psychology (NEP) and *broad* evolutionary psychology (BEP). Advocates of NEP follow the groundbreaking work of Cosmides and Tooby, arguing that the mind is like a Swiss Army knife loaded with specific mental tools that evolved in our Pleistocene past to solve specific problems of survival. In response to (a) and (b) above, adherents to NEP argue that

(a) the mind is a host of specialized, domain-*specific* mental modules, and (b) the Pleistocene epoch is *the* time-period in which the basic psychological structure of the modern human mind was solidified in our genetic makeup.

In contrast to NEP, advocates of BEP consider alternative approaches to Cosmides and Tooby's Pleistocene epoch-forming, Swiss Army knife model of the mind, and want to argue that (a) the mind probably does not contain the myriad of specialized, domain-specific mental modules as the NEPers would have us believe, but relies more upon domain-*general* mental capacities that have evolved to handle the various and sundry problems a human faces (Samuels 1998). They also want to claim that (b) although the Pleistocene epoch is a significant time-period in our evolutionary past, it is by no means a single environment, nor is it the only environment that has shaped the modern mind (Daly and Wilson 1999; Laland and Brown 2002).

So, is it possible to determine which camp of thinkers has the more accurate picture concerning our mental architecture and its evolution? In this paper, I take up the challenge of trying to adjudicate between NEP and BEP by offering a

Robert Arp, "The Environments of Our Hominin Ancestors, Tool Usage, and Scenario Visualization," *Biology & Philosophy* (2006), 1, 95–117. With kind permission from Springer Science and Business Media.

I wish to thank George Terzis, Kent Staley, Richard Blackwell, Brian Cameron, Susan Arp, Kim Sterelny, and a referee from *Biology & Philosophy* for their comments on earlier versions of this paper.

theory of mental segregation and integration I call *scenario visualization* that is rooted in problem solving tasks our early human ancestors would have faced in their environments. The evidence for scenario visualization and problem is found in the types of tools these early peoples utilized. After a presentation of Cosmides and Tooby's NEP approach, as contrasted with one BEP approach put forward by Steven Mithen (1996, 1998, 2001), I formulate my theory concerning scenario visualization. In essence, scenario visualization is an advance upon Mithen's account of cognitive fluidity, which itself (i.e., Mithen's account) is an advance upon Cosmides and Tooby's model of the mind as being composed of encapsulated mental modules. Then, I present archeological evidence of tool making to show that scenario visualization emerged to deal with the many visually-related problems encountered by our hominin ancestors in their ever-changing environments.

NEP, the Pleistocene, and the Emergence of Modularity

According to advocates of NEP, the adaptive problems in the Pleistocene environment occasioned the emergence of psychological modules *designed* to handle the various and sundry problems of such an environment. These modules are envisioned as domains of specificity, handling only one kind of adaptive problem to the exclusion of others. Modules are encapsulated in this sense, and do not share information with one another. Like the various kinds of tools in a Swiss Army knife, the various mental modules are supposed to solve specific problems; but, they do so to the exclusion of each other. The scissors of the Swiss Army knife are not functionally related to the Phillip's-head screwdriver, which is not functionally related to the toothpick, etc.

However, there seems to be a fundamental limitation in the NEPers' reasoning, especially if the environment in which the domain-specific module has been selected is supposed to have remained fairly stable. Cosmides and Tooby (1987: 28) note that these domain-specific modules have evolved "for solving *long-enduring* (my italics) adaptive problems", and Hirschfeld and Gelman (1994: 21) characterize a module as a "stable response to a set of *recurring* (my italics)

and complex problems faced by the organism." Now Daly and Wilson (1999), Foley (1995), and Boyd and Silk (1997) have shown that the Pleistocene did not consist of a single hunter-gatherer type of environment, but was actually a constellation of environments that presented a host of challenges to the mind. So, the first problem for advocates of NEP has to do with the possibility of the environment in which a particular module evolved being stable enough for the module to have evolved. In other words, Daly *et al.*'s criticism of NEP is that the environments in which our Pleistocene ancestors lived were *too* varied and *too* erratic for the Swiss Army knife blades to be solidified in our genetic makeup.

Even if we tempered Daly *et al.*'s claims regarding the multitude of environments faced by our ancestors and grant that the Pleistocene consisted of a more unitary Stone-age-hunter-gatherer-life-out-on-the-savannah kind of existence (like the one Cosmides and Tooby would have us believe), then we have the further problem of the possibility of some stable and routine module being able to handle the *un*stable and *non*-routine events occurring in some environment. When routine perceptual and knowledge structures fail, or when atypical environments present themselves, it is *then* that we need to be innovative in dealing with this novelty. If mental modules are encapsulated, and are designed to perform certain routine functions, how can this modularity account for novel circumstances? The problem for NEP can be phrased in the form of a disjunction. Either (a) the environment was not stable enough to occasion the emergence of domain-specific modules, as is part of the thrust of Daly *et al.*'s criticism. Or (b) the environment was stable, allowing for domain-specific mental modules to emerge; but then the environment changed, making it such that the modules specified for the old environment would no longer be helpful in the new environment.

Now, imagine the ever-changing environments of the Pleistocene epoch. The environmental shifts had dramatic effects on modularity, since now the specific content of the information from the environment in a particular module was no longer relevant. The information that was formerly suited for life in a certain environment could no longer be relied upon in the new environmental niches. Appeal to modularity alone would have led to certain death and extinction of many mammalian species. The successful progression from typical kinds of environments to other *atypical* kinds of environments would have required some other kind of mental capacity to emerge in the minds of our hominin ancestors that creatively could handle the new environments. Mere mental associations, or trial-and-error kinds of mental activities, would not be enough since the environments were wholly new, and there would have been no precedent by or through which one could form mental associations utilizing past information. Mental associations deal with the familiar. What is one to do when encountering the wholly unfamiliar? Although important, modules have their limitations, since they do all of their associative work in routine environments. What happens if an environment radically changes, making the information that a particular module characteristically selects in a familiar environment no longer relevant in a wholly new environment? A radical re-adaptation and re-adjustment would be needed – one that transcends the limitations of the routine. This totally new environment would require that one be creative or innovative in solving environmental problems so as to survive. But how is it that one would have been creative in solving the environmental problems of the Pleistocene?

Here, it is important to draw a distinction between what Mayer (1995: 4) refers to as *routine* problem solving and *nonroutine creative* problem solving. In routine problem solving one recognizes many possible solutions to a problem, given that the problem was solved through one of those solutions in the past. We also can engage in activities that are more abstract and creative. We can invent new tools based upon mental blueprints, synthesize concepts that, at first glance, seemed wholly disparate or unrelated, and devise novel solutions to problems. If a person decided to pursue a wholly new way to solve a problem by, say, inventing some kind of tool, then we would have an instance of nonroutine creative problem solving. Nonroutine creative problem solving involves finding a solution to a problem that has not been solved previously. The invention of a new tool would be an example of nonroutine creative problem solving because the inventor did not possess a way to solve the problem already. The significant question becomes, then: How is it

that humans evolved the ability to engage in forms of nonroutine creative problem solving, especially given that the Pleistocene environment in which our hominin ancestors existed either was really a constellation of ever-changing environments (Daly *et al.*'s criticism), or was a single environment is filled with a myriad of nonroutine problems that seem only to be able to be handled creatively?

One BEP Response: Mithen's Cognitive Fluidity

This is where Steven Mithen has made an advance upon advocates of NEP by introducing *cognitive fluidity*, an idea that explains creative response to novel environments. He sees the evolving mind as going through a three-step process (1996: 64, 1998). The first step begins prior to 6 million years ago when the primate mind was dominated by what he calls a *general intelligence*. This general intelligence consisted of an all-purpose, trial-and-error learning mechanism devoted to multiple tasks. Learning was slow, errors were frequent, and behaviors were imitated, much like the mind of the chimpanzee.

The second step coincides with the evolution of the *Australopithecine* line, and continues all the way through the *Homo* lineage to *H. neandertalensis*. In this second step, multiple specialized intelligences, or modules, emerge alongside general intelligence. Learning within these modules was faster, and more complex activities could be performed. Compiling data from fossilized skulls, tools, foods, and habitats, Mithen concludes that *H. habilis* probably had a general intelligence, as well as modules devoted to social intelligence (because they lived in groups), natural history intelligence (because they lived off of the land), and technical intelligence (because they made stone tools). Neandertals and archaic *H. sapiens* would have had all of these modules, including a primitive language module, because their skulls exhibit bigger frontal and temporal areas. According to Mithen, the neandertals and archaic *H. sapiens* would have had the Swiss Army knife mind that advocates of NEP speak about.

Mithen rightly criticizes the NEPers who think that the essential ingredients of mind evolved during the Pliestocene epoch. The potential variety of problems encountered in generations subsequent to the Pleistocene is too vast for a much more limited Swiss-army-knife mental repertoire; there are just too many situations for which nonroutine creative problem solving would have been needed in order to not simply survive, but flourish and dominate the earth. The emergence of distinct mental modules *during the Pleistocene* as being adequate to account for learning, negotiating, and problem solving *in our world today* cannot be correct (Mithen 1996: 45–46). Pinker (2002: 40–41, 220–221) thinks that there are upwards of 15 different domains, and various other evolutionary psychologists have their chosen number of mental domains (e.g., Shettleworth 2000; Plotkin 1997; Palmer and Palmer 2002). However, there is potentially an infinite number of problems to be faced on a regular basis by animals as they negotiate environments. It does not seem that there would be a way for 15, 20, 25 – or even 1000 – domains to handle all of these potential problems. That we negotiate environments so well shows that we have some capacity to handle the various and sundry potential nonroutine problems that arise in our environments.

Here is where the third step in Mithen's evolution of the mind comes into play. In this final step, which coincides with the emergence of modern humans, the various modules are working together with a flow of knowledge and ideas between and among them (1996: 154, 1998). The information and learning from one module can now influence another, resulting in an almost limitless capacity for imagination, learning, and problem solving. The working together of the various mental modules as a result of *cognitive fluidity* is consciousness, for Mithen, and represents the most advanced form of mental activity (1996: 188–192, 1998).

Mithen goes on to note that his model of cognitive fluidity can account for human creativity in terms of problem solving, art, ingenuity, and technology. Mithen's idea of cognitive fluidity helps to explain our conscious ability to creatively solve nonroutine problems because the potential is always there to make innovative, previously unrelated connections between ideas or perceptions, given that the information between and among modules has the capacity to be mixed together, or intermingle. This is not to say that the information will in fact mix together by an individual. This is just to say that there is always the potential for such a mental process to occur in our species.

Scenario Visualization, Segregation, and Integration

Mithen's account of cognitive fluidity allows for the free movement of information between and among modules. This is important for mental activities, like imagination, requiring the simultaneous utilization of several modules. So for example, Mithen would think that totemic anthropomorphism associated with animals in, say, a totem pole made up of part-human/part-animal figures, derives from the free flow of information between a natural history module dealing specifically with animals and their characteristics, and a social module dealing specifically with people and their characteristics (see Mithen 1996: 164–167). A totem carved out of wood is the *material* result of the free flow of information between the natural history and social modules.

Mithen's account is unsatisfactory, however, because he makes consciousness to be a passive thing. On his account, consciousness is just a flexible fluidity, a free-flowing of information between and among modules. This does not seem to be the full account of consciousness. When we are engaged in conscious activity, we are *doing* something. The fundamental insight derived from Kant (1929), and reiterated by numerous philosophers, psychologists, and neuroscientists, is that consciousness is an active process (e.g., Kandel *et al.* 2000: 412; Sekuler and Blake 2002: 123–124; Kanizsa 1976, 1979). Consider the illustration in figure 27.1 below. We immediately recognize the space in the middle as an octagon; however, the reason why we can seems to be because our visual perception is constructive. The mind brings something to the diagram and fills in the blank (literally!), in generating the image of the octagon.

I want to bolster Kant's fundamental insight and suggest that a certain kind of conscious activity I call *scenario visualization* involves activities of segregation and integration, as well as the flexibility of free-flowing information (Mithen's cognitive fluidity). These psychological properties of segregation and integration are akin to the visual

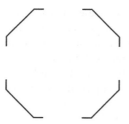

Figure 27.1. A fill-in-the-blank octagon.

processes that actively segregate and integrate the information concerning the lines and spaces in figure 27.1 above, so as to produce a coherent picture of the octagon.

What I mean by scenario visualization is a conscious process that entails: *selecting visual information from mental modules; forming a coherent and organized visual cognition; and the subsequent transforming and projecting of that visual cognition into some suitable imagined scenario, for the purpose of solving some problem posed by the environment in which one inhabits by the usage of a tool.*

As an active, mental process scenario visualization would include the following steps: (a) reception of visual stimulus cues from a relevant external environment, indicating that a problem is present; (b) identification of a goal to be achieved or problem to be solved in some external environment; (c) selection of visual images that appear to be relevant to the solution from several possible choices of visual images; (d) integration of the visual images concerning possible scenarios into organized and coherent visual scenes; (e) integration of the visual images concerning the imagined problem solving tool into an organized and coherent perception, vis-à-vis the imagined possible scenarios; (f) projection of visual images into imagined scenarios to judge the potential viability or appropriateness of a particular problem solving tool to a problem; (g) recollection of the particular goal of the project from memory; and (h) recognition that a particular problem solving tool is appropriate as a solution in the relevant environment that prompted the process of scenario visualization in the first place.

The key feature of scenario visualization is the mind's ability to select and integrate visual images from mental modules, as well as project and transform these images in possible situations,

circumstances or scenarios for the purpose of solving some problem in an environment. As a conscious process, visualization is distinct from the psychological processes of simply forming a visual image or recalling a visual image from memory; these activities can be performed by non-human primates, mammals, and possibly other animals. Visualization requires a mind that is more active in the *utilization* of visual images and/or memories from modules against an environmental backdrop. It is not the having of visual images that is important; it is what the mind does in terms of actively segregating and integrating visual information relative to environments that really matters.

Certain forms of problem solving offer clear examples that are expressive of the intimate relationship between consciousness and the visual system (see Tye 1991). Smith (1995: 136–137) has noted that when we problem solve, we engage in several mental functions: first, we generate some kind of mental representation of the goal to be achieved; next, we select the appropriate means for achieving this goal; then, we execute the planned strategy; finally, in the midst of executing the strategy, we monitor how successful we are. The formation of visual images, and then being able to manipulate, switch around, and/or project those images in various imagined visual scenarios, can play a role in conscious creative problem solving. I suggest that this entails a process of scenario visualization. Scenario visualizing requires forming a visual image, and depending upon what is being visualized, may require recalling a visual image from memory. Each of the above problem solving steps mentioned by Smith requires a visual image of some kind.

We are the only kind of species that can visualize in this more complete way, and what I am suggesting further is threefold: First, humans have the unique ability to select among visual images, integrate and organize visual information to form a coherent visual cognition, go beyond the present in order to project visual images into future scenarios, as well as transform the visual images within a variety of imagined environments – this is scenario visualization. We construct tools to do work in some environment. We need some kind of environmental setting in which to construct an artifact precisely because the artifact, presumably, is going to serve some purpose in some environment. In order to survive in un-stable and changing environments, our hominin ancestors evolved a capacity to deal with this instability, whereby they could visually anticipate the kinds of tools needed for a variety of settings.

Second, our capacity to scenario visualize is a central feature of conscious behavior, an idea that comports well with: Sternberg's (2001) notion of consciousness entailing the setting up of future goals; Carruthers's (2002) idea that humans are the only kinds of beings able to generate – and then reason with – novel suppositions or ima-ginary scenarios; Arnheim (1969) and Kosslyn and Koenig's (1995: 146) position that the very elements of reasoning – thoughts, concepts, abstractions, and words – seem to require visual imagery, and the further use of that imagery in creative and imaginative ways; Gardenfors (2000), Lakoff (1987), Rosch (1975, 1981), and Gray's (1999) views that the conscious mind utilizes basic Gestalt-type mental images and pre-conceptual structures when forming certain concepts and producing linguistic expressions; and Crick and Koch's (1999: 324) claim that "conscious seeing" requires the brain's ability to "form a conscious representation of the visual scene that it then can use for many different actions or thoughts".

Third, scenario visualization emerged as a natural consequence of our evolutionary history, which includes the development of a complex nervous system in association with environmental pressures that occasioned the evolution of such a function. In attempts to recreate tools from the Mousterian and Upper Paleolithic era, archeo-logists like Mithen (1996: 171, 1998), Wynn (1979, 1981, 1991, 1993), and Wynn and McGrew (1989) have shown that the construction of such tools would require several mental visualizations, as well as numerous revisions of the material. Such visualizations likely included the abilities to, at least, identify horizontal or vertical lines within a distracting frame, select an image from several possible choices, distinguish a target figure embedded in a complex background, construct an image of a future scenario, project an image onto that future scenario, as well as recall from memory the particular goal of the project. If an advanced form of toolmaking acts as a mark of consciousness, then given complex and changing Pleistocene environments, as well as the scenario visualization that is necessary to produce tools

so as to survive these environments, what I am suggesting is that visual processing most likely was the primary way in which this consciousness emerged on the evolutionary scene.

Mithen thinks that the kinds of unique beha-viors we engage in are the result of a free flow of information between and among modules. This cannot be the full story. My claim is that scenario visualization emerged as a conscious property of the brain to act as a kind of metacognitive process that segregates and integrates relevant visual information from psychological modules, in performing certain functions in novel envir-onments. More accurately, we scenario visualize, viz., selectively attend to visual information from certain modules, and actively integrate that visual information from those modules so as to attain mental coherence. If consciousness were *merely* free flow of information, there would be no mental coherency; the information would be chaotic and directionless, and not really inform-ative at all. It would be more like meaningless data that free-floated around. However, data needs to be segregated and integrated so that it can become informative for the cognizer. Segre-gation and integration of visual information from mental modules are the jobs of scenario visualization.

For example, that the visual images in the social module pertaining to human behaviors and the visual images in the natural history module pertaining to animal behaviors are put together in anthropomorphic animal totemism means that they had to be *segregated* or *selected out from* other modules as relevant. Other modu-lar images would be bracketed out as irrelevant, as the images in these two modules would be *focused* upon. However, it is not just *that* chan-nels have been opened between these modules, so that their specified information can intermix. Cognitive fluidity is necessary; but, something more active needs to occur when the idea of anthropomorphic animal totemism is brought to mind. The modules pertaining to such an idea must be synthesized, so that something coher-ent results. Another way to say this is that the information from both modules is integrated, allowing for something sublimated (to use a Hegelian notion), or innovative to emerge anew as a result of the process. Fodor (1998) expresses a similar claim about integration:

Even if early man had modules for "natural intelligence" and "technical intelligence," he couldn't have become modern man just by adding what he knew about fires to what he knew about cows. The trick is in thinking out what happens when you put the two together; you get *steak au poivre* by integrating knowledge bases, not by merely summing them. (p. 159)

Consider that toolmaking, as much as language, characterizes our apparent human uniqueness among species in the animal kingdom. There is ample evidence of advanced forms of toolmaking in our past – specifically, those that began with the Mousterian industry – that require a mind having the capacity to visualize (Wynn 1979, 1981, 1991, 1993; Isaac 1986; Pelegrin 1993; Mithen 1998). The most basic step in constructing a stone tool has to do with simply striking a flake from a cobble. We have been able to get chimpanzees to imitate this behavior in captivity, but there is no evidence of apes in the wild performing this rudimentary procedure (Griffin 1992: 113, 248; Stanford 2000: 111–114). To strike a sequence of flakes in such a way that each one aids in the removal of others, however, demands much more control of the brain, as well as a hand equipped with a variety of grips. The various steps in the process must be evaluated for their own merits, and previous steps must be seen in light of future steps as well. It would seem that such stone working cannot be operated by an inflexible and mechanical trial-and-error, or imitative mental routine, because there are too many possibilities at every strike of the stone. When we consider that our early hominid ancestors not only had to select certain materials that were appropriate to solve some problem, but also utilized a diverse set of techniques, and went through a number of steps involving an array of stages that resulted in a variety of tool types, then it becomes apparent that a fairly advanced form of mental activity had to occur. Thus, Wynn (1993: 396–397) claims that tool behavior "entails problem solving, the ability to adjust behavior to a specific task at hand, and, for this, rote sequences are not enough". This mental complexity has caused McNabb and Ashton (1995) to refer to our toolmaking ancestors as "thoughtful flakers".

Tool-usage and Early Hominin Environments

In the Introduction, I noted that there is a debate among evolutionary psychologists as to (a) the type and number of mental modules the human mind contains, as well as (b) the exact time-period or time-periods when these mental modules were solidified in our psyche. Also, I noted that all evolutionary psychologists are in agreement with the fact that certain environmental selection forces were present in our early hominid past, and that these forces contributed to the mind's formation. Further, forming an accurate picture of what those selection forces were like is integral to our understanding of the mental mechanisms that have survived the process. At the same time, once we have an understanding of the environmental challenges faced by our ancestors, we can get a better picture of what our mental architecture has evolved to look like. I will now utilize my idea of scenario visualization, in conjunction with pieces of archeological evidence concerning tool-usage and creative problem solving, in order to offer my own hypothesis regarding the evolution of our mental architecture.

Along with bipedalism, it is generally agreed by biologists, anthropologists, archeologists, and other researchers that a variety of factors contributed to the evolution of the human brain. These include, but are not limited to, the following: diversified habitats; social systems; protein from large animals; higher amounts of starch; delayed consumption of food; food sharing; language; and toolmaking (Martin 1990; Aiello and Dunbar 1993; Gibson and Ingold 1994; Relethford 1994; Aboitiz 1996; Aiello 1997; Deacon 1997; Donald 1997; Allman 2000; Roth 2000). It is not possible to get a complete picture of the evolution of the brain without looking at all of these factors, as brain development is involved in a complex coevolution with physiology, environment, and social circumstances. However, I wish to focus on toolmaking as essential in the evolution of the brain and visual system. I do this for three reasons.

First, toolmaking is the mark of intelligence that distinguishes the *Australopithecine* species from the *Homo* species in our evolutionary past. *H. habilis* was the first toolmaker, as the Latin name – *handy-man* – suggests. Second, tools offer us indirect, but compelling, evidence that psychological states emerged from brain states. In trying to simulate ancient toolmaking techniques, archeologists have discovered that certain tools only can be made according to mental templates, as Wynn (1979, 1981, 1991, 1993), Isaac (1986), and Pelegrin (1993) have demonstrated. Finally, as I will show, the evolution of toolmaking parallels the evolution of the visual system from non-cognitive visual processing to conscious cognitive visual processing in terms of scenario visualization. We now turn to each of these three points.

The first stone artifacts, Oldowan stones, were found between 3 and 2 mya, and are associated with *H. habilis*. They consist of choppers, bone breakers, and flakes that likely were used to break open the bones of animals to extract the protein-rich marrow. The key innovation has to do with the technique of chopping stones to create a chopping or cutting edge. Typically, many flakes were struck from a single *core* stone, using a softer hammer stone to strike the blow. Here, we have the first instance of *making* a tool to *make* another tool, and it is arguable that this technique is what distinguishes ape-men from apes. Another way to put this is that chimps and *Australopithecines* used the tools they made, but did not use these tools to make other tools. So, the distinction is between a tool-*user* who has made a certain tool A to serve some function (e.g., *Australopithecines* and chimps), and a tool-*maker* who has made a certain tool A to serve the specific function of making another tool B to serve some function (e.g., *H. habilis*).

The Acheulian tool industry consisted of axes, picks, and cleavers. It first appeared around 1.5 mya, and is associated with *H. erectus*. The key innovations are the shaping of an entire stone to a stereotyped tool form, as well as chipping the stone from both sides to produce a symmetrical (bifacial) cutting edge. This activity required manual dexterity, strength, and skill. However, the same tools were being used for a variety of tasks such as slicing open animal skins, carving meat, and breaking bones.

The Acheulian industry stayed in place for over a million years. The next breakthrough in tool technology was the Mousterian industry that arrived on the scene with the *H. neandertalensis* lineage, near the end of the archaic *Homo* lineage, around 200,000 ya. Mousterian techniques involved a careful preliminary shaping of the stone core from which the actual blade is struck off, either first by shaping a rock into a rounded surface before striking off the raised area as a wedge-shaped flake, or by shaping the core as a long prism of stone before striking off triangular flakes from its length. This was an innovation in tool technology because of its more complex three-stage process of constructing (a) the basic core stone, (b) the rough blank, and (c) the refined finalized tool.

Also, such a process enabled various kinds of tools to be created, since the rough blank could follow a pattern that ultimately became either cutting tools, serrated tools, flake blades, scrapers, or lances. Further, these tools had wider applications as they were being used with other material components to form handles and spears, and they were being used as tools to make other tools, such as wooden and bone artifacts.

By 40,000 ya, some 60,000 years after anatomically modern *H. sapiens* evolved, we find instances of human art in the forms of beads, tooth necklaces, cave paintings, stone carvings, and figurines. This period in tool manufacture is known as the Upper Paleolithic, and it ranges from 40,000 ya to the advent of agriculture around 12,000 ya. This toolmaking era shows a remarkable proliferation of tool forms, tool materials, and much greater complexity of toolmaking techniques. Sewing needles and fish hooks made of bone and antlers first appear, along with flaked stones for arrows and spears, burins (chisel-like stones for working bone and ivory), multibarbed harpoon points, and spear throwers made of wood, bone, or antler (Wynn 1979, 1981, 1991, 1993; Isaac 1986; Wynn and McGrew 1989; Pelegrin 1993; Mithen 1996, 1998, 2001).

Before we go any further, it is necessary to distinguish four levels of visual processing in the visual system. The first level is a *non-cognitive* visual processing that occurs at the lowest level of the visual hierarchy associated with the eye, LGN, and primary visual cortex. At this level, the animal is wholly unaware of the processing, as the brain receives the disparate pieces of basic information in the visual field concerned with lines, shapes, distance, depth, color, etc., of an object in the visual field (Julesz 1984; Merikle and Daneman 1998; Kandel *et al.* 2000: 349–356, 507–522).

The second level of visual processing is a *cognitive* visual processing that occurs at a higher level of visual awareness associated with the *what* and *where* unimodal areas. When it is said that an animal visually perceives *what* an object looks like or *where* an object is located, this means that the animal is cognitively aware of or cognitively attends to that object in the visual field (van Essen *et al.* 1992; Goodale *et al.* 1994). The move from non-cognitive visual processing to cognitive visual processing is a move from the purely neurobiological to the psychological dimension associated with the brain's activities. Words like *cognition, awareness,* and *perception* all refer to similar psychological abilities of an animal.

The third level of visual processing is a *cognitive* visual processing that occurs at an even higher level of visual awareness concerned with the *integration* of the disparate pieces of unimodal visual information in the unimodal association areas. There are times when an animal must determine both what an object is and where it is located, and this level of visual processing makes such a determination possible (van Essen and Gallant 1994; Desimone and Ungerleider 1989).

The fourth level of visual processing is a *conscious cognitive* visual processing that occurs at the highest level of the visual hierarchy associated with the multimodal areas, frontal areas, and most probably the summated areas of the cerebral cortex. This is the kind of visual processing associated with human awareness and experience, and evidence for this level comes from reports made by individuals, as well as from observing human behavior (Roth 2000: 84–86; Kandel *et al.* 2000: 396–402, 1317–1319).

We can think of the four levels of visual processing in relation to the various species in the animal kingdom. All vertebrate species in the phylum chordata with a rudimentary visual system (mammals, birds, reptiles, amphibians and fishes) exhibit non-cognitive visual processing of some kind. These same vertebrate species also exhibit cognitive visual processing to some degree or another (Stamp Dawkins 1993; Marten and Psarakos 1994; Byrne 1995; Parker 1996; Pearce 1997). However, within the order primates, human beings alone seem to exhibit conscious cognitive visual processing to a full degree, while the other primates may do so to a lesser degree.

The question now becomes, What level of visual processing did our early ancestors achieve? Given the neural connections of present-day chimps, and the eye-socket formations and endocasts of *Australopithecines*, it is clear that both have (had) non-cognitive visual processing. Also, given that chimps are aware of and attend to visual stimuli, it is likely that they, along with *Australopithecines*, have (had) cognitive visual processing. But, can these species be said to have *conscious* cognitive visual processing? Possibly, to a certain extent. But, what counts against such species having *full* consciousness is a lack of toolmaking. There seems to be a direct connection between advanced forms of toolmaking and conscious visual processing.

I have suggested that advanced forms of toolmaking require *conscious* visual processing in terms of scenario visualization. Such an activity entails that a mind be able to *visualize* the many different aspects of the toolmaking process. The key feature is the mental ability to go beyond the present in order to project visual images from mental modules into future, possible situations, circumstances and/or scenarios, as well as transform or manipulate the visual images against a variety of imagined backgrounds. My claim is that in the same way that visual integration performs the function of segregating and integrating visual module areas, so too, consciousness emerged as a property of the brain to act as a kind of *metacognitive* process that not only is the intermixing and interplaying of visual information from psychological modules (Mithen's cognitive fluidity), but segregates and integrates relevant visual information, in performing certain functions in novel environments. Put another way, just as the visual system working in tandem with other parts of the brain actively fills in the needed space so as to attain a coherent picture of the octagon, I utilized in figure 27.1 a few pages back, so too consciousness, in terms of scenario visualization, actively segregates and integrates visual information so as to attain a coherent "picture" utilizing psychological modules. Again, in the Kantian spirit, the mind is an active thing, and consciousness represents the most complex activity of the mental hierarchy.

The Evolution of the Javelin

In what follows, I trace the development of the multi-purposed javelin from its meager beginnings as a stick, through the modification of the stick into the spear, to the specialization of the spear as a javelin, equipped with a launcher. This tool is illustrative of the emergence of conscious visual awareness in terms of scenario visualization that tells a concrete evolutionary story. This story is meant to be presented as a coherent and plausible account of how it is that scenario visualization would have emerged in our evolutionary past and, like most evolutionary accounts, is not meant to be an account for which we have *decisive* evidence.

Step 1: the stick

It seems that we can take present-day chimpanzee activities to be representative of early hominin life, and we can see that chimps in their native jungle environments do indeed use tools. The chimps also use these branches to hit in self-defense, or in attack. This is probably what our ancestors did while in their African environments as well. The kind of activities chimps engage in when they use tools can be categorized as trial-and-error learning, or imitative learning. If we watch baby chimps, they try to imitate the actions of older chimps, including the usage of tools. Researchers have tried to get chimps to make tools to make other tools, the way in which early *H. habilis* likely made tools to make tools, by flaking and edging, but they cannot do it (Tomasello *et al.* 1993; Byrne 1995: 89–93, 2001: 162–164). So, it seems that chimps form visual images, and can even recall visual images from memory, when they use tools. However, they do not visualize. Their tool-usage merely is imitative, and wholly lacking in innovation.

When the climate changed and our ancestors moved from the jungles to forage and kill food out on the various environments of Africa, they eventually constructed javelins that they could throw from a distance in order to kill prey. One could continue to hit prey or a predator with a stick until it dies, as was done in the jungle environment. This may work for some prey and predators, but what about the ones that are much bigger than you, like wooly mammoths and

saber-toothed tigers? Imagine being stuck out on the savannah with a stick as your only tool of defense against these animals. Stated simply, you would need to become more creative in your toolmaking just to survive.

The progression from stick to thrown javelin went through its own evolution that is indicative of the advance from *cognitive* visual processing in terms of forming visual images, to *conscious cognitive* visual processing in terms of visualizing. The kind of toolmaking that our early *Homo* ancestors engaged in was likely to be little more than trial-and-error or imitative learning that was passed on from generation to generation, the same way certain activities are passed on from one chimp generation to the next. Flakes were constructed; so too, sticks were constructed. Apparently, however, it never occurred to these species to place one of their flaked stones on the edge of a stick.

Step 2: the spear

By the Mousterian era (200,000 ya), archaic *H. sapiens* and *H. neandertalensis* were going through a three-step stone-forming process, allowing for the possibility that a variety of tools be constructed in the outcome. Also, such stone flakes were placed on the end of sticks as spears. It is safe to say that the variety of tools constructed is evidence that they were visualizing future scenarios in which these tools could be used; otherwise, *what would be the point of constructing a variety of tools in the first place?* Chimps use the same medium of sticks or rocks to either hit, throw, or smash. However, the construction of a variety of tools indicates that they have a variety of purposes. What is a purpose, other than the formation of a visual image, the projection of that visual image onto some future scenario, and the intent to carry out or act on such visualization? The variety of tools is the material result of purposive visualization. Following Wynn, Mithen (1996) notes that a mind with an ability to "think about hypothetical objects and events" is

> absolutely essential for the manufacture of a stone tool like the handaxe. One must form a mental image of what the finished tool is to look like before starting to remove flakes from the stone nodule. Each strike follows from a hypothesis as to its effect on the shape of the tool. (p. 36)

Step 3: the javelin

With the arrival of modern humarn on the evolutionary scene (i.e., *H. sapiens sapiens*), we find evidence of a variety of types of javelins, spears, and javelin-launchers. Archeologists like Wynn (1993) and Mithen (1996: 171) have shown that the construction of a javelin would require several mental visualizations, as well as numerous revisions of the material, so as to attain its optimal performance. Such visualizations likely included the abilities to: (a) identify horizontal or vertical lines within a distracting frame; (b) select an image from several possible choices; (c) distinguish a target figure embedded in a complex background; (d) construct an image of a future scenario; (e) project an image onto that future scenario; and (f) recall from memory the particular goal of the project in the first place.

Different types of javelins with different shaped heads and shafts were constructed, depending upon the kind of kill or defense anticipated. If one tried to simply walk up to and hit a large animal, one likely would have been killed. In fact, this is probably what happened on more than one occasion to the early hominin working out of the environmental framework of the jungle in this totally new environmental framework of the savannah. Eventually our ancestors, such as *H. neandertalensis*, developed the spear; however, there is evidence suggesting that they could only develop spears, and not javelins (Wynn 1991; Mithen 1996: 147–150). *H. sapiens sapiens* developed javelins – equipped with launchers – that could be used in creative ways to not only throw from a distance, but also to spear at close range, hack, and cut (Wynn 1993; Mithen 1996: 171). So, the emergence of the javelin and its myriad uses would seem to indicate the presence of a different kind of mind that could creatively form, recall, re-adjust, select, and integrate visual scenes and scenarios for the purposes of surviving and flourishing in either *constantly changing* or *novel* environments.

Given the concrete evidence of fossilized tools, Sperber (1994), Mithen (1996), Donald (1997), and Pinker (1997) speculate that *H. sapiens sapiens* were clearly conscious, whereas *Australopithecines* clearly did not have consciousness. This is consistent with my claim that consciousness involves scenario visualization, and that such conscious cognitive visualization emerged with the production of more complex tools.

Below is a diagram (figure 27.2) that depicts the mental processes involved in the construction of a javelin by a member of the species *H. sapiens sapiens* who lived out in the savannah around 40,000 ya. This illustration is supposed to represent the slower, intelligent processes associated with consciously selecting and integrating the free flow of visual information between mental modules, so as to construct a certain kind of javelin in order to solve some adaptive problem. In this case, the problem to be solved has to do with easily and efficiently killing a large antelope for the purposes of skinning it and using its body parts for food and warmth during the approaching winter months. I ask you to imagine that this is the *very first instance* of one of our ancestors coming up with the idea of the javelin, with the intention of manufacturing it. At first, s/he has no prior knowledge of the javelin; but through the process of scenario visualization, s/he eventually "puts two and two together", and devises the mental blueprints for the manufacture of the javelin. In other words, this is supposed to a schematization of nonroutine creative problem solving at work.

In the first figure, the hunter has separate visual images associated with antelope characteristics, the manufacture of the bi-faced hand-axe, as well as with how projectiles move through the air. The hunter also has visual images associated with all kinds of other pieces of information like the faces of the members of his/her group, a mental map of the immediate area, some intuitive sense of mechanics and biology, etc. In accordance with Mithen's idea of cognitive fluidity, the information between and among these mental spheres has the potential to intermix, and is represented by the dotted-line bubbles. Notice that, consistent with the data presented by developmental and evolutionary psychologists (e.g., Spelke 1991; Gardner 1993, 1999; Pinker 1994, 1997, 2002; Palmer and Palmer 2002), there are several mental modules (dotted-line bubbles) that make up the hunter's mind. In the second figure, scenario visualization is beginning as the animal, biological, technology, and intuitive physics modules are bracketed off or segregated from the other mental modules. In the third figure, visualization is continuing because the hunter is

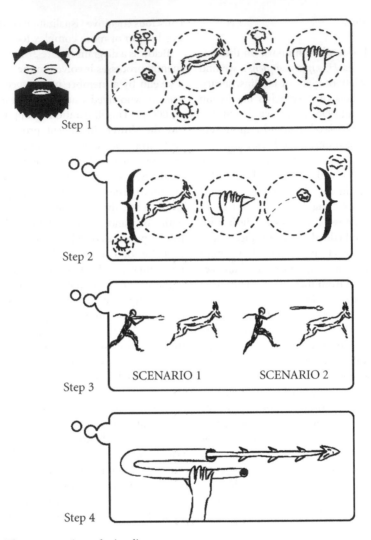

Figure 27.2. The construction of a javelin.

manipulating, inverting, and transforming the images as they are projected into future imagined scenarios. In the fourth figure, these modules are actively integrated so that a wholly new coherent and organized image is formed that can become implemented in the actual production of the javelin.

Some Objections, Questions, and Clarifications

Someone may object at this point that (1) not everyone visualizes when they solve problems,

and/or that (2) blind persons, who cannot visualize in the way I have defined the term, surely have the ability to solve problems. There are also questions that may arise as to (3) whether scenario visualization is at work in the *use* of a javelin by a hunter in some environment, once it has been manufactured, as well as (4) how scenario visualization, as a conscious process, is related to the conscious process of goal-formation. I will respond briefly to these objections and questions.

(1) It seems implausible that no one *ever* visualizes when trying to solve a problem. There is a debate concerning whether people use visual

images or some other form of semantic reasoning when they problem solve (see Tye 1991). I am not suggesting that people *always* visualize or *never* use semantic forms of reasoning, or other forms of reasoning, when solving problems. I simply am pointing out that there exists this capacity to scenario visualize in our species as a whole and that, at times, people utilize it to solve problems in innovative ways. In fact, whether one utilizes scenario visualization most likely will depend upon the type of problem with which one is confronted. There are some problems – for example, certain mathematical problems – that can be solved without the use of scenario visualization. Other problems, like spatial relation or depth perception problems, may require scenario visualization. The kinds of problems with which our ancestors were confronted *most likely were of the spatial relation and depth relation types* and so, the capacity to scenario visualize would have been useful for their survival. Our ancestors were not solving math equations; they were negotiating environments primarily with the use of their visual systems.

(2)   I am trying to give an account of how it is that *the human species as a whole, with their visual systems intact*, evolved the ability to solve *vision-related* problems associated with their environments. So, my account skirts the issue of a blind person's capacity to problem solve because such a person does not have an intact visual system, and so s/he does not solve vision-related problems. Blind persons assuredly have the capacity to problem solve in sophisticated and innovative ways. Louis Braille, the man who invented Braille as a means for blind persons to communicate information with the usage of bumps on pieces of paper, is a prime example (see Davidson 1971). However, he obviously could not scenario visualize (maybe he could scenario *tactilize*?). The human species as a whole has evolved the capacity to engage in scenario visualization, even though certain members of our species do not have this capacity because of blindness. This makes sense since humans, in general, are not coded genetically for blindness – they are coded for sight.

(3)   Above, I presented a schematization associated with scenario visualization and the *manufacture* of the javelin. Questions may arise as to whether scenario visualization is at work in the *use* of a javelin by a hunter in some environment,

once it has been manufactured. I have tried to present scenario visualization as representative of a slower, step-by-step mental process that would be going on in someone's head as they were engaged in the actual manufacturing of a javelin. However, it is conceivable that scenario visualization, as a creative response to novelty in an environment, is at work while a hunter is actually using a tool. Situations arise that call upon one to be mentally creative or innovative in some unexpected circumstance. For example, let's say the hunter is running after the antelope with the intention of stopping to launch the javelin and, just before s/he does so, the antelope turns around and actually charges the hunter. In this case, we can imagine the hunter using the javelin to cut, hack, or slice in self-defense. The import of Mithen's cognitive fluidity position is that the information from mental modules has the potential to intermix; the import of my scenario visualization view is that the information can be segregated, integrated, and organized so that one can respond to environmental problems creatively. So, we can envision scenario visualization at work in the manufacture, as well as the use, of a tool such as a javelin.

(4)   Scenario visualization is only one aspect of consciousness. There are several other aspects of consciousness, including intentionality, indexicality, and qualia-based perceptual awareness (experience), to name just a few (see Chalmers 1996; Crick and Koch 1999; Roth 2000). Earlier in this paper, I mentioned Sternberg's (2001) idea that consciousness comprises the ability to coherently form a belief or set up a goal that a human being can ultimately act upon, through selectivity and filtering of information. When one visualizes in order to solve a problem, one must not only have some idea of the environment in which the solution to the problem presents itself, but one must also have some idea of the *goal* to be achieved though solving the problem in that environment. My suggestion is that the aspect of conscious behavior regarding belief/goal formation works with the aspect of conscious behavior regarding scenario visualization in order to solve vision-related problems creatively. How do these two work together? I believe these two aspects of conscious behavior mutually inform one another in a vision-related, problem solving process.

To start with, some visual cue causes one to form a belief regarding some goal to be achieved. The goal to be achieved then causes one to select visual images that *seem to be relevant* to the solution to the problem at hand. I say "seem to be relevant" because, at first, the images are not integrated or organized fully in one's mind. In other words, the solution utilizing the certain selected visual images is not seen clearly. This would be kind of like a *hunch* concerning the relevance of certain visual images to the solution of some problem. The integrative aspect of scenario visualization then goes to work, attempting a variety of possible visual scenarios through manipulating and adjusting the selected visual images. Again, this integration occurs against a backdrop of some kind of environment, since the solution to the problem must be believed to be relevant in some situation. Once a visual scenario comes into view clearly or is clarified in one's mind as being appropriate to solve some problem in an environment, the visual scenario then informs the goal that has been set up at the beginning of the entire process. A solution is then believed to be the accurate one to pursue, and one sets out to actually solve the problem through constructing some tool, devising some plan, etc.

Conclusion

In this paper, I have given an account of how our ancestors evolved a conscious ability I call *scenario visualization* that enabled them to construct novel tools so as to survive and flourish in the ever-changing and complex environments in which they lived. I presented the ideas and arguments put forward by NEPers that the mind evolved certain mental abilities as adaptive responses to environmental pressures in the Pleistocene epoch. In critiquing the NEP approach, I presented Mithen's idea of cognitive fluidity, and agreed with Mithen that cognitive fluidity acts as a necessary condition for novel tool production, but disagreed that cognitive fluidity alone will suffice for such an activity. I then argued that the flexible exchange of information between and among modules *together with* scenario visualization is what accounted for our ancestors' abilities to construct the novel tools needed to survive and

flourish in the environments in which they lived. Finally, through the archeological evidence presented by Mithen and Wynn, I traced the development of the multi-purposed javelin, from its meager beginnings as a stick, in order to illustrate the emergence of scenario visualization.

References

Aboitiz F. 1996. Does bigger mean better? evolutionary determinants of brain size and structure. *Brain Behav. Evol.* 47: 225–245.

Aiello L. 1997. Brain and guts in human evolution: the expensive tissue hypothesis. *Braz. J. Genet.* 20: 141–148.

Aiello L. and Dunbar R. 1993. Neocortex size, group size and the evolution of language. *Curr. Anthropol.* 34: 184–193.

Allman J. 2000. *Evolving Brains.* Scientific American Library, New York.

Arnheim R. 1969. *Visual Thinking.* University of California Press, Berkeley.

Boyd R. and Silk J. 1997. *How Humans Evolved.* Norton, New York.

Byrne R. 1995. *The Thinking Ape: Evolutionary Origins of Intelligence.* Oxford University Press, Oxford.

Byrne R. 2001. Social and technical forms of primate intelligence. In: DeWaal F. (ed.), *Tree of Origin: What Primate Behavior Can Tell Us About Human Social Evolution.* Harvard University Press, Cambridge, MA, pp. 145–172.

Carruthers P. 2002. Human creativity: its evolution, its cognitive basis, and its connections with childhood pretence. *Brit. J. Philos. Sci.* 53: 1–29.

Chalmers D. 1996. *The Conscious Mind: In Search of a Fundamental Theory.* Oxford University Press, New York.

Cosmides L. and Tooby J. 1987. From evolution to behavior: Evolutionary psychology as the missing link. In: Dupre J. (ed.), *The Latest on the Best: Essays on Evolution and Optimality.* MIT Press, Cambridge, MA, pp. 27–36.

Cosmides L. and Tooby J. 1992. The psychological foundations of culture. In: Barkow J., Cosmides L. and Tooby J. (eds), *The Adapted Mind.* Oxford University Press, New York, pp. 19–136.

Cosmides L. and Tooby J. 1994. Origins of domain specificity: The evolution of functional organization. In: Hirschfeld L. and Gelman S. (eds), *Mapping the Mind: Domain Specificity in Cognition and Culture.* Cambridge University Press, Cambridge, UK, pp. 148–165.

Crick F. and Koch C. 1999. The problem of consciousness. In: Damasio A. (eds), *The Scientific*

American Book of the Brain. Scientific American, New York, pp. 311–324.

Daly M. and Wilson M. 1999. Human evolutionary psychology and animal behaviour. *Anim. Behav.* 57: 509–519.

Davidson M. 1971. *Louis Braille: The Boy Who Invented Books for the Blind*. Scholastic, Inc., New York.

Deacon T. 1997. *The Symbolic Species: The Co-evolution of Language and the Brain*. W.W. Norton and Company, New York.

Desimone R. and Ungerleider L. 1989. Neural mechanisms of visual processing in monkeys. In: Boller F. and Grafman J. (eds), *Handbook of Neuropsychology*. Elsevier Publishers, New York, pp. 267–299.

Donald M. 1997. The mind considered from a historical perspective. In: Johnson D. and Erneling C. (eds), *The Future of the Cognitive Revolution*. Oxford University Press, New York, pp. 355–365.

Fodor J. 1998. Review of Steven Mithen's the prehistory of the mind. In: Fodor J. (ed.), *In Critical Condition: Polemical Essays on Cognitive Science and the Philosophy of Mind*. MIT Press, Cambridge, MA, pp. 153–160.

Foley R. 1995. The adaptive legacy of human evolution: a search for the environment of evolutionary adaptedness. *Evol. Anthropol.* 4: 192–203.

Gardenfors P. 2000. *Conceptual Spaces: The Geometry of Thought*. MIT Press, Cambridge, MA.

Gardner H. 1993. *Multiple Intelligences: The Theory in Practice*. Basic Books, New York.

Gardner H. 1999. *Intelligence Reframed: Multiple Intelligences for the 21st Century*. Basic Books, New York.

Gibson K. and Ingold T. (eds) 1994. *Tools, Language and Cognition in Human Evolution*. Cambridge University Press, Cambridge, UK.

Goodale M. *et al.* 1994. Separate neural pathways for the visual analysis of object shape in perception and prehension. *Curr. Biol.* 4: 604–606.

Gray C. 1999. The temporal correlation hypothesis of visual feature integration: still alive and well. *Neuron* 24: 31–47.

Griffin D. 1992. *Animal Minds*. University of Chicago Press, Chicago.

Hirschfeld L. and Gelman S. 1994. Toward a topography of mind: An introduction to domain specificity. In: Hirschfeld L. and Gelman S. (eds), *Mapping the Mind: Domain Specificity in Cognition and Culture*. Cambridge University Press, Cambridge, UK, pp. 3–35.

Isaac G. 1986. Foundation stones: Early artifacts as indicators of activities and abilities. In: Bailey G. and Callow P. (eds), *Stone Age Prehistory*. Cambridge University Press, Cambridge, UK, pp. 221–241.

Julesz B. 1984. Toward an axiomatic theory of preattentive vision. In: Edelman G. (eds), *Dynamic Aspects of Neocortical Function*. Wiley Publishers, New York, pp. 585–612.

Kandel E., Schwartz J. and Jessell T. 2000. *Principles of Neural Science*. McGraw-Hill, New York.

Kanizsa G. 1976. Subjective contours. *Sci. Am.* 234: 48.

Kanizsa G. 1979. *Organization in Vision: Essays on Gestalt Perception*. Praeger, New York.

Kant I. 1929. *Critique of Pure Reason*. N. Kemp-Smith (trans), Macmillan, New York.

Kosslyn S. and Koenig O. 1995. *Wet Mind: The New Cognitive Neuroscience*. The Free Press, New York.

Lakoff G. 1987. *Women, Fire, and Dangerous Things*. University of Chicago Press, Chicago.

Laland K. and Brown G. 2002. *Sense and Nonsense: Evolutionary Perspectives on Human Behaviour*. Oxford University Press, Oxford.

Marten K. and Psarakos S. 1994. Evidence of self-awareness in the bottlenose dolphin (*Tursiops truncates*). In: Parker S. (eds), *Self-awareness In Animals and Humans*. Cambridge University Press, New York, pp. 361–379.

Martin R. 1990. *Primate Origins and Evolution*. Chapman & Hall, London.

Mayer R. 1995. The search for insight: Grappling with Gestalt psychology's unanswered questions. In: Sternberg R. and Davidson J. (eds), *The Nature of Insight*. MIT Press, Cambridge, MA, pp. 3–32.

McNabb J. and Ashton N. 1995. Thoughtful flakers. *Camb. Archeol. J.* 5: 289–301.

Merikle P. and Daneman M. 1998. Psychological investigations of unconscious perception. *J. Consciousness Stud.* 5: 5–18.

Mithen S. 1996. *The Prehistory of the Mind: The Cognitive Origins or Art and Science*. Thames and Hudson, London.

Mithen S. 1998. Handaxes and ice age carvings: Hard evidence for the evolution of consciousness. In: Hameroff S. (eds), *Toward a Science of Consciousness: The Second Tucson Discussions and Debates*. MIT Press, Cambridge, MA, pp. 281–296.

Mithen S. 2001. Archeological theory and theories of cognitive evolution. In: Hodder I. (ed.), *Archeological Theory Today*. Polity Press, Cambridge, UK, pp. 98–121.

Palmer J. and Palmer A. 2002. *Evolutionary Psychology: The Ultimate Origins of Human Behavior*. Allyn and Bacon, Needham Heights, MA.

Parker S. 1996. Apprenticeship in tool-mediated extractive foraging: the origins of imitation, teaching and self-awareness in great apes. *J. Comparative Psychol.* 106: 18–34.

Pearce J. 1997. *Animal Learning and Cognition*. Psychology Press, Erlbaum, Hillsdale, NJ.

Pelegrin J. 1993. A framework for analyzing stone tool manufacture and a tentative application to some early stone industries. In: Berthelet A. and

Chavaillon J. (eds), *The Use of Tools by Human and Non-human Primates*. Clarendon Press, Oxford, pp. 302–314.

Pinker S. 1994. *The Language Instinct*. W. Morrow and Company, New York.

Pinker S. 1997. *How the Mind Works*. W. W. Norton & Company, New York.

Pinker S. 2002. *The Blank Slate: The Modern Denial of Human Nature*. Penguin Books, New York.

Plotkin H. 1997. *Evolution in Mind: An Introduction to Evolutionary Psychology*. Harvard University Press, Cambridge, MA.

Relethford J. 1994. *The Human Species: An Introduction to Evolutionary Anthropology*. Mayfield Publishing, Mountain View, CA.

Rosch E. 1975. Cognitive representations of semantic categories. *J. Exp. Psychol. Gen.* 104: 192–253.

Rosch E. 1981. Prototype classification and logical classification: The two systems. In: Scholnick E. (ed.), *New Trends of Conceptual Representation*. Erlbaum, Hillsdale, NJ, pp. 73–85.

Roth G. 2000. The evolution and ontogeny of consciousness. In: Metzinger T. (ed.), *Neural Correlates of Consciousness*. MIT Press, Cambridge, MA, pp. 77–98.

Samuels R. 1998. Evolutionary psychology and the massive modularity hypothesis. *Brit. J. Philos. Sci.* 49: 575–602.

Scher S. and Rauscher F. 2003. Nature read in truth or flaw. Locating alternatives in evolutionary psychology. In: Scher S. and Rauscher F. (eds), *Evolutionary Psychology: Alternative Approaches*. Kluwer Academic Publishers, Boston, pp. 1–30.

Sekuler R. and Blake R. 2002. *Perception*. McGraw Hill, New York.

Shettleworth S. 2000. Modularity and the evolution of cognition. In: Heyes C. and Huber L. (eds), *The Evolution of Cognition*. MIT Press, Cambridge, MA, pp. 43–60.

Smith S. 1995. Fixation, incubation, and insight in memory and creative thinking. In: Smith S., Ward T. and Finke R. (eds), *The Creative Cognition Approach*. MIT Press, Cambridge, MA, pp. 135–156.

Spelke E. 1991. Physical knowledge in infancy: Reflections on Piaget's theory. In: Carey S. and Gelman S.

(eds), *Epigenesis of Mind: Studies in Biology and Culture*. Erlbaum, Hillsdale, NJ, pp. 131–169.

Sperber D. 1994. The modularity of thought and the epidemiology of representations. In: Hirschfeld L. and Gelman S. (eds), *Mapping the Mind: Domain Specificity in Cognition and Culture*. Cambridge University Press, Cambridge, UK, pp. 39–67.

Stamp Dawkins M. 1993. *Through Our Eyes Only? The Search for Animal Consciousness*. Oxford University Press, Oxford.

Stanford C. 2000. *Significant Others: The Ape–Human Continuum and the Quest for Human Nature*. Basic Books, New York.

Sternberg R. 2001. *Complex Cognition: The Psychology of Human Thought*. Oxford University Press, New York.

Tomasello M. *et al.* 1993. Imitative learning of actions on objects by children, chimpanzees, and enculturated chimpanzees. *Child Dev.* 64: 1688–1705.

Tye M. 1991. *The Imagery Debate*. MIT Press, Cambridge, MA.

van Essen D. and Gallant J. 1994. Neural mechanisms of form and motion processing in the primate visual system. *Neuron* 13: 1–10.

van Essen D. *et al.* 1992. Information processing in the primate visual system: an integrated systems perspective. *Science* 255: 419–423.

Wilson D. 2003. Evolution, morality, and human potential. In: Scher S. and Rauscher F. (eds), *Evolutionary Psychology: Alternative Approaches*. Kluwer Academic Publishers, Boston, pp. 55–70.

Wynn T. 1979. The intelligence of later Acheulian hominids. *Man* 14: 371–391.

Wynn T. 1981. The intelligence of oldowan hominids. *J. Hum. Evol.* 10: 529–541.

Wynn T. 1991. Tools, grammar and the archeology of cognition. *Camb. Archeol. J.* 1: 191–206.

Wynn T. 1993. Layers of thinking in tool behavior. In: Gibson K. and Ingold T. (eds), *Tools, Language and Cognition in Human Evolution*. Cambridge University Press, Cambridge, UK, pp. 389–406.

Wynn T. and McGrew W. 1989. An ape's view of the Oldowan man. *Man* 24: 383–398.

PART XII

DESIGN AND CREATIONISM

Introduction

Evolutionary theory has challenged the design argument for God's existence, essentially because the order and complexity exhibited by living things can be explained through basic naturalistic principles such as variation, inheritance, population increase, struggle for existence, differential survival, differential production, and natural selection. Thus, the explanation of adaption has no need for a god-hypothesis. This does imply that the theory of natural selection is incompatible with God's existence, though its reliance on chance in dictating long-term evolutionary outcome makes natural selection incompatible with the existence of an omniscient agent who exploited the process to produce humans "in his own image." Natural selection is too unreliable a process to be counted on by an omniscient god to produce intelligent life, let alone us.

Since the publication of Darwin's *Origin*, this "no need for the god hypothesis" has caused theists committed to the "inerrancy" (freedom from error) of the Bible to reject the theory of natural selection. In the first selection in this section (chapter 28), Donald Prothero traces the history of what is known as *creationism* and its relationship to American politics up to, and including, the *Kitzmiller v. Dover* trial and consequent ruling against teaching creationism in Pennsylvania public schools. Creationists believe not only that the universe was designed and created by an all-powerful, intelligent being, but they also believe that this idea should be taught alongside evolutionary theory and the rest of science in biology classrooms. Most creationists are anti-evolutionists of one stripe or another, but their arguments against it invariably reveal their misunderstandings and misinterpretations of science.

The proposal that creationism (or its latest variant) be taught as science strikes most biologists and philosophers of biology as impossible for several reasons. First, the creationist's god is beyond the purview of publicly testable empirical knowledge (or, if it is open to empirical examination, such an existence has been decisively disproved). More important, there is no clear, consistent research program underwritten by creationist hypotheses. The reader is encouraged to return to Part I: Basic Principles and Proofs of Darwinism, as well as the Further Reading in Philosophy of Science material at the end of the general introduction, for more on how Darwin's theory makes the role of an omnipotent designer dispensable in science, if it ever had a role to begin with.

Modern-day creationists, who style themselves advocates of *intelligent design* instead of *creationists*, have attempted to resurrect a version of the design argument by claiming that certain biological organism or systems – the bacterial flagellum, the blood clot cascade, for that matter Darwin's favorite example, the eye – are irreducibly complex. Irreducible complexity cannot be achieved by gradual accumulation of small improvements and could therefore only have been created by a far-sighted designer.

In the second selection in this section (chapter 29), intelligent-design creationist Michael Behe argues that the parts and processes associated with the bacterial flagellum are irreducibly complex, leading us to the conclusion that it must have been designed by some mind. Behe uses the analogy to the typical spring-loaded bar mousetrap with its numerous parts: the mouse trap must come together *wholesale*; otherwise, it would not work *piecemeal*. Likewise, there are natural parts and processes that must be brought together all at once (and wholesale) as an irreducibly complex unit through the influence of some designer god.

In our final selection (chapter 30), Kenneth Miller demonstrates how events have overtaken Behe's example, and shown that in this instance he is as mistaken about the irreducible complexity of the bacterial flagellum, as he is about several other cases of alleged irreducible complexity. The bacterial flagellum has been demonstrated to be *reducibly* complex (and, in some cases, simple).

Further Reading

Ayala, F. (2007). *Darwin and intelligent design*. Minneapolis, MN: Fortress Press.

Berra, T. (1990). *Evolution and the myth of creationism*. Stanford: Stanford University Press.

Bridgham, J., Carroll, S., & Thornton, J. (2006). Evolution of hormone-receptor complexity by molecular ex-ploitation. *Science, 312,* 97–101.

Coyne, J. (1996). God in the details. *Nature, 383,* 227–228.

Dembski, W., & McDowell, S. (2008). *Understanding intelligent design: Everything you need to know in plain language*. Eugene, OR: Harvest House Publishers.

Dembski, W., & Ruse, M. (2007). *Debating design: From Darwin to DNA*. Cambridge: Cambridge University Press.

DeRosier, D. (1998). The turn of the screw: The bacterial flagellar motor. *Cell, 93,* 17–20.

Faugy, D., & Farrel, K. (1999). A twisted tale: The origin and evolution of motility and chemotaxis in prokaryotes. *Microbiology, 145,* 279–280.

Hume, D. (1779/1947). *Dialogues concerning natural religion*. Indianapolis: Bobbs-Merrill.

Kitzmiller v. Dover Area School District. 400 F. Supp. 2d 707. Case No. 04cv2688 (M.D. Pa 2005). Available at: http://www.pamd.uscourts.gov/kitzmiller/kitzmiller_3 42. pdf.

Krem, M., & Di Cera, E. (2002). Evolution of enzyme cascades from embryonic development to blood coagulation. *Trends in Biochemical Sciences, 27,* 67–74.

McDonald, J. (2008). A reducibly complex mousetrap. Available at: http://udel.edu/~mcdonald/mousetrap. html.

Murphy, N., & Schloss, J. (2008). Biology and religion. In M. Ruse (ed.), *The Oxford handbook of philosophy of biology* (pp. 545–569). Oxford: Oxford University Press.

National Academy of Sciences (2008). *Science, evoltion, and creationism*. Washington, DC: National Academies Press.

Pennock, R. (2000). *Tower of Babel: The evidence against the new creationism*. Cambridge, MA: MIT Press.

Pennock, R. (ed.) (2001). *Intelligent design creationism and its critics: Philosophical, theological, and scientific perspectives*. Cambridge, MA: MIT Press.

Pennock, R. (2002). Should creationism be taught in public schools? *Science and Education, 11,* 111–133.

Petto, A., & Godfrey, L. (eds.) (2008). *Scientists confront creationism: Intelligent design and beyond*. New York: W.W. Norton.

Rosenberg, A., & McShea, D. (2008). *Philosophy of biology: A contemporary introduction* (Chapter 5). London: Routledge.

Ruse, M. (2003). *Darwin and design: Does evolution have a purpose?* Cambridge, MA: Harvard University Press.

Scott, E. (2005). *Evolution vs. creationism: An introduction*. Berkeley: University of California Press.

Shanks, N. (2004). *God, the devil, and Darwin: A critique of intelligent design theory*. Oxford: Oxford University Press.

Shanks, N., & Joplin, K. (1999). Redundant complexity: A critical analysis of intelligent design in biochemistry. *Philosophy of Science, 66,* 268–298.

Trachtenberg, S., Gilad, R., & Geffen, N. (2003). The bacterial linear motor of *Spiroplasma melliferum* BC3: From single molecules to swimming cells. *Molecular Microbiology 47,* 671–697.

Weber, B. (1999). Irreducible complexity and the problem of biochemical emergence *Biology and Philosophy, 14,* 593–605.

28

Science and Creationism

Donald Prothero

Mything Links

These bits of information from ancient times, which have to do with themes that have supported human life, built civilizations, and informed religions over the millennia, have to do with deep inner problems, inner mysteries, inner thresholds of passage, and if you don't know what the guide-signs are along the way, you have to work it out for yourself.

Joseph Campbell, *The Power of Myth*

Nearly every culture on Earth has some form of a creation story or myth that it uses to explain its place in the universe and its relationships to its god or gods. As Joseph Campbell wrote in *The Power of Myth* (1988), these stories are essential for a culture to understand itself and its role in the cosmos and for individuals to know what their gods and their culture expect of them. At one time, myths served the role of explaining how the world came to be, usually with the subtext that it explained how that culture fit within the universe. In our modern technological scientific age, we tend to scoff at the stories that were believed by the Sumerians, Norse, and Greeks, but in their time, they served both as a metaphor and allegory for their place in the universe and a rational explanation for how things came to be.

Many creation stories have common elements or themes that are universal across cultures and time. They often have elements of birth or eggs in them because these are very powerful symbols of the creation of life in our world. In some versions of the Japanese creation myths, a jumbled mass of elements appeared in the shape of the egg, and later in the story, Izanami gives birth to the gods. In the beginning of one of the Greek myths, the bird Nyx lays an egg that hatches into Eros, the god of love. The shell pieces become Gaia and Uranus. In Iroquois legend, Sky Woman fell from a floating island in the sky because she was pregnant and her husband pushed her out. After she landed, she gave birth to the physical world. There are many Hindu creation stories. In one of them, the god Brahma created the primal waters as the womb for a small seed, which grew into a golden egg. Brahma split it apart and made the heavens from one half and the Earth and all her creatures from the other. The Chinook Indians of the Pacific Northwest were created out of a great egg laid by the Thunderbird. Similar stories of a cosmic egg are known from Chinese, Finnish, Persian, and Samoan mythology.

Many stories often have a mother and father figure that are responsible for the creation. The mother figure is often some form of "Mother

Donald Prothero, *Evolution: What the Fossils Say and Why It Matters* (Columbia, 2007), chapter 2.

Earth," and her fertility is symbolic of the earth's fertility. The Greek creation myths, for example, have the world arising from the mating of the earth goddess Gaia, the sky god Uranus, and their union created the pantheon of Greek gods, who in turn created the physical universe. In Japan, Izanagi and Izanami mated, and the mother goddess Izanami gave birth to three children, Amaterasu, the sun, Tsukiyumi, the moon, and Susano-o, their unruly son. The Australian aborigines believed in the Sun-Mother who created all the animals, plants, and bodies of water at the suggestion of the Father of All Spirits. Primordial parents are found in the myths of many other cultures, including the Egyptians, Cook Islanders, Tahitians, and the Luiseño and Zuni Indians.

Most creation myths, such as the Hebrew, Greek, and Japanese myths mentioned above, as well as the Sumerian–Babylonian myths discussed below, have some form of chaos or nothingness at the beginning that is organized or separated into sky and earth by their gods. Other myths, however, imagine a world that existed before our present world, and one of their gods from the earlier world brings our universe into existence. The Bushmen of Africa, for example, imagined a world where people and animals lived together in peace and harmony. Then The Great Master and Lord of All Life, Kaang, planned a wondrous land above theirs and planted a great tree that spread over it. At the base of the tree, he dug a hole and brought the people and animals up from below. In the Hopi myth, there were past worlds beneath ours. When life became unbearable in those worlds, the people and animals climbed up the pine trees to reach new, unspoiled worlds where they could live. This ladder is endless, so some creatures may still be climbing out of this world and into the next. The Navajo creation myth is similar but instead of climbing pine trees from one world to the next, they climb through a great hollow reed.

The theme of humans breaking some sort of divine edict from the gods and causing pain and suffering by their disobedience is also common. In addition to the Adam and Eve in the Garden of Eden story, there is the Greek story of Pandora, who was given to Epimetheus as a gift from Zeus, along with a box she was not allowed to open. But Zeus also gave her curiosity, so that when she did open the box she released all the sins and troubles of the world. The African Bushmen were told by their gods not to build fire, and when they disobeyed, their peaceful relationships with animals were destroyed forever. According to the Australian Aborigines, the Sun-Mother created the animals, which she demanded must live peacefully together. But envy overcame them, and they began to quarrel. She came back to Earth and gave them a chance to change into any shape they wanted, resulting in the strange combination of animals in Australia. But because the animals had disobeyed the Sun Mother's instructions, she created two humans that would rule over the animals and dominate them.

The theme of a great flood that destroys nearly all of life is common to nearly all mythologies. In addition to the Sumerian story of Ziusudra and Babylonian story of Utnapishtim (described below) and the Hebrew legend of Noah (probably derived from the Sumerian or Babylonian account), the Greeks talked of Deucalion, who survived the great flood and seeded the land with the humans after the floodwaters receded. There are similar flood legends in Norse, Celtic, Indian, Aztec, Chinese, Mayan, Assyrian, Hopi, Romanian, African, Japanese, and Egyptian mythology. Scholars suggest that this may be because most cultures that live near large bodies of water (which nearly all do except those in the mountains) have experienced some catastrophic flood in their distant past. It also wiped out much of their culture and tradition, so that flood achieves legendary status when its story is told generation after generation. Only a few cultures or religions, including the Jainists of India and the Confucianists of China, have no creation myth whatsoever.

This brief thematic summary does not do justice to the details and the imagery of the original myths nor to the power of the language in which they were written. If you have never done so, I strongly recommend that you pick up a book of comparative mythology or examine some of the many texts that are now available on the Internet. Through all this discussion, we have seen how mythologies often reflect universal themes about human existence and about how humans fit into the universe. None of these stories is necessarily "true" or "false" – they are products of their own cultures and were essential to those people in giving their world a context. All humans hunger for an understanding of their origins, so they generate some sort of story to explain it. Once that story has been passed from generation to generation, it acquires its own sort of reality or "truth," and it is important to the members of that culture so that they understand their own role in the world and their relationship to their gods.

As Michael Shermer (1997:30) sums it up, "Does all this mean that the biblical creation and re-creation stories are false? To even ask the question is to miss the point of the myths, as Joseph Campbell (1949, 1982) spent a lifetime making clear. These flood myths have deeper meanings tied to re-creation and renewal. Myths are not about truth. Myths are about the human struggle to deal with the great passages of time and life – birth, death, marriage, the transitions from childhood to adulthood to old age. They meet a need in the psychological or spiritual nature of humans that has absolutely nothing to do with science. To turn a myth into science, or science into a myth, is an insult to myths, an insult to religion, and an insult to science. In attempting to do this, the creationists have missed the significance, meaning and sublime nature of myths. They took a beautiful story of creation and re-creation and ruined it."

The Genesis of Genesis

When on high heaven was not named,
And the earth beneath did not yet bear a name,
And the primeval Apsu, who begat them,
And chaos, Tiamat, the mother of them both
Their waters were mingled together,
And no field was formed, no marsh was to be seen;
When of the gods none had been called into being,
And none bore a name, and no destinies were
   ordained;
Then were created the gods in the midst of heaven,
Lahmu and Lahamu were called into being . . .
Ages increased . . .

Enuma Elish, about 3000 BC

The origin of the Hebrew creation stories in the Bible has been studied for nearly 200 years and is well known and accepted by most Bible scholars. In the 1860s and 1870s, archeologists excavated several ancient Sumerian cities in Mesopotamia (what is now Iraq) and found clay tablets written

in cuneiform. This is the oldest written language on Earth, created with marks in soft clay made by a wedge-shaped stylus. Some of the stories date back at least to 4000 BC, and most were recycled by the mythology of the Akkadian, Babylonian, and Assyrian cultures that replaced the Sumerians in Mesopotamia. The longest and best known of these stories is the *Enuma Elish* (in Babylonian, the first two words of the story, translated "When on high . . ."), which describes a creation epic that bears remarkable similarity to Genesis 1, including a formless void and chaos with gods dividing the waters from the land and naming the creatures. Since the story predates any of the Hebrew creation stories by centuries, there is little doubt that the early Hebrews were influenced by this powerful epic accepted by all Mesopotamian civilizations for over two millennia. Psalm 74 also borrows heavily from the *Enuma Elish*, where Yahweh destroys the Leviathan and splits its head open in an almost word-for-word copy of the way in which Marduk, the chief god of Babylon, splits open the head of Tiamat, the goddess of the ocean.

Another source is *The Epic of Gilgamesh*, which dates to about 2750 BC. The Sumerians had a hero called Ziusudra (called Atrahasis by the Akkadians and Utnapishtim by the Babylonians), who is warned by the earth goddess Ea to build a boat because the god Ellil was tired of the noise and trouble of humanity and planned to wipe them out with a flood. When the floodwaters receded, the boat was grounded on the mountain of Nisir. After Ziusudra's boat was stuck for seven days, he released a dove, which found no resting place and returned. He then released a swallow that also returned, but the raven that was released the next day did not return. Ziusudra then sacrificed to Ea on the top of Mount Nisir. The story is nearly identical to that of Noah's flood, not only in its plot and structure, but also in the details of its phrasing. Only the characters' and gods' names and a few details have been changed to suit the differences between the monotheistic Hebrew culture and the polytheistic cultures of the Sumerians, Akkadians, and Babylonians.

Two centuries of detailed study by scholars has also revealed the way in which the Bible was put together. In its original Hebrew, the Old Testament (especially the first five books, or Pentateuch) show unmistakable signs of different

authors writing different parts, and then someone later patching the whole thing together.
Someone reading a later translation (especially
the outdated King James translation) cannot pick
up these differences easily, but they are obvious to
those who read Hebrew. In high school, I was
troubled by the contradictions between what I
learned in my Presbyterian Sunday School and
what I had learned from science; I decided to
find out about the Bible myself. Not only did
I read numerous books about Biblical scholarship,
but I also learned to read Hebrew, so I could deci-
pher Genesis on my own, making my own judgment
about translations. In college, I also learned ancient
Greek, and I can still read the New Testament in the
original text and recognize when someone is mis-
translating or misinterpreting the original.

To Hebrew scholars, the most obvious signs of
different authorship are their choices of certain
phrases and words, especially the word they use
for God. One source is known as the "J" source,
after *Jahveh*, a common name for God. This
name is also spelled and pronounced "Yahweh"
or "YHWH" for those who dare not speak God's
name (since early written Hebrew had no vowels
or even the modern system of vowel points,
only the consonants are used). This name was
mispronounced and misspelled as "Jehovah" by
later authors. The authors of the J document
were priests of the southern kingdom of Judah,
who wrote sometime between 848 BC and the
Assyrian destruction of Israel in 722 BC. They use
terms such as "Sinai," "Canaanites," and phrases
such as "find favor in the sight of," "call on the
name of," and "bring *out* from the land of
Egypt." The J authors were probably religious
leaders associated with Solomon's temple, very
concerned with delineating the guiding hand of
Jahweh in their history but not so concerned
with the miraculous.

The second main source is known as the "E"
document, after their name for God, *Elohim*,
"powerful ones" in Hebrew. The priests who
composed the E document were interested in different issues, used a different set of phrases, and
can be traced to the northern kingdom of Israel,
sometime between 922 BC and the Assyrian con-
quest in 722 BC. The E authors use such terms as
"Horeb" instead of Sinai, "Amorites" instead of
Canaanites, and the phrase "bring *up* from the land
of Egypt." Most scholars think that the E authors

were Ephraimite priests, who were more interested
in the righteousness that God requires of his
people. When people sinned they must repent.
Moses is the central focus of these accounts,
along with miraculous aspects of their history.

A third source, the "P" source, or Priestly
Code, was apparently written by Aaronid priests
around the time of the Babylonian captivity in
587 BC. It is the youngest of the sources of the
Old Testament. The P source emphasizes the
role of Aaron and diminishes the role of Moses
in the early books of the Bible. This source
frequently uses long lists and is characterized by
long boring interruptions to the narrative and cold
unemotional descriptions. To Hebrew scholars,
the P source is also distinctive in its low, clumsy,
inelegant literary style. The P source views God
as distant and transcendental, acting and com-
municating only through the priesthood. According to P, God is just but also unmerciful, using
brutal, abrupt punishment when laws are broken.

Sometime during the reign of King Josiah
around 622 BC, the Hebrews began combining
these different traditions along with other sources
(such as the "D" source of the Deuteronomic
code). All of these documents date from the
period before Judah was captured, Jerusalem
burned, the Temple destroyed, and the Hebrews
dragged off to captivity by Nebuchadnezzar of
Babylon in 587 BC.

Verse by verse, scholars can tease apart the
way in which each book of the Old Testament
was woven together (see Friedman 1987 or Pelikan
2005). As a result, the Bible is full of internal contradictions that make it impossible for anyone who
reads it closely to take it literally, but only makes
sense in the context of different sources being
blended together. For example, Genesis 1 (largely
from the P source) gives the order of creation as
plants, animals, man, and woman, but Genesis 2
(from the J source) gives it as man, plants, animals,
and woman. According to Genesis 1 : 3–5, on the
first day, God created light, then separated light and
darkness, but according to Genesis 1 : 14–19, the
sun (which separates night and day) wasn't created
until the fourth day.

Genesis 6–7 gives the story of Noah twice, once
from the J source and once from the P source, with
verses from the two sources intermingled so that
they sometimes contradict each other. Genesis
6 : 5–8 are from the J source, but Genesis 6 : 9–22

are from the P source. Then Genesis 7 : 1–5 are from the J source, and Genesis 7 : 6–24 are alternately from the J and P sources every other line or so (Friedman 1987:54). This leads to many contradictions, such as Genesis 7 : 2 (from the J source), saying that Noah took seven pairs of each clean beast in the ark, but Genesis 7 : 8–15 (from the P source), saying he took only one pair of each beast in the ark. In Genesis 7 : 7, Noah and his family finally enter the ark, and then in Genesis 7 : 13 they enter it all over again (the first verse from the J source, the second from the P source). According to Genesis 6 : 4, there were Nephilim (giants) on the earth before the Flood, then Genesis 7 : 21 says that all creatures other than Noah's family and those on the ark were annihilated – but Numbers 13 : 33 says there were Nephilim after the Flood.

Many more examples could be cited, but the basic point is clear: the Bible is a composite of multiple sources that did not always agree on details. This was no problem to the ancient Hebrew culture, which used the Bible for inspiration but was not concerned with literal consistency. It is a big problem for modern fundamentalists (most of whom have never read the Bible in the original Hebrew or Greek, so they are in no position to argue) who believe that every word of the Bible is literally true. Most nonfundamentalist Christians, Catholics, Jews, and Muslims have accepted what scholarship has shown us about the origin of the Bible and use it as a book for understanding their relationship to their God but not as a science textbook or literal account of history. As Joseph Campbell and many other later authors have pointed out, these religious stories are important to believers for their meaning and symbolism and connection to the inner mysteries of life, not as detailed literal accounts of events. Only in our modern scientific age, with its obsession with literalism and detail, have fundamentalists made such a gross error about the spirit and meaning of the Scriptures (see Vawter 1983 and other papers in Frye 1983).

What Is Creationism?

It not infrequently happens that something about the earth, about the sky, about other elements of this world, about the motion and rotation or even the magnitude and distances of the stars, about definite eclipses of the sun and moon, about the passage of years and seasons, about the nature of animals, of fruits, of stones, and of other such things, may be known with the greatest certainty by reasoning or by experience, even by one who is not a Christian. It is too disgraceful and ruinous, though, and greatly to be avoided, that he [the non-Christian] should hear a Christian speaking so idiotically on these matters, and as if in accord with Christian writings, that he might say that he could scarcely keep from laughing when he saw how totally in error they are. In view of this and in keeping it in mind constantly while dealing with the book of Genesis, I have, insofar as I was able, explained in detail and set forth for consideration the meanings of obscure passages, taking care not to affirm rashly some one meaning to the prejudice of another and perhaps better explanation.

St. Augustine, *The Literal Interpretation of Genesis 1 : 19–20*

The United States is home to a unique and peculiar form of religious extremism known as creationism. As a movement, it has almost no following in Canada, Europe, Asia, or most of the rest of the world, but in America it has had a long influence on science education and public understanding of evolution. As a result, most Americans still don't understand or accept the evidence of evolution.

Ironically, the creationist movement is not only a uniquely American phenomenon, but it is also the latest form of backlash against the inevitable forces of change and modernity. For most of the past 2000 years, people did not question the literal accounts of creation in the first books of Genesis. Even as early as AD 426, however, the great Christian philosopher St. Augustine wrote (quoted above) that the Genesis account of creation was an allegory and should not be interpreted literally, as adherence to a literal reading of Genesis might discredit the faith.

As more and more scientific discoveries were made, some of the literalistic readings of the Bible had to be rethought. Once people accepted that the spherical Earth revolved around the sun, it was no longer plausible to think that Joshua had made the sun stand still, that the Earth was flat, or that it was the center of the universe, as described in the Bible. By the mid-1700s, enough facts about nature had accumulated that many educated people doubted the literal accounts of the Bible. During the "French enlightenment" of the mid-1700s, writers such as Diderot, Voltaire, and Rousseau rejected the Church's dogmas, and in 1749, the great naturalist Georges-Louis Leclerc, the Count of Buffon, even speculated that the Earth was 75,000 years old, that life evolved, and that humans and apes were closely related.

By the early 1800s, the idea that Genesis was a literal account of Earth's history was widely questioned by educated people, especially in England, France, and Germany. As a backlash to this widespread skepticism, a number of ministers and naturalists tried to write accounts that reconciled nature with the Bible (the *Bridgewater Treatises*) or tried to use the apparent design and perfection in nature as evidence of a divine Designer (natural theology). But literal belief in Genesis was already widely discredited long before Darwin published his book in 1859.

Darwin, of course, changed the terms of the debate entirely, polarizing the Western world into those accepting evolution and those rejecting it. At first, the argument was intense as the shock of Darwin's ideas began to sink in, but by the time Darwin died in 1882, the fact that life had evolved was no longer controversial in any European scientific or intellectual community. Darwin's ideas had become so respectable when he died that he was buried with honors in the Scientists' Corner of Westminster Abbey, right next to Isaac Newton and many other famous British scientists.

Most American scholars and scientists also came to terms with Darwin by the 1880s or created their own form of compromise between their own religious beliefs and the idea that life had evolved. For example, in 1880, the editor of one American religious weekly estimated that "perhaps a quarter, perhaps a half of the educated ministers in our leading Evangelical denominations" believed "that the story of the creation and the fall of man, told in Genesis, is no more the record of actual occurrences than is the parable of the Prodigal Son" (Numbers 1992:3). At the same time, a skeptical analytical approach known as "Higher Criticism" was being applied to the Bible itself, and scholars (especially in

Germany) were able to show by careful analysis of the original texts and their language that the Old Testament is a composite of several schools of thought in Hebrew history, not the words of Moses and the Prophets.

Higher criticism alarmed the devout Biblical literalists even more than Darwinism and evolution, so in 1878, ministers met in the First Niagara Bible Conference. Beginning in 1895 and concluding by 1910, they had published 90 pamphlets that were known as *The Fundamentals* of their faith (hence the term *fundamentalist*). Most of *The Fundamentals* concerned the miracles of Jesus, his virgin birth, his bodily resurrection, his death on the cross to atone for our sins, and finally, that the Bible is the directly inspired word of God. Fundamentalism was largely a reaction to the "Higher Criticism" of the Bible, and its early proponents were not quite as strongly against evolution, because it was already widely accepted not only by scientists but also by most ministers. A. C. Dixon, the first editor of *The Fundamentals,* wrote that he felt "a repugnance to the idea than an ape or an orang-outang was my ancestor" but was willing "to accept the humiliating fact, if proved" (Numbers 1992:39). Reuben A. Torrey, who edited the last two volumes of *The Fundamentals,* acknowledged "for purely scientific reasons" that a man could "believe thoroughly in the absolute infallibility of the Bible and still be an evolutionist of a certain type" (Numbers 1991:39). Although the early fundamentalists were not happy with evolution, they were willing to live with it; they were not as stridently opposed to the idea as they would be a generation later. More importantly, evolution was accepted by most of the science textbooks of the time, so even if the parents were fundamentalists who rejected evolution, their children accepted it. Even in the conservative Baptist South, evolution was taught without much resistance in many educational institutions (Numbers 1992:40).

Twentieth-Century Creationism

Congress shall make no law respecting an establishment of religion.

First Amendment to the U.S.
Constitution, 1789

"Creation science" ... *is simply not science.*
Judge William Overton,
McLean vs. Arkansas

The first two decades of the twentieth century were a time of global turmoil, with the progressive politics of Presidents Theodore Roosevelt and Woodrow Wilson, the bloodshed of World War I, and the great influenza epidemic of 1918. Then came the "Roaring Twenties" and a national conservative backlash. It was also a "Return to Normalcy" as Warren Harding promised when he won the presidency in 1920. With the conservative backlash came Prohibition. This did nothing to stop alcohol consumption in the United States, but it did make profitable careers for gangsters and moonshiners and the owners of illegal speakeasies. Another conservative movement, however, was the backlash against evolution by the resurgent fundamentalist movement. The movement was led by William Jennings Bryan, one of the most popular and powerful political figures in the United States, who had run for President on the Democratic ticket three times and lost. By the 1920s, however, Bryan was in his sixties, in failing health, and beginning to promote conservative causes that were becoming popular in the 1920s. Bryan campaigned vigorously for laws to outlaw the teaching of evolution. By the end of the 1920s, more than 20 states had debated such laws, and five (Tennessee, Mississippi, Arkansas, Oklahoma, and Florida) had banned or curtailed the teaching of evolution in their public schools. It went so far that the U.S. Senate debated, but never passed, a resolution that banned radio broadcasts favorable to evolution.

Ironically, Bryan himself was not a Biblical literalist. He confided to a friend shortly before his death that he had no objection to evolution, as long as it didn't include man (Numbers 1992:43). He was also less than literal about the meaning of Genesis 1, subscribing to the common "day-age" theory that each "day" in Genesis 1 was actually a long period of geologic time, or "age." Nevertheless, he became the national spokesman for a witch hunt that hounded many biologists out of their jobs in southern universities and destroyed the careers of many other scientists.

The climax of the creationist movement in the 1920s was the infamous Scopes Monkey Trial of 1925, long called the "Trial of the Century" until the O. J. Simpson trial eclipsed it in notoriety. Not only was it a titanic struggle between two of the giants of the time, Bryan and the legendary defense attorney Clarence Darrow, but it was also one of the first trials to be covered live on radio and in newsreels, beginning the modern trend toward celebrity trial journalism. Among the press covering the trial was none other than the famous satirist and essayist H. L. Mencken, who wrote numerous savage columns and editorials for the *Baltimore Sun*, ridiculing the Biblical literalism and backward habits and racism of the South.

The trial itself was originally planned as a publicity stunt by the town fathers of Dayton Tennessee. Anxious to garner attention and to provide a test case to challenge the recently passed Tennessee Butler Act, or "monkey laws" that banned the teaching of evolution, the civic leaders recruited a local high school teacher, John T. Scopes, to be their guinea pig. Scopes volunteered to take time off from teaching gym to teach biology for one day so that he could test the law, although later he admitted that he wasn't sure he had actually taught anything about evolution. He did, however, use the classic textbook, *Hunter's Civic Biology*, which mentioned evolution prominently. Once the trial was underway, Darrow's defense plans collapsed because Judge John T. Raulston would not allow the testimony of any of the expert scientific witnesses that Darrow had brought. The judge ruled that the case only concerned whether Scopes had broken the law, and witnesses challenging the law itself were irrelevant. In desperation, Darrow turned this defeat into one of the greatest legal tour de forces in history. He baited Bryan into taking the stand as an expert witness on the Bible. Under a blistering cross-examination (vividly portrayed in the famous play and movie *Inherit the Wind*), Darrow got Bryan to admit to many of the logical absurdities of a literalistic interpretation of the Bible. Bryan could not explain how Joshua had gotten the sun (and therefore the earth) to stand still or where Cain had gotten his wife (when there were only supposed to be four people on earth, Adam and Eve and Cain and Abel), or many other problems with literal interpretation of the Bible. Even more devastating, Bryan admitted under oath that the "days" of

Genesis were not 24-hour days but could be long geological "ages," a revelation that shocked most of his fundamentalist followers. Soon, fundamentalism and the "Monkey Law" itself were subject to ridicule. Bryan died a week after the trial, which occurred during a torrid heat wave and was very stressful for his failing health. More importantly, the fundamentalist monkey laws had taken a bad beating in the press and in the public eye, and most Americans were embarrassed that our nation had been portrayed as so scientifically backward.

The trial itself, however, was inconclusive. The judge mistakenly levied a $100 fine on Scopes that was supposed to be levied by the jury, so his verdict was thrown out on this technicality. As a result, it was not possible to take the case to higher courts and have a verdict examined on appeal. Scopes never had to pay the judge's fine. Eventually, Scopes went to college and became a successful oil geologist. And the Tennessee monkey law stayed on the books for decades and was not declared unconstitutional until 1968. In that year, Susan Epperson, a young biology teacher in Arkansas, got her case heard before the Supreme Court, who then struck down all laws forbidding the teaching of evolution – 43 years after the Scopes trial!

By 1929 the Great Depression had changed the mood of the country, and creationism was no longer in the forefront. Fundamentalists were more concerned about issues like sex education, and no further legal cases challenging the monkey laws were filed, although the old laws remained on the books. Instead, the fundamentalists focused their attention on making sure evolution vanished from the biology textbooks, which it did shortly after the Scopes trial (due to pressure by a few determined creationists on textbook publishers and on local school boards). Creationism and evolution existed in an uneasy truce until the Soviets shook America with the launch of Sputnik in 1957. Then Americans were shocked to discover how far behind our science and technology had fallen, and by 1958, the Republican Congress and Eisenhower administration poured big money back into scientific research and science education. Science also became more and more respected by the American public, especially after the technological advances of World War II, the atomic bomb, and eventually

the space race. Federal funding from scientific research went from 0.02 percent of the gross national product during the Hoover administration (1929–1933) to 1.5 percent of the GNP by 1960. With this new emphasis on science came biology textbooks that reflected the new ideas in evolution represented by the Neo-Darwinian synthesis of the 1940s and 1950s. The new generation of science textbook authors was not as cowed by creationist pressure to dilute or eliminate coverage of evolution when the nation was facing a crisis in science education, in large part caused by the lackadaisical coverage of science in public schools.

The newly Darwinized biology textbooks caused the creationists to rise out of their inactivity. In 1961, Henry Morris and John C. Whitcomb published *The Genesis Flood*, which represented a whole new approach of creationists to not only discredit evolution but also geology. By 1963, they had founded the Creation Research Society near San Diego, followed by the Institute for Creation Research (ICR), which has been the main base of operations for fundamentalist creationists ever since. Through their books, debate appearances, and public speeches, they raised awareness of their literalist views to a new level, although they had no impact on the community of science yet.

Creationists, however, still faced one major hurdle: the Constitution and the legal system. By 1968, the Supreme Court had struck down all the old antievolutionary "monkey laws," and the creationists no longer had the backing of conservative legislatures as they had in the 1920s. Because they could no longer legally exclude evolution from the classroom, they tried a tactic of demanding equal time for their ideas. However, in court case after court case, they were turned back, because their ideas were clearly religious in origin, with no scientific content, and the Constitution prohibits the government from establishing a state religion or favoring one religion over another. Led by fundamentalist lawyer Wendell Bird, the creationists changed tactics yet again. They began calling their ideas "scientific creationism" and claimed that their ideas were as scientific as evolution and deserved equal time in science classes. Of course, this is simply "bait and switch," because the creationist literature is full of references to God and the Bible.

They even published two editions of the same textbook, one of which was labeled "Public School Edition" and deleted the overt references to God and Bible, but otherwise the text was the same.

Their main spokesmen seemed to be talking out of both sides of their mouths. In public, they argue that "creation science" is good science, but when speaking to a religious audience, they let their fundamentalist beliefs show. For example, Henry Morris (1972, preface) writes, "Creation, on the other hand, is a scientific theory which does fit all the facts of true science, as well as God's revelation in the Holy Scriptures." On page 58 of the same book, he writes "we conclude that special creation theory is the best theory, strictly on the scientific merits of the case." Yet the ICR's principal debater and spokesman, Duane Gish, wrote (1973:40) "we cannot discover by scientific investigations anything about the creative processes used by the Creator," and (p. 8), "creation is, of course, unproven and unprovable by the methods of experimental science. Neither can it qualify as a scientific theory."

The climax came when Arkansas and Louisiana passed bills that mandated "equal time" for creationism in science classes, and these laws were promptly challenged in federal court. The ACLU, challenging the Arkansas law, put not only distinguished scientists and philosophers of science on the witness stand, but also a variety of ministers and theologians, as well as parents of children in the school district. In fact, the lead plaintiff challenging the law was a minister, the Reverend Bill McLean of Little Rock. The witnesses showed example after example of how there was no difference between "creation science" and religion, and how the nature of science forbids any belief system that twists the facts to fit their preexisting conclusions. The creationist case was further hampered by the fact that they had no credible scientific witnesses to bring to the stand. One of their star witnesses, the maverick British astrophysicist Chandra N. Wickramasinghe, openly scoffed at the idea of creation science. On January 5, 1982, Judge William R. Overton gave his ruling on *McLean vs. Arkansas*. Judge Overton saw through the thin disguise of creation science and ruled that the Arkansas law "was simply and purely an effort to introduce the Biblical version of creation into the public school curricula." According to Overton, the law "left no

doubt that the major effect of the Act is the advancement of particular religious beliefs." The law that required balanced treatment "lacks legitimate educational value because 'creation science' as defined in that section is simply not science." In 1985, Federal Judge Adrian Duplantier ruled in a summary judgment (thus not requiring even a trial or witnesses) that Louisiana's equal time law was also unconstitutional. In the 1987 *Edwards vs. Aguillard* case, the United States Supreme Court, on a 7–2 vote, upheld the decisions of the lower courts, and the creationists lost their last legal battle in this round.

For about ten years, the creationists stayed away from the courts and stopped trying to force their way into education by legal means. Instead, they focused their energies on school boards and textbook publishers. Every week, those of us in the front lines of the creationism battle heard news of another school district that was under pressure to teach creationism or put antievolutionary stickers in biology textbooks. Most of these battles were eventually decided against the creationists, but they are a determined and well-funded minority that has nothing but time, energy, and money to push their cause, while most scientists are too busy doing legitimate research to pay attention to the problem.

"Intelligent Design" – or "Breathtaking Inanity"?

The evidence, so far at least and laws of Nature aside, does not require a Designer. Maybe there is one hiding, maddeningly unwilling to be revealed. But amid much elegance and precision, the details of life and the Universe also exhibit haphazard, jury-rigged arrangements and much poor planning. What shall we make of this: an edifice abandoned early in construction by the architect?

Carl Sagan, *Pale Blue Dot*

The label "scientific creationist" was seen as a fraud by Judges Overton and Duplantier and by the Supreme Court. So the creationists resorted to a new strategy: "intelligent design" (commonly abbreviated ID). In order to find a way to make their ideas constitutional and legal, they had to try to eliminate any signs of religion from their dogmas, not simply dress up Biblical ideas as "scientific creationism." In the 1990s, a new generation of creationists came up with a different strategy that focused on the apparent "design" in nature, arguing that it requires some sort of "intelligent designer." Led by Berkeley lawyer Phillip Johnson, Lehigh biochemist Michael Behe, and former Baylor professor William Dembski, they published a number of books that promoted their views. They argued that nature was full of things that not only showed intelligent design but also were "irreducibly complex" and could not have evolved by chance. They pointed to a number of examples, such as the flagellum and the eye, which they believed could not be explained by chance events or by gradual evolution.

In most ways, their arguments are recycled from over two centuries ago, when many devout naturalists ascribed to the school of thought known as "natural theology." (Ministers were often naturalists back then because they had lots of time for studying nature as evidence of God's handiwork, and there were no professional scientists.) The most famous advocate of natural theology was the Reverend William Paley, who in 1802 wrote *Natural Theology,* the classic treatment of the subject. His most famous metaphor is the "watchmaker" analogy. If you were to find a watch on a beach, you would immediately recognize that it was "intricately contrived" and infer that it had a watchmaker. To Paley, the "intricate contrivances" of nature were evidence that there was a Divine Watchmaker, namely God.

In its day, the natural theology school of thought was very influential, and Darwin himself knew Paley's book almost by heart. Yet the basic arguments had been discredited even before the time of Paley. In 1779, the Scottish philosopher David Hume published *Dialogues Concerning Natural Religion,* which demolished the whole argument from design. Using dialogues between characters to voice different points of view, Hume puts the standard natural theology arguments in the mouth of a character called Cleanthes, then he tears them down in the words of a skeptic named Philo. Philo notes that pointing to the design in nature is a faulty analogy because we have no standard to compare our world to, and it is possible to imagine a world much better designed than the one in which we live. Even if we concede that the world looks

designed, it does not follow that the designer is the Judaeo-Christian God. It could have been the god of another religion or culture, or the work of a committee of gods, or a juvenile god who makes mistakes. Jews and Christians simply assumed that if there was a Designer, it must be their God, but there is no compelling evidence to show that it wasn't some other god.

More importantly, evidence was already in existence in Hume's and Paley's times that did not reflect well on the Divine Designer. For all the examples of beauty or symmetry in nature, one could also point to many examples where nature is poorly designed or jury-rigged so that it just barely works, or where nature shows astonishing cruelty that does not reflect a caring, compassionate God. Stephen Jay Gould pointed to examples such as *Lampsilis,* the freshwater clam that sticks a brood sac full of eggs out of its shell that looks vaguely like a fish. It's not a very good fishing lure, but it's good enough to get fish to bite it and transfer the eggs to their gills, where they are passed on to another generation. Similarly, the anglerfish has a crude fringe on the tip of a spine above its eyes that looks vaguely fishlike, and when flicked around, is just good enough to lure prey close enough to be gulped down. Again, it's not a very good facsimile of a fish, but it's good enough to lure prey within reach. Gould's favorite example is the panda's "thumb". Pandas, like most cats, dogs, bears, and members of the order Carnivora, have all five fingers united in a paw, yet pandas are almost the only carnivore that eats plants (bamboo). Consequently, pandas have modified a wrist bone, the radial sesamoid, into a crude thumb-like device, which is not jointed and not very flexible or strong, but just strong enough to allow pandas to strip off the leaves from the bamboo as they eat. Once again, a clumsy, poorly designed jury-rigged device – good enough to allow the survival of pandas (although we're now driving them to extinction due to habitat destruction in China), but evidence of a very clumsy Designer at best.

Examples of poor or at least very puzzling design can be accumulated endlessly. Many cave-dwelling fish and salamanders have the rudiments of eyes, but they are completely blind. If God specially created these creatures to live in totally dark caves, why bother to give them nonfunctional eyes in the first place? Even more peculiar is the course of the recurrent laryngeal nerve, which connects the brain to the larynx and allows us to speak. In mammals, this nerve avoids the direct route between brain and throat and instead descends into the chest, loops around the aorta near the heart, then returns to the larynx. That makes it seven times longer than it needs to be! For an animal like the giraffe, it traverses the entire neck twice, so it is fifteen feet long (fourteen feet of which are unnecessary!). Not only is this design wasteful, but it also makes a animal more susceptible to injury. Of course, the bizarre pathway of this nerve makes perfect sense in evolutionary terms. In fish and early mammal embryos, the precursor of the recurrent laryngeal nerve attached to the sixth gill arch, deep in the neck and body region. Fish still retain this pattern, but during later human embryology, the gill arches are modified into the tissues of our throat region and pharynx. Parts of the old fish-like circulatory system were rearranged, so the aorta (also part of the sixth gill arch) moved back into the chest, taking the recurrent laryngeal nerve (looped around it) backward as well.

In fact, the more one looks at nature, the more one finds examples of clumsy or jury-rigged design because, unlike a Divine Designer, evolution does not require perfection. Any solution that ensures the survival of an organism long enough to breed is sufficient. We humans are classic examples of an organism not optimally designed to our current lifestyles. Our backs and our feet are not well adapted to walking upright, as those of us who suffer with back and foot pain know. Our knees are poorly constructed and easily damaged, as those who have had knee surgery can attest. Our eyes are designed backward, with several layers of cells and tissues blocking and distorting the light hitting the retina in the back of our eye before the light finally reaches the photoreceptor cells on the very bottom layer. We have vestigial organs, such as our tiny tailbones, tonsils, and appendix, the latter two of which no longer perform an important function but can become infected and be deadly to us. These only make sense if they were inherited from ancestors who had functioning versions of these organs. Our genome is full of nonfunctional DNA, including inactive pseudogenes that were active in our ancestors. Humans,

like most primates, cannot make vitamin C and must get it from their diet. We still carry all the genes for making vitamin C but no longer use them, probably because our primate ancestors got it from their fruit-rich diets instead. Finally, ask any ID advocate: why did God give men nonfunctional nipples?

ID creationists may want to think twice before pointing to God's handiwork as evidence of a benevolent God, because it is full of examples of not only poor or incompetent design but also outright cruelty. The most famous example is the family of wasps known as the Ichneumonidae, which consist of about 3300 species who all reproduce in a distinctive way. The female wasp stings a prey animal with her ovipositor and lays her eggs inside the paralyzed prey. After the eggs hatch, the larvae slowly eat the living prey animal from the inside, destroying the less essential parts first and only eating the essential parts (and killing the host) at the very end, when they are ready to hatch out of its dead shell. (Shades of the creepy extraterrestrials in the movie *Alien*.) The Victorians were horrified when this example became well known and were at a loss as to how to square this fact of nature with their idea of a benevolent God who looks after the tiny sparrow and cares about all of His creation. As Charles Darwin wrote,

> I cannot persuade myself that a beneficent and omnipotent God would have designedly created parasitic wasps with the express intention of their feeding within the living bodies of Caterpillars.

But that has long been a problem for those who would believe in a God who is all-knowing and all-powerful. If so, why does He allow innocents to suffer and die? Why can't He stop great natural disasters? (As I write this on December 26, 2005, we mark the one-year anniversary of the Indian Ocean tsunami, which killed almost a million innocent people). This is the classic "problem of pain" (*theodicy*) that has always tortured Christian apologetics, but many skeptics consider it good evidence against a Divine Designer who watches His handiwork closely. As Darwin himself put it (in an 1856 letter to Huxley): "What a book a devil's chaplain might write on the clumsy, wasteful, blundering low and horridly cruel works of nature!"

Reading the ID creationists closely, you find that they don't offer any new scientific ideas or a true alternative theory of life competing with evolution. All they argue is that some parts of nature seem too complex for them to imagine an evolutionary explanation.

This is the classic "god of the gaps" approach: concede to science that which it has already explained but reserve to supernatural forces that which hasn't been explained – yet. Back in the Middle Ages, people thought that God made the heavens run and the stars and planets move until Copernicus, Galileo, Newton, and Kepler showed that it could all be explained by natural laws and processes without divine intervention. So theology retreated from explaining that part of nature, and it has been retreating ever since. But nature is always full of things that we have not explained. Explaining the unexplained is the goal of science – to continue solving those unsolved mysteries, not to stop and throw up our hands and say, "Oh, well, I can't think of an explanation now, so God must have done it." As Michael Shermer (2005:182) points out, the ID approach is actually quite arrogant: if the ID creationist can't think of a natural explanation, then they are asserting that *no* scientist can either, and the problem cannot be solved. Needless to say, this is not how science operates: giving up on hypotheses and testable explanations, shrugging our shoulders, and going home while saying "God works in mysterious ways."

ID creationists actually concede that they don't really have a real alternative theory to evolution. Leading ID creationist Paul Nelson said at a meeting at Biola College in Los Angeles in 2004: "Easily the biggest challenge facing the ID community is to develop a full-fledged theory of biological design. We don't have such a theory right now, and that's a problem. Without a theory, it's very hard to know where to direct your research focus. Right now, we've got a bag of powerful intuitions, and a handful of notions such as 'irreducible complexity' and 'specified complexity' – but, as yet, no general theory of biological design." Nor is their "research program" legitimate. During cross-examination in the Dover, Pennsylvania, ID creationism trial, Behe was forced to confess, "There are no peer-reviewed articles by anyone advocating for intelligent design supported by pertinent experiments or calculations which provide detailed rigorous accounts of how intelligent design of any biological systems occurred." Behe also conceded that there were no peer-reviewed articles supporting some of the other claims that systems (such as the blood-clotting cascade, the immune system, and the bacterial flagellum) were irreducibly complex or intelligently designed. Their entire literature consists of books and articles published by their own supporters, or for the general trade book market, where there are no scientific standards of peer review. (The one exception I'm aware of is discussed later in this chapter.)

But all this talk about intelligent design is actually a smokescreen for what is still fundamentally a religious dogma. For public consumption, the ID advocates may say that the designer need not be the Judaeo-Christian God but could also be an alien or some other supernatural entity. Dembski claims that "scientific creationism has prior religious commitments whereas intelligent design does not." But in reality, they are nearly all evangelical Christian men, who clearly have used intelligent design as a smokescreen for their real agenda: get religion into science classrooms and evolution out – or weaken it, at least. In public, they try to hide these religious convictions, but when speaking to their fellow fundamentalists, they let their true colors show. In an article in the Christian magazine *Touchstone,* Dembski wrote, "Intelligent design is just the Logos theology of John's Gospel restated in the idiom of information theory." In 1999, Dembski wrote, "Any view of the sciences that leaves Christ out of the picture must be seen as fundamentally deficient. ... The conceptual soundness of a scientific theory cannot be maintained apart from Christ." On February 6, 2000, Dembski told the National Religious Broadcasters: "Intelligent Design opens the whole possibility of us being created in the image of a benevolent God. ... The job of apologetics is to clear the ground, to clear obstacles that prevent people from coming to the knowledge of Christ. ... And if there's anything that I think has blocked the growth of Christ as the free reign of the Spirit and people accepting the Scripture and Jesus Christ, it is the Darwinian naturalistic view." At the same conference, Phillip Johnson said, "Christians in the twentieth century have been playing defense.

They've been fighting a defensive war to defend what they have, to defend as much of it as they can. It never turns the tide. What we're trying to do is something entirely different. We're trying to go into enemy territory, their very center, and blow up the ammunition dump. What is their ammunition dump in this metaphor? It is their version of creation." In 1996, Johnson said, "This isn't really, and never has been, a debate about science. . . . It's about religion and philosophy." One of the ID creationist authors, Jonathan Wells, is a follower of the Reverend Sun-Myung Moon and his Unification Church (which is vehemently antievolutionary). As Wells wrote, "When Father chose me (along with about a dozen other seminary graduates) to enter a PhD program in 1978, I welcomed opportunity to prepare myself for battle."

Ironically, most ID creationists accept some microevolutionary change and conventional geology and the great age of the earth, and regard the "young-earth" literalist creationists of the ICR or Answers in Genesis as irrelevant dinosaurs, relicts of the past. In 2005, Dembski actually debated the dean of the old-guard creationists, Henry Morris, where he said, "Thus, in its relation to Christianity, intelligent design should be viewed as a ground-clearing operation that gets rid of the intellectual rubbish that for generations has kept Christianity from receiving serious consideration."

Even though the ID creationists pretend to be dispassionately following the truth, when you look closely at their internal documents, it is clear that they are waging outright warfare on science by whatever dirty tactics and PR techniques that are necessary. Brown and Alston (2007) in their book *Flock of Dodos: Behind Modern Creationism, Intelligent Design, and the Easter Bunny* detail some of the more dishonest activities of the Discovery Institute, and print in full the infamous "Wedge Document" of the ID creationists, which details their devious political and PR strategy to force their viewpoints on the American scientific community and educational system. As Brown and Alston summarize, the Discovery Institute "is willing to mischaracterize the results achieved by real scientists in order to achieve short-lived propaganda victories, and it is willing to continue to do so even after these real scientists object and

even after it has apologized and promised to stop doing so. Above all, it is willing to cloak its true socio-political goals behind a consciously-crafted veil of dispassionate scientific inquiry, even while denouncing science itself. If the Discovery Institute tells a lie, it does so in order to advance the Truth. Because the Discovery Institute fights for morality, it is above morality. Indeed, the intent of the Discovery Institute is simple enough. Con men are rarely complicated" (pp. 136–137).

If the words of the ID creationists were not evidence enough, we can always heed the warning of "Deep Throat" (in *All the President's Men*): "Follow the money." The ID movement is largely based at the Center for the Renewal of Science and Culture (CRC), a part of the Discovery Institute in Seattle. The CRC receives most of its funding from right-wing evangelical and religious organizations and from rich individuals and foundations whose expressed goals are to promote evangelical Christianity. These include $750,000 from the Ahmanson Foundation, whose executor, Howard Ahmanson Jr., said that his goal was "the total integration of biblical law into our lives." The McClellan Institute gave $450,000 to promote "the infallibility of the Scripture"; they give grants to organizations "committed to furthering the Kingdom of Christ." The Stewardship Foundation gives $200,000 a year, and their goal is "to contribute to the propagation of the Christian Gospel by evangelical and missionary work." Most of the 22 organizations funding the CRC were politically and religiously conservative, according to the *New York Times*. The *Times* also reported that the CRC received $4.1 million in 2003 and spends about $50,000 to $60,000 a year on about 50 researchers. That money buys a lot of air time on radio and TV, funding for their advocates to speak and debate around the country and publish their books, get hearings at various conservative school boards – and file lawsuits promoting their ideas.

By 2005, ID creationism had reached a peak in publicity, when they made the cover of *Time* magazine and got the endorsement of President Bush. They also got their ideas heard by the conservative Kansas State Board of Education (which endorsed them) and tried to push their ideas onto the Dover, Pennsylvania, school board (which also tried to follow them until sued by the parents in the district). As I write these words, however, they have just suffered a great defeat in the case of *Kitzmiller et alia vs. the Dover Area School District.* Federal Judge John E. Jones III, a traditional Christian appointed by President Bush in 2002 (and *not* a liberal activist judge), saw through their smokescreen and ruled that ID creationism was clearly an unconstitutional establishment of a particular religion in public schools. His 139-page ruling was very thorough and detailed. Judge Jones castigated the evangelical Christian school board that rammed through the creationist curriculum and drew the lawsuit by the parents of the students. In his words, "The breathtaking inanity of the board's decision is evident when considered against the factual backdrop which has now been fully revealed through the trial. The students, parents, and teachers of the Dover Area School District deserved better than to be dragged into this legal maelstrom, with its resulting utter waste of monetary and personal resources." Judge Jones was particularly irritated by the hypocrisy of the ID creationists, who attempt to sound secular when the Constitution is involved, but crowed about their religious motives when not in court. "The citizens of the Dover area were poorly served by the members of the board who voted for the intelligent design policy. It is ironic that several of these individuals who so staunchly and proudly touted their religious convictions in public would time and again lie to cover their tracks and disguise the real purpose behind the intelligent design policy." And in another passage, "We find that the secular purposes claimed by the board amount to a pretext for the board's real purpose, which was to promote religion in the public school classroom." Still later he wrote, "Any asserted secular purposes by the board are a sham and are merely secondary to a religious objective."

The judge also pointed out that the ID advocates had tried to paint evolution as atheism, which was preposterous. "Both defendants and many of the leading proponents of intelligent design make a bedrock assumption which is utterly false. Their presupposition is that evolutionary theory is antithetical to a belief in the existence of a supreme being and to religion in general." Finally, the judge was puzzled by the fact that there was no real theory behind intelligent design creationism, just criticisms of evolutionary biology and a vague god of the gaps idea. In Judge Jones' words, "Defendants' asserted secular purpose of improving science education is belied by the fact that most if not all of the board members who voted in favor of the biology curriculum change concealed that they still do not know, or nor have they ever known, precisely what intelligent design is."

Brown and Alston (2007) dissect the absurdities of the Dover trial. Their first chapter gives a detailed account of the trial, and uses the court transcripts and the creationists' own words to expose their lying and dishonesty. Brown and Alston quote extensively from the confused and convoluted testimony of William Buckingham, the creationist school board chairman who openly promoted his religious motivations before trial, then lied under oath repeatedly to cover his tracks, apparently at the instructions of the lawyers from the Discovery Institute. As Brown and Alston sum it up, "to know William Buckingham is to know the millions of our fellow Americans who are ignorant not only of the theory they'd like to discredit, but also of the pseudo-theory with which they'd like to replace it; who, knowing full well that they lack the basic data to make a decision between the two, do so anyway and loudly at that; and who lie through their teeth when asked exactly what it is that motivates them to do these sorts of things in the first place. William Buckingham lied because he believed it was necessary to do so in order to preserve the truth as he saw it – that literalized Christianity is the one true religion, and that Darwinism is its greatest threat" (p. 26).

Most analysts believe that the Dover verdict is the death knell for future attempts by ID creationists to win victories by legal means, because in most cases courts follow precedents established by other courts (especially if they are thorough and well-reasoned). Ironically, the point is largely moot in Dover, Pennsylvania, because in November 2005, the citizens of Dover (embarrassed by all the negative publicity) voted the conservatives off the school board and voted in a new school board that was opposed to teaching intelligent design in its schools. Naturally, this new school board did not wish to appeal the judge's ruling, but applauded it – but they were still stuck with the legal bills that the folly of the old school board had generated.

The Monkey Business of Creationism

Creation is, of course, unproven and unprovable by the methods of experimental science. Neither can it qualify as a scientific theory.

Duane Gish, 1973

Creation isn't a theory. The fact that God created the universe is not a theory – it's true. However, some of the details are not specifically nailed down in Scripture. Some issues – such as creation, a global Flood, and a young age for the earth – are determined by Scripture, so they are not theories. My understanding from Scripture is that the universe is in the order of 6,000 years old. Once that has been determined by Scripture, it is a starting point that we build theories upon.

Kurt Wise, 1995

Ultimately, creationism has nothing to do with science, except that the creationists want to replace a valid scientific idea with their own religious dogmas. It is all about politics and power and promoting their cherished ideas whatever the cost. Creationists don't do normal science, don't publish their antievolutionary ideas in peer-reviewed scientific journals, don't present their results at legitimate scientific meetings, and more importantly, don't even begin to follow the basic precept of science: *there is no final truth, and all ideas must be subject to testing and falsification.* Creationists have their conclusions already determined. The ICR even makes their members swear a loyalty oath that predetermines their conclusions. No real scientist would ever do this, since in real science, the conclusions must remain tentative and subject to change. Creationists will do whatever it takes to twist and jumble and distort the evidence to support their case. Indeed, the term "creation science" is an oxymoron, a contradiction in terms, like "jumbo shrimp." Creationists are not really doing science as long as their conclusions are predetermined, and they are unwilling to test and falsify their conclusions. The quotes from Duane Gish above and cited earlier in the chapter clearly confess this.

As Shermer (1997:131) pointed out, the creationists have much in common with the Neo-Nazi Jew-hating Holocaust deniers, who refuse to acknowledge that millions of Jews were killed by the Nazis. Like the creationists, Holocaust deniers pretend to be legitimate objective scholars and deny their underlying motives in public, but in private they reveal their true anti-Semitic hatreds that drive them to distort and deny the truth. The principal strategy of Holocaust deniers is to find small errors in scholarship of historians (or scientists, in the case of creationists) and imply the entire field is wrong, as if scholars never disagreed or made mistakes. Holocaust deniers often quote other people out of context (Nazis, Jews, other Holocaust scholars) to make them seem as if they are supporting the deniers' position; creationists do the same to evolutionists' publications. The existence of a debate over details is used by the Holocaust deniers to suggest that the Holocaust didn't happen, or the scholars can't get their stories straight, and creationists do the same to the legitimate scientific debate among evolutionary biologists. However, as Shermer says, the Holocaust deniers can at least be partially right in that the number of Jews killed may be revised, but the creationists cannot. Once you introduce supernaturalism to the debate, it is no longer scientific.

Because they have lost every battle in the courts, creationists resort to other tactics: pressuring school boards, intimidating textbook publishers, harassing people who oppose them, and disguising their religious motives by such flimsy ruses as intelligent design. Because their unscientific ideas could never pass peer review in scientific journals, nor make it into the university curriculum, they publish their own books and journals, and create their own educational institutions to reflect their dogmas. Because their ideas would not withstand the scrutiny of peer review in scientific meetings, they seldom attend real scientific meetings but preach to the choir instead.

The exceptions to these statements prove the rule. In August 2005, an ID creationist article on the "Cambrian explosion" appeared in the obscure *Journal of the Biological Society of Washington*. According to reports, the peer reviews were scathing and recommended rejection of the article, but the editor had creationist sympathies and let it be published anyway. Once the rest of the editorial board and the Smithsonian scientists became aware of what had been slipped past them, they repudiated the article, and the editor resigned. To my knowledge, this is the only openly creationist paper that has ever appeared in a legitimate scientific peer-reviewed journal – and only because the editor was sympathetic to their cause and violated journal policy by overruling his reviewers. The other papers published by Fritz, Baumgardner, Austin, and creationist "flood geologists" don't appear in peer-reviewed scientific journals, only in the creationists' own publications. Any of their writings that do appear in a legitimate peer-reviewed journal concern some minor issue (such as the polystrate trees of Yellowstone, or the fossil concentrations in some places), and nowhere does the author reveal his creationist agenda in the research.

Likewise, creationists do not present their arguments at legitimate professional meetings of respected scientific societies; they use stealth tactics instead. Forrest and Gross (2004) describe the sneaky efforts of an ID creationist, Paul Chien, to organize a conference in China, ostensibly about the amazing Precambrian and Cambrian fossils that have been found there. But when respected scientists, such as Dr. David Bottjer of the University of Southern California and Dr. Nigel Hughes of University of California Riverside, arrived, they found that the meeting was funded by the Discovery Institute and full of ID creationist speakers. The whole conference was a deliberate ruse to get the papers of legitimate scientists published alongside those of creationists and to lend creationists some respectability.

Whenever they do try to engage the scientific community, creationists do so through a debate format. At first, this seems like a fair strategy, because for many fields we have a long tradition of using debate to explore evidence and clarify ideas. But, in fact, the debate format does nothing to sort out the evolution/creation dispute, except to resolve who has better rhetorical and debating skills. Creationists are very skilled at this, since they do it all the time and have a lot of practice. By contrast, scientists never actually engage in a true formal debate (complete with a pro and con position, moderator, rebuttals, etc.) at scientific meetings. In addition, creationists

dictate the terms of the debate by constantly attacking their evolutionist opponent with one charge after another, jumping from astronomy to thermodynamics to paleontology to biology to anthropology. The scientist opposing the creationist debater cannot possibly answer all of the misconceptions and distortions of complex concepts that they have introduced in the short debate format, because they can't teach an audience the actual science as fast as the creationists can distort it. When the evolutionist debater tries to go on the offensive, the creationist quickly dodges the question and continually tries to make his (mostly religious) audience believe that they must believe the creationist or become an atheist. When a scientist with good debating skills (especially one with religious convictions who can't be called an atheist) pins them down, creationists crumble because their knowledge of their favorite scientific subjects is superficial and learned by rote, so they really don't understand what they are talking about. But their skill in debating is such that they seldom get pinned down or rattled for very long. Most scientists won't even bother to debate them, because it's a no-win situation; everyone who attends the debate has already made up their minds. In addition, most scientists are poorly trained at debating, and we don't want to treat them as scientific peers (they aren't) and dignify their arguments with the pretense of a debate. Plus, we all have much better things to do, such as real scientific research. Consequently, creationists taunt scientists and claim that they are afraid to defend evolution.

Stephen Jay Gould said it best,

> Debate is an art form. It is about the winning of arguments. It is not about the discovery of truth. There are certain rules and procedures to debate that really have nothing to do with establishing fact – which they are very good at. Some of those rules are: never say anything positive about your own position because it can be attacked, but chip away at what appear to be the weaknesses in your opponent's position. They are good at that. I don't think I could beat the creationists at debate. I can tie them. But in courtrooms they are terrible, because in courtrooms you cannot give speeches. In a courtroom you have to answer direct questions about the positive status of your belief. We destroyed

them in Arkansas. On the second day of the two-week trial, we had our victory party. (from a 1985 Caltech lecture, quoted in Shermer 1997: 153)

My own experience with debating creationists was truly eye-opening. In October 1983, I was asked by Stanley Weinberg, head of the Committees of Correspondence (predecessor of the National Center for Science Education, which battles creationism) to represent downstate Illinois for the Committee. I had responsibility for fighting the creationists over the entire state except Chicago, all while I was doing research and teaching a full load of geology classes at Knox College in Galesburg, Illinois. Luckily, for most of my term it was "all quiet on the downstate front," except for when ICR's top debater Duane Gish made a tour through the area. He was invited to a debate at Purdue University, just across the Illinois border in West Lafayette, Indiana. When no faculty at Purdue would take him on, I agreed to their invitation, just for the experience and to say I'd done it once.

First, though, I talked with people who had debated (and beaten him) before. A week before the debate, I went to see Gish give an unopposed lecture at the University of Illinois in Champaign-Urbana. At an open lecture with no opponent, he gives his standard canned speech and slides. Since attendance was free, it drew a crowd of hundreds of bright, skeptical college students who hissed and booed and heckled him regularly at his outrageous misstatements and distortions. For me, it was valuable, because I saw his slides and talk in advance, and I'd been told that he was a robot. He never changes a line of his memorized script or the slide sequence, never acknowledges his opponent, and never realizes that his arguments have just been demolished. (In 1995, I saw Gish debate again, and his patter had not changed in 12 years, except that the slides looked noticeably faded.) I knew that I was going to get the first and third half-hours of our first two hours (in a four-hour debate!), so I prepared to attack and discredit his positions before he even got to them. Sure enough, he never noticed that I had done so and did not change a line of his standard litany, even though its credibility was already demolished. I was even tempted to steal his inevitable lame jokes and

use them first to see if he noticed and was forced to change. I realized, however, that putting a picture of a chimpanzee on the screen and saying (as he does in every lecture), "How did that picture of my grandson get in there?" wasn't much of a laugh line for a 29-year-old unmarried evolutionist.

More startling than this, however, was the behavior of the creationists who organized the debate. They bused in hundreds of people from all the nearby churches, yet discouraged others from attending by charging heavily for Purdue students, who have a lot better things to do on campus on a Saturday night that don't cost a lot of money. As a result, the audience was already 95 percent creationists when I arrived after driving for hours with my five students from Illinois who rode with me. During introductions, the organizers had the gall to say, "We couldn't get Stephen Jay Gould or Niles Eldredge to debate Dr. Gish, so we got . . ." implying that I was some unknown sucker from the boonies of Illinois. Of course, this is their way of taunting evolutionists by suggesting that Gould and Eldredge are afraid of Gish, but they never considered the possibility that it was an insult to me as a scientist. Then, after I had demolished Gish during the first two hours (many people came up to me and said so, and said they had been converted from creationism), they went to a question-and-answer format. After the break, they handed me a stack of 3 × 5 inch index cards with questions from the audience, and they put questions like "What are your religious beliefs?" and "Are you a sexual pervert?" and "Are you going to hell?" at the top. Clearly, the debate was never about science from the beginning, and the mostly fundamentalist audience didn't come to hear about science, but only to cheer their champion and to pity the soul of the poor damned evolutionist. Gish, of course, shuffled the cards until he found sympathetic questions, which he used to add points that he had not discussed in the first two hours.

Even though I beat Gish badly, I decided not to debate him again because it was a waste of time preaching to the converted, and I didn't want to continually dignify their position. But in 2002, I was invited to take part in a panel debating evolution on public television in Los Angeles. Two creationists from ICR were the opponents, and I spent the entire debate canceling out their outrageous lies and distortions. Luckily, they made the mistake of mentioning the fossil record, so I had a huge advantage over them. I'd learned from past experience never to let their misstatements go unchallenged, so I interrupted and cut them off as soon as they said their lies. My debating partner was a lawyer from the ACLU, and he ended up winning the debate by calmly pointing out again and again the simple fact that creationism is religion and forbidden from public school science classes by the Constitution.

What startles you most about the creationist debaters is that they never learn anything new or come up with different arguments. They repeat the same old, tired lines over and over again like a mantra, seemingly unaware or unwilling to admit that their argument has long been discredited. For example, nearly every creationist debater will mention the second law of thermodynamics and argue that complex systems like the earth and life cannot evolve, because the second law seems to say that everything in nature is running down and losing energy, not getting more complex. But that's NOT what the second law says; every creationist has heard this but refuses to acknowledge it. The second law *only applies to closed systems,* like a sealed jar of heated gases that gradually cools down and loses energy. But the earth is *not* a closed system – it continually gets new energy from the sun, and this (through photosynthesis) is what powers life and makes it possible for life to become more complex and evolve. It seems odd that the creationists continue to misuse the second law of thermodynamics when they have been corrected over and over again, but the reason is simple: it sounds impressive to their audience with limited science education, and if a snow job works, you stay with it.

Gish is particularly dishonest in this regard. If he is beaten in a debate in one city and forced to admit that an argument is not true, he will still use the same invalid argument the next night in front of a different audience, since they didn't see him recant the argument the previous night (Arthur 1996; Petto 2005). Gish has been repeatedly caught in lies and deliberate deceptions (Arthur 1996; see www.holysmoke.org/gish.htm), yet he refuses to change even one line of his deceptive and discredited ideas. How honest or truthful can a debater be if he cynically uses an argument he knows has been proven wrong on the next unsuspecting audience?

Likewise, creationists will use spurious arguments from probability to claim that evolution cannot occur, again counting on their followers to be impressed with math and statistics. Creationists will cite the improbability that a monkey could type the works of Shakespeare as an analogy for evolution building complex systems by random chance. Gish's favorite analogy (borrowed from maverick astronomer Fred Hoyle) concerns the improbability that a hurricane blowing through a junkyard could assemble a 707. But these analogies are completely off base. Evolution is not "random chance" but a process whereby natural selection weeds out unfavorable variations and greatly improves the likelihood of events. A better analogy is a monkey with a word processor, whose program (like your spell checker) automatically deletes or fixes mistakes, so that even by typing random keys, the monkey will eventually assemble a recognizable string of words. Richard Dawkins (1986, 1996) has provided many interesting examples and computer models that show just how easily this can be done.

Besides, as anyone who really understands probability knows, you can't make this kind of argument after the fact. If you do so, then *any complex sequence of events is extremely improbable, even though they actually occur.* A good analogy is the one I used in the Gish debate. I asked the audience of several hundred to estimate the probability *after the fact* that all of the events that had happened in their lives would actually happen, and the probability that among all those unlikely events, they would all end up in this room at this particular moment. Naturally, the improbability of this event is enormous. I pointed out to the audience that by Gish's probability arguments, they could not exist!

Most of the standard shopworn creationist arguments are debunked elsewhere, so I will not discuss them here. Suffice it to say that if a real scientist never learned anything new in 50 years of research and never changed their position once they'd been proven wrong, they wouldn't last long in the scientific community.

References

Alters, Brian and Sandra Alters. 2001. *Defending Evolution*. Sudbury, Mass.: Jones and Bartlett.

Arthur, Joyce. 1996. Creationism: Bad science or immoral pseudoscience? (an expose of creationist Dr. Duane Gish). Available at: http://mypage.direct.ca/w/writer/gish.html

Berra, Tim. 1990. *Evolution and the Myth of Creationism*. Stanford, Calif.: Stanford University Press.

Brockman, John (ed.). 2006. *Intelligent Thought: Science Versus the Intelligent Design Movement*. New York: Vintage.

Brown, B. and J. P. Alston. 2007. *Flock of Dodos: Behind Modern Creationism, Intelligent Design, and the Easter Bunny*. Cambridge, England: Cambridge House.

Campbell, Joseph. 1949. *The Hero with a Thousand Faces*. Washington DC: Bollingen Foundation.

Dawkins, R. 1986. *The Blind Watchmaker*. New York: W.W. Norton.

Dawkins, R. 1996. *Climbing Mount Improbable*. New York: W. W. Norton.

Dembski, William. 1999. *The Design Inference: Eliminating Chance Through Small Probabilities*. Cambridge: Cambridge University Press.

Eldredge, Niles. 1982. *The Monkey Business: A Scientist Looks at Creationism*. New York: Pocket Books.

Eldredge, Niles. 2000. *The Triumph of Evolution and the Failure of Creationism*. New York: W. H. Freeman.

Forrest, B. and P. R. Gross. 2004. *Creationism's Trojan Horse: The Wedge of Intelligent Design*. New York: Oxford University Press.

Franz, Marie-Louise von. 1972. *Creation Myths*. Zurich, Switzerland: Spring.

Friedman, Richard. 1987. *Who Wrote the Bible?* New York: Harper & Row.

Frye, Roland Mushat (ed.). 1983. *Is God a Creationist? The Religious Case Against Creation-Science*. New York: Charles Scribner.

Futuyma, Douglas. 1983. *Science on Trial: The Case for Evolution*. New York: Pantheon.

Gish, Duane. 1973. *Evolution? The Fossils Say No!* Bethany House Publishing.

Godfrey, Laurie (ed.). 1983. *Scientists Confront Creationism*. New York: W. W. Norton.

Gould, Stephen. 1992. *The Panda's Thumb: More Reflections in Natural History*. New York: W. W. Norton.

Graves, Robert and Raphael Patai. 1963. *Hebrew Myths: The Book of Genesis*. New York: McGraw-Hill.

Heidel, Alexander. 1942. *The Babylonian Genesis*. Chicago: University of Chicago Press.

Heidel, Alexander. 1946. *The Gilgamesh Epic and Old Testament Parallels*. Chicago: University of Chicago Press.

Humes, Edward. 2007. *Monkey Girl: Evolution, Education, Religion, and the Battle for America's Soul*. New York: Ecco.

Isaak, Mark. 2006. *The Counter-Creationism Handbook*. Berkeley: University of California Press.

Kitcher, Philip. 1982. *Abusing Science: The Case Against Creationism*. Cambridge, Mass.: MIT Press.

Larson, Edward. 1985. *Trial and Error: The American Controversy Over Creation and Evolution*. New York: Oxford University Press.

McGowan, Chris. 1984. *In the Beginning: A Scientist Shows Why the Creationists Are Wrong*. Buffalo, NY: Prometheus.

Miller, Kenneth. 1999. *Finding Darwin's God: A Scientist's Search for Common Ground Between God and Evolution*. New York: Harper Collins.

Morris, Henry. 1972. The Remarkable Birth of Planet Earth. Bethany House Publishing.

New York Times. 2005. Politicized Scholars Put Evolution on the Defensive. By Jodi Wilgoren, August 21.

Numbers, Ronald. 1992. *The Creationists: The Evolution of Scientific Creationism*. New York: Knopf.

Olasky, Marvin and John Perry. 2005. *Monkey Business: The True Story of the Scopes Trial*. New York: B&H.

Pelikan, Jaroslav. 2005. *Whose Bible is it? A History of the Scriptures Through the Ages*. New York: Viking.

Pennock, Robert. 1999. *Tower of Babel: The Evidence Against the New Creationism*. Cambridge, Mass.: MIT Press.

Perakh, M. 2004. *Unintelligent Design*. Buffalo, NY: Prometheus.

Pigliucci, M. 2002. *Denying Evolution: Creationism, Scientism, and the Nature of Science*. Sunderland, Mass.: Sinauer.

Ruse, Michael. 1982. *Darwinism Defended*. New York: Addison-Wesley.

Ruse, Michael. 1988. *But is it Science? The Philosophical Questions in the Creation/Evolution Controversy*. Buffalo, NY: Prometheus.

Ruse, Michael. 2003. *Darwin and Design: Does Evolution Have a Purpose?* Cambridge, Mass.: Harvard University Press.

Ruse, Michael. 2005. *The Evolution-Creation Struggle*. Cambridge, Mass.: Harvard University Press.

Sagan, Carl. 1997. *Pale Blue Dot*. New York: Ballantine Books. University Press.

Sarna, Nahum. 1966. *Understanding Genesis: The Heritage of Biblical Israel*. New York: Schocken.

Scott, E. C. 2005. *Evolution vs. Creationism: An Introduction*. Berkeley: University of California Press.

Shanks, N. 2004. *God, the Devil, and Darwin: A Critique of Intelligent Design Theory*. Oxford: Oxford University Press.

Shermer, Michael. 1997. *Why People Believe Weird Things: Pseudoscience, Superstition, and Other*

Confusions of Our Time. New York: W.H. Freeman Press.

Shermer, Michael. 2004. *The Science of Good and Evil: Why People Cheat, Gossip, Care, Share, and Follow the Golden Rule.* New York: Henry Holt.

Shermer, Michael. 2006. *Why Darwin Matters: Evolution and the Case Against Intelligent Design.* New York: Henry Holt/Times Books.

Shulman, Seth. 2007. *Undermining Science: Suppression and Distortion in the Bush Administration.* Berkeley: University of California Press.

Smith, Cameron M. and Charles Sullivan. 2007. *The Top Ten Myths About Evolution.* Amherst, NY: Prometheus.

Smith, Homer. 1952. *Man and His Gods.* New York: Little, Brown and Company.

Kurt Wise. 1995. Fossil expert says . . . think weird! *Creation* 18(1): 42–44.

Young, M. and T. Edis (eds.). 2005. *Why Intelligent Design Fails: A Scientific Critique of the New Creationism.* Piscataway, NJ: Rutgers University Press.

29

Irreducible Complexity:
Obstacle to Darwinian Evolution

Michael J. Behe

A Sketch of the Intelligent Design Hypothesis

In his seminal work *On the Origin of Species*, Darwin hoped to explain what no one had been able to explain before – how the variety and complexity of the living world might have been produced by simple natural laws. His idea for doing so was, of course, the theory of evolution by natural selection. In a nutshell, Darwin saw that there was variety in all species: For example, some members of a species are bigger than others, some faster, some brighter in color. He knew that not all organisms that are born will survive to reproduce, simply because there is not enough food to sustain them all. So Darwin reasoned that the ones whose chance variation gives them an edge in the struggle for life would tend to survive and leave offspring. If the variation could be inherited, then over time the characteristics of the species would change, and over great periods of time, perhaps great changes could occur.

It was an elegant idea, and many scientists of the time quickly saw that it could explain many things about biology. However, there remained an important reason for reserving judgment about whether it could actually account for all of biology: the basis of life was as yet unknown. In Darwin's day, atoms and molecules were still theoretical constructs – no one was sure if such things actually existed. Many scientists of Darwin's era took the cell to be a simple glob of protoplasm, something like a microscopic piece of Jell-O. Thus the intricate molecular basis of life was utterly unknown to Darwin and his contemporaries.

In the past hundred years, science has learned much more about the cell and, especially in the past fifty years, much about the molecular basis of life. The discoveries of the double helical structure of DNA, the genetic code, the complicated, irregular structure of proteins, and much else have given us a greater appreciation for the elaborate structures that are necessary to sustain life. Indeed, we have seen that the cell is run by machines – literally, machines made of molecules. There are molecular machines that enable the cell to move, machines that empower it to transport nutrients, machines that allow it to defend itself.

In light of the enormous progress made by science since Darwin first proposed his theory, it is reasonable to ask if the theory still seems to be a good explanation for life. In *Darwin's Black Box: The Biochemical Challenge to Evolution* (Behe 1996), I argued that it is not. The main difficulty

Michael Behe, "Irreducible Complexity: Obstacle to Darwinian Evolution," in W. Dembski and M. Ruse (eds.) *Debating Design: From Darwin to DNA Black Box: The Biochemical Challenge to Evolution* (Cambridge, 2004), pp. 352–370.

for Darwinian mechanisms is that many systems in the cell are what I termed "irreducibly complex." I defined an irreducibly complex system as: a single system that is necessarily composed of several well-matched, interacting parts that contribute to the basic function, and where the removal of any one of the parts causes the system to effectively cease functioning (Behe 2001). As an example from everyday life of an irreducibly complex system, I pointed to a mechanical mousetrap such as one finds in a hardware store. Typically, such traps have a number of parts: a spring, a wooden platform, a hammer, and other pieces. If one removes a piece from the trap, it can't catch mice. Without the spring, or hammer, or any of the other pieces, one doesn't have a trap that works half as well as it used to, or a quarter as well; one has a broken mousetrap, which doesn't work at all.

Irreducibly complex systems seem very difficult to fit into a Darwinian framework, for a reason insisted upon by Darwin himself. In the *Origin*, Darwin wrote that "[i]f it could be demonstrated that any complex organ existed which could not possibly have been formed by numerous, successive, slight modifications, my theory would absolutely break down. But I can find out no such case" (Darwin 1859, 158). Here Darwin was emphasizing that his was a gradual theory. Natural selection had to improve systems by tiny steps, over a long period of time, because if things improved too rapidly, or in large steps, then it would begin to look as if something other than natural selection were driving the process. However, it is hard to see how something like a mousetrap could arise gradually by something akin to a Darwinian process. For example, a spring by itself, or a platform by itself, would not catch mice, and adding a piece to the first non-functioning piece wouldn't make a trap either. So it appears that irreducibly complex biological systems would present a considerable obstacle to Darwinian evolution.

The question then becomes, are there any irreducibly complex systems in the cell? Are there any irreducibly complex molecular machines? Yes, there are many. In *Darwin's Black Box*, I discussed several biochemical systems as examples of irreducible complexity: the eukaryotic cilium, the intracellular transport system, and more. Here I will just briefly describe the bacterial flagellum

(DeRosier 1998; Shapiro 1995), since its structure makes the difficulty for Darwinian evolution easy to see. The flagellum can be thought of as an outboard motor that bacteria use to swim. It was the first truly rotary structure discovered in nature. It consists of a long filamentous tail that acts as a propeller; when it is spun, it pushes against the liquid medium and can propel the bacterium forward. The propeller is attached to the drive shaft indirectly through something called the hook region, which acts as a universal joint. The drive shaft is attached to the motor, which uses a flow of acid or sodium ions from the outside to the inside of the cell to power rotation. Just as an outboard motor has to be kept stationary on a motorboat while the propeller turns, there are proteins that act as a stator structure to keep the flagellum in place. Other proteins act as bushings to permit the drive shaft to pass through the bacterial membrane. Studies have shown that thirty to forty proteins are required to produce a functioning flagellum in the cell. About half of the proteins are components of the finished structure, while the others are necessary for the construction of the flagellum. In the absence of almost any of the proteins – in the absence of the parts that act as the propeller, drive shaft, hook, and so forth – no functioning flagellum is built.

As with the mousetrap, it is quite difficult to see how Darwin's gradualistic process of natural selection sifting random mutations could produce the bacterial flagellum, since many pieces are required before its function appears. A hook by itself, or a driveshaft by itself, will not act as a propulsive device. But the situation is actually much worse than it appears from this cursory description, for several reasons. First, there is associated with the functioning of the flagellum an intricate control system, which tells the flagellum when to rotate, when to stop, and sometimes when to reverse itself and rotate in the opposite direction. This allows the bacterium to swim toward or away from an appropriate signal, rather than in a random direction that could much more easily take it the wrong way. Thus the problem of accounting for the origin of the flagellum is not limited to the flagellum itself but extends to associated control systems as well.

Second, a more subtle problem is how the parts assemble themselves into a whole. The analogy to an outboard motor fails in one respect: an outboard motor is generally assembled under the direction of a human – an intelligent agent who can specify which parts are attached to which other parts. The information for assembling a bacterial flagellum, however (or, indeed, for assembling any biomolecular machine), resides in the component proteins of the structure itself. Recent work shows that the assembly process for a flagellum is exceedingly elegant and intricate (Yonekura *et al.* 2000). If that assembly information is absent from the proteins, then no flagellum is produced. Thus, even if we had a hypothetical cell in which proteins homologous to all of the parts of the flagellum were present (perhaps performing jobs other than propulsion) but were missing the information on how to assemble themselves into a flagellum, we would still not get the structure. The problem of irreducibility would remain.

Because of such considerations, I have concluded that Darwinian processes are not promising explanations for many biochemical systems in the cell. Instead, I have noted that, if one looks at the interactions of the components of the flagellum, or cilium, or other irreducibly complex cellular system, they look like they were designed – purposely designed by an intelligent agent. The features of the systems that indicate design are the same ones that stymie Darwinian explanations: the specific interaction of multiple components to accomplish a function that is beyond the individual components. The logical structure of the argument to design is a simple inductive one: whenever we see such highly specific interactions in our everyday world, whether in a mousetrap or elsewhere, we unfailingly find that the systems were intentionally arranged – that they were designed. Now we find systems of similar complexity in the cell. Since no other explanation has successfully addressed them, I argue that we should extend the induction to subsume molecular machines, and hypothesize that they were purposely designed.

Misconceptions about What a Hypothesis of Design Entails

The hypothesis of Intelligent Design (ID) is quite controversial, mostly because of its philosophical and theological overtones, and in the years since *Darwin's Black Box* was published a number of scientists and philosophers have tried to refute its main argument. I have found these rebuttals to be unpersuasive, at best. Quite the opposite, I think that some putative counterexamples to design are unintentionally instructive. Not only do they fail to make their case for the sufficiency of natural selection, they show clearly the obstacle that irreducible complexity poses to Darwinism. They also show that Darwinists have great trouble recognizing problems with their own theory. I will examine two of those counterexamples in detail a little later in this chapter. Before I do, however, I will first address a few common misconceptions that surround the biochemical design argument.

First of all, it is important to understand that a hypothesis of Intelligent Design has no quarrel with evolution per se – that is, evolution understood simply as descent with modification, but leaving the mechanism open. After all, a designer may have chosen to work that way. Rather than common descent, the focus of ID is on the *mechanism* of evolution – how did all this happen, by natural selection or by purposeful Intelligent Design?

A second point that is often overlooked but should be emphasized is that Intelligent Design can happily coexist with even a large degree of natural selection. Antibiotic and pesticide resistance, antifreeze proteins in fish and plants, and more may indeed be explained by a Darwinian mechanism. The critical claim of ID is not that natural selection doesn't explain *anything*, but that it doesn't explain *everything*.

My book, *Darwin's Black Box*, in which I flesh out the design argument, has been widely discussed in many publications. Although many issues have been raised, I think the general reaction of scientists to the design argument is well and succinctly summarized in the recent book *The Way of the Cell*, published by Oxford University Press and authored by the Colorado State University biochemist Franklin Harold. Citing my book, Harold writes, "We should reject, as a matter of principle, the substitution of intelligent design for the dialogue of chance and necessity (Behe 1996); but we must concede that there are presently no detailed Darwinian accounts of the evolution of any biochemical system, only a variety of wishful speculations" (Harold 2001, 205).

Let me emphasize, in reverse order, Harold's two points. First, as other reviewers of my book have done,[1] Harold acknowledges that Darwinists have no real explanation for the enormous complexity of the cell, only hand-waving speculations, more colloquially known as "just-so stories." I had claimed essentially the same thing six years earlier in *Darwin's Black Box* and encountered fierce resistance – mostly from internet fans of Darwinism who claimed that, why, there were hundreds or thousands of research papers describing the Darwinian evolution of irreducibly complex biochemical systems, and who set up web sites to document them.[2]

As a sufficient response to such claims, I will simply rely on Harold's statement quoted here, as well as the other reviewers who agree that there is a dearth of Darwinian explanations. After all, if prominent scientists who are no fans of Intelligent Design agree that the systems remain unexplained, then that should settle the matter. Let me pause, however, to note that I find this an astonishing admission for a theory that has dominated biology for so long. That Darwinian theory has borne so little fruit in explaining the molecular basis of life – despite its long reign as the fundamental theory of biology – strongly suggests that it is not the right framework for understanding the origin of the complexity of life.

Harold's second point is that there is some principle that forbids us from investigating Intelligent Design, even though design is an obvious idea that quickly pops into your mind when you see a drawing of the flagellum or other complex biochemical system. What principle is that? He never spells it out, but I think the principle probably boils down to this: design appears to point strongly beyond nature. It has philosophical and theological implications, and that makes many people uncomfortable. Because they think that science should avoid a theory that points so strongly beyond nature, they want to rule out intelligent design from the start.

I completely disagree with that view and find it fainthearted. I think science should follow the evidence wherever it seems to lead. That is the only way to make progress. Furthermore, not only Intelligent Design, but *any* theory that purports to explain how life occurred will have

philosophical and theological implications. For example, the Oxford biologist Richard Dawkins has famously said that "Darwin made it possible to be an intellectually-fulfilled atheist" (Dawkins 1986, 6). A little less famously, Kenneth Miller has written that "[God] used evolution as the tool to set us free" (Miller 1999, 253). Stuart Kauffman, a leading complexity theorist, thinks Darwinism cannot explain all of biology: "Darwinism is not enough. . . . [N]atural selection cannot be the sole source of order we see in the world" (Kauffman 1995, viii). But Kauffman thinks that his theory will somehow show that we are "at home in the universe." The point, then, is that all theories of origins carry philosophical and theological implications. There is no way to avoid them in an explanation of life.

Another source of difficulty for some people concerns the question, how could biochemical systems have been designed? A common misconception is that designed systems would have to be created from scratch in a puff of smoke. But that isn't necessarily so. The design process may have been much more subtle. In fact, it may have contravened no natural laws at all. Let's consider just one possibility. Suppose the designer is indeed God, as most people would suspect. Well, then, as Kenneth Miller points out in his book, *Finding Darwin's God*:

> The indeterminate nature of quantum events would allow a clever and subtle God to influence events in ways that are profound, but scientifically undetectable to us. Those events could include the appearance of mutations . . . and even the survival of individual cells and organisms affected by the chance processes of radioactive decay. (Miller 1999, 241)

Miller doesn't think that guidance is necessary in evolution, but if it were (as I believe), then a route would be open for a subtle God to design life without overriding natural law. If quantum events such as radioactive decay are not governed by causal laws, then it breaks no law of nature to influence such events. As a theist like Miller, that seems perfectly possible to me. I would add, however, that such a process would amount to Intelligent Design, not Darwinian evolution. Further, while we might not be able to detect quantum manipulations, we may nevertheless be able to conclude confidently that the final structure was designed.

Misconceptions Concerning Supposed Ways around the Irreducibility of Biochemical Systems

Consider a hypothetical example where proteins homologous to all of the parts of an irreducibly complex molecular machine first had other individual functions in the cell. Might the irreducible system then have been put together from individual components that originally worked on their own, as some Darwinists have proposed? Unfortunately, this picture greatly oversimplifies the difficulty, as I discussed in *Darwin's Black Box* (Behe 1996, 53). Here analogies to mousetraps break down somewhat, because the parts of a molecular system have to find each other automatically in the cell. They can't be arranged by an intelligent agent, as a mousetrap is. In order to find each other in the cell, interacting parts have to have their surfaces shaped so that they are very closely matched to each other, as pictured in Figure 29.1. Originally, however, the individually acting components would not have had complementary surfaces. So all of the interacting surfaces of all of the components would first have to be adjusted before they could function together. And only then would the new function of the composite

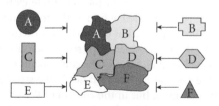

Figure 29.1 The parts of an irreducibly complex molecular machine must have surfaces that are closely matched to each other to allow specific binding. This drawing emphasizes that even if individually acting proteins homologous to parts of a complex originally had separate functions, their surfaces would not be complementary to each other. Thus the problem of irreducibility remains even if the separate parts originally had individual functions. (The blocked arrows indicate that the original protein shapes are not suitable to bind other proteins in the molecular machine.)

system appear. Thus, I emphasize strongly, *the problem of irreducibility remains, even if individual proteins homologous to system components separately and originally had their own functions.*

Another area where one has to be careful is in noticing that some systems that have extra or redundant components may have an irreducibly complex *core*. For example, a car with four spark plugs might get by with three or two, but it certainly can't get by with none. Rat traps often have two springs, to give them extra strength. The trap can still work if one spring is removed, but it can't work if both springs are removed. Thus in trying to imagine the origin of a rat trap by Darwinian means, we still have all the problems we had with a mousetrap. A cellular example of redundancy is the hugely complex eukaryotic cilium, which contains about 250 distinct protein parts (Dutcher 1995). The cilium has multiple copies of a number of components, including multiple microtubules and dynein arms. Yet a working cilium needs at least one copy of each in order to work, as I pictured in my book (Behe 1996, 60). Thus, like the rat trap's, its gradual Darwinian production remains quite difficult to envision. Kenneth Miller has pointed to the redundancy of the cilium as a counterexample to my claim of its irreducibility (Miller 1999, 140–5). But redundancy only delays irreducibility; it does not eliminate it.

Finally, rather than showing how their theory could handle the obstacle, some Darwinists are hoping to get around irreducible complexity by verbal tap dancing. At a debate between proponents and opponents of Intelligent Design sponsored by the American Museum of Natural History in April 2002, Kenneth Miller actually claimed (the transcript is available at the web site of the National Center for Science Education) that a mousetrap isn't irreducibly complex because subsets of a mousetrap, and even each individual part, could still "function" on their own. The holding bar of a mousetrap, Miller observed, could be used as *a toothpick*, so it still has a "function" outside the mousetrap. Any of the parts of the trap could be used as a paperweight, he continued, so they all have "functions." And since any object that has mass can be a paperweight, then any part of anything has a function of its own. *Presto*, there is no such thing as irreducible complexity! Thus the acute problem for gradualism that any child can see in systems like the mousetrap is smoothly explained away.

Of course, the facile explanation rests on a transparent fallacy, a brazen equivocation. Miller uses the word "function" in two different senses. Recall that the definition of irreducible complexity notes that removal of a part "causes the *system* to effectively cease functioning." Without saying so, in his exposition Miller shifts the focus from the separate function of the intact *system* itself to the question of whether we can find a different use (or "function") for some of the *parts*. However, if one removes a part from the mousetrap that I have pictured, it can no longer catch mice. The *system* has indeed effectively ceased functioning, so the *system* is irreducibly complex, just as I have written. What's more, the functions that Miller glibly assigns to the parts – paperweight, toothpick, key chain, and so forth – have little or nothing to do with the function of the system – catching mice (unlike the mousetrap series proposed by John McDonald, to be discussed later) – so they give us no clue as to how the system's function could arise gradually. Miller has explained precisely nothing.

With the problem of the mousetrap behind him, Miller then moved on to the bacterial flagellum – and again resorted to the same fallacy. If nothing else, one has to admire the breathtaking audacity of verbally trying to turn another severe problem for Darwinism into an advantage. In recent years, it has been shown that the bacterial flagellum is an even more sophisticated system than had been thought. Not only does it act as a rotary propulsion device, it also contains within itself an elegant mechanism used to transport the proteins that make up the outer portion of the machine from the inside of the cell to the outside (Aizawa 1996). Without blinking, Miller asserted that the flagellum is not irreducibly complex because some proteins of the flagellum could be missing and the remainder could still transport proteins, perhaps independently. (Proteins similar – but not identical – to some found in the flagellum occur in the type III secretory system of some bacteria. See Hueck 1998). Again, he was equivocating, switching the focus from the function of the system, acting as a rotary propulsion machine, to the ability of a subset of the system to transport proteins across a membrane. However, taking away the parts of the flagellum certainly destroys the ability of the system to act as a rotary propulsion machine, as I have argued. Thus, contra Miller, the flagellum is indeed irreducibly complex. What's more, the function of transporting proteins has as little directly to do with the function of rotary propulsion as a toothpick has to do with a mousetrap. So discovering the supportive function of transporting proteins tells us precisely nothing about how Darwinian processes might have put together a rotary propulsion machine.

The Blood Clotting Cascade

Having dealt with some common misconceptions about intelligent design, in the next two sections I will examine two systems that were proposed as serious counterexamples to my claim of irreducible complexity. I will show not only that they fail, but also how they highlight the seriousness of the obstacle of irreducible complexity.

In *Darwin's Black Box*, I argued that the blood clotting cascade is an example of an irreducibly complex system (Behe 1996, 74–97). At first glance, clotting seems to be a simple process. A small cut or scrape will bleed for a while and then slow down and stop as the visible blood congeals. However, studies over the past fifty years have shown that the visible simplicity is undergirded by a system of remarkable complexity. (Halkier 1992). In all, there are over a score of separate protein parts involved in the vertebrate clotting system. The concerted action of the components results in the formation of a weblike structure at the site of the cut, which traps red blood cells and stops the bleeding. Most of the components of the clotting cascade are involved not in the structure of the clot itself, but in the control of the timing and placement of the clot. After all, it would not do to have clots forming at inappropriate times and places. A clot that formed in the wrong place, such as in the heart or brain, could lead to a heart attack or stroke. Yet a clot that formed even in the right place, but too slowly, would do little good.

The insoluble weblike fibers of the clot material itself are formed of a protein called fibrin. However, an insoluble web would gum up blood flow before a cut or scrape happened, so fibrin exists in the bloodstream initially in a soluble, inactive form called fibrinogen. When the closed circulatory system is breached, fibrinogen is activated by having a piece cut off from one end of two of the three proteins that comprise it. This exposes sticky sites on the protein, which allows them to aggregate. Because of the shape of the fibrin, the molecules aggregate into long fibers that form the meshwork of the clot. Eventually, when healing is completed, the clot is removed by an enzyme called plasmin.

The enzyme that converts fibrinogen to fibrin is called thrombin. Yet the action of thrombin itself has to be carefully regulated. If it were not, then thrombin would quickly convert fibrinogen to fibrin, causing massive blood clots and rapid death. It turns out that thrombin exists in an inactive form called prothrombin, which has to be activated by another component called Stuart factor. But by the same reasoning, the activity of Stuart factor has to be controlled, too, and it is activated by yet another component. Ultimately, the component that usually begins the cascade is tissue factor, which occurs on cells that normally do not come in contact with the circulatory system. However, when a cut occurs, blood is exposed to tissue factor, which initiates the clotting cascade.

Thus in the clotting cascade, one component acts on another, which acts on the next, and so forth. I argued that the cascade is irreducibly complex because, if a component is removed, the pathway is either immediately turned on or permanently turned off. It would not do, I wrote, to postulate that the pathway started from one end, fibrinogen, and then added components, since fibrinogen itself does no good. Nor is it plausible even to start with something like fibrinogen and a nonspecific enzyme that might cleave it, since the clotting would not be regulated and would be much more likely to do harm than good.

So said I. But Russell Doolittle – an eminent protein biochemist, a professor of biochemistry at the University of California–San Diego, a member of the National Academy of Sciences, and a lifelong student of the blood clotting system – disagreed. As part of a symposium discussing my book and Richard Dawkins' *Climbing Mount Improbable* in the *Boston Review*, which is published by the Massachusetts Institute of Technology, Doolittle wrote an essay discussing the phenomenon of gene duplication – the process by which a cell maybe provided with an extra copy

of a functioning gene. He then conjectured that the components of the blood clotting pathway, many of which have structures that are similar to each other, arose by gene duplication and gradual divergence. This is the common view among Darwinists. Professor Doolittle went on to describe a then-recent experiment that, he thought, showed that the cascade is not irreducible after all. Professor Doolittle cited a paper by Bugge and colleagues (1996a) entitled "Loss of Fibrinogen Rescues Mice from the Pleiotropic Effects of Plasminogen Deficiency." Of that paper, he wrote:

> Recently the gene for plaminogen [sic] was knocked out of mice, and, predictably, those mice had thrombotic complications because fibrin clots could not be cleared away. Not long after that, the same workers knocked out the gene for fibrinogen in another line of mice. Again, predictably, these mice were ailing, although in this case hemorrhage was the problem. And what do you think happened when these two lines of mice were crossed? For all practical purposes, the mice lacking both genes were normal! Contrary to claims about irreducible complexity, the entire ensemble of proteins is not needed. Music and harmony can arise from a smaller orchestra. (Doolittle 1997)

(Again, fibrinogen is the precursor of the clot material itself. Plasminogen is the precursor of plasmin, which removes clots once their purpose is accomplished.) So if one knocks out either one of those genes of the clotting pathway, trouble results; but, Doolittle asserted, if one knocks out both, then the system is apparently functional again. That would be a very interesting result, but it turns out to be incorrect. Doolittle misread the paper.

The abstract of the paper states that "[m]ice deficient in plasminogen and fibrinogen are phenotypically indistinguishable from fibrinogen-deficient mice." In other words, the double mutants have all the problems that the mice lacking just fibrinogen have. Those problems include inability to clot, hemorrhaging, and death of females during pregnancy. Plasminogen deficiency leads to a different suite of symptoms – thrombosis, ulcers, and high mortality. Mice missing both genes were "rescued" from the ill

Table 29.1   *Effects of knocking out genes for blood clotting components*

Missing Protein	Symptoms	Reference
Plasminogen	Thrombosis, high mortality	Bugge *et al.* 1995
Fibrinogen	Hemorrhage, death in pregnancy	Suh *et al.* 1995
Plasminogen/fibrinogen	Hemorrhage, death in pregnancy	Bugge *et al.* 1996a
Prothrombin	Hemorrhage, death in pregnancy	Sun *et al.* 1998
Tissue factor	Hemorrhage, death in pregnancy	Bugge *et al.* 1996b

effects of plasminogen deficiency only to suffer the problems associated with fibrinogen deficiency.[3] The reason for this is easy to see. Plasminogen is needed to remove clots that, left in place, interfere with normal functions. However, if the gene for fibrinogen is also knocked out, then clots can't form in the first place, and their removal is not an issue. Yet if clots can't form, then there is no functioning clotting system, and the mice suffer the predictable consequences.

Clearly, the double-knockout mice are not "normal." They are not promising evolutionary intermediates.

The same group that produced the mice missing plasminogen and fibrinogen has also produced mice individually missing other components of the clotting cascade – prothrombin and tissue factor. In each case, the mice are severely compromised, which is *exactly* what one would expect if the cascade is irreducibly complex (Table 29.1).

What lessons can we draw from this incident? The point is certainly not that Russell Doolittle misread a paper, which anyone might do. (Scientists, as a rule, are not known for their ability to write clearly, and Bugge and colleagues were no exception.) Rather, the main lesson is that irreducible complexity seems to be a much more severe problem than Darwinists recognize, since the experiment Doolittle himself chose to demonstrate that "music and harmony can arise from a smaller orchestra" showed exactly the opposite. A second lesson is that gene duplication is not the panacea that it is often made out to be. Professor Doolittle knows as much about the structures of the clotting proteins and their genes as anyone on Earth, and he is convinced that many of them arose by gene duplication and exon shuffling. Yet that knowledge did not prevent him from proposing utterly nonviable mutants as

possible examples of evolutionary intermediates. A third lesson is that, as I had claimed in *Darwin's Black Box*, there are no papers in the scientific literature detailing how the clotting pathway could have arisen by Darwinian means. If there were, Doolittle would simply have cited them.

Another significant lesson that we can draw is that, while the majority of academic biologists and philosophers place their confidence in Darwinism, that confidence rests on no firmer grounds than Professor Doolittle's. As an illustration, consider the words of the philosopher Michael Ruse:

> For example, Behe is a real scientist, but this case for the impossibility of a small-step natural origin of biological complexity has been trampled upon contemptuously by the scientists working in the field. They think his grasp of the pertinent science is weak and his knowledge of the literature curiously (although conveniently) outdated.
>
> For example, far from the evolution of clotting being a mystery, the past three decades of work by Russell Doolittle and others has thrown significant light on the ways in which clotting came into being. More than this, it can be shown that the clotting mechanism does not have to be a one-step phenomenon with everything already in place and functioning. One step in the cascade involves fibrinogen, required for clotting, and another, plaminogen [*sic*], required for clearing clots away. (Ruse 1998)

And Ruse goes on to quote Doolittle's passage from the *Boston Review* that I quoted earlier. Now, Ruse is a prominent Darwinist and has written many books on various aspects of Darwiniana. Yet, as his approving quotation of Doolittle's mistaken reasoning shows (complete with his copying of Doolittle's typo-misspelling of "plaminogen"),

Ruse has no independent knowledge of how natural selection could have put together complex biochemical systems. As far as the scientific dispute is concerned, Ruse has nothing to add.

Another such example is seen in a recent essay in *The Scientist*, "Not-So-Intelligent Design," by Neil S. Greenspan, a professor of pathology at Case Western Reserve University, who writes (Greenspan 2002), "The Design advocates also ignore the accumulating examples of the reducibility of biological systems. As Russell Doolittle has noted in commenting on the writings of one ID advocate . . ." Greenspan goes on to cite approvingly Doolittle's argument in the *Boston Review*. He concludes, with unwitting irony, that "[t]hese results cast doubt on the claim by proponents of ID that they know which systems exhibit irreducible complexity and which do not." But since the results are precisely the opposite of what Greenspan supposed, the shoe is now on the other foot. This incident casts grave doubt on the claim by Darwinists – both biologists and philosophers – that they know that complex cellular systems are explainable in Darwinian terms. It demonstrates that Darwinists either cannot or will not recognize difficulties for their theory.

The Mousetrap

The second counterargument to irreducibility I will discuss here concerns not a biological example but a conceptual one. In *Darwin's Black Box*, I pointed to a common mechanical mousetrap as an example of irreducible complexity. Almost immediately after the book's publication, some Darwinists began proposing ways in which the mousetrap could be built step by step. One proposal that has gotten wide attention, and that has been endorsed by some prominent scientists, was put forward by John McDonald, a professor of biology at the University of Delaware, and can be seen on his web site.[4] McDonald's main point was that the trap that I pictured in my book consisted of five parts, yet he could build a trap with fewer parts.

I agree. In fact, I said exactly the same thing in my book. I wrote:

> We need to distinguish between a *physical* precursor and a *conceptual* precursor. The trap described above is not the only system that can immobilize a mouse. On other occasions my family has used a glue trap. In theory at least, one can use a box propped open with a stick that could be tripped. Or one can simply shoot the mouse with a BB gun. However, these are not physical precursors to the standard mousetrap since they cannot be transformed, step-by-Darwinian-step, into a trap with a base, hammer, spring, catch, and holding bar. (Behe 1996, 43)

Thus the point is not that mousetraps can be built in different ways, with different numbers of pieces. (My children have a game at home called "Mousetrap," which has many, many pieces and looks altogether different from the common mechanical one.) Of course they can. The only question is whether a particular trap can be built by "numerous, successive, slight modifications" to a simple starting point – without the intervention of intelligence – as Darwin insisted that his theory required.

The McDonald traps cannot. The structure of the second trap, however, is not a single, small, random step away from the first. First notice that the one-piece trap is not a simple spring – it is shaped in a very special way. In fact, the shape was deliberately chosen by an intelligent agent, John McDonald, to act as a trap. Well, one has to start somewhere. But if the mousetrap series is to have any relevance at all to Darwinian evolution, then intelligence can't be involved at any further point.

Yet intelligence saturates the whole series. Consider what would be necessary to convert the one-piece trap to the "two-piece" trap. One can't just place the first trap on a simple piece of wood and have it work as the second trap does. Rather, the two protruding ends of the spring first have to be reoriented. What's more, two staples are added to hold the spring onto the platform so that it can be under tension in the two-piece trap. So we have gone not from a one-piece to a two-piece trap, but from a one-piece to a four-piece trap. Notice also that the placement of the staples in relation to the edge of the platform is critical. If the staples were moved a quarter-inch from where they are, the trap wouldn't work. Finally, consider that, in order to have a serious analogy to the robotic processes of the cell, we can't have an intelligent human setting the mousetrap –

the first trap would have to be set by some unconscious charging mechanism. So, when the pieces are rearranged, the charging mechanism too would have to change for the second trap.

It's easy for us intelligent agents to overlook our role in directing the construction of a system, but nature cannot overlook any step at all, so the McDonald mousetrap series completely fails as an analogy to Darwinian evolution. In fact, the second trap is best viewed not as some Darwinian descendant of the first but as a completely different trap, designed by an intelligent agent, perhaps using a refashioned part or two from the first trap.

Each of the subsequent steps in the series suffers from analogous problems, which I have discussed elsewhere.[5]

In his endorsement of the McDonald mousetrap series, Kenneth Miller wrote: "If simpler versions of this mechanical device [the mousetrap] can be shown to work, then simpler versions of biochemical machines could work as well . . . and this means that complex biochemical machines could indeed have had functional precursors."[6] But that is exactly what it doesn't show – if by "precursor" Miller means "Darwinian precursor." On the contrary, McDonald's mousetrap series shows that even if one does find a simpler system to perform some function, that gives one no reason to think that a more complex system performing the same function could be produced by a Darwinian process starting with the simpler system. Rather, the difficulty in doing so for a simple mousetrap gives us compelling reason to think it cannot be done for complex molecular machines.

Future Prospects of the Intelligent Design Hypothesis

The misconceived arguments by Darwinists that I have recounted here offer strong encouragement to me that the hypothesis of Intelligent Design is on the right track. After all, if well-informed opponents of an idea attack it by citing data that, when considered objectively, actually demonstrate its force, then one is entitled to be confident that the idea is worth investigating.

Yet it is not primarily the inadequacy of Darwinist responses that bodes well for the design hypothesis. Rather, the strength of design derives mainly from the work-a-day progress of science. In order to appreciate this fact, it is important to realize that the idea of Intelligent Design arose not from the work of any individual but from the collective work of biology, particularly in the last fifty years. Fifty years ago, the cell seemed much simpler, and in our innocence it was easier then to think that Darwinian processes might have accounted for it. But as biology progressed and the imagined simplicity vanished, the idea of design became more and more compelling. That trend is continuing inexorably. The cell is not getting any simpler; it is getting much more complex. I will conclude this chapter by citing just one example, from the relatively new area of proteomics.

With the successful sequencing of the entire genomes of dozens of microorganisms and one vertebrate (us), the impetus has turned toward analyzing the cellular interactions of the proteins that the genomes code for, taken as a whole. Remarkable progress has already been made. Early in 2002, an exhaustive study of the proteins comprising the yeast proteome was reported. Among other questions, the investigators asked what proportion of yeast proteins work as groups. They discovered that nearly fifty percent of proteins work as complexes of a half-dozen or more, and many as complexes of ten or more (Gavin *et al.* 2002).

This is not at all what Darwinists had expected. As Bruce Alberts wrote earlier in the article "The Cell as a Collection of Protein Machines":

> We have always underestimated cells. Undoubtedly we still do today. But at least we are no longer as naive as we were when I was a graduate student in the 1960s. Then most of us viewed cells as containing a giant set of second-order reactions. . . .
>
> But, as it turns out, we can walk and we can talk because the chemistry that makes life possible is much more elaborate and sophisticated than anything we students had ever considered. Proteins make up most of the dry mass of a cell. But instead of a cell dominated by randomly colliding individual protein molecules, we now know that nearly every major process in a cell is carried out by assemblies of 10 or more protein molecules. And, as it carries out its biological functions, each of these protein assemblies interacts

with several other large complexes of proteins. Indeed, the entire cell can be viewed as a factory that contains an elaborate network of interlocking assembly lines, each of which is composed of a set of large protein machines. (Alberts 1998)

The important point here for a theory of Intelligent Design is that molecular machines are not confined to the few examples that I discussed in *Darwin's Black Box*. Rather, most proteins are found as components of complicated molecular machines. Thus design might extend to a large fraction of the features of the cell, and perhaps beyond that into higher levels of biology.

Progress in twentieth-century science has led us to the design hypothesis. I expect progress in the twenty-first century to confirm and extend it.

Notes

1   For example, the microbiologist James Shapiro of the University of Chicago declared in *National Review* that "[t]here are no detailed Darwinian accounts for the evolution of any fundamental biochemical or cellular system, only a variety of wishful speculations" (Shapiro 1996, 65). In *Nature*, the University of Chicago evolutionary biologist Jerry Coyne stated, "There is no doubt that the pathways described by Behe are dauntingly complex, and their evolution will be hard to unravel.... [W]e may forever be unable to envisage the first protopathways" (Coyne 1996, 227). In a particularly scathing review in *Trends in Ecology and Evolution*, Tom Cavalier-Smith, an evolutionary biologist at the University of British Columbia, nonetheless wrote, "For none of the cases mentioned by Behe is there yet a comprehensive and derailed explanation of the probable steps in the evolution of the observed complexity. The problems have indeed been sorely neglected – though Behe repeatedly exaggerates this neglect with such hyperboles as 'an eerie and complete silence'" (Cavalier-Smith 1997, 162). The Evolutionary biologist Andrew Pomiankowski, writing in *New Scientist,* agreed: "Pick up any biochemistry textbook, and you will find perhaps two or three references to evolution. Turn to one of these and you will be lucky to find anything better than 'evolution selects the fittest molecules for their biological function'" (Pomiankowski 1996, 44). In *American Scientist*, the Yale molecular biologist Robert Dorit averred, "In a narrow sense, Behe is correct when he argues that we do not yet fully understand the evolution of the flagellar

motor or the blood clotting cascade" (Dorit 1997, 474).

2   A good example is found on the "World of Richard Dawkins" web site, maintained by a Dawkins fan named John Catalano at <www.world-of-dawkins.com/Catalano/box/published.htm>. It is to this site that the Oxford University physical chemist Peter Atkins was referring when he wrote in a review of *Darwin's Black Box* for the "Infidels" web site: "Dr. Behe claims that science is largely silent on the details of molecular evolution, the emergence of complex biochemical pathways and processes that underlie the more traditional manifestations of evolution at the level of organisms. Tosh! There are hundreds, possibly thousands, of scientific papers that deal with this very subject. For an entry into this important and flourishing field, and an idea of the intense scientific effort that it represents (see the first link above) [*sic*]" (Atkins 1998).

3   Bugge and colleagues (1996a) were interested in the question of whether plasminogen had any role in metabolism other than its role in clotting, as had been postulated. The fact that the direct effects of plasminogen deficiency were ameliorated by fibrinogen deficiency showed that plasminogen probably had no other role.

4   <http://udel.edu/~mcdonald/oldmousetrap.html>. Professor McDonald has recently designed a new series of traps that can be seen at <http://udel.edu/~mcdonald/mousetrap.html>. I have examined them and have concluded that they involve his directing intelligence to the same degree.

5   M. J. Behe, "A Mousetrap Defended: Response to Critics." <www.crsc.org>

6   <http://biocrs.biomed.brown.edn/Darwin/DI/Mousetrap.html>

References

Aizawa, S. I. 1996. Flagellar assembly in Salmonella typhimurium. *Molecular Microbiology* 19: 1–5.

Alberts, B. 1998. The cell as a collection of protein machines: Preparing the next generation of molecular biologists. *Cell* 92: 291–4.

Atkins, P. W. 1998. Review of Michael Behe's *Darwin's Black Box.* <www.infidels.org/library/modern/peter_atkins/behe_html>.

Behe, M. J. 1996. *Darwin's Black Box: The Biochemical Challenge to Evolution*. New York: The Free Press.

Behe, M. J. 2001. Reply to my critics: A response to reviews of *Darwin's Black Box: The Biochemical Challenge to Evolution. Biology and Philosophy* 16: 685–709.

Bugge, T. H., M. J. Flick, C. C. Daugherty, and J. L. Degen. 1995. Plasminogen deficiency causes severe thrombosis but is compatible with development and reproduction. *Genes and Development* 9: 794–807.

Bugge, T. H., K. W. Kombrinck, M. J. Flick, C. C. Daugherty, M. J. Danton, and J. L. Degen. 1996a. Loss of fibrinogen rescues mice from the pleiotropic effects of plasminogen deficiency. *Cell* 87: 709–19.

Bugge, T. H., Q. Xiao, K. W. Kombrinck, M. J. Flick, K. Holmback, M. J. Darnton, M. C. Colbert, D. P. Witte, K. Fujikawa, E. W. Davie, and J. L. Degen. 1996b. Fatal embryonic bleeding events in mice lacking tissue factor, the cell-associated initiator of blood coagulation. *Proceedings of the National Academy of Sciences (USA)* 93: 6258–63.

Cavalier-Smith, T. 1997. The blind biochemist. *Trends in Ecology and Evolution* 12: 162–3.

Coyne, J. A. 1996. God in the details. *Nature* 383: 227–8.

Darwin, C. 1859. *The Origin of Species.* New York: Bantam Books.

Dawkins, R. 1986. *The Blind Watchmaker.* New York: Norton.

DeRosier, D. J. 1998. The turn of the screw: The bacterial flagellar motor. *Cell* 93: 17–20.

Doolittle, R. F. A delicate balance. *Boston Review,* February/March 1997, pp. 28–9.

Dorit, R. 1997. Molecular evolution and scientific inquiry, misperceived. *American Scientist* 85: 474–5.

Dutcher, S. K. 1995. Flagellar assembly in two hundred and fifty easy-to-follow steps. *Trends in Genetics* 11: 398–404.

Gavin, A. C., *et al.* 2002. Functional organization of the yeast proteome by systematic analysis of protein complexes. *Nature* 415: 141–7.

Greenspan, N. S. 2002. Not-so-intelligent design. *The Scientist* 16: 12.

Halkier, T. 1992. *Mechanisms in Blood Coagulation Fibrinolysis and the Complement System.* Cambridge: Cambridge University Press.

Harold, F. M. 2001. *The Way of the Cell.* Oxford: Oxford University Press.

Hueck, C. J. 1998. Type III protein secretion systems in bacterial pathogens of animals and plants. *Microbiology and Molecular Biology Reviews* 62: 379–433.

Kauffman, S. A. 1995. *At Home in the Universe: The Search for Laws of Self-Organization and Complexity.* New York: Oxford University Press.

Miller, K. R. 1999. *Finding Darwin's God: A Scientist's Search for Common Ground between God and Evolution.* New York: Cliff Street Books.

Pomiankowski, A. 1996. The God of the tiny gaps. *New Scientist,* September 14, pp. 44–5.

Ruse, M. 1998. Answering the creationists: Where they go wrong and what they're afraid of. *Free Inquiry,* March 22, p. 28.

Shapiro, J. 1996. In the details... what? *National Review,* September 16, pp. 62–5.

Shapiro, L. 1995. The bacterial flagellum: From genetic network to complex architecture. *Cell* 80: 525–7.

Sub, T. T., K. Holmback, N. J. Jensen, C. C. Daugherty, K. Small, D. I. Simon, S. Potter, and J. L. Degen. 1995. Resolution of spontaneous bleeding events but failure of pregnancy in fibrinogen-deficient mice. *Genes and Development* 9: 2020–33.

Sun, W. Y., D. P. Witte, J. L. Degen, M. C. Colbert, M. C. Burkart, K. Holmback, Q. Xiao, T. H. Bugge, and S. J. Degen. 1998. Prothrombin deficiency results in embryonic and neonatal lethality in mice. *Proceedings of the National Academy of Sciences USA* 95: 7597–602.

Yonekura, K., S. Maki, D. G. Morgan, D. J. DeRosier, F. Vonderviszt, K. Imada, and K. Namba. 2000. The bacterial flagellar cap as the rotary promoter of flagellin self-assembly. *Science* 290: 2148–52.

30

The Flagellum Unspun: *The Collapse of "Irreducible Complexity"*

Kenneth R. Miller

Almost from the moment *On the Origin of Species* was published is 1859, the opponents of evolution have fought a long, losing battle against their Darwinian foes. Today, like a prizefighter in the late rounds losing badly on points, they've placed their hopes on one big punch – a single claim that might smash through the overwhelming weight of scientific evidence to bring Darwin to the canvas once and for all. Their name for this virtual roundhouse right is "Intelligent Design."

In the last several years, the Intelligent Design (ID) movement has attempted to move against the standards of science education in several American states, most famously in Kansas and Ohio (Holden 1999; Gura 2002). The principal claim made by adherents of this view is that they can detect the presence of "Intelligent Design" in complex biological systems. As evidence, they cite a number of specific examples, including the vertebrate blood clotting cascade, the eukaryotic cilium, and most notably, the eubacterial flagellum (Behe 1996a; Behe 2002).

Of all these examples, the flagellum has been presented so often as a counterexample to evolution that it might well be considered the "poster child" of the modern anti-evolution movement. Variations of its image now appear on web pages of anti-evolution groups such as the Discovery Institute, and on the covers of "Intelligent Design" books such as William Dembski's *No Free Lunch* (Dembski 2002a). To anti-evolutionists, the high status of the flagellum reflects the supposed fact that it could not possibly have been produced by an evolutionary pathway.

There is, to be sure, nothing new or novel in an anti-evolutionist pointing to a complex or intricate natural structure and professing skepticism that it could have been produced by the "random" processes of mutation and natural selection. Nonetheless, the "argument from personal incredulity," as such sentiments have been appropriately described, has been a weapon of little value in the anti-evolution movement. Anyone can state at any time that he or she cannot imagine how evolutionary mechanisms might have produced a certain species, organ, or structure. Such statements, obviously, are personal – and they say more about the limitations of those who make them than they do about the limitations of Darwinian mechanisms.

The hallmark of the Intelligent Design movement, however, is that it purports to rise above the level of personal skepticism. It claims to have found a *reason* why evolution could not have produced a structure like the bacterial

Kenneth Miller, "The Flagellum Unspun: The Collapse of 'Irreducible Complexity'," in W. Dembski and M. Ruse (eds.) *Debating Design: From Darwin to DNA Black Box: The Biochemical Challenge to Evolution* (Cambridge, 2004), pp. 81–97.

flagellum – a reason based on sound, solid scientific evidence.

Why does the intelligent design movement
regard the flagellum as unevolvable? Because it
is said to possesses a quality known as "irreducible
complexity." Irreducibly complex structures, we
are told, could not have been produced by
evolution – or, for that matter, by any natural
process. They do exist, however, and therefore
they must have been produced by something. That
something could only be an outside intelligent
agency operating beyond the laws of nature –
an intelligent designer. That, simply stated, is
the core of the new argument from design, and
the intellectual basis of the Intelligent Design
movement.

The great irony of the flagellum's increasing
acceptance as an icon of the anti-evolutionist
movement is that fact that research had demolished its status as an example of irreducible
complexity almost at the very moment it was
first proclaimed. The purpose of this chapter is
to explore the arguments by which the flagellum's
notoriety has been achieved, and to review the
research developments that have now undermined
the very foundations of those arguments.

The Argument's Origins

The flagellum owes its status principally to
Darwin's Black Box (Behe 1996a), a book by
Michael Behe that employed it in a carefully
crafted anti-evolution argument. Building upon
William Paley's well-known "argument from
design," Behe sought to bring the argument two
centuries forward into the realm of biochemistry. Like Paley, Behe appealed to his readers
to appreciate the intricate complexity of living
organisms as evidence for the work of a designer.
Unlike Paley, however, he raised the argument
to a new level, claiming to have discovered a scientific principle that could be used to prove that
certain structures could not have been produced
by evolution. That principle goes by the name of
"irreducible complexity."

An irreducibly complex structure is defined
as "a single system composed of several well-
matched, interacting parts that contribute to
the basic function, wherein the removal of any
one of the parts causes the system to effectively

cease functioning" (Behe 1996a, 39). Why would
such systems present difficulties for Darwinism?
Because they could not possibly have been produced by the process of evolution:

> An irreducibly complex system cannot be pro-
> duced directly by numerous, successive, slight
> modifications of a precursor system, because
> any precursor to an irreducibly complex system
> that is missing a part is by definition nonfunc-
> tional. . . . Since natural selection can only choose
> systems that are already working, then if a bio-
> logical system cannot be produced gradually it
> would have to arise as an integrated unit, in
> one fell swoop, for natural selection to have
> anything to act on. (Behe 1996b)

The phrase "numerous, successive, slight modifications" is not accidental. The very same words
were used by Charles Darwin in the *Origin of
Species* in describing the conditions that had to
be met for his theory to be true. As Darwin
wrote, if one could find an organ or structure
that could not have been formed by "numerous,
successive, slight modifications," his "theory would
absolutely break down" (Darwin 1872, 191). To
anti-evolutionists, the bacterial flagellum is now
regarded as exactly such a case – an "irreducibly
complex system" that "cannot be produced
directly by numerous successive, slight modifications." A system that could not have evolved –
a desperation punch that just might win the fight
in the final round, a tool with which the theory
of evolution might be brought down.

The Logic of Irreducible Complexity

Living cells are filled, of course, with complex
structures whose detailed evolutionary origins
are not known. Therefore, in fashioning an argument against evolution one might pick nearly any
cellular structure – the ribosome, for example –
and claim, correctly, that its origin has not been
explained in detail by evolution.

Such arguments are easy to make, of course, but
the nature of scientific progress renders them far
from compelling. The lack of a detailed current
explanation for a structure, organ, or process
does not mean that science will never come up
with one. As an example, one might consider

the question of how left-right asymmetry arises in vertebrate development, a question that was beyond explanation until the 1990s (Belmonte 1999). In 1990, one might have argued that the body's left-right asymmetry could just as well be explained by the intervention of a designer as by an unknown molecular mechanism. Only a decade later, the actual molecular mechanism was identified (Stern 2002), and any claim one might have made for the intervention of a designer would have been discarded. The same point can be made, of course, regarding any structure or mechanism whose origins are not yet understood.

The utility of the bacterial flagellum is that it seems to rise above this "argument from ignorance." By asserting that it is a structure "in which the removal of an element would cause the whole system to cease functioning" (Behe 2002), the flagellum is presented as a "molecular machine" whose individual parts must have been specifically crafted to work as a unified assembly. The existence of such a multipart machine therefore provides genuine scientific proof of the actions of an intelligent designer.

In the case of the flagellum, the assertion of irreducible complexity means that a minimum number of protein components, perhaps thirty, are required to produce a working biological function. By the logic of irreducible complexity, these individual components should have no function until all thirty are put into place, at which point the function of motility appears. What this means, of course, is that evolution could not have fashioned those components a few at a time, since they do not have functions that could be favored by natural selection. As Behe wrote, "natural selection can only choose among systems that are already working" (Behe 2002), and an irreducibly complex system does not work unless all of its parts are in place. The flagellum is irreducibly complex, and therefore, in must have been designed. Case closed.

Answering the Argument

The assertion that cellular machines are irreducibly complex, and therefore provide proof of design, has not gone unnoticed by the scientific community. A number of detailed rebuttals have appeared in the literature, and many have pointed out the poor reasoning of recasting the classic argument from design in the modern language of biochemistry (Coyne 1996; Miller 1996; Depew 1998; Thornhill and Ussery 2000). I have suggested elsewhere that the scientific literature contains counterexamples to any assertion that evolution cannot explain biochemical complexity (Milller 1999, 147), and other workers have addressed the issue of how evolutionary mechanisms allow biological systems to increase in information content (Adami, Ofria, and Collier 2000; Schneider 2000).

The most powerful rebuttals to the flagellum story, however, have not come from direct attempts to answer the critics of evolution. Rather, they have emerged from the steady progress of scientific work on the genes and proteins associated with the flagellum and other cellular structures. Such studies have now established that the entire premise by which this molecular machine has been advanced as an argument against evolution is wrong – the bacterial flagellum is not irreducibly complex. As we will see, the flagellum – the supreme example of the power of this new "science of design" – has failed its most basic scientific test. Remember the claim that "any precursor to an irreducibly complex system that is missing a part is by definition nonfunctional"? As the evidence has shown, nature is filled with examples of "precursors" to the flagellum that are indeed "missing a part," and yet are fully functional – functional enough, in some cases, to pose a serious threat to human life.

The Type III Secretory Apparatus

In the popular imagination, bacteria are "germs" – tiny microscopic bugs that make us sick. Microbiologists smile at that generalization, knowing that most bacteria are perfectly benign, and that many are beneficial – even essential – to human life. Nonetheless, there are indeed bacteria that produce diseases, ranging from the mildly unpleasant to the truly dangerous. Pathogenic, or disease-causing, bacteria threaten the organisms they infect in a variety of ways, one of which is by producing poisons and injecting them directly into the cells of the body. Once inside, these toxins break down and destroy the host cells, producing illness, tissue damage, and sometimes even death.

In order to carry out this diabolical work, bacteria not only must produce the protein toxins that bring about the demise of their hosts, but also must efficiently inject them across the cell membranes and into the cells of their hosts. They do this by means of any number of specialized protein secretory systems. One, known as the type III secretory system (TTSS), allows gram-negative bacteria to translocate proteins directly into the cytoplasm of a host cell (Heuck 1998). The proteins transferred through the TTSS include a variety of truly dangerous molecules, some of which are known as "virulence factors," that are directly responsible for the pathogenic activity of some of the most deadly bacteria in existence (Heuck 1998; Büttner and Bonas 2002).

At first glance, the existence of the TTSS, a nasty little device that allows bacteria to inject these toxins through the cell membranes of their unsuspecting hosts, would seem to have little to do with the flagellum. However, molecular studies of proteins in the TTSS have revealed a surprising fact: the proteins of the TTSS are directly homologous to the proteins in the basal portion of the bacterial flagellum. These homologies extend to a cluster of closely associated proteins found in both of these molecular "machines." On the basis of these homologies, McNab (1999) has argued that the flagellum itself should be regarded as a type III secretory system. Extending such studies with a detailed comparison of the proteins associated with both systems, Aizawa has seconded this suggestion, noting that the two systems "consist of homologous component proteins with common physico-chemical properties" (Aizawa 2001, 163). It is now clear, therefore, that a smaller subset of the full complement of proteins in the flagellum makes up the functional transmembrane portion of the TTSS.

Stated directly, the TTSS does its dirty work using a handful of proteins from the base of the flagellum. From the evolutionary point of view, this relationship is hardly surprising. In fact, it is to be expected that the opportunism of evolutionary processes would mix and match proteins in order to produce new and novel functions. According to the doctrine of irreducible

complexity, however, this should not be possible. If the flagellum is indeed irreducibly complex, then removing just one part, let alone ten or fifteen, should render what remains "by definition nonfunctional." Yet the TTSS is indeed fully functional, even though it is missing most of the parts of the flagellum. The TTSS may be bad news for us, but for the bacteria that possess it, it is a truly valuable biochemical machine.

The existence of the TTSS in a wide variety of bacteria demonstrates that a small portion of the "irreducibly complex" flagellum can indeed carry out an important biological function. Because such a function is clearly favored by natural selection, the contention that the flagellum must be fully assembled before any of its component parts can be useful is obviously incorrect. What this means is that the argument for intelligent design of the flagellum has failed.

Counterattack

Classically, one of the most widely repeated charges made by anti-evolutionists is that the fossil record contains wide "gaps" for which transitional fossils have never been found. Therefore, the intervention of a creative agency – an intelligent designer – must be invoked to account for each gap. Such gaps, of course, have been filled with increasing frequency by paleontologists – the increasingly rich fossil sequences demonstrating the origins of whales are a useful example (Thewissen, Hussain, and Arif 1994; Thewissen et al. 2001). Ironically, the response of anti-evolutionists to such discoveries is frequently to claim that things have only gotten worse for evolution. Where previously there had been just one gap, as a result of the transitional fossil there are now two (one on either side of the newly discovered specimen).

As word of the relationship between the eubacterial flagellum and the TTSS has begun to spread among the "design" community, the first hints of a remarkably similar reaction have emerged. The TTSS only makes problems worse for evolution, according to this response, because now there are two irreducibly complex systems to deal with. The flagellum is still irreducibly complex – but so is the TTSS. So now there are two systems for evolutionists to explain instead of just one.

Unfortunately for this line of argument, the claim that one irreducibly complex system might contain another is self-contradictory. To understand this, we need to remember that the entire point of the design argument, as exemplified by the flagellum, is that only the entire biochemical machine, with all of its parts, is functional. For the Intelligent Design argument to stand, this must be the case, since it provides the basis for their claim that only the complete flagellum can be favored by natural selection, not any of its component parts.

However, if the flagellum contains within it a smaller functional set of components such as the TTSS, then the flagellum itself cannot be irreducibly complex – by definition. Since we now know that this is indeed the case, it is obviously true that the flagellum is not irreducibly complex.

A second reaction, which I have heard directly after describing the relationship between the secretory apparatus and the flagellum, is the objection that the TTSS does not tell us how either it or the flagellum evolved. This is certainly true, although Aizawa has suggested that the TTSS may indeed be an evolutionary precursor of the flagellum (Aizawa 2001). Nonetheless, until we have produced a step-by-step account of the evolutionary derivation of the flagellum, one may indeed invoke the argument from ignorance for this and every other complex biochemical machine.

However, in agreeing to this, one must keep in mind that the doctrine of irreducible complexity was intended to go one step beyond the claim of ignorance. It was fashioned in order to provide a rationale for claiming that the bacterial flagellum could not have evolved, even in principle, because it is irreducibly complex. Now that a simpler, functional system (the TTSS) has been discovered among the protein components of the flagellum, the claim of irreducible complexity has collapsed, and with it any 'evidence" that the flagellum was designed.

The Combinatorial Argument

At first glance, William Dembski's case for Intelligent Design seems to follow a distinctly different strategy in dealing with biological complexity. His recent book, *No Free Lunch* (Dembski 2002a), lays out this case, using information theory and mathematics to show that life is the result of Intelligent Design. Dembski makes the assertion that living organisms contain what he calls "complex specified information" (CSI), and he claims to have shown that the evolutionary mechanism of natural selection cannot produce CSI. Therefore, any instance of CSI in a living organism must be the result of intelligent design. And living organisms, according to Dembski, are chock-full of CSI.

Dembski's arguments, couched in the language of information theory, are highly technical and are defended, almost exclusively, by reference to their utility in detecting information produced by human beings. These include phone and credit card numbers, symphonies, and artistic woodcuts, to name just a few. One might then expect that Dembski, having shown how the presence of CSI can be demonstrated in man-made objects, would then turn to a variety of biological objects. Instead, he turns to just one such object, the bacterial flagellum.

Dembski then offers his readers a calculation showing that the flagellum could not possibly have evolved. Significantly, he begins that calculation by linking his arguments to those of Behe, writing: "I want therefore in this section to show how irreducible complexity is a special case of specified complexity, and in particular I want to sketch how one calculates the relevant probabilities needed to eliminate chance and infer design for such systems" (Dembski 2002a, 289). Dembski then tells us that an irreducibly complex system, like the flagellum, is a "discrete combinatorial object." What this means, as he explains, is that the probability of assembling such an object can be calculated by determining the probabilities that each of its components might have originated by chance, that they might have been localized to the same region of the cell, and that they would have been assembled in precisely the right order. Dembski refers to these three probabilities as Porig, Plocal, and Pconfig, and he regards each of them as separate and independent (Dembski 2002a, 291).

This approach overlooks the fact that the last two probabilities are actually contained within the first. Localization and self-assembly of complex protein structures in prokaryotic cells are properties generally determined by signals built into the primary structures of the proteins themselves. The same is probably true for the amino acid sequences of the thirty or so protein components of the flagellum and the approximately twenty proteins involved in the flagellum's assembly (McNab 1999; Yonekura *et al.* 2000). Therefore, if one gets the sequences of all the proteins right, localization and assembly will take care of themselves.

To the ID enthusiast, however, this is a point of little concern. According to Dembski, evolution still could not construct the thirty proteins needed for the flagellum. His reason is that the probability of their assembly falls below what he terms the "universal probability bound." According to Dembski, the probability bound is a sensible allowance for the fact that highly improbable events do occur from time to time in nature. To allow for such events, he argues that given enough time, any event with a probability larger than 10^{-150} might well take place. Therefore, if a sequence of events, such as a presumed evolutionary pathway, has a calculated probability less than 10^{-150}, we may conclude that the pathway is impossible. If the calculated probability is greater than 10^{-150}, it is possible (even if unlikely).

When Dembski turns his attention to the chances of evolving the thirty proteins of the bacterial flagellum, he makes what he regards as a generous assumption. Guessing that each of the proteins of the flagellum have about 300 amino acids, one might calculate that the chance of getting just one such protein to assemble from "random" evolutionary processes would be 20^{-300}, since there are 20 amino acids specified by the genetic code. Dembski, however, concedes that proteins need not get the *exact* amino acid sequence right in order to be functional, so he cuts the odds to just 20^{-30}, which he tells his readers is "on the order of 10^{-39} (Dembski 2002a, 301). Since the flagellum requires thirty such proteins, he explains that thirty such probabilities "will all need to be multiplied to form the origination probability" (Dembski 2002a, 301). That would give us an origination probability for the flagellum of 10^{-1170}, far below the universal probability bound. The flagellum could not have evolved, and now we have the numbers to prove it. Right?

Assuming Impossibility

I have no doubt that to the casual reader, a quick glance over the pages of numbers and symbols in Dembski's books is impressive, if not downright intimidating. Nonetheless, the way in which he calculates the probability of an evolutionary origin for the flagellum shows how little biology actually stands behind those numbers. His computation calculates only the probability of spontaneous, random assembly for each of the proteins of the flagellum. Having come up with a probability value on the order of 10^{-1170}, he assures us that he has shown the flagellum to be unevolvable. This conclusion, of course, fits comfortably with his view that "[t]he Darwinian mechanism is powerless to produce irreducibly complex systems" (Dembski 2002a, 289).

However complex Dembski's analysis, the scientific problem with his calculations is almost too easy to spot. By treating the flagellum as a "discrete combinatorial object" he has shown only that it is unlikely that the parts of the flagellum could assemble spontaneously. Unfortunately for his argument, no scientist has ever proposed that the flagellum, or any other complex object, evolved in that way. Dembski, therefore, has constructed a classic "straw man" and blown it away with an irrelevant calculation.

By treating the flagellum as a discrete combinatorial object, he has assumed in his calculation that no subset of the thirty or so proteins of the flagellum could have biological activity. As we have already seen, this is wrong. Nearly a third of those proteins are closely related to components of the TTSS, which does indeed have biological activity. A calculation that ignores that fact has no scientific validity.

More importantly, Dembski's willingness to ignore the TTSS lays bare the underlying assumption of his entire approach to the calculation of probabilities and the detection of "design." *He assumes what he is trying to prove.*

According to Dembski, the detection of "design" requires that an object display complexity that could not be produced by what he calls "natural causes." In order to do that, one must first examine all of the possibilities by which an object, such as the flagellum, might have been generated naturally. Dembski and Behe, of course, come to the conclusion that there are no such natural causes. But how did they determine that? What is the scientific method used to support such a conclusion? Could it be that their assertions of the lack of natural causes simply amount to an unsupported personal belief? Could it be that there are such causes but that they simply happened not to think of them? Dembski actually seems to realize that this is a serious problem. He writes: "Now it can happen that we may not know enough to determine all the relevant chance hypotheses. Alternatively, we might think we know the relevant chance hypotheses, but later discover that we missed a crucial one. In the one case a design inference could not even get going; in the other, it would be mistaken" (Dembski 2002a, 123, note 80).

What Dembski is telling us is that in order to "detect" design in a biological object, one must first come to the conclusion that the object could not have been produced by any "relevant chance hypotheses" (meaning, naturally, evolution). Then, and only then, are Dembski's calculations brought into play. Stated more bluntly, what this really means is that the "method" first involves *assuming the absence* of an evolutionary pathway leading to the object, followed by a calculation "proving" the impossibility of spontaneous assembly. Incredibly, this a priori reasoning is exactly the sort of logic upon which the new "science of design" has been constructed.

Not surprisingly, scientific reviewers have not missed this point – Dembski's arguments have been repeatedly criticized on this issue and on many others (Charlesworth 2002; Orr 2002; Padian 2002).

Designing the Cycle

In assessing the design argument, therefore, it only *seems* as though two distinct arguments have been raised for the unevoluability of the flagellum. In reality, those two arguments, one invoking irreducible complexity and the other specified complex information, both depend upon a single scientifically insupportable position – namely, that we can look at a complex biological object and determine with absolute certainty that none of its component parts could have been first selected to perform other functions. Now the discovery of extensive homologies between the type

III secretory system and the flagellum has shown just how wrong that position was.

When anti-evolutionary arguments featuring the bacterial flagellum rose to prominence, beginning with the 1996 publication of *Darwin's Black Box* (Behe 1996a), they were predicated upon the assertion that each of the protein components of the flagellum was crafted, in a single act of design, to fit the specific purpose of the flagellum. The flagellum was said to be unevolvable because the entire complex system had to be assembled first in order to produce any selectable biological function. This claim was broadened to include all complex biological systems, and to assert further that science would never find an evolutionary pathway to any of these systems. After all, it hadn't so far, at least according to one of the principal advocates of "design":

> There is no publication in the scientific literature – in prestigious journals, specialty journals, or books – that describes how molecular evolution of any real, complex, biochemical system either did occur or even might have occurred. (Behe 1996a, 185)

As many critics of intelligent design have pointed out, that statement is simply false. Consider, as just one example, the Krebs cycle, an intricate biochemical pathway consisting of nine enzymes and a number of cofactors that occupies center stage in the pathways of cellular metabolism. The Krebs cycle is "real," "complex," and "biochemical." Does it also present a problem for evolution? Apparently it does, according to the authors of a 1996 paper in the *Journal of Molecular Evolution*, who wrote:

> "The Krebs cycle has been frequently quoted as a key problem in the evolution of living cells, hard to explain by Darwin's natural selection: How could natural selection explain the building of a complicated structure in toto, when the intermediate stages have no obvious fitness functionality? (Melendez-Hevia, Wadell, and Cascante 1996)

Where Intelligent Design theorists threw up their hands and declared defeat for evolution, however, these researchers decided to do the hard scientific work of analyzing the components of the cycle and seeing if any of them might have been selected for other biochemical tasks. What they found should be a lesson to anyone who asserts that evolution can act only by direct selection for a final function. In fact, nearly all of the proteins of the complex cycle can serve different biochemical purposes within the cell, making it possible to explain in detail how they evolved:

> In the Krebs cycle problem the intermediary stages were also useful, but for different purposes, and, therefore, its complete design was a very clear case of opportunism. . . . the Krebs cycle was built through the process that Jacob (1977) called "evolution by molecular tinkering," stating that evolution does not produce novelties from scratch: It works on what already exists. The most novel result of our analysis is seeing how, with minimal new material, evolution created the most important pathway of metabolism, achieving the best chemically possible design. In this case, a chemical engineer who was looking for the best design of the process could not have found a better design than the cycle which works in living cells. (Melendez-Hevia, Wadell, and Cascante 1996)

Since this paper appeared, a study based on genomic DNA sequences has confirmed the validity of its approach (Huynen, Dandekar, and Bork 1999). By contrast, how would Intelligent Design have approached the Krebs cycle? Using Dembski's calculations as our guide, we would first determine the amino acid sequences of each of the proteins of the cycle, and then calculate the probability of their spontaneous assembly. When this is done, an origination probability of less than 10^{-400} is the result. Therefore, applying "design" as a predictive science would have told both groups of researchers that their ultimately successful studies would have been fruitless, since the probability of spontaneous assembly falls below the "universal probability bound."

We already know, however, the reason that such calculations fail. They carry a built-in assumption that the component parts of a complex biochemical system have no possible function beyond serving the completely assembled system itself. As we have seen, this assumption is false. The Krebs cycle researchers knew

better, of course, and were able to produce two important studies describing how a real, complex, biochemical system might have evolved – the very thing that design theorists once claimed did not exist in the scientific literature.

The Failure of Design

It is no secret that concepts like "irreducible complexity" and "Intelligent Design" have failed to take the scientific community by storm (Forrest 2002). Design has not prompted new research studies, new breakthroughs, or novel insights on so much as a single scientific question. Design advocates acknowledge this from time to time, but they often claim that this is because the scientific deck is stacked against them. The Darwinist establishment, they say, prevents them from getting a foot in the laboratory door.

I would suggest that the real reason for the cold shoulder given "design" by the scientific community – particularly by life science researchers – is that time and time again its principal scientific claims have turned out to be wrong. Science is a pragmatic activity, and if your hypothesis doesn't work, it is quickly discarded.

The claim of irreducible complexity for the bacterial flagellum is an obvious example of this, but there are many others. Consider, for example, the intricate cascade of proteins involved in the clotting of vertebrate blood. This has been cited as one of the principal examples of the kind of complexity that evolution cannot generate, despite the elegant work of Russell Doolittle (Doolittle and Feng 1987; Doolittle 1993) to the contrary. A number of proteins are involved in this complex pathway, as described by Behe:

> When an animal is cut, a protein called Hagemann factor (XII) sticks to the surface of cells near the wound. Bound Hagemann factor is then cleaved by a protein called HMK to yield activated Hagemann factor. Immediately the activated Hagemann factor converts another protein, called prekallikrein, to its active form, kallikrein. (Behe 1996a, 84)

How important are each of these proteins? In line with the dogma of irreducible complexity, Behe argues that each and every component must be in place before the system will work, and he is perfectly clear on this point:

> [N]one of the cascade proteins are used for anything except controlling the formation of a clot. Yet in the absence of any of the components, blood does not clot, and the system fails. (Behe 1996a, 86)

As we have seen, the claim that every one of the components must be present in order for clotting to work is central to the "evidence" for design. One of those components, as these quotations indicate, is factor XII, which initiates the cascade. Once again, however, a nasty little fact gets in the way of Intelligent Design theory. Dolphins lack factor XII (Robinson, Kasting, and Aggeler 1969), yet their blood clots perfectly well. How can this be, if the clotting cascade is indeed irreducibly complex? It cannot, of course, and therefore the claim of irreducible complexity is wrong for this system as well. I would suggest, therefore, that the real reason for the rejection of "design" by the scientific community is remarkably simple – the claims of the Intelligent Design movement are contradicted time and time again by the scientific evidence.

The Flagellum Unspun

In any discussion of the question of "Intelligent Design," it is absolutely essential to determine what is meant by the term itself. If, for example, the advocates of design wish to suggest that the intricacies of nature, life, and the universe reveal a world of meaning and purpose consistent with an overarching, possibly Divine, intelligence, then their point is philosophical, not scientific. It is a philosophical point of view, incidentally, that I share, along with many scientists. As H. Allen Orr pointed out in a recent review:

> Plenty of scientists have, after all, been attracted to the notion that natural laws reflect (in some way that's necessarily poorly articulated) an intelligence or aesthetic sensibility. This is the religion of Einstein, who spoke of "the grandeur of reason incarnate in existence" and of the scientist's religious feeling [that] takes the form of a rapturous amazement at the harmony of natural law. (Orr 2002)

This, however, is not what is meant by "Intelligent Design" in the parlance of the new anti-evolutionists. Their views demand not a universe in which the beauty and harmony of natural law has brought a world of vibrant and fruitful life into existence, but rather a universe in which the emergence and evolution of life is expressly made impossible by the very same rules. Their view requires that behind each and every novelty of life we find the direct and active involvement of an outside Designer whose work violates the very laws of nature that He had fashioned. The world of Intelligent Design is not the bright and innovative world of life that we have come to know through science. Rather, it is a brittle and unchanging landscape, frozen in form and unable to adapt except at the whims of its designer.

Certainly, the issue of design and purpose in nature is a philosophical one that scientists can and should discuss with great vigor. However, the notion at the heart of today's Intelligent Design movement is that the direct intervention of an outside designer can be demonstrated by the very existence of complex biochemical systems. What even they acknowledge is that their entire scientific position rests upon a single assertion – that the living cell contains biochemical machines that are irreducibly complex. And, they assert, the bacterial flagellum is the prime example of such a machine.

Such an assertion, as we have seen, can be put to the test in a very direct way. If we are able to find contained within the flagellum an example of a machine with fewer protein parts that serves a purpose distinct from motility, the claim of irreducible complexity is refuted. As we have also seen, the flagellum does indeed contain such a machine, a protein-secreting apparatus that carries out an important function even in species that lack the flagellum altogether. A scientific idea rises or falls on the weight of the evidence, and the evidence in the case of the bacterial flagellum is abundantly clear.

As an icon of the anti-evolutionist movement, the flagellum has fallen.

The very existence of the type III secretory system shows that the bacterial flagellum is not irreducibly complex. It also demonstrates, more generally, that the claim of "irreducible complexity" is scientifically meaningless, constructed as it is upon the flimsiest of foundations –

the assertion that because science has not yet found selectable functions for the components of a certain structure, it never will. In the final analysis, as the claims of Intelligent Design fall by the wayside, its advocates are left with a single remaining tool with which to battle against the rising tide of scientific evidence. That tool may be effective in some circles, of course, but the scientific community will be quick to recognize it for what it really is – the classic argument from ignorance, dressed up in the shiny cloth of biochemistry and information theory.

When three leading advocates of Intelligent Design were recently given a chance to make their case in an issue of *Natural History* magazine, they each concluded their articles with a plea for design. One wrote that we should recognize "the design inherent in life and the universe" (Behe 2002), another that "design remains a possibility" (Wells 2002), and the third "that the natural sciences need to leave room for design" (Dembski 2002b). Yes, it is true. Design does remain a possibility, but not the type of "Intelligent Design" of which they speak.

As Darwin wrote, there is grandeur in an evolutionary view of life, a grandeur that is there for all to see, regardless of their philosophical views on the meaning and purpose of life. I do not believe, even for an instant, that Darwin's vision has weakened or diminished the sense of wonder and awe that one should feel in confronting the magnificence and diversity of the living world. Rather, to a person of faith it should enhance the sense of the Creator's majesty and wisdom (Miller 1999). Against such a backdrop, the struggles of the Intelligent Design movement are best understood as clamorous and disappointing double failures – rejected by science because they do not fit the facts, and having failed religion because they think too little of God.

References

Adami, C., C. Ofria, and T. C. Collier. 2000. Evolution of biological complexity. *Proceedings of the National Academy of Sciences* 97: 4463–8.

Aizawa, S.-I. 2001. Bacterial flagella and type III secretion systems. *FEMS Microbiology Letters* 202: 157–64.

Behe, M. 1996a. *Darwin's Black Box*. New York: The Free Press.

Behe, M. 1996b. Evidence for intelligent design from biochemistry. Speech given at the Discovery Institute's God and Culture Conference, August 10, 1996, Seattle, Washington. Available at <http://www.arn.org/docs/behe/mb_idfrombiochemistry.htm>.

Behe, M. 2002. The challenge of irreducible complexity. *Natural History* 111 (April): 74.

Büttner, D., and U. Bonas. 2002. Port of entry – the Type III secretion translocon. *Trends in Microbiology* 10: 186–91.

Belmonte, J. C. I. 1999. How the body tells right from left. *Scientific American* 280 (June): 46–51.

Charlesworth, B. 2002. Evolution by design? *Nature* 418: 129.

Coyne, J. A. 1996. God in the details. *Nature* 383: 227–8.

Darwin, C. 1872. *On the Origin of Species*, 6th ed. London: Oxford University Press.

Dembski, W. 2002a. No free lunch: Why specified complexity cannot be purchased without intelligence. Lanham, MD: Rowman and Littlefield.

Dembski, W. 2002b. Detecting design in the natural sciences. *Natural History* 111 (April): 76.

Depew, D. J. 1998. Intelligent design and irreducible complexity: A rejoinder. *Rhetoric and Public Affairs* 1: 571–8.

Doolittle, R. F. 1993. The evolution of vertebrate blood coagulation: A case of yin and yang. *Thrombosis and Heamostasis* 70: 24–8.

Doolittle, R. F., and D. F. Feng. 1987. Reconstructing the evolution of vertebrate blood, coagulation from a consideration of the amino acid sequences of clotting proteins. *Cold Spring Harbor Symposia on Quantitative Biology* 52: 869–74.

Forrest, B. 2002. The newest evolution of creationism. *Natural History* 111 (April): 80.

Gura, T. 2002. Evolution critics seek role for unseen hand in education. *Nature* 416: 250.

Heuck, C. J. 1998. Type III protein secretion systems in bacterial pathogens of animals and plants. *Microbiology and Molecular Biology Reviews* 62: 379–433.

Holden, C. 1999. Kansas dumps Darwin, raises alarm across the United States. *Science* 285: 1186–7.

Huynen, M. A., T. Dandekar, and P. Bork. 1999. Variation and evolution of the citric-acid cycle: A genomic perspective. *Trends in Microbiology* 7: 281–91.

McNab, R. M. 1999. The bacterial flagellum: Reversible rotary propellor and type III export apparatus. *Journal of Bacteriology* 181: 7149–53.

Melendez-Hevia, E., T. G. Wadell, and M. Cascante. 1996. The puzzle of the Krebs citric acid cycle: Assembling the pieces of chemically feasible reactions, and opportunism in the design of metabolic pathways during evolution. *Journal of Molecular Evolution* 43: 293–303.

Miller, K. R. 1999. *Finding Darwin's God*. New York: HarperCollins.

Miller, K. R. 1996. A review of *Darwin's Black Box*. *Creation/Evolution* 16: 36–40.

Orr, H. A. 2002. The return of intelligent design. *The Boston Review* (Summer 2002): 53–6.

Padian, K. 2002. Waiting for the watchmaker. *Science* 295: 2575–4.

Pennock, R. T. 2001. *Intelligent Design Creationism and Its Critics: Philosophical, Theological, and Scientific Perspectives*. Cambridge, MA: MIT Press.

Robinson, A. J., M. Kropatkin, and P. M. Aggeler. 1969. Hagemann factor (factor XII) deficiency in marine mammals. *Science* 166: 1420–2.

Schneider, T. D. 2000. Evolution of biological information. *Nucleic Acids Research* 28: 2794–9.

Stern, C. 2002. Embryology: Fluid flow and broken symmetry. *Nature* 418: 29–30.

Thewissen, J. G. M., S. T. Hussain, and M. Arif. 1994. Fossil evidence for the origin of aquatic locomotion in archaeocete whales. *Science* 286: 210–12.

Thewissen, J. G. M., E. M. Williams, L. J. Roe, and S. T. Hussain. 2001. Skeletons of terrestrial cetaceans and the relationship of whales to artiodactyls. *Nature* 413: 277–81.

Thomhill, R. H., and D. W. Ussery. 2000. A classification of possible routes of Darwinian evolution. *Journal of Theoretical Biology* 203: 111–16.

Wells, J. 2002. Elusive icons of evolution. *Natural History* 111 (April): 78.

Yonekura, K., S. Maki, D. G. Morgan, D. J. DeRosier, F. Vonderviszt, K. Imada, and E. Namba. 2000. The bacterial flagellar cap as the rotary promoter of flagellin self-assembly. *Science* 290: 2148–52.

Printed and bound by CPI Group (UK) Ltd, Croydon, CR0 4YY

27/10/2024

14580392-0003